Advances in Intelligent Systems and Computing

Volume 1389

The series “Advances in Intelligent Systems and Computing” contains publications on theory, applications, and design methods of Intelligent Systems and Intelligent Computing. Virtually all disciplines such as engineering, natural sciences, computer and information science, ICT, economics, business, e-commerce, environment, healthcare, life science are covered. The list of topics spans all the areas of modern intelligent systems and computing such as: computational intelligence, soft computing including neural networks, fuzzy systems, evolutionary computing and the fusion of these paradigms, social intelligence, ambient intelligence, computational neuroscience, artificial life, virtual worlds and society, cognitive science and systems, Perception and Vision, DNA and immune based systems, self-organizing and adaptive systems, e-Learning and teaching, human-centered and human-centric computing, recommender systems, intelligent control, robotics and mechatronics including human-machine teaming, knowledge-based paradigms, learning paradigms, machine ethics, intelligent data analysis, knowledge management, intelligent agents, intelligent decision making and support, intelligent network security, trust management, interactive entertainment, Web intelligence and multimedia.

The publications within “Advances in Intelligent Systems and Computing” are primarily proceedings of important conferences, symposia and congresses. They cover significant recent developments in the field, both of a foundational and applicable character. An important characteristic feature of the series is the short publication time and world-wide distribution. This permits a rapid and broad dissemination of research results.

Indexed by DBLP, EI Compendex, INSPEC, WTI Frankfurt eG, zbMATH, Japanese Science and Technology Agency (JST).

All books published in the series are submitted for consideration in Web of Science.

More information about this series at http://www.springer.com/series/11156

Wojciech Zamojski · Jacek Mazurkiewicz ·
Jarosław Sugier · Tomasz Walkowiak ·
Janusz Kacprzyk
Editors

Theory and Engineering of Dependable Computer Systems and Networks

Proceedings of the Sixteenth International Conference on Dependability of Computer Systems DepCoS-RELCOMEX, June 28 – July 2, 2021, Wrocław, Poland

Springer

Editors
Wojciech Zamojski
Wrocław University of Science and Technology
Wrocław, Poland

Jacek Mazurkiewicz
Wrocław University of Science and Technology
Wrocław, Poland

Jarosław Sugier
Wrocław University of Science and Technology
Wrocław, Poland

Tomasz Walkowiak
Wrocław University of Science and Technology
Wrocław, Poland

Janusz Kacprzyk
Polish Academy of Sciences
Systems Research Institute
Warsaw, Poland

ISSN 2194-5357 ISSN 2194-5365 (electronic)
Advances in Intelligent Systems and Computing
ISBN 978-3-030-76772-3 ISBN 978-3-030-76773-0 (eBook)
https://doi.org/10.1007/978-3-030-76773-0

This Springer imprint is published by the registered company Springer Nature Switzerland AG
The registered company address is: Gewerbestrasse 11, 6330 Cham, Switzerland

Preface

We are pleased to present the proceedings of the Sixteenth International Conference on Dependability of Computer Systems DepCoS-RELCOMEX which is scheduled, as at the time of writing this preface, to be held online in Wrocław, Poland, from June 28 to July 2, 2021.

One year ago, when preparing the previous proceedings, we were just facing the rising tide of the COVID pandemic spreading across the globe. Being not able to predict the outcome of the rapidly changing situation in the world and the degree to which Europe will be affected, writing the preface we still hoped to organize the conference in its long-established venue—the Brunów Palace—and to see all our participants in face-to-face discussions, as it had been our tradition and pride since 2006. After the proceedings went to press, it soon turned out that this was not possible and we had to organize the 2020 event online with open admission. Despite technical difficulties and limited experience in this area (also among our participants), the sessions based on remote access successfully accomplished conference goals, with the papers presented by the authors live. Although the limited possibilities of online contacts could not fully match the benefits of a real-life traditional meeting in Brunów, we strived to achieve the most of what the conditions would allow, taking as a bonus the additional possibility to host external attendees who could join our open meetings. Considering the circumstances, the conference—first of all thanks to active reaction, discipline and support which we received from our participants—was realized successfully on the planned dates.

This positive experience with virtual organization of the previous event encouraged us to organize the 16th Conference online again because it was the only way to respect the annual DepCoS-RELCOMEX tradition. The first event was organized in 2006 at the Faculty of Electronics, Wrocław University of Science and Technology, by the Institute of Computer Engineering, Control and Robotics (CECR), and since then, new editions have been taking place every year; now, the organizer is the Department of Computer Engineering, successor of the institute. The conference inspiration came from the heritage of the other two cycles of events: RELCOMEX (1977 – 89) and Microcomputer School (1985 – 95) which were organized by the Institute of Engineering Cybernetics (the previous name of CECR) under the leadership of Prof. Wojciech Zamojski, now also DepCoS Chairman. This volume of "Advances in Intelligent Systems and Computing" is the latest in the series of conference proceedings which were first published by the IEEE Computer Society (2006–09), then by Wrocław University of Science and Technology Publishing House (2010–12) and finally by Springer Nature in the AISC volumes no. 97 (2011), 170 (2012), 224 (2013), 286 (2014), 365 (2015), 479 (2016), 582 (2017), 761 (2018), 987 (2019) and 1173 (2020). Springer Nature is one of the largest and most prestigious scientific publishers, with the AISC title—one of the fastest growing book series in its program—being submitted for indexing in CORE Computing Research & Education Database, ISI Conference Proceedings Citation Index (now run by Clarivate), Ei Compendex, DBLP, Scopus, Google Scholar, SpringerLink and many other indexing services around the world.

The mission of the DepCoS-RELCOMEX conference is to focus on diverse issues which are constantly arising in performability and dependability analysis of contemporary computer systems and networks. Being probably the most complex technical systems ever engineered by man (and also—the most dynamically evolving ones), their organization cannot be any longer interpreted only as a structure built on the base of technical resources (hardware) but their evaluation must take into account a unique blend of interacting people (their needs and behaviors), networks (together with mobile properties, cloud organization, Internet of Everything) and a large number of users dispersed geographically and producing an unimaginable number of applications. An ever-growing number of research methods being continuously developed for such analyses apply the newest results of artificial intelligence (AI) and computational intelligence (CI). Selection of papers in these proceedings illustrates broad variety of multi-disciplinary topics which should be considered in contemporary dependability explorations.

Presenting the reader this proceedings, we would like to thank everyone who participated in organization of the conference and in preparation of the volume: authors, members of the Program and the Organizing Committees, and all who helped in this difficult time. But especially, this volume would not be possible without invaluable contribution of the following 32 reviewers: Ali Al-Dahoud, Andrzej Białas, Ilona Bluemke, Wojciech Bożejko, Eugene Brezhniev, Dariusz Caban, Frank Coolen, Manuel Gil Perez, Zbigniew Gomółka, Ireneusz Jóźwiak, Vyacheslav Kharchenko, Urszula Kużelewska, Alexey Lastovetsky, Henryk Maciejewski, Jan Magott, Jacek Mazurkiewicz, Marek Młyńczak, Yiannis Papadopoulos, Ewaryst Rafajłowicz, Rafał Scherer, Czesław Smutnicki, Robert Sobolewski, Janusz Sosnowski, Jarosław Sugier, Kamil Szyc, Victor Toporkov, Tomasz Walkowiak, Max Walter, Marek Woda, Min Xie, Wojciech Zamojski and Wlodek Zuberek. Their work and detailed comments have helped to select and refine the conference submissions, and not mentioned anywhere else in this book, their support deserves even more recognition in the introduction.

Finally, we would like to thank all authors who decided to publish and discuss their research on the DepCoS-RELCOMEX platform. We express our hope that the included papers will contribute to further progress in design, analysis and engineering of dependable computer systems and networks, creating a valuable source material for scientists, researchers, practitioners and students who work in these areas.

Wojciech Zamojski
Jacek Mazurkiewicz
Jarosław Sugier
Tomasz Walkowiak
Janusz Kacprzyk

Organization

Sixteenth International Conference on Dependability of Computer Systems DepCoS-RELCOMEX

Wrocław, Poland, June 28 – July 2, 2021

Program Committee

Wojciech Zamojski (Chairman)	Wrocław University of Science and Technology, Poland
Ali Al-Dahoud	Al-Zaytoonah University, Amman, Jordan
Andrzej Białas	Research Network ŁUKASIEWICZ, Institute of Innovative Technologies EMAG, Katowice, Poland
Ilona Bluemke	Warsaw University of Technology, Poland
Wojciech Bożejko	Wrocław University of Science and Technology, Poland
Eugene Brezhniev	National Aerospace University "KhAI", Kharkov, Ukraine
Dariusz Caban	Wrocław University of Science and Technology, Poland
De-Jiu Chen	KTH Royal Institute of Technology, Stockholm, Sweden
Frank Coolen	Durham University, UK
Mieczysław Drabowski	Cracow University of Technology, Poland
Francesco Flammini	University of Linnaeus, Sweden
Manuel Gill Perez	University of Murcia, Spain
Franciszek Grabski	Gdynia Maritime University, Gdynia, Poland

Aleksander Grakowskis	Transport and Telecommunication Institute, Riga, Latvia
Ireneusz Jóźwiak	Wrocław University of Science and Technology, Poland
Igor Kabashkin	Transport and Telecommunication Institute, Riga, Latvia
Janusz Kacprzyk	Polish Academy of Sciences, Warsaw, Poland
Vyacheslav S. Kharchenko	National Aerospace University "KhAI", Kharkov, Ukraine
Krzysztof Kołowrocki	Gdynia Maritime University, Poland
Leszek Kotulski	AGH University of Science and Technology, Krakow, Poland
Henryk Krawczyk	Gdansk University of Technology, Poland
Urszula Kużelewska	Bialystok University of Technology, Białystok, Poland
Alexey Lastovetsky	University College Dublin, Ireland
Jan Magott	Wrocław University of Science and Technology, Poland
Henryk Maciejewski	Wrocław University of Science and Technology, Poland
Jacek Mazurkiewicz	Wrocław University of Science and Technology, Poland
Marek Młyńczak	Wrocław University of Science and Technology, Poland
Yiannis Papadopoulos	Hull University, UK
Ewaryst Rafajłowicz	Wrocław University of Science and Technology, Poland
Przemysław Rodwald	Polish Naval Academy, Gdynia, Poland
Elena Savenkova	Peoples' Friendship University of Russia, Moscow, Russia
Rafał Scherer	Częstochowa University of Technology, Poland
Mirosław Siergiejczyk	Warsaw University of Technology, Poland
Czesław Smutnicki	Wrocław University of Science and Technology, Poland
Robert Sobolewski	Bialystok University of Technology, Poland
Janusz Sosnowski	Warsaw University of Technology, Poland
Jarosław Sugier	Wrocław University of Science and Technology, Poland
Victor Toporkov	Moscow Power Engineering Institute (Technical University), Russia
Tomasz Walkowiak	Wrocław University of Science and Technology, Poland
Max Walter	Siemens, Germany
Tadeusz Więckowski	Wrocław University of Science and Technology, Poland

Bernd E. Wolfinger	University of Hamburg, Germany
Min Xie	City University of Hong Kong, Hong Kong SAR, China
Irina Yatskiv	Transport and Telecommunication Institute, Riga, Latvia
Włodzimierz Zuberek	Memorial University, St.John's, Canada

Organizing Committee

Chair

Wojciech Zamojski	Wrocław University of Science and Technology, Poland

Members

Jacek Mazurkiewicz	Wrocław University of Science and Technology, Poland
Jarosław Sugier	Wrocław University of Science and Technology, Poland
Tomasz Walkowiak	Wrocław University of Science and Technology, Poland
Tomasz Zamojski	Wrocław University of Science and Technology, Poland
Mirosława Nurek	Wrocław University of Science and Technology, Poland

Contents

Application of Assumption Modes and Effects Analysis to XMECA

Ievgen Babeshko[1], Kostiantyn Leontiiev[2], Vyacheslav Kharchenko[1], Andriy Kovalenko[3](✉), and Eugene Brezhniev[1,4]

[1] National Aerospace University "KhAI", Kharkiv, Ukraine
e.babeshko@khai.edu, v.kharchenko@csn.khai.edu
[2] Research and Production Corporation Radiy, Kropyvnytskyi, Ukraine
research@radiy.com
[3] Kharkiv National University of Radio Electronics, Kharkiv, Ukraine
andriy.kovalenko@nure.ua
[4] Radics LLC, Kropyvnytskyi, Ukraine
radics@radics.tech

Abstract. Failure modes, effects and criticality analysis (FMECA) is a well-known risk assessment method used to diagnose potential failure modes of a product or system being designed. This method is much based on expert experience and aims to develop improvement strategies so as to reduce the risk of possible failures. Expert decisions and assumptions affect analysis results significantly, and when several experts are involved into analysis, obtaining final results becomes even more complicated and expert-dependent. XMECA is a generic extension of FMECA that can be applied to analysis not only failures, but other aspects related to safety and security analysis like intrusions (intrusion modes, effects and criticality analysis, IMECA) etc. This paper suggests a novel approach to support XMECA called assumption modes and effect analysis. This approach refers to a technique used to minimize risks involved in making assumptions done by expert during performance of XMECA. Possible assessment scenario are described, case study is provided.

Keywords: FMECA · Safety assessment · Expert decisions · Risk assessment

1 Introduction

Failure modes, effects and criticality analysis (FMECA) technique is a standardized systematic method for identifying modes of failure and covering their local and global effects on the product (system) or process [1]. In case of product or system, it can be applied to either their hardware or software [2]. Also, some FMECA standards define pre-tailored FMECA processes, i.e. specific groups of interrelated FMECA activities that are designed to mitigate specific types of failure/fault risks [3]. IMECA is a FMECA modification intended to assess vulnerabilities that can be used in addition to standardized FMECA for safety-related domains where each vulnerability may lead to a failure in the

W. Zamojski et al. (Eds.): DepCoS-RELCOMEX 2021, AISC 1389, pp. 1–11, 2021.
https://doi.org/10.1007/978-3-030-76773-0_1

case of successful intrusion [4–8]. Unified approach that combines FMECA, IMECA and other assessment methods is referred as XMECA [9].

In spite of the fact that FMECA is known for almost more than 70 years, performed analysis shows that it is a challenge to define and classify the outputs of FMECA applied to modern complex products and systems so as to process them efficiently [10]. In addition, several researchers confirm that FMECA is typically carried out manually by human experts, or, in other cases, human experts are still required for the validation of the automatic FMECA updates, emphasizing that expert evaluations during analysis are quite subjective and could lead to ambiguousness of the obtained assessment results [11–13].

Several studies confirm that different experts involved in FMECA process produce different types of assessment information that could be complete/incomplete, precise/imprecise, known/unknown, etc. [14, 15]. Literature reviews on FMECA confirm different FMECA shortcomings and disadvantages, but do not pay attention to assessment of expert performing analysis [16–18].

In our previous works [19, 20] we analyzed possible experts' errors and the uncertainty degree of their decisions. Expert impact is relevant not only for FMECA, but for other assessment methods (see Table 1).

Table 1. Expert impact on safety assessment.

Assessment method	Expert support (limitations)	Percent of operations
XMECA	Selection of critical elements, failure modes, criticality assessment (task dimension)	Over 50%
Software fault injection testing	Selection of statements (operators), error types, criticality assessment (task dimension and technological complexity)	Over 30%
Markov models	Definition of states, failure and recovery rates (task dimension)	Over 70%
Common cause failure	Definition of diversity types and metrics (absence of representative statistics, testing complexity)	Over 50%

The rest of the paper is organized as follows. In the next section we briefly discuss the FMECA model, provide its formal notation and present several examples of FMECA tables from different experts. Section 3 presents the suggested expert evaluation for FMECA, covering both cases for equal and different qualification of experts. We discuss assumption modes and effect analysis in Sect. 4 and conclude our work in Sect. 5.

2 FMECA Model

FMECA-table could be represented as FT list, which consists of set of T tuples:

$$FT = < f_i, m_i = \{m_{ij}\}, e_i = \{e_{ij}\}, p_i = \{p_{ij}\}, s_i = \{s_{ij}\}, j = 1, \ldots, k_i >_{i=1}^{F}, \quad (1)$$

where f_i is a failed element (failure cause); e_i is a set of failure effects (consequences); p_i, s_i are respectively failure probability that can be defined using fuzzy scale (for instance, «high» - «medium» - «low») or as value in range 0..1, and severity, that can be defined using fuzzy scale; c_i is a failure criticality defined as function of fuzzy variables φ, $c_i = \varphi(p_i, s_i)$; m_i is a set of failure modes; number of considered failure modes of element i is k_i; total number of failure modes,

$$k = k_1 + k_2 + \ldots + k_F. \quad (2)$$

Relation between f_i, m_i, and e_i is shown on Fig. 1, relation between s_i, p_i and c_i, is shown on Fig. 2.

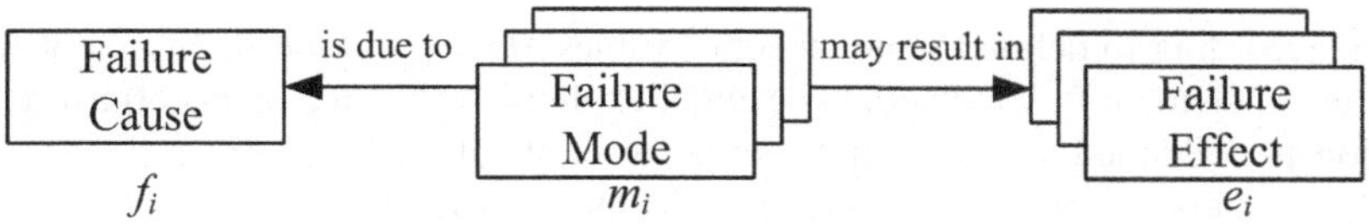

Fig. 1. Relation between failure cause, modes and effects.

Number of rows in a FMECA table $F^* = F$, if $k_1 = k_2 = \ldots = k_F = 1$; in general case $F^* = K$.

During performance of FMECA, an expert sequentially defines:

- elements f_i (module components, program operators, process operations etc.), failures of which must be taken into account (or considered), i.e. $f_i \in \Delta F$, $\Delta F \subset MF$, where MF is a set of components; ΔF is a subset of components taken into account;
- failure modes m_{ij} of element f_i, that have to be taken into account, i.e.

$$m_{ij} \in \Delta M_i, \Delta M_i \subset MM_i, \quad (3)$$

where ΔM_i is a set of element f_i failures taken into account; MM_i is a set of all element f_i failures; - effects e_{ij} of failure mode m_{ij} of element f_i, which are to be described, i.e.

$$e_{ij} \in \Delta E_i, \Delta E_i \subset ME_i, \quad (4)$$

where ΔE_i is a set of failure effects defined by expert for a particular failure mode m_{ij} of an element f_{ii}; ME_i is a set of all possible effects for particular failure mode of this element; probability and severity of failure mode m_{ij} of element f_i; probability p_{ij} is being chosen according to defined scale on set of values $MP = \{p'_h\}$; severity s_{ij} is being

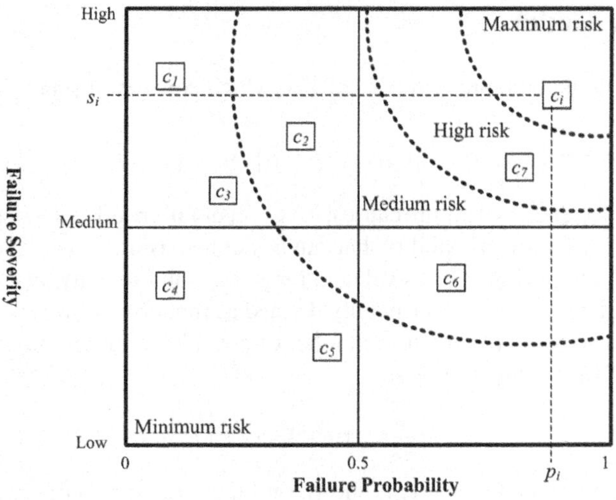

Fig. 2. Relation between failure severity and probability.

chosen according to defined scale on set of values $MS = \{s'_g\}$; criticality c_{ij} of failure mode m_{ij} of element f_i, which could be either explicitly evaluated by expert using given function φ, or defined by expert on his own on set of values $MC = \{c'_g\}$.

Three examples of FMECA prepared by three different experts are provided in Tables 2, 3 and 4. These tables are used as input in the following section.

Table 2. FMECA example prepared by expert 1.

Name	Type	Failure mode	Failure effect	Failure probability	Failure severity
FU06	Fuse	Fail to open	–	5.0E−07	Medium
		Slow to open	–	4.0E−07	Low
		Premature open	Voltage disconnection	1.0E−07	High
C12	Capacitor	Short circuit	Voltage drop	3.0E−08	High
		Open circuit	–	1.8E−08	Medium
R18	Resistor	Short circuit	Voltage drop	9E−09	Medium
		Open circuit	Open input path	5.4E−08	High

Table 3. FMECA example prepared by expert 2.

Name	Type	Failure mode	Failure effect	Failure probability	Failure severity
FU06	Fuse	Fail to open	–	5.0E−07	Medium
		Slow to open	–	4.0E−07	Low
C12	Capacitor	Short circuit	Voltage drop	3.0E−08	High
		Open circuit	–	1.8E−08	Medium
		Reduced value up to 0.5x	–	6.0E−09	Low

Table 4. FMECA example prepared by expert 3.

Name	Type	Failure mode	Failure effect	Failure probability	Failure severity
FU06	Fuse	Fail to open	–	5.0E−07	Medium
		Slow to open	–	4.0E−07	Low
		Premature open	Voltage disconnection	1.0E−07	High
C12	Capacitor	Short circuit	Voltage drop	3.0E−08	High
		Reduced value up to 0.5x	–	6.0E−09	Low
		Increased value up to 2x	–	6.0E−09	Low

3 Expert Evaluation of FMECA

3.1 Evaluation in Case of Equal Qualification (Self-assessment) of Experts

If FMECA assessment is being performed by group of Q experts that have identical qualification or qualification is not considered, so as to assess opinions of different experts it is required:

- to analyze kinds of divergences associated with different elements of model (1);
- to produce final version for each divergence;
- to obtain integrated version of FMECA;
- to perform analysis of it and provide eventual safety assessment.

When FMECA assessment is being performed by group consisting of Q experts the following divergences are possible:

- different sets of elements which failures have to be considered f_i are defined, i.e. given set $M\Delta F$ of sets $\Delta F^{(q)}$, $q = 1,\ldots, Q$, where $\Delta F^{(q)}$ is a set of elements which failures are considered by q-th expert;

- different sets of failure modes m_{ij} of element f_i which have to be considered are defined, i.e. given set $M\Delta M_i$ of sets $\Delta M_i^{(q)}$, for all q, $\Delta M_i^{(q)} c\ MM_i$, where $\Delta M_i^{(q)}$ is a set of element failure modes f_i, considered by q-th expert;
- different sets of effects e_{ij} of failure mode m_{ij} of element f_i, that have to be described are defined, i.e. given set $M\Delta E_i$ of sets $\Delta E_i^{(q)}$, for all q, $\Delta E_i^{(q)} c\ ME_i$, where $\Delta E_i^{(q)}$ is a set of failure effects of element f_i, considered by q-th expert;
- different probabilities of failure modes m_{ij} of element f_i are defined, i.e. given set $M\Delta P_{ij}$ of sets $\Delta P_{ij}^{(q)}$, for all q, $\Delta P_{ij}^{(q)} c\ MP$, where $\Delta P_{ij}^{(q)}$ is a set of probabilities of failure modes m_{ij} of element f_i, considered by q-th expert;
- different severities of failure modes m_{ij} of element f_i, i.e. given set $M\Delta S_i$ of sets $\Delta S_i^{(q)}$, for all q, $\Delta S_i^{(q)} c\ MS$, where $\Delta S_i^{(q)}$ is a set of failure severities of element f_i, considered by q-th expert;
- different criticalities of failure modes m_{ij} of element f_i, i.e. given set $M\Delta C_i$ of sets $\Delta C_i^{(q)}$, for all q, $\Delta C_i^{(q)} c\ MC$, criticality is either evaluated explicitly by q-th expert using specified function φ, or is defined by expert on his own. These two cases can be treated separately.

Doing so, the following assumptions are considered to be fair: firstly, sets $M\Delta F$, $M\Delta M_i$ and $M\Delta E_i$ entirely cover all possible expert opinions, and, secondly, scales (values) for assessment of failure probabilities, severities and criticality *MP*, *MS* and *MC* are common and can't be changed during assessment.

Three assessment scenarios based on expert opinions are available: conservative (ScC), when the most comprehensive list of failure modes is being generated, and consequences and risks assessment is performed in a pessimistic way; optimistic (ScO), when the minimal list of failure modes based on intersection of sets of failure modes is being generated, and during consequences and risks assessment best values are chosen; weighted (ScW), when common subset of failures is generated and then complemented by modes discovered and selected by two or more experts; consequences and risks assessment is based on average values in this case.

Scenario ScC. In this case the following assessments are provided:

- set of elements, which have to be included into FMECA table according to (1):

$$M\,\Delta F(ScC) = \mathrm{U}\Delta F^{(q)}, q = 1, \ldots, Q;$$

- sets of failure modes for all elements $f_i \varepsilon\ M\Delta F(ScC)$, which have to be considered:

$$M\,\Delta M_i(ScC) = \mathrm{U}\Delta M_i^{(q)}, q = 1, \ldots, Q;$$

- sets of failure effects e_{ij} of mode m_{ij} of element f_i, which have to be considered:

$$M\,\Delta E_i(ScC) = \mathrm{U}\Delta E_i^{(q)}, q = 1, \ldots, Q;$$

- failure probabilities of mode m_{ij} of element f_i are to be evaluated by equation:

$$p_{ij}(ScC) = \max\{\Delta P_{ij}^{(q)}\}, q = 1, \ldots, Q;$$

- failure severities of mode m_{ij} of element f_i are described by equation:

$$s_{ij}(ScC) = \max\{\Delta S_{ij}^{(q)}\}, q = 1, \ldots, Q;$$

- failure criticalities of mode m_{ij} of element f_i could be evaluated using equation:

$$c_{ij}(ScC) = \max\{\Delta C_{ij}^{(q)}\}, q = 1, \ldots, Q.$$

Scenario ScO. In this case the following equations are used:

- set of elements, which failures have to be included in FMECA table according to (1) is determined by equation:

$$M\,\Delta F(ScO) = \cap \Delta F^{(q)}, q = 1, \ldots, Q;$$

- sets of failure modes for all elements $f_i \in M\Delta F(ScC)$, which have to be considered:

$$M\,\Delta M_i(ScO) = \cap \Delta M_i^{(q)}, q = 1, \ldots, Q;$$

- sets of failure consequences e_{ij} of mode m_{ij} of element f_i, which have to be considered:

$$M\,\Delta E_i(ScO) = \cap \Delta E_i^{(q)}, q = 1, \ldots, Q;$$

- probabilities of failure modes m_{ij} of element f_i are to be obtained using equation:

$$p_{ij}(ScO) = \min\{\Delta P_{ij}^{(q)}\}, q = 1, \ldots, Q;$$

- severities of failure modes m_{ij} of element f_i are described by equation:

$$s_{ij}(ScO) = \min\{\Delta S_{ij}^{(q)}\}, q = 1, \ldots, Q;$$

- failure criticalities of mode m_{ij} of element f_i are to be evaluated using equation:

$$c_{ij}(ScO) = \min\{\Delta C_{ij}^{(q)}\}, q = 1, \ldots, Q.$$

Scenario ScW. In this case the following assessments are provided:

- set of elements, which failures are to be included in FMECA table according to (1):

$$M\,\Delta F(ScW) = \cap \Delta F^{(q)} \mathrm{U} \Delta F^{(q)*}, q = 1, \ldots, Q,$$

  where $\Delta F^{(q)*}$ is a set of elements, which failures are considered by two or more experts;
- sets of failure modes for all elements $f_i \in M\Delta F(ScC)$, that have to be considered:

$$M\,\Delta M_i(ScW) = \cap \Delta M_i^{(q)} \mathrm{U} \Delta M_i^{(q)*}, q = 1, \ldots, Q,$$

  where $\Delta M_i^{(q)*}$ is a set of elements' failure modes considered by two or more experts;

- sets of failure consequences e_{ij} of mode m_{ij} of element f_i, which have to be considered:

$$M\Delta E_i(ScW) = \cap \Delta E_i^{(q)} \cup \Delta E_i^{(q)*}, q = 1, \ldots, Q,$$

where $\Delta E_i^{(q)*}$ is a set of elements' failure consequences considered by two or more experts;
- probabilities of failure modes m_{ij} of element f_i are to be evaluated as ceiling function applied to the average:

$$p_{ij}(ScW) = \text{avermax}\{\Delta P_{ij}^{(q)}\}, q = 1, \ldots, Q;$$

- severities of failure modes m_{ij} of element f_i are described by equation:

$$s_{ij}(ScW) = \text{avermax}\{\Delta S_{ij}^{(q)}\}, q = 1, \ldots, Q;$$

- failure criticalities of mode m_{ij} of element f_i are to be evaluated using equation:

$$c_{ij}(ScW) = \text{avermax}\{\Delta C_{ij}^{(q)}\}, q = 1, \ldots, Q.$$

3.2 Evaluation in Case of Different Qualification (Self-assessment) of Experts

If FMECA is being performed by group of Q experts, whose qualification is different and is considered, to assess opinions of different experts it is required to add MScD sets to discussed above ScC, ScO, ScW. We consider the following scenarios.

Scenarios ScDT. Scenarios from this group are based on discarding of assessments provided by experts that have minimal or specified qualification, and further going to implementation of one of scenarios ScC, ScO, ScW, that are converted into scenarios ScTC, ScTO, ScTW, when qualification of experts is not considered anymore.

Scenarios ScDW. Scenarios of this group are based on ScC in generation of sets *MΔF(ScC)*, *MΔM$_i$(ScC)* and *MΔE$_i$(ScC)* and further weighted failure probability, severity and criticality assessments with rounding using ceiling function.

4 Assumption Modes and Effect Analysis. Case Study

The following situations are to be considered when analysis is being performed by several experts:

- different sets of elements: sets of elements provided by different experts can be merged;
- different sets of failure modes: two scenarios of merging are possible: optimistic (intersection) and conservative (union);
- different sets of failure effects: preference relation could be used so as to choose more critical effects.

Table 5. Assumption modes and effect example

Assumption	Mode	Effect
Absolute expert trust	Incomplete analysis	Incorrect assessment
Expert qualification	Incorrect choice of failure modes	Chosen more failure modes than required
		Chosen less failure modes that required
		Chosen wrong failure modes
	Incorrect choice of failure effects	Overestimated effect
		Underestimated effect
		Wrong effect

Table 6. FMECA table after ScC application

Name	Type	Failure mode	Failure effect	Failure probability	Failure severity
FU06	Fuse	Fail to open	–	5.0E−07	High
		Slow to open	–	4.0E−07	High
		Premature open	Voltage disconnection	1.0E−07	High
C12	Capacitor	Short circuit	Voltage drop	3.0E−08	High
		Open circuit	–	1.8E−08	Medium
		Reduced value up to 0.5x	–	6.0E−09	Low
		Increased value up to 2x	–	6.0E−09	Low
R18	Resistor	Short circuit	Voltage drop	9E−09	Medium
		Open circuit	Open input path	5.4E−08	High

Table 7. FMECA table after ScO application

Name	Type	Failure mode	Failure effect	Failure probability	Failure severity
FU06	Fuse	Fail to open	–	5.0E−07	High
		Slow to open	–	4.0E−07	High
C12	Capacitor	Short circuit	Voltage drop	3.0E−08	High

An example of assumption modes and effect analysis table is provided in Table 5.

Applying conservative scenario described above to FMECA tables provides in Tables 2, 3 and 4, we obtain results provided in Table 6.

After application of optimistic scenario described above to FMECA tables provides in Tables 2, 3 and 4, we obtain results provided in Table 7.

Using weighted scenario described above to FMECA tables provides in Tables 2, 3 and 4, we obtain results provided in Table 8.

Table 8. FMECA table after ScW application

Name	Type	Failure mode	Failure effect	Failure probability	Failure severity
FU06	Fuse	Fail to open	–	5.0E−07	High
		Slow to open	–	4.0E−07	High
		Premature open	Voltage disconnection	1.0E−07	High
C12	Capacitor	Short circuit	Voltage drop	3.0E−08	High
		Open circuit	–	1.8E−08	Medium
		Reduced value up to 0.5x	–	6.0E−09	Low

5 Conclusion and Future Work

This work systematizes safety assessment procedures using XMECA procedures on unified set-theory model basis considering expert based approach and technique application. Main difficulty of expert assessment is processing non-quantitative data to make decision concerning failure types, modes and effects.

Algorithms and different scenarios of XMECA assessment by involvement of group of experts with equal or different qualification (self-assessment) have been developed and discussed. This allows to decrease the risks of incorrect (incomplete) assessments.

Future research will be connected with development more formalized technique and tool for processing expert reports for FMECA and FIT based safety assessment.

References

1. IEC: IEC 60812:2018. Failure modes and effects analysis (FMEA and FMECA) (2018). https://webstore.iec.ch/publication/26359. Accessed 04 Jan 2021
2. VTT Industrial Systems: Failure mode and effects analysis of software-based automation systems (2002). http://www.julkari.fi/bitstream/handle/10024/124480/stuk-yto-tr190.pdf. Accessed 04 Jan 2021

3. Jackson, A.B., Jackson, T., Jackson, K.B.: Chronology of continuous improvement of the world's best FMECA standard. In: 2020 Annual Reliability and Maintainability Symposium (RAMS), Palm Springs, CA, USA, pp. 1–6 (2020)
4. Babeshko, E., Kharchenko, V., Gorbenko, A.: Applying F(I)MEA-technique for SCADA-based industrial control systems dependability assessment and ensuring. In: 2008 Third International Conference on Dependability of Computer Systems DepCoS-RELCOMEX, Szklarska Poreba, Poland (2008)
5. Androulidakis, I., Kharchenko, V., Kovalenko, A.: Imeca-based technique for security assessment of private communications: technology and training. Inf. Secur. Int. J. **35**(1), 99–120 (2016)
6. Kharchenko, V., Illiashenko, O., Brezhnev, E., Boyarchuk, A., Golovanevskiy, V.: Security informed safety assessment of industrial FPGA-based systems. In: Proceedings of the Probabilistic Safety Assessment and Management Conference, Honolulu, Hawaii (2014)
7. Schmittner, C., Gruber, T., Puschner, P., Schoitsch, E.: Security application of failure mode and effect analysis (FMEA). In: Bondavalli, A., Di Giandomenico, F. (eds.) SAFECOMP 2014. LNCS, vol. 8666, pp. 310–325. Springer, Cham (2014). https://doi.org/10.1007/978-3-319-10506-2_21
8. Kharchenko, V., Gorbenko, A., Sklyar, V., Phillips, C.: Green computing and communications in critical application domains: challenges and solutions. In: The International Conference on Digital Technologies, Zilina, pp. 191–197 (2013)
9. Illiashenko, O., Kharchenko, V., Chuikov, Y.: Safety analysis of FPGA-based systems using XMECA for V-model of life cycle. Radioelectron. Comput. Syst. **80**, 141–147 (2016)
10. Çetin, E.N.: FMECA applications and lessons learnt. In: 2015 Annual Reliability and Maintainability Symposium (RAMS), Palm Harbor, FL, USA (2015)
11. Lee, Y.-S., Kim, H.-C., Cha, J.-M., Kim, J.-O.: A new method for FMECA using expert system and fuzzy theory. In: 2010 9th International Conference on Environment and Electrical Engineering, Prague, Czech Republic (2010)
12. Liu, H.-C., Chen, X.-Q., You, J.-X., Li, Z.: A new integrated approach for risk evaluation and classification with dynamic expert weights. IEEE Trans. Reliab. **70**, 163–174 (2020)
13. Colli, M., Sala, R., Pirola, F., Pinto, R., Cavalieri, S., Wæhrens, B.V.: Implementing a dynamic FMECA in the digital transformation era. IFAC-PapersOnLine **52**, 755–760 (2019)
14. Chin, K.-S., Wang, Y.-M., Poon, G.K.K., Yang, J.-B.: Failure mode and effects analysis using a group-based evidential reasoning approach. Comput. Oper. Res. **36**(6), 1768–1779 (2009)
15. Liu, H.-C.: FMEA Using Uncertainty Theories and MCDM Methods. Springer, Singapore (2016). https://doi.org/10.1007/978-981-10-1466-6
16. Liu, H.-C., Chen, X.-Q., Duan, C.-Y., Wang, Y.-M.: Failure mode and effect analysis using multi-criteria decision making methods: a systematic literature review. Comput. Ind. Eng. **135**, 881–897 (2019)
17. Liu, H.-C., Liu, L., Liu, N.: Risk evaluation approaches in failure mode and effects analysis: a literature review. Expert Syst. Appl. **40**, 828–838 (2013)
18. Dai, W., Maropoulos, P., Cheung, W., Tang, X.: Decision-making in product quality based on failure knowledge. Int. J. Product Lifecycle Manag. **5**(2/3/4), 143–163 (2011)
19. Yasko, A., Babeshko, E., Kharchenko, V.: FMEDA-based NPP I&C systems safety assessment: toward to minimization of experts' decisions uncertainty. In: 24th International Conference on Nuclear Engineering, Charlotte, North Carolina, USA (2016)
20. Leontiiev, K., Babeshko, I., Kharchenko, V.: Assumption modes and effect analysis of XMECA: expert based safety assessment. In: 2020 IEEE 11th International Conference on Dependable Systems, Services and Technologies (DESSERT), Kyiv, Ukraine, pp. 90–94 (2020)

Improving Effectiveness of the Risk Management Methodology in the Revitalization Domain

Andrzej Bialas(✉)

Research Network ŁUKASIEWICZ, Institute of Innovative Technologies EMAG, Leopolda 31, 40-189 Katowice, Poland
andrzej.bialas@emag.lukasiewicz.gov.pl

Abstract. The paper concerns the EU RFCS SUMAD (Sustainable Use of Mining Waste Dump) project. It presents the concept of an extension of the risk management tool which can be applied to plan the revitalization process of post-mining areas, such as waste dumps. The tool will support decision makers in the revitalization planning phase by the selection of the most advantageous revitalization activities for the considered waste dump and the assumed land use. The proposed tool is based on three pillars: Risk Reduction Assessment (RRA), Cost-Benefits Analysis (CBA) and Qualitative Criteria Analysis (QCA), used to work out aggregated information for a decision maker. The extension of this tool, discussed in the paper, embraces a new functionality called the performance evaluation subsystem. It allows to monitor the revitalized object after the implementation of the revitalization plan. Three categories of performance indicators are designed: environmental, financial and social, and four types of the indicators with respect to the monitored values. Different modes of indicators feeding are discussed. The presented model was validated on some diversified examples of indicators.

Keywords: Risk management · Measures and indicators · Key performance indicators · Cost-Benefits Analysis · Qualitative Criteria Analysis · Post-mining areas revitalization

1 Introduction

The paper presents research dealing with the international project SUMAD (Sustainable Use of Mining Waste Dump). It is an international, interdisciplinary project embracing geotechnics, geology, ecology and IT issues. The project aim is to explore possible future uses of areas spoiled by coal-mining activities. The identification of the right revitalization methodology for the given area will be achieved by risk management and physical or numerical modeling with respect to geotechnical, sustainability, environmental, socio-economic, and long-term management challenges.

The author's organization, a partner of the project consortium, provides the software component called SUMAD RMT (Risk Management Tool) [1]. The component is to

W. Zamojski et al. (Eds.): DepCoS-RELCOMEX 2021, AISC 1389, pp. 12–21, 2021.
https://doi.org/10.1007/978-3-030-76773-0_2

support decision makers in planning revitalization processes of post-mining sites, especially spoil heaps. The revitalization plans have to take into account different factors representing risk management, financial and non-financial constraints. The SUMAD methodology is based on the knowledge and experience in the field of advanced risk management acquired from the earlier performed ValueSec [2] and CIRAS [3] projects, but concerning other domains of applications than revitalization.

The SUMAD risk management methodology currently implemented in the tool according to the project requirements offers decision support in the revitalization process planning. The question is what can be beyond and how this methodology can be improved.

The objective of the paper is to improve the SUMAD methodology by applying effectiveness measures and indicators to assess results of revitalization decisions in certain time perspective and to allow revitalization improvements.

Section 2 presents the SUMAD methodology, currently implemented within the SUMAD RMT. Section 3 discusses its extensions beyond the revitalization planning process. Section 4 presents models of the proposed solutions which can be implemented as new options of SUMAD RMT, and Sect. 5 – the model validation. The final section concludes the performed research.

2 The SUMAD Methodology and Its Implementation

The SUMAD methodology is based on the interdisciplinary approach to risk management, including technical, ecological and geotechnical issues with a view on financial and non-financial limitations of the applied revitalization methods considered "safety measures" here.

SUMAD RMT is designed to support strategic decisions related to the revitalization objectives of the given site, i.e. the selection of revitalization methods, which properly reduce risks, are economically effective and are free of different non-financial constraints. The given method embraces a set of different revitalization techniques.

In the SUMAD methodology a certain number of revitalized methods (called revitalization alternatives) are analyzed and one of them is selected by the decision maker for implementation.

Three types of analyses are possible with the use of SUMAD-RMT:

- RRA – Risk Reduction Assessment; focuses on the risk existing before and after revitalization;
- CBA – Cost-Benefits Analysis; considers investment cost, operation cost and future benefits before and after revitalization;
- QCA – Qualitative Criteria Analysis; deals with non-financial constraints, like societal, ethical, political, technological, environmental parameters, etc., existing before and after revitalization.

Figure 1 shows a general scheme of SUMAD operations, starting from opening a revitalization project until making the decision related to the selection of the most advantageous revitalization method for implementation. Initially, the current state of the existing heap ("as is") is identified with respect to the inherent values of risks, current costs/benefit parameters and non-financial constraints. Considering the planned land use and possible revitalization strategies, the decision maker analyzes the risk picture and defines a certain number of revitalization alternatives. Next, he/she performs the RRA, CBA and QCA analyses for each alternative, obtaining risk pictures after the implementation of each alternative. It allows to compare the pictures with respect to the project criteria and select the most advantageous revitalization alternative for implementation. As a result, a revitalization plan can be elaborated and next implemented outside the SUMAD project.

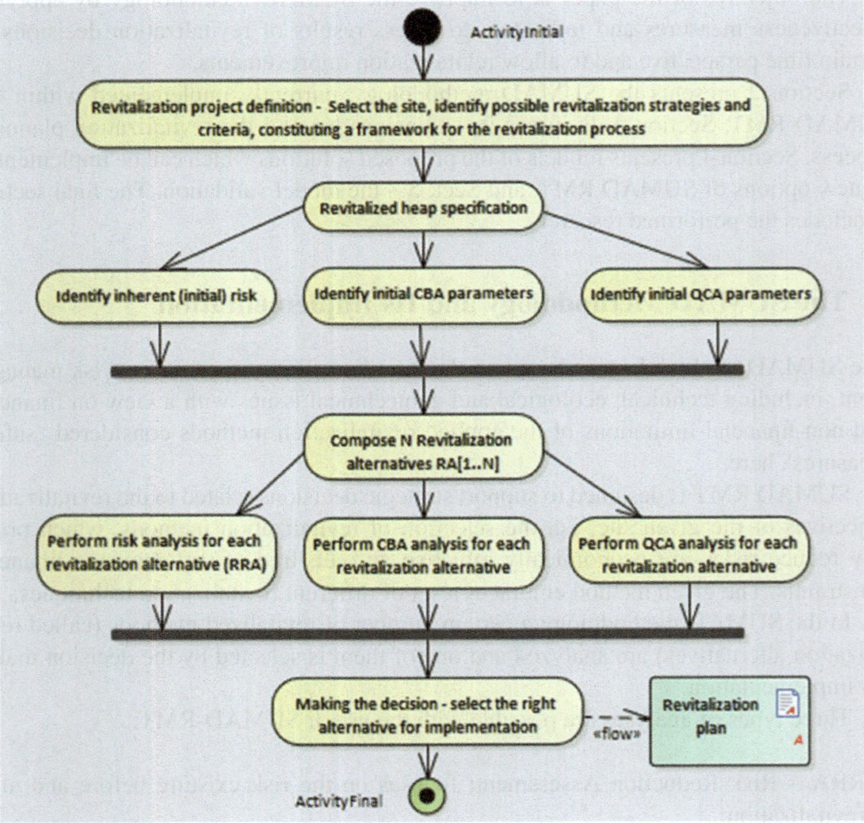

Fig. 1. Main SUMAD activities in the revitalization planning.

More detailed information about SUMAD RMT is included in the paper [1].

3 Extension of the SUMAD Methodology

The ISO 14000 standard concerning the Environment Management Systems (EMS) [4], points out the importance of the performance evaluation and the continual improvement of the implemented systems. The SUMAD project concerns ecological aspects as well, but the methodology embraced by the SUMAD project ends with the revitalization planning stage and does not embrace the above issues.

The author proposes to extend the methodology beyond the planning stage by introducing a new functionality to SUMAD RMT related to the performance indicators. It will allow to monitor, measure, evaluate and improve the revitalization project results in a longer period of time after the project implementation. Apart from this the activities existing in the SUMAD monitoring can be oriented towards the performance indicators.

The monitored parameters are very diversified with respect to their quantity, static/dynamic character, complexity, level of aggregation, source of information, etc. A general framework for the performance indicators management will be proposed as the extension of SUMAD RMT.

Performance indicators (PI), also called effectiveness measures, key performance indicators (KPI) or performance measurement indicators (PMIs), are used in different management systems, including business management. They are defined by experts for different domains of application and sometimes are standardized.

ISO 14000 includes general requirements for monitoring, measurement, analysis and evaluation, extended by the [5] guidelines. Experts, implementing EMSes, identify the environmental aspects (i.e. elements of activities or products or services that can interact with the environment) of the given organization, define the specific performance indicators and continual improvement processes, e.g. [6].

The paper [7] concerns environmental performance evaluation in industry. A conceptual tool was presented for this domain of application. Eight evaluation categories relevant for the industry domain were identified:

1. Amount of input-output materials (material balance);
2. Consumption/production of energy (energy balance);
3. Destination of solid, liquid and gaseous emissions;
4. Environmental impact assessment;
5. Environmental costs (material, energy and emissions);
6. Legal compliance and stakeholder requirements;
7. Surrounding environment conditions;
8. Applied measures to prevent pollution.

Apart from the categories, a calculation scheme for total and relative costs of the environmental aspects was defined. The tool was validated on the yoghurt caps production process.

The paper [8] presents the EMS performance indicators for the construction industry. Three main evaluation categories were identified: regulatory compliance, auditing activities and resources consumption, and some sub-categories specific for this domain.

Some examples of the KPIs for EMS are discussed on the web page [9]. The proposed main categories are: Natural Resources, e.g. water, gas usage, Emissions and Waste, e.g. weight to landfill, Incidents, Proactive Measures, e.g.: risk reduction measures implemented.

The paper [10] discusses KPIs in implementing sustainable strategies for the value creation process. These KPIs are considered, similarly to SUMAD, in three categories:

environmental, social and economic, and in a longer time horizon. For this reason they form a good foundation for the SUMAD performance indicators, discussed later.

Key performance areas (KPA) related to the mining activity were discussed in the paper [11]. KPAs are understood as the areas of performance that are reflected explicitly or implicitly in the vision and strategies of the organization. Each KPA embraces a set of KPIs. The paper discusses many examples of KPIs, including environmental, safety and social aspects. They concern mining processes, though some measures can be applied in the proposed SUMAD performance evaluation subsystem.

The guideline [5] introduces the environmental performance indicator (EPI), which can be divided into two main categories:

- management performance indicator (MPI), providing information about the management activities to influence an organization's environmental performance,
- operational performance indicator (OPI), providing information about the environmental performance of an organization's operational process.

The EPIs are focused on the needs of the organization interacting with the environment, and it would be difficult to apply them directly to the SUMAD project, as the context is different here. There is no organization but a heap which, once revitalized, may interact with the environment for a long time.

No performance evaluation system for the heap revitalization domain was encountered. It is a typical situation for a specific domain. In such a case performance measurement evaluation systems are elaborated from scratch.

The SUMAD project domain is specific and for this reason specific indicators should be defined in the future by the domain experts. The paper focuses only on the project of software solutions – mechanisms allowing to define different types of them. The tool will be open for different evaluation categories.

4 Model of the Proposed Performance Evaluation Subsystem

Two issues should be considered during the development of the performance evaluation subsystem as an optional functionality of the developed SUMAD RMT tool:

- the classification scheme of the performance indicators allowing to define the evaluation categories for the application domain,
- the additional software functionality responsible for data management, measurements implementation, feeding and presentation.

4.1 Classification Scheme of Performance Indicators

It is assumed that software should be flexible enough to accept any classification of measures worked out for the specific project domain. The UML (Unified Modeling Language) composition diagram (Fig. 2) presents three main categories of performance indicators: environmental (EPI), financial (FPI) and social (SPI). They are related to the RRA, CBA and QCA categories respectively.

Each category may have many indicators of different character expressed by the UML multiplicity 0..*, defined for the given application. The composition of the right set of performance measures for a given revitalization project is a task for the domain experts. Some examples are shown in the paper.

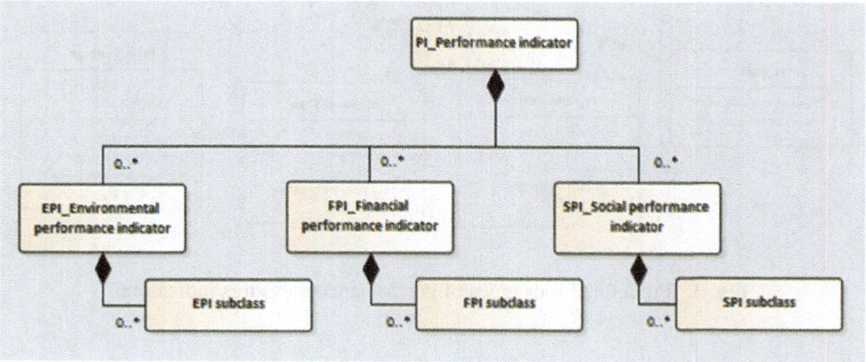

Fig. 2. Hierarchy of performance indicators.

4.2 Software Support for Performance Evaluation

It was assumed that:

- currently SUMAD RMT can manage many revitalization projects, each for one heap,
- a revitalized heap is considered a “protected asset” in the risk management methodology,
- SUMAD RMT embraces one process (one phase of the life-cycle model) for each revitalization project, i.e. revitalization planning;
- the second process is proposed here – monitoring the revitalized heap,
- the considered here performance evaluation subsystem will be related to this process,
- the proposed functionality implies the extensions of the currently developed SUMAD RMT software:
  - predefined data management module should be extended by performance measures representation,
  - user module, esp. the decision maker dashboard should be extended to present measurement-related data.

The SUMAD performance measures can be of different types, shown in Fig. 3:

- the `PI_Basic` type indicator presents the value of the measurement variable and measurement unit,
- the `PI_NoMoreThan` and `PI_NoLessThan` do the same and compare the value with the right reference levels issuing warning/alerting messages,

– the PI_InTheRange indicator helps to keep the value in the Min-Max range, issuing proper messages.

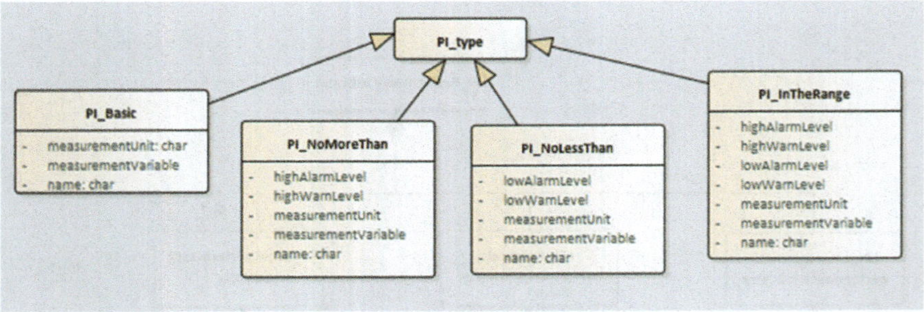

Fig. 3. Predefined four types of performance measures/indicators.

Three types of feeding sources for PIs (shown in Fig. 4) are proposed:

– the Manual_FS type is designed for information introduced manually by operators using the named window; it will be used mostly for SPI indicators;
– the Automatic_FS type is designed for data transferred (semi-)automatically from data bases or external equipment, including IoT devices, sensors;
– the Derived_FS type is designed to compose complex measures coming from many external sources; the outcome measurement variable is calculated on the input variables basis.

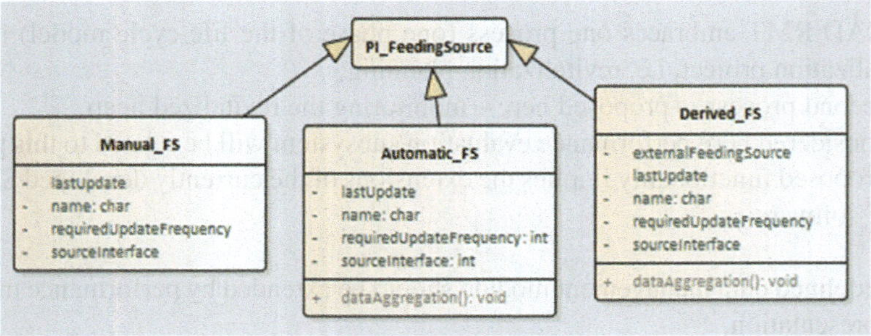

Fig. 4. Types of feeding sources for indicators.

5 Model Validation

The presented concept should be validated at the model stage before its implementation in software. Let us assume a certain revitalized heap, which is monitored to check its behaviour in time, and the effects of revitalization.

The validation embraces three different cases (Fig. 5):

1. Assessing the ground behaviour of the revitalized object, using information about events or incidents (their number, category and consequences) registered in the database within a certain time period.
2. Monitoring the energy production by a photovoltaic or wind power station.
3. A satisfaction survey of the people living around the revitalized object.

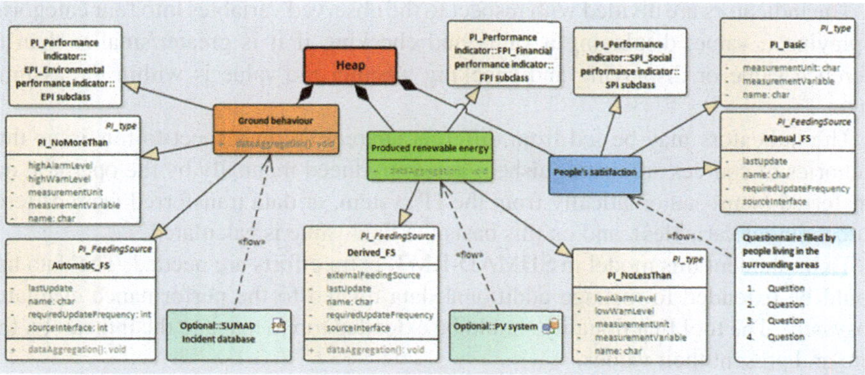

Fig. 5. Example of the heap revitalization performance measures.

The first case concerns the EPI indicator, configured in SUMAD-RMT as "Ground behaviour". It is fed automatically (`Automatic_FS`) from the event/incident data base and controlled whether the number of incidents of a selected category exceeds the reference value (`PI_NoMoreThan`). The operator obtains warnings, or even alerts. It is possible to monitor the aggregated consequences of incidents, etc. in the similar way. The "Ground behaviour" indicator is based on the historical data. Assuming that a certain number of sensors in the heap exists, the monitoring can be extended to the real-time data. The `Automatic_FS` mode of operation can be substituted by `Derived_FS`. Data transferred from many sensors can be aggregated and presented as the picture of the current ground behaviour.

The second case concerns financial gains related to the monetary value of the energy produced by the photovoltaic power plant. The FPI indicator is used. The monetary value of the produced energy is calculated on the basis of different parameters obtained from the power plant production management system (`Derived_FS`). The business goal is to maximize the production value. In this case the `NoLessThan` type of indicator is applied. Assuming that the amount (in kWh) of produced energy in a certain time period should be monitored, some changes in the performance evaluation subsystem are made. In this case the EPI indicator will be used, and the `Automatic_FS` source type.

The third case concerns "soft issues". A periodical satisfaction survey is provided. The people living around the revitalized object answer some questions. The filled questionnaires are summarized and the conclusions are manually introduced (Manual_FS) to the performance evaluation subsystem by the operator (SCI indicator). The PI_Basic type indicator is suitable in this case.

6 Conclusions

The paper presents the model of the performance evaluation subsystem for SUMAD-RMT. The basic three categories of performance indicators, i.e. environmental, financial and social, are distinguished expressing issues considered in the SUMAD project.

The indicators are divided with respect to the observed variables into four categories: displaying a value, displaying a value and checking if it is greater/smaller than the reference value or displaying and checking whether the value is within the assumed range.

The indicators may be fed from different sources. With respect to this issue three categories of sources are distinguished: data introduced manually by the operator, data transferred (semi-)automatically from the IT system, or data transferred from different sources (input variables), and on this basis the final value is calculated.

To implement this model in SUMAD-RMT, extra efforts are needed. The data base should be extended to manage additional data related to the performance evaluation subsystem. The tool functionality should be extended too to manage the indicators, feed them and present their values.

The model was validated on some sample indicators, showing the possibility to define a broad range of diversified indicators.

Acknowledgements. The SUMAD project leading to this application has received funding from the EU Research Fund for Coal and Steel under grant agreement No. 847227.

References

1. Bialas, A.: Risk management approach for revitalization of post-mining areas. In: Zamojski, W., Mazurkiewicz, J., Sugier, J., Walkowiak, T., Kacprzyk, J. (eds.) Theory and Applications of Dependable Computer Systems: Proceedings of the Fifteenth International Conference on Dependability of Computer Systems DepCoS-RELCOMEX, June 29–July 3, 2020, Brunów, Poland. Advances in Intelligent Systems and Computing, vol. 1173, pp. 71–81 (2020). https://doi.org/10.1007/978-3-030-48256-5
2. ValueSec. https://cordis.europa.eu/project/rcn/97989/factsheet/en. Accessed Feb 2021
3. Ciras. http://cirasproject.eu/. Accessed Feb 2021
4. ISO 14001:2015 Environmental management systems — Requirements with guidance for use
5. ISO 14031:2013 Environmental management — Environmental performance evaluation — Guidelines
6. Biswas, P.: ISO 14001:2015 Clause 9 Performance evaluation. https://isoconsultantkuwait.com/2019/06/23/iso-140012015-clause-9-performance-evaluation/. Accessed Jan 2021
7. Correa Maceno, M.M., Pawlowsky, U., Machado Scurupa, K.: Environmental performance evaluation – a proposed analytical tool for an industrial process application. J. Cleaner Prod. **172**, 1452–1464 (2018)

8. Tam, V.W.Y., Tam, C.M., Zeng, S.X., Chan, K.K.: Environmental performance measurement indicators in construction. Build. Environ. **2**(41), 164–173 (2006)
9. KPI examples for ISO14000: https://14000store.com/articles/iso-14001-key-performance-indicators/. Accessed Feb 2021
10. Hristov, I., Chirico, A.: The role of sustainability Key Performance Indicators (KPIs) in implementing sustainable strategies. Sustainability **11**(20), 1–19 (2019). https://doi.org/10.3390/su11205742
11. Dougall, A.W., Mmola, T.M.: Identification of key performance areas in the Southern African surface mining delivery environment. The Southern African Institute of Mining and Metallurgy Surface Mining (2014)

Automated Music Generation Using Recurrent Neural Networks

Mateusz Czyz and Michal Kedziora(✉)

Faculty of Computer Science and Management, Wroclaw University of Science and Technology, Wroclaw, Poland
michal.kedziora@pwr.edu.pl

Abstract. The paper aims to devise a set of machine learning models based on recurrent neural networks with emphasis on utilizing LSTM layers. These models are meant to be able to generate musical features such as melody notes or chords in sequence, or in other words generate music. Authors has decided to implement methods for music notation generation. Moreover, the paper contains a thorough description of the preprocessing of the obtained dataset along with the used ML technology and the latest research in related fields. In the paper, the authors elaborate on the process of training the devised models and example results of prediction done by the neural networks.

Keywords: Neural networks · LSTM · Recurrent neural networks · Music · Machine learning

1 Introduction

Nowadays, increasingly music is being created and performed. There is an increase in demand for music, even if this music is not meant to be remembered for years, but rather authentic and fresh for the sake of performance. Musicians create remakes and covers of popular songs and take inspiration from them to create their own music on the basis of someone else's pieces. What is more, music can often be constructed on popular and predictable patterns, which opens perspectives for algorithmic creation of such music or generation by artificial intelligence methods [1]. When considering music, which is meant to be easy to perform, the best choice is generating sheet music. Some attempts at generating polyphonic sheet music have already been made and were successful [9]. Yet, such notation is quite constrained and this nuisance can be overcome by aiming at generating only lead sheets, which are defining enough structure to relieve a human of most composing work and still leave much field for improvisation.

Regarding Machine Learning, the type of data and the way it is processed has a large influence on the type of model which can handle processing it and how well it will perform when the process of learning is over. Data organised spatially is likely to be learned best by convolutional neural networks, as demonstrated, for example, by MNIST dataset various model comparisons regarding error rate,

W. Zamojski et al. (Eds.): DepCoS-RELCOMEX 2021, AISC 1389, pp. 22–31, 2021.
https://doi.org/10.1007/978-3-030-76773-0_3

where convolutional neural networks are at the very top of the ranking [2]. For sequential data, like time series, especially music, recurrent neural networks are well suited [5]. Within this category, various specialized architectures serve different applications best [6]. For example, advanced RNNs based upon GRU and LSTM cells with multiple inputs, multiple outputs and multiple hidden layers perform very well in natural language processing tasks like translation [7].

The objective of this paper is to construct a method within the field of artificial intelligence, capable of producing sequences of musical features, like chords and melody notes. The second goal is to process the obtained dataset into a form capable to be fed into these models, which shall then be trained on this devised training set. After training, the prediction mechanism of the constructed models shall be used to generate sequences of music features like melody notes with their rhythm values and pitches, along with chords or chord symbols.

The remainder of this paper is structured as follows: The second chapter provides insight into pieces of research that have been conducted over the years in this field. The next chapter serves as an explanation how and why specific steps were taken upon the dataset during preprocessing and preparing the data to be used in the models. In Research Methodology chapter, there is a characterization of the conducted research and an insight into the results.

2 Related Work

One of the first attempts at generating lead sheets is [14]. In this paper, Weil et al. describe what a lead sheet is and propose a way to generate such a notation given input audio data. The authors propose four different modules that communicate with each other to establish beats, estimate melody and harmony (chords), and render a lead sheet containing these pieces of information. The chords are encoded in the form of a root, resulting in a variable with 24 possible values. The resulting lead sheets are encoded in a very convenient format of Band-In-A-Box, which offers all necessary information encoding with an addition of lead sheet printout. This format has been a candidate for this paper output lead-sheet format but has been replaced by the more popular MusicXML, based on XML, which can be interpreted and displayed by the most popular free score editing software, MuseScore. In [11], as in many other papers, there is clearly stated that recently developed models fail to capture and generate any musical data that is structured in a convincing way. Otherwise, such music is evident to have been generated by a machine and far less pleasant than anything composed by a human. Roy et al. deals with this problem by utilizing Markov models with constraints that can be imposed by the user on the structure of the generated pieces. This method is, therefore, half-automized and takes advantage of the user's musical knowledge. The authors prevail at generating convincingly structured lead sheets by generating variations on input themes.

Most work regarding music generating involves RNN, which is known to give satisfactory results when handling and generating time sequences. Some of it, however, exploit CNN, which is not the most obvious choice. Yet yields some promising results. One such paper is [15], where Yang et al. generate a melody one bar after another using this model class. Moreover, the devised model is capable of generating melodies both from scratch and conditioned on several types of input data, like a chord sequence or priming melody. This paper does not specify a method for lead sheet generation but rather a multitrack MIDI music. Nevertheless, the results can easily be viewed as a lead sheet as long as the underlying chords are voiced and enhanced by the performer. There has also been a substantial breakthrough in the field of generating chords for existing melodies. The two most recent solutions utilize Markov models and a probabilistic context-free grammar (PCFG) [12] as well as BiLSTM networks [8]. LSTM and BiLSTM networks fall under the category of RNN, which are best suited for learning, and generating time sequences and therefore had been chosen as a model for this paper. The most important of all the related works is that of Cedric de Boom et al. [4]. Said model is mostly based on LSTM and BiLSTM networks. What is quite unique about the proposed approach is that the process of learning consists of two steps: first, chord patterns and rhythm patterns are learned, and then they are fed into two BiLSTM layers and, along with the melody, fed into LSTM once more. This solution implicitly assumes that the melody depends on the chord and rhythmic structure of a piece of music.

3 Data Preprocessing

The dataset used in this research is the Wikifonia dataset. It consists of over 6700 lead sheets. The data format used to encode music in the Wikifonia dataset is MusicXML. It is a commonly known format designed and maintained specifically for the purpose of allowing elasticity and precision in musical notation.

The first step of preprocessing was eliminating polyphony. Melodies in all devised models are defined as a sequence of notes, where at any given moment there is exactly one note to be played. This note can either have a pitch or be defined as a rest. However, in the dataset, not all pieces are defined this way. Therefore, a way of extracting the melody from multiple notes stacked upon each other must be conceived. Discovering which part of a polyphonic harmony is considered a melody at any given time is a complex problem that is hard to solve in an analytical way. In this paper, an advantage has been taken of the fact that in musicXML all elements are defined sequentially and so whenever multiple voices are present in a given instant, the first defined note of the polyphony is retained.

Coming up next was splitting notes that were connected by ties. It affects the music quite considerably, because without ties, the sound of a piece is recognisably different. It is necessary, however, due to the fact that bar lines are learned along with the actual music. As previously mentioned, there can be only one note at any given moment. If a bar line is encoded in the same model feature as melody notes, then there can be no note that lasts while passing from one bar to the next. The next step taken was unfolding the repetitions. To that end, there

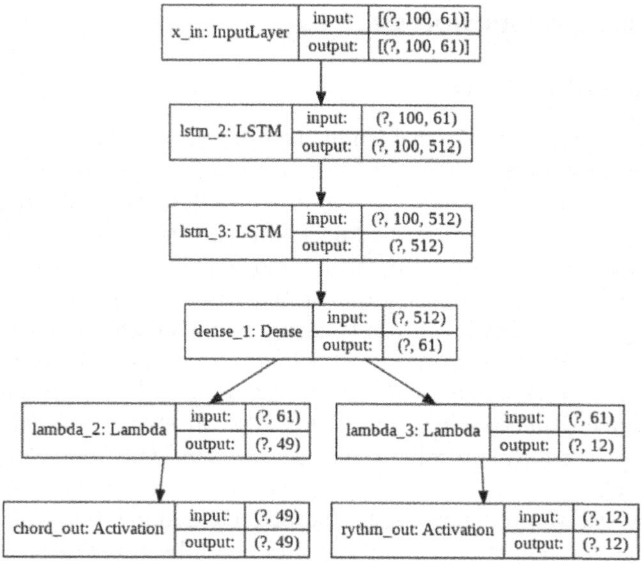

Fig. 1. Architecture of the model used for chord and rhythm prediction

is a way to mark which fragment is to be performed multiple times and specify the exact places where to start and end the repetition. Such notation naturally makes the performance of a piece consist of a number of bars more than there are on a lead sheet. To model music exactly the way it is to be performed and to avoid modeling repetition signs, such repetitions have to be unfolded.

There is a variety of chord types in the dataset, most of which are quite rare and occur in one or two pieces. Modelling such chords with that much of an underrepresentation would be challenging and no method of balancing the dataset was applied to augment the data in that direction. That problem was handled by mapping more exotic and specialized chords to more basic and more common versions of those same chords. Firstly, the total number of occurrences of each chord type has been checked. After diagram analysis, most chords have been mapped to one of four modes: major, minor, augmented, and diminished as those are the most basic types of chord types and other chords can be mapped to one of them without making too much assumptions. There has been a similar case with rhythm, although there is no possible mapping that one can easily do, because replacing the rhythmic value of a note is impossible when each note is processed on its own and not regarding its vicinity. Therefore, to avoid modeling very rare and exotic rhythmic values, all pieces which contain rhythm types that are not among the 11 most common ones in the dataset, have been removed from the dataset.

4 Research Methodology

4.1 Extracting Features

The data has been cleaned by this moment in the research, but continues to be encoded in an intermediate with unnecessary information still encoded therein. Therefore, all features that will be modeled need to be extracted. The three features that are modeled in this project are: chords, rhythm, and melody. In this step, a sequence of bars with chords and notes inside them is transformed into three vectors, each containing information about the pitch and rhythm type of the note being played at any given timestep and what chord needs to be played as a harmony [3]. Previous assumptions about chord mapping and rhythm type removal define how many possible states can a single variable in each vector have at any given timestep [10]. For simplicity, all chord roots are mapped to their enharmonic equivalent with either no accidentals or one sharp [13]. This operation yields 12 possible chord roots. As mentioned before, there are also 4 modes that the input chords are mapped to. Taking that into account, a single variable in a chord vector can have 48 possible states. Moreover, barlines are encoded explicitly, so another state has to be dedicated for that purpose, resulting in a 49-state chord variable. There are 11 rhythm types that a given note can have, so with the barline included, the rhythm vector consists of 12-state variables. For potential compatibility with the MIDI format, as done by de Boom et al., the melody vector element's size has been defined as 128 (from C-1 at 8.18 Hz to G#9 at 13289.75 Hz), and with two separate, special states reserved for the rest and barline, the resulting variable can potentially have 130 states.

4.2 Creating Sequences

Working with LSTMs requires the input data to be given as sequences of input vectors. Therefore, before training, all preprocessed and properly encoded musical files had to be segmented into training sequences. A training sequence is defined by one or more matrices, where each row represents a single one-hot vector with a given feature encoded inside. Moreover, the number of rows is fixed in advance. The order in which these vectors are set in is persisted in the source music file from which it is derived. For each training sequence, there also exists a value of the immediate successor, which is used in the process of network learning.

4.3 Experiments

Four models have been constructed. The first model is designed to encapsulate information about the harmonic and rhythmic structure of a piece and does not vary from generator to generator except the very last one. The chord and rhythm model consists of eight layers. For simplicity, the input layer is already merged with chord and rhythm vectors. Input is then passed to two LSTM layers that are supposed to learn the relations in data. Then, the output of the RNN layer is fed into a dense layer for prediction. After that, the output is split into two vectors, both of which are then treated with the softmax function. That way, the predicted chords and rhythms are output separately (Fig. 1).

The second model is the most complex model for learning melodies (Fig. 2). It consists of two inputs, one being merged chords and rhythms and the second being the melody, and seven other layers. The concatenated chords and rhythms are fed into a couple of BiLSTM layers that look through the chord and rhythmic structure of the training sequence, then the output of the second BiLSTM layer is concatenated with the melody input and fed through another couple of LSTM layers. At the end, there is a dense layer and a softmax activation layer used for prediction of melody. The second melody model differs from the previous one only by replacing BiLSTM layers with normal LSTM. This model is supposed to speed up the learning process and reduce the number of parameters.

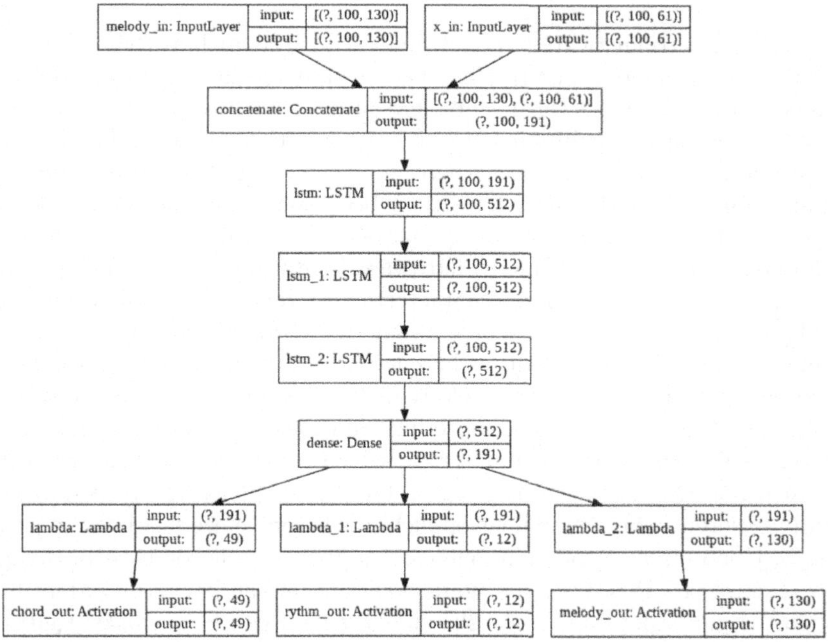

Fig. 2. Architecture of the model used for simultaneous prediction of all extracted features

The last model is the simplest one and is used for all feature simultaneous prediction. At the beginning, all inputs are concatenated, then fed through three consecutive LSTM layers and modelled by a dense layer. The output of this layer is then split into three separate vectors. To each of them, a softmax is applied. During the research, all models were trained on a set of approximately 70% of all training sequences, while the remaining 30% have been left out to enable unbiased validation of the devised models. The validation process has been performed with categorical accuracy as a metric implemented in Keras' metrics module. It is a simple, yet trustworthy metric, which depicts how often predictions match the actual real data, expressed as percentage. All models have achieved the same result of approximately 75% accuracies in rhythm and chord prediction and about 60% in melody prediction. Similar results can be read from values of the loss function chosen as the loss function for all outputs of all networks. The learning curve has not been illustrated in this paper, as training any of the given networks for more than one epoch has proven to be the cause of overfitting. During the learning of each model, the loss function over a batch has been monitored, and early stopping after one full epoch was necessary. For each batch of training sequences after this point, the accuracy of each batch was rising to values as high as 95% without any improvement, or even a decrease on the validation set.

5 Results

First, the most sophisticated model generating the melody has been tested (BiLSTM-based model). The test has been conducted by generating 100 pieces per model and analysing the results. Every model requires a kind of seed to start generating its own subsequent notes. In our research, a random seed from the dataset was chosen and from there, a sequence of 50 feature vectors were generated for each seed, which have then been split from the original training vector. A model, which has as many parameters as this one will always need a rich dataset to learn, but due to hardware limitations and a smaller learning set, the results of this model are characterised by relatively small variance. The first observation, which can be made about the obtained sequences is that only two pieces out of 100 have fully coherent barlines from beginning to the end. For example, given that only the melody feature has been generated as a barline and the two others have a normal note-like value, then the barline would have to be noted as having some rhythmic value, which contradicts any intuition, rules of musical notation or music in general. Two of the generated pieces, however, are fully coherent, and a fragment of such a coherent piece has been illustrated in Fig. 3. Upon closer inspection, one can see that the chords used in this piece follow a simple I - IV - V - I harmonic structure in the key of C-major all way through the piece, not only in the shown excerpt. It can be surmised that it has been conditioned on the same harmonic pattern. This chord progression can be found in music in pieces throughout music, common also with a minor chord on the sixth step of the major scale. The rhythmic structure is monotonous and repetitive, however, it is correct in terms of meter. Repetitions in music are common and very useful, however, this whole piece consists of only two bars repeated throughout its entirety, which resembles more a sample beat than a piece of music. It can be therefore concluded, as a proof of concept, that the devised training set and first model (predicting chords and rhythm) are sufficient to learn the basic harmonic and rhythmic structures of music, however, introducing more diversity by augmenting the training data is highly advisable

```
( F_major   ,half            ,F4   )
( F_major   ,quarter         ,rest )
( F_major   ,quarter         ,C5   )
( |         ,|               ,|    )
( F_major   ,dotted_quarter  ,C5   )
( F_major   ,eighth          ,A4   )
( F_major   ,quarter         ,A4   )
( F_major   ,quarter         ,F4   )
( |         ,|               ,|    )
( G_major   ,half            ,A#4  )
( G_major   ,quarter         ,G4   )
( G_major   ,quarter         ,C5   )
```

Fig. 3. A fragment of music generated by the BiLSTM-based model

to avoid overfitting towards the key of F-major and repeating same, simple patterns of length of 2 bars during a whole piece without any diversity. What could also help improve the results is introducing an element of randomness into each subsequent generated element with a small probability to provoke some changes. Melody prediction done by this model is characterized by a narrow spectrum of used notes, however there is a decent level of diversity of used notes, even with some trace amounts of rests. It generally tends to stay within the C-major scale with occasional accidentals such as in bar 3 of the excerpt. While it could be a matter of conditioning, the model generates music rather dissonant and entirely not obvious without profound and radical modifications and extensions of the outlined chords. The second tested model was LSTM. This model behaves worse than the same model with BiLSTM layers, which was to be expected. In this model, only one out of 100 melodies is fully coherent regarding barlines and it is most probable that the sole reason for this is the simplicity of the generated chord/rhythm pattern. Notes' pitches vary by a negligible amount and pitches are generally unpleasantly placed atop specific chords. It is worth mentioning that the chord/rhythmic pattern has been generated by the same model, which provided the structure for the previous melody model. Long-term dependencies in this model's melody prediction are difficult to encounter.

The next tested model was "Three features model". Surprisingly, the model that was expected to have given the worst results turned out to be the most reliable. The monolithic model, which predicts all three features simultaneously, performs at a 51% success rate when success is defined as assembling a fully coherent piece. It utilizes a full spectrum of chord modes utilizing even some augmented chords. The melodies tend not to be overtly complex, yet still are interestingly put atop the accompanying chords. This models' results actually suggest that the chord symbols have been oversimplified and that this model would benefit greatly from increasing the number of chords to which the input chords are mapped. Other ways to improve the readability of this music would be to rewrite the chords in a postprocessing phase to fit the melody better by specifying extensions and substituting a major chord with a suspended fourth or a suspended second maintaining the same root.

6 Conclusion and Future Work

In this paper, we accomplished the main goal by devising a set of Recurrent Neural Networks that have been trained on a dataset processed specifically for that purpose. This set consists of sequences of music features, whose distributions inside the training set have been modeled. After devising and inference of the constructed models, they have been validated on a previously secluded set of data that has not been utilised during the models' training. Immediately thereafter, a set of hundred new musical sequences consisting of 50 feature vectors each have been generated by conditioning the model on the random sequences contained within the training data. All generated results have been explored and described, with an emphasis on outlining possible ways to improve upon them. Along with that, a thorough examination of the obtained datasets has been conducted, the technology used for each step has been put forward, and each choice has been justified by listing possible alternatives and giving reasons for not choosing the other path. The results achieved through the models' predictions in this thesis are very promising, as all models were able to, to some degree, learn about the rules of musical notation without specifying them explicitly, for example that the rhythmical values of notes in each bar have to follow a certain regularity dictated by the time signature of a specific fragment.

There is still much room for improvement. Among the ways to improve upon the achieved results, firstly, is to enhance the training set by providing transposed sequences. Secondly, broadening the set of possible chords encoded in the chord feature would eliminate situations, where chords that are innately impossible to be mapped into a major or minor mode, are forced to a major mode. Furthermore, for data simplified compared to these assembled by Cedric de Boom et al., the model described by them as the worst, promises better results, as it has far fewer parameters to train. This simplification is inherently connected with less information capacity, which means less sophisticated relations in the data should be able to be learned. Modeling such sophisticated relations may, however, be unnecessary for the results to be satisfactory, and therefore, such unified models could be further researched in the future. Another solution would be to model more features by more complicated models, for example, ties that have been ignored in this paper, may be modeled to increase the similarity of generated data to real-world data. Instead of modelling the rhythmical values directly, this can be done by modelling the "duration" element situated inside all "note" elements. Notes connected with ties can have their durations added up and modelled this way.

References

1. Briot, J.-P., Hadjeres, G., Pachet, F.-D.: Deep learning techniques for music generation–a survey. arXiv preprint arXiv:1709.01620 (2017)
2. Chandra, A.L.: Mcculloch-pitts neuron – mankind's first mathematical model of a biological neuron (2018)

3. Ciaburro, G., Joshi, P.: Python Machine Learning CookBook, 2nd edn. Packt, Birmingham (2019)
4. De Boom, C., Van Laere, S., Verbelen, T., Dhoedt, B.: Rhythm, chord and melody generation for lead sheets using recurrent neural networks. In: Joint European Conference on Machine Learning and Knowledge Discovery in Databases, pp. 454–461. Springer (2019)
5. Dieleman, S., van den Oord, A., Simonyan, K.: The challenge of realistic music generation: modelling raw audio at scale. Adv. Neural Inf. Process. Syst. **31**, 7989–7999 (2018)
6. Kedziora, M., Gawin, P., Szczepanik, M., Jozwiak, I.: Malware detection using machine learning algorithms and reverse engineering of android java code. Int. J. Network Secur. Appl. (IJNSA) **11** (2019)
7. Kumar, N.S., Amencherla, M., Vimal, M.G.: Emotion recognition in sentences - a recurrent neural network approach. In: IFIP Advances in Information and Communication Technology book series (IFIPAICT), vol. 578 (2020)
8. Lim, H., Rhyu, S., Lee, K.: Chord generation from symbolic melody using blstm networks. In: 18th International Society for Music Information Retrieval Conference (ISMIR 2017), pp. 621–627 (2017)
9. Liu, H.-M., Yang, Y.-H.: Lead sheet generation and arrangement by conditional generative adversarial network. In: 2018 17th IEEE International Conference on Machine Learning and Applications (ICMLA), pp. 722–727. IEEE (2018)
10. (Hayden) Liu, Y.: Python Machine Learning by Example, 2nd Edn. Packt, Birmingham (2019)
11. Pachet, F., Papadopoulos, A., Roy, P.: Sampling variations of sequences for structured music generation. In: 18th International Society for Music Information Retrieval Conference (ISMIR 2017), pp. 167–173 (2017)
12. Tsushima, H., Nakamura, E., Itoyama, K., Yoshii, K.: Function-and rhythm-aware melody harmonization based on tree-structured parsing and split-merge sampling of chord sequences. In: ISMIR, pp. 502–508 (2017)
13. Vasilev, I., Slater, D., Spacagna, G., Roelants, P., Zocca, V.: Python Deep Learning, 2nd edn. Packt, Birmingham (2019)
14. Weil, J., Sikora, T., Durrieu, J.-L., Richard, G.: Automatic generation of lead sheets from polyphonic music signals. In: 10th International Society for Music Information Retrieval Conference (ISMIR 2009), pp. 603–608 (2009)
15. Yang, L.-C., Chou, S.-Y., Yang, Y.-H.: MidiNet: a convolutional generative adversarial network for symbolic-domain music generation. In: 18th International Society for Music Information Retrieval Conference (ISMIR 2017), pp. 324–331 (2017)

Non-exhaustive Verification in Integrated Model of Distributed Systems (IMDS) Using Vagabond Algorithm

Wiktor B. Daszczuk

Institute of Computer Science,
Warsaw University of Technology, Nowowiejska Street 15/19, 00-665 Warsaw, Poland
wbd@ii.pw.edu.pl

Abstract. Model checking is one of the leading techniques in systems verification, yet it suffers from combinatorial explosion. Heuristic non-exhaustive search techniques allow overcoming this limitation to some extent. These algorithms: A*, Ant Colony and Genetic, use some heuristics to speed up finding required states (for example: deadlock). As the algorithms work on a part of the reachability space of the verified system, a number of outgoing transitions from a state and their characteristics are used in the heuristics.

The new algorithm "Vagabond", presented in the paper, takes into account the progress of processes expressed in Integrated Model of Distributed Systems (IMDS). The formalism is uniform, but the two projections highlight communication deadlocks in the node view and resource deadlocks in the agent view. Moreover, the combination of IMDS with the Vagabond algorithm allows distinguishing between deadlock and termination, while they are both discontinuation of all processes. This feature is unique.

The mentioned non-exhaustive search algorithms use sets of several parameters which values are not obvious. We decided to use as few parameters as possible, preferably none.

Keywords: Deadlock detection · Integrated Model of Distributed Systems · Non-exhaustive search · Model checking heuristic

1 Introduction

Deadlock is mostly observed software malfunction. Even if other errors occur, such as sequence error or invariant violation, they lastly express themselves as deadlocks, i.e., states with no future in which all processes are stuck. Many deadlock detection techniques are developed, including on-line and off-line, model checking and Petri net static analysis etc.

In off-line methods, also known as static ones, the reachability space of the verified system against deadlocks is checked. These methods are successful, but they have some disadvantages:

W. Zamojski et al. (Eds.): DepCoS-RELCOMEX 2021, AISC 1389, pp. 32–43, 2021.
https://doi.org/10.1007/978-3-030-76773-0_4

- Space explosion problem – resulting from cooperating parallel tasks, often asynchronous. The elaboration of reachability space is time-consuming (typically exponential) and takes a large amount of memory.
- Most methods, especially those addressed to cycling systems, do not distinguish deadlock from distributed termination.

The latter problem is defeated by our formalism – Integrated Model of Distributed Systems (IMDS [1]). Communication deadlock, resource deadlock and distributed termination are defined as general temporal formulas, not related to the structure of the verified system. Verification is possible for systems having an arbitrary shape, containing both cycling and terminating processes. The formulas for distributed termination checking distinguish deadlock from termination.

Some methods fight space explosion: symbolic model checking, abstraction, etc. There are also methods that give an approximate solution, based on non-exhaustive heuristics for finding deadlocks. Some of them are discussed in Sect. 2. Finding a deadlock proves its existence, but a deadlock not found does not guarantee its absence. However, nowadays we are used to inaccurate methods, for example, machine learning can give both false positive and false negative solutions [2].

There are a few non-exhaustive verification methods. Three of them are widely described in the literature: ant colony optimization - ACO [3], genetic algorithms - GA [4], and A* search [3]. In some articles, those methods are combined. All three methods construct a set of reachable paths, and check if at least one of them leads to a deadlock. They were originally designed for optimization purposes, therefore some of their features do not match model checking, like drawing the genome: it does not necessarily produce a path leading to a reachable state. Also, the mentioned articles apply sets of various parameters without any procedure to elaborate their values. For example in genetic algorithms, some parameters like population size, chromosome length, gene crossing rate, mutation rate, fitness function, etc. are often taken in an arbitrary way. In ant algorithms, the parameters include ants set cardinality, pheromone evaporation factor, parameters for pheromone deposit, etc.

The number of paths constructed in the mentioned algorithms often exceeds the number of choices that can be made in the initial part of a verified system, therefore the starting part of many paths (we call it *leader*) is equal. A deadlock-free leader running through the initial activities of a system can occur in all paths.

The method developed originally for verification is Statistical Model Checking [5], see Sect. 2.

Deadlock is often defined as a state with no outgoing transitions. Obviously, it is a total deadlock. However, sometimes termination of processes in a verified system is treated in a similar way, as a state with no outgoing transitions. For example, successful termination of the solitaire game is found just as if it were a deadlock [6]. The drawback of such an approach is that if a real deadlock may happen in the system, it would not be distinguished from termination. Of course, a designer may inspect the suspicious state and decide if it is a deadlock or termination, but this makes automated deadlock detection impossible.

The advantages of the mentioned non-exhaustive algorithms are:

- The global reachability space is not elaborated.
- A path leading to a deadlock is a counterexample leading from the initial state to the deadlock found.

Non-exhaustive verification is shown on some benchmarks, typically Dijkstra's philosophers problem, with a number of philosophers as high as possible, to show the strength of presented methods. Also, other benchmarks are used, but all of them use *global state* in their logic and contain intentional deadlocks. However, in verification practice, the systems must be checked which are unknown to be free from deadlock. This is the reason the verification is applied in industrial systems development. The papers on non-exhaustive verification do not deal with such systems, they do not refer to the problem of how long the verification should be performed if a deadlock is not found yet.

This is the reason why we started our work on needed an efficient non-exhaustive algorithm for deadlock detection. In our work on exhaustive distributed deadlock detection, we distinguish resource deadlock from communication deadlock, and we also identify distributed termination. We aimed to preserve these features in a non-exhaustive search. Namely, the requirements for the algorithm were the following:

- Identify both communication and resource deadlocks, and distinguish among them.
- Check distributed termination (in the present paper, we deal only with deadlock, due to paper size limit).
- A set of parameters as small as possible, preferably none.
- Avoiding examination of leader subpaths.

The contribution of this paper is the new verification algorithm called Vagabond, which fulfills all the requirements outlined above. The former two requirements are fulfilled thanks to preserving information on features of distributed nodes and agents of IMDS in reachability space constructed for verification. Thanks to this, general temporal formulas, not dependent on the shape of a verified system, are elaborated. The latter two requirements are satisfied by the construction of the algorithm, in which a vagabond process wanders through reachability space.

The paper is organized as follows: Related work on non-exhaustive deadlock detection is presented in Sect. 2. The IMDS formalism is covered in Sect. 3: basic definition, processes extraction, and deadlock definition. The new Vagabond algorithm is presented in Sect. 4 in 3 versions: for communication deadlock detection and resource deadlock detection. Some experiments with the new algorithm are covered in Sect. 5. The limitations of the Vagabond algorithm are discussed in Sect. 6. Sect. 7 concludes the paper.

2 Related Work

In this section, we use the term *state* as it is used in the literature on discrete systems. Note that we call the entire system status a *configuration* in IMDS formalism (Sect. 3.1), while *state* is reserved for the local status of a node in a distributed environment.

Static methods of deadlock detection contain model checking [7] and Petri net siphon analysis [8]. In general, they find a situation from which no escape is possible. However, reachability space explosion is the main obstacle in static verification. There are multiple methods which purpose is to decrease this feature, for example, symbolic model checking, bounded model checking, compositional model checking, local model checking, partial order reduction, abstraction, and slicing [7].

Non-exhaustive search fights the explosion problem as the cost of unsure result. Three main methods are used: Ant Colony Optimization (ACO), A* graph search, Genetic Algorithms (GA), and Statistical Model Checking (SMC).

ACO is defined for optimization rather than for searching [3, 9]. A verification process works like an ant that finds the target, i.e., deadlock or termination, and other ants are supposed to follow the pheromone track. ACO is ideal for finding the shortest counterexample if a deadlock is found. Yet, during the verification, it is difficult to determine which path more likely leads to a deadlock than another path.

A* uses the heuristic evaluation function, associated with each node or transition [3, 6]. The heuristic must be admissible, i.e., it must underestimate the cost. The algorithm prefers paths that seem to be better, following nodes or transitions with a smaller value of the evaluation function, which should lead to the goal faster. It is hard to specify a function that prefers paths leading to an unknown deadlock. In [6], such a heuristic is proposed for specification in synchronous CCS, which is inadequate for modeling distributed systems in our opinion.

GA generates paths of execution as chromosomes [4, 10]. The genes are choices of transitions. The above problems with evaluating transitions in ACO and A* are similar to the troubles with the fitness function of the chromosomes in GA. If we investigate for a deadlock, finding it simply finishes the algorithm and no further path evaluation is needed. If we search for a counterexample having given features (for example length), a more subtle evaluation is needed. For example, reinforcement learning applied to GA is described in [11].

Statistical model checking [5] can investigate a representative part of reachability space, which is most often visited. It applies simulation-based techniques to avoid exhaustive state space exploration, and provides probabilistic guarantees for property satisfaction. Any statistical model checking toolset is built by combining:

- monitoring procedure to decide whether a finite execution satisfies the property under consideration,
- statistical model checking algorithm,
- a tool that allows to describe a system and generate sets of executions.

3 Model Checking in Integrated Model of Distributed Systems

3.1 Basic IMDS Definition

Integrated Model of Distributed Systems (IMDS) is based on an observation of the operation of distributed systems. A *node* staying in a given state receives a message, which fires an *action* of the node. As a result, the node changes its state and issues the next message. Although such behavior has some phases, we treat the execution of an action atomically, i.e. a relation $\Lambda \subset ((M \times P) \times (M \times P))$, where M is a finite set of *messages* $M = \{m_1, m_2, \ldots\}$ and P is a finite set of *states* $P = \{p_1, p_2, \ldots\}$.

To describe sequences of messages passed through and forth the nodes, we introduce a notion of *agents*. An agent describes a sequential calculation in a distributed environment, performed on the nodes while the agent migrates between them. To model terminating systems, we allow agent processes to terminate. Terminating action does not contain an output message, so it has a form $\lambda = ((m,p),(p'))$.

The current situation in a system (we do not call it a *state*, because this term is addressed to nodes) is created from the current states of all nodes and messages pending at these nodes. The set of all states and messages is called *configuration* T. A system starts from its initial configuration T_0. Every action $\lambda \in \Lambda$ converts its input configuration $T_{inp}(\lambda)$ to its output configuration $T_{out}(\lambda)$.

The behavior of a distributed system is described by a global reachability graph called Labeled Transition System (LTS). This graph covers all possible paths the system may follow. The graph is composed of configurations and actions. Formally:

$$\begin{aligned} &LTS = <Q, q_0, W> \\ &Q = \{q_0 = T_0, q_1 = T_1, \ldots\} \quad (vertices); \\ &q_0 = T_0 (initial\ vertex) \\ &W = \{(T, \lambda, T') | \lambda \in \Lambda, T = T_{inp}(\lambda), T' = T_{out}(\lambda)\} \quad (transitions) \end{aligned} \tag{1}$$

3.2 Processes in IMDS

For verification, especially for finding deadlocks and checking termination, processes must be defined in a distributed system. We can associate resident processes with system nodes: a process is a sequence of actions executed on a node. In the same system, we can associate processes with agents: a process is a sequence of actions threaded by messages of an agent. The difference is in the manner of grouping actions. We may say that a system is decomposed into *node processes* or into *agent processes*. For technical purposes, we define processes as *sets* of actions (of nodes or of agents, respectively), rather than sequences. Having P_i a set of states of the ith node, and M_j a set of messages of the jth agent, we get:

$$\begin{aligned} B_i &= \{\lambda \in \Lambda | \lambda = ((m,p),(m',p')) \vee \lambda = ((m,p),(p')), p, p' \in P_i\} \\ C_j &= \{\lambda \in \Lambda | \lambda = ((m,p),(m',p')) \vee \lambda = ((m,p),(p')), m, m' \in M_j\} \end{aligned} \tag{2}$$

We call the two decompositions *node view* and *agent view* of a system.

3.3 Communication Deadlock and Resource Deadlock

In general, deadlock is a situation in which all processes wait for events that cannot happen. A node process is in deadlock if it will never execute any action. However, if no message is pending at a node, is it simply idle. Therefore, a deadlock requires messages to be pending at deadlocked nodes. In practice, node deadlock (also called communication deadlock, as nodes communicate by means of messages) deadlock occurs when no action is enabled while there pending messages in the system.

An agent process is deadlocked if it will never execute an action, but it is not terminated (i.e., its message is pending at some node). As agents communicate by setting node states, it is a resource deadlock. Technically every communication deadlock is a resource deadlock, but a counterexample generated for node deadlock follows node processes while the other one follows agent processes. Note that if a partial deadlock is concerned (i.e. a deadlock regarding a subset of processes), resource deadlock does not necessarily mean a communication deadlock, but we deal only with total deadlocks in this paper.

4 The Vagabond Algorithm

4.1 Communication Deadlock (Node Deadlock)

We start the description of our new algorithm with communication deadlock detection. The principles are:

- A path from the initial configuration is constructed. The path consists of configurations, randomly choosing an action among actions enabled in a configuration. In a configuration T an action $\lambda = ((m,p),(m',p'))$ is enabled if $\{m,p\} \subset T$.
- If a deadlock configuration is reached, the algorithm stops.
- If a configuration belonging to the path is found, it denotes a loop. In such situation, a path is truncated by a random number of configurations.

Note that the algorithm is extremely simple and does not refer to any parameters. However, one parameter is needed to successfully run it: how long must we construct the path if no deadlock is found. Our first attempt was to limit the number of algorithm steps (number of path extensions) to N:

For node deadlocks:

$$N = \mathrm{card}(\Lambda) * \left\lceil \log_2\left(\prod_{k=1}^{n} \mathrm{card}(\Lambda s_k)\right)\right\rceil * \lceil \log_2(n) \rceil * \lceil \log_2(m) \rceil \tag{3}$$

For agent deadlocks:

$$N = \mathrm{card}(\Lambda) * \left\lceil \log_2\left(\prod_{k=1}^{m} \mathrm{card}(\Lambda a_k)\right)\right\rceil * \lceil \log_2(n) \rceil * \lceil \log_2(m) \rceil \tag{4}$$

where Λs_k is a set of actions of the node process s_k, Λa_k is a set of actions of the agent process a_k, n and m are respectively the number of nodes and the number on agents. The first factor simply takes a number of actions. The factor about Λs_k or Λa_k assumes that actions communicate pairwise, but not every pair. The logarithm follows that the more states, the smaller part of their pairs communicate. The rightmost two factors reflect that nodes communicate pairwise and agents communicate pairwise, two of them in every action.

However, in some examples, we found some false negatives, i.e., existent deadlock not found. We introduced a parameter D, which is the multiplying factor. We suggest $D = 4$, as typically it is enough. Greater D may be used, but the verification may last for days in such a case.

```
Vagabond (set of all nodes S):
insert initial configuration T0 to the path Φ;
do
{
  take a configuration T finishing the path Φ;
  if (no s∈S enabled and a message pends in every s∈S in T)
    halt(deadlock);
  else
  {
    draw an action λ∈Λ_T;
    if (λ leads to a configuration T∈Φ)
                    // a cycle detected - backtracking
    {
      cut a random number of configurations from path Φ;
      while
      (last configuration Tlast≠T0 in path Φ has one enabled action)
        cut Tlast from path Φ;
      exclude an action drawn previously in new Tlast from next draw;
    }
    else append T to the path Φ;
  }
}
while not(D*N steps completed);
halt(no deadlock);
```

4.2 Resource Deadlock (Agent Deadlock)

Recall node deadlock (Sect. 3.3): no action is enabled while there are pending messages at the nodes. The algorithm for agents differs: agents enablement is tested instead of nodes. Agents are enabled if their actions are enabled. A terminated agent is not assumed to be disabled.

4.3 Heuristics

Finding deadlocks and checking termination using the Vagabond algorithm is effective, but we think it needs to be expanded. Heuristics used in the mentioned algorithms (ACO, A*, GA) concern a number of transitions outgoing from given states, and on type of transition. In IMDS, system processes are preserved in reachability space and their behavior may be a hint in choosing the next transition in a path.

In communication deadlock detection, the heuristic can be used to speed up finding a deadlock. A disabled process can be enabled only by some other process. Therefore, the actions leading to configurations with fewer node processes enabled are chosen with higher probability than other actions. However, this slows down the search because output configurations of all the enabled actions should be elaborated. A similar rule concerns resource deadlock detection (the number of enabled agents is taken). In some cases (for example, 12 philosophers, see Sect. 5) the deadlock cannot be found in 4*N steps without the heuristic.

5 Experimental Results

The non-exhaustive search was examined in the literature using a set of benchmarks. However, most of the benchmarks cannot be used in our case because they use global or non-local states, while our methodology concerns distributed systems. Below we show verification of three non-trivial examples.

5.1 Philosophers

The most commonly used benchmark is Dijkstra's philosophers. It was transformed into a distributed version, in which the philosophers *ph* (agents) use their own nodes called chairs, and forks also reside on their own nodes. We verified the system with 4, 5, 10, 11, and 12 philosophers. For *X* philosophers, there are 2*X* nodes (*chairs* and *forks*) and *X* agents (*ph*). The results are collected in Table 1, compared to exhaustive verification (except 5 philosophers and butlers, which gives memory overrun). Time of exhaustive verification includes the elaboration of the reachability graph. Individual times are given as [s], [m:s] or [h:m:s] (in all tables).

Table 1. Philosophers – deadlock.

Number of philosophers	Type, dd = deadlock	Exhaustive verif. time	Vagabond verif. time
4	Agents dd	17	0
5	Agents dd	1:06:07	0
10	Agents dd		23
11	Agents dd		14:51
12	Agents dd		1:14:09

Additionally, three deadlock-free solutions are verified: for *X* philosophers *X-asym* – asymmetric – one of the philosophers takes forks in inverse order; *X-butlers* – with two additional butler nodes; *X-2atOnce* – taking 2 forks or none. The results of the verification are presented in Table 2. Both node view and agent view are verified exhaustively (except for 5 nodes with butlers, which exceeds available memory) and non-exhaustively. Exhaustive verification time is given totally for the elaboration of reachability space and the evaluation of the proper temporal formula.

5.2 Karlsruhe Production Cell

The Karlsruhe production cell was verified in numerous papers [12], however mainly using synchronous formalisms. It consists of 8 cooperating controllers, with elements circulating between them. The cell was tested for 1 produced element, and this gives expected deadlock resulting from the cell design. For more elements, the cell works without a deadlock. The verification of a system with 6 elements takes almost an hour.

Table 2. Philosophers – deadlock-free.

System	Type	Exhaustive verif. time	Vagabond verif. time
4-2atOnce	Nodes	2:46	9:40
4-2atOnce	Agents	2:46	32:25
4-asym	Nodes	3	2:48
4-asym	Agents	3	2:46
4-butlers	Nodes	2:58:50	1:24
4-butlers	Agents	2:58:50	3:41
5-2atOnce	Nodes	4:48:34	18:44
5-2atOnce	Agents	4:48:22	1:21:49
5-asym	Nodes	11:14	13:02
5-asym	Agents	11:15	12:43
5-butlers	Nodes		5:32
5-butlers	Agents		10:10

We have modified the system to model a situation in which an element can be dropped the belt due to synchronization error. We model such a situation by an "artificial" agent deadlock. For several circulating elements the system can be verified exhaustively. We enlarged the number of elements to 100, then to 150, which much exceeds the possibility of the exhaustive search. The Vagabond algorithm finds total deadlock in 15 and 38 s, respectively. Verification for 200 elements caused a memory overrun error in the Dedan verifier.

5.3 Train Scheduling

The third verified system is train scheduling, described in [13]. The system consists of a network of tracks, with traveling 8 trains. The original article describes a deadlock if the trains occupy all stations belonging to a given subset. The Vagabond algorithm finds this deadlock in 7 s for communication deadlock, and 23 s for resource deadlock. The authors of [13] show how to avoid a deadlock using a solution similar to Dijkstra's butlers: the two butlers allow 7 trains to enter the hazardous set. However, this solution cannot be implemented in a distributed way, because the butlers' states are accessible globally.

We constructed distributed butlers governing the critical regions. The reachability space of such a modified system exceeds the available memory if an attempt to perform the exhaustive search is applied. The Vagabond algorithm justifies deadlock freeness in more than 11 h. We suggest that the time is acceptable, taking into account that $D*N$ is almost 5 million (for agent verification) and that reachability space exceeds available memory and thus cannot be checked exhaustively.

6 Limitations

Exhaustive search gives proof of a verified feature. Non-exhaustive search in its nature may leave given features unverified. The methods mentioned in related work leave some uncertainty. If we treat finding deadlock as a positive result, false-positive cannot happen. If a deadlock is found, it is proved. However, false-negative may happen, i.e., existing deadlock may not be found. We never have proof that the whole reachability space has been searched, even if we tested more states than the fully connected graph of possible configurations:

$$N_{max} = \prod_{k=1}^{n} card(\Lambda s_k) * \prod_{k=1}^{m} card(\Lambda a_k) \quad (5)$$

If we want to have the sureness that all reachable configurations are searched, we must build full reachability space and mark every visited configuration. This makes heuristic search useless.

For exhaustive search, memory size is the main limit. We may use reduction techniques mentioned in Sect. 2. All those treatments shift the limit of the size of verified models, but it always exists. Also, the time of building the reachability space is important.

In non-exhaustive model checking we switch memory limits to time limits. The question is: how long can we wait for a result. The longer we may wait, the greater is our sureness that the result is proper. For greater confidence in a result, we should expand the verification time.

In our algorithm, we use a function for N (Eqs. 3 and 4) which grows much slower than N_{max} (Eq. 5), due to the use of logarithms. The use of logarithms results from the assumption that in a set of many processes, the dependencies between them do not form a large network of relationships. Yet, it is obvious that in some cases this function would give to small N, as in general $Nmax$ grows exponentially with the number of nodes and agents, while N grows logarithmically. Therefore, we should enlarge the D factor to be more sure about verification results.

It is difficult to compare the results of our experiments with other algorithms described in the literature. First, IMDS formalism is addressed to distributed systems, which is why each modeled system is constructed of distributed nodes communicating via messages. For individual nodes making their own decisions, the global system state is unavailable. Every benchmark should be converted to the asynchronous model without a global state, which is not always possible.

Second, all benchmarks used in the mentioned articles [3, 4, 6, 9–11, 14] (and others) show how quickly a deadlock can be found in a system that is known to contain a deadlock. In practice, however, users are interested in checking that their systems are secure, rather than checking how many philosophers can fall into a deadlock. None of these articles relates to how long it takes to say that the system being verified is free of deadlocks.

Table 3 compares various heuristic-based methods of deadlock detection with their maximum number of philosophers tested.

Table 3. Philosophers.

Paper	Modeling language	Technique	No. of philosophers	Verification time
[3]	CCS	ACO	15	?
[3]	CCS	A*	12	?
[4]	C	GA	(52% runs) 17	2:57
[4]	C	GA (mutation only)	(52% runs) 17	2:16
[6]	CCS	A*	12	?
[6]	CCS	Greedy	40	?
[10]	GTS[a]	GA	30	40:09
[10]	GTS[a]	A*	8	24:37
[11]	GTS[a]	GA + machine learning	100	56
[11]	GTS[a]	GA	20	23
[11]	GTS[a]	A*	10	3:57
[14]	Promela	ACO	8	1:12:14
This paper	IMDS	Vagabond	12	1:14:09

[a]GTS - Graph Transformation System

7 Conclusions and Further Work

In the paper, the non-exhaustive vagabond algorithm for deadlock and termination verification is presented. The algorithm is tailored especially for distributed IoT systems consisting of a number of nodes running a set of cooperating agents. The algorithm was tested on several examples, both falling into a deadlock and safe from a deadlock. Also, some student exercises with a result unknown in advance were investigated.

The algorithm confirms our assumption: it has only one parameter D controlling the number of steps performed during the heuristic search. Also, in many cases, it searches leader subpaths less frequently than the rest of the reachability space.

Our algorithm is designed for IMDS specification, but we believe that it may be applied to other formalisms. However, IMDS formalism is the only one that differentiates deadlock from termination independently from the shape of a verified system, and which can identify communication deadlock in node view and resource deadlock in agent view [15].

The algorithm uses a vagabond, which may be started multiple times in parallel. As the runs are independent, various distributed implementations are possible. Elements of other approaches: ACO, GA, A*, and machine learning may increase the performance of vagabond. We plan to apply these heuristics in our future research.

References

1. Daszczuk, W.B.: Specification and verification in integrated model of distributed systems (IMDS). MDPI Comput. **7**, 1–26 (2018). https://doi.org/10.3390/computers7040065
2. Batista, G.E.A.P.A., Prati, R.C., Monard, M.C.: A study of the behavior of several methods for balancing machine learning training data. ACM SIGKDD Explor. Newsl. **6**(1), 20–29 (2004). https://doi.org/10.1145/1007730.1007735
3. Francesca, G., Santone, A., Vaglini, G., Villani, M.L.: Ant colony optimization for deadlock detection in concurrent systems. In: 2011 IEEE 35th COMPSAC, Munich, Germany, 18–22 July 2011. pp. 108–117. IEEE (2011). https://doi.org/10.1109/COMPSAC.2011.22.
4. Godefroid, P., Khurshid, S.: Exploring very large state spaces using genetic algorithms. In: Katoen, J.P., Stevens, P. (eds.) Tools and Algorithms for the Construction and Analysis of Systems, pp. 266–280. Springer Berlin Heidelberg, Berlin, Heidelberg (2002). https://doi.org/10.1007/3-540-46002-0_19
5. Legay, A., Delahaye, B., Bensalem, S.: Statistical model checking: an overview. In: Barringer, H., Falcone, Y., Finkbeiner, B., Havelund, K., Lee, I., Pace, G., Roşu, G., Sokolsky, O., Tillmann, N. (eds.) Runtime Verification, pp. 122–135. Springer, Heidelberg (2010). https://doi.org/10.1007/978-3-642-16612-9_11
6. Gradara, S., Santone, A., Villani, M.L.: DELFIN+: an efficient deadlock detection tool for CCS processes. J. Comput. Syst. Sci. **72**, 1397–1412 (2006). https://doi.org/10.1016/j.jcss.2006.03.003
7. Baier, C., Katoen, J.-P.: Principles of Model Checking. MIT Press, Cambridge (2008)
8. Abdul-Hussin, M.H.: Elementary siphons of Petri nets and deadlock control in FMS. J. Comput. Commun. **3**, 1–12 (2015). https://doi.org/10.4236/jcc.2015.37001
9. Chicano, F., Alba, E.: Ant colony optimization with partial order reduction for discovering safety property violations in concurrent models. Inf. Process. Lett. **106**, 221–231 (2008). https://doi.org/10.1016/j.ipl.2007.11.015
10. Yousefian, R., Rafe, V., Rahmani, M.: A heuristic solution for model checking graph transformation systems. Appl. Soft Comput. **24**, 169–180 (2014). https://doi.org/10.1016/j.asoc.2014.06.055
11. Pira, E., Rafe, V., Nikanjam, A.: Deadlock detection in complex software systems specified through graph transformation using Bayesian optimization algorithm. J. Syst. Softw. **131**, 181–200 (2017). https://doi.org/10.1016/j.jss.2017.05.128
12. Lewerentz, C., Lindner, T. (eds.): Formal Development of Reactive Systems: Case Study Production Cell. Springer, Heidelberg (1995). https://doi.org/10.1007/3-540-58867-1
13. Mazzanti, F., Ferrari, A., Spagnolo, G.O.: Towards formal methods diversity in railways: an experience report with seven frameworks. Int. J. Softw. Tools Technol. Transf. **20**, 263–288 (2018). https://doi.org/10.1007/s10009-018-0488-3
14. Alba, E., Chicano, F.: Ant colony optimization for model checking. In: Díaz, R.M., Pichler, F., Arencibia, A.Q. (eds.) Computer Aided Systems Theory – EUROCAST 2007, pp. 523–530. Springer, Heidelberg (2007). https://doi.org/10.1007/978-3-540-75867-9_66
15. Daszczuk, W.B.: Communication and resource deadlock analysis using IMDS formalism and model checking. Comput. J. **60**, 729–750 (2017). https://doi.org/10.1093/comjnl/bxw099

Automatic Multi-class Classification of Polish Complaint Reports About Municipal Waste Management

Alicja Dąbrowska, Robert Giel(✉), and Sylwia Werbińska-Wojciechowska

Wroclaw University of Science and Technology, Wroclaw, Poland
robert.giel@pwr.edu.pl

Abstract. One of the many application areas for machine learning tools is waste management, where mainly classification of different waste materials and prediction of waste generation can be found. The conducted literature review indicated that there is a lack of papers concerning waste-related text classification.

In this research, we investigated the automatic text classification of complaint reports written in Polish that were sent to the municipal waste management system operating in one of the biggest Polish cities, Wroclaw. The analyzed problem regards a multi-class Machine Learning classification. Waste-related vocabulary, Polish language, and unlabeled dataset are the main difficulties in the considered problem.

The results showed that the automatic classification using Machine Learning algorithms achieved higher accuracy than the standard procedure based on people's choices currently applied in the reporting system.

Keywords: Machine learning · Polish language · Complaints · Text classification · Waste management · Multi-class classification

1 Introduction

Nowadays, there are many developments in the field of Machine Learning that may be widely used for various application areas. One of these areas is waste management. There can be found papers concerning the prediction of waste generation [1–3] and filling level of containers [4–6]. In [1], socio-economic and demographic parameters were considered using Artificial Neural Network and Decision Trees. Authors of [3] were trying to predict weekly and daily waste generation at the building scale through the combination of machine learning and area estimation techniques. Another work [7] concerning forecasting waste generation developed three models: collection zone, socio-economic stratification, population, and generated waste quantity were considered factors. There also can be found papers focused on clustering. In [8], based on socio-economic parameters and containers information such as location, volume, and waste weight, waste generation type profiles were created. The authors of [9] predicted trash bin status based on their information. Besides clustering and quantity of waste prediction,

W. Zamojski et al. (Eds.): DepCoS-RELCOMEX 2021, AISC 1389, pp. 44–52, 2021.
https://doi.org/10.1007/978-3-030-76773-0_5

there are many papers focused on classification. The authors of [10] proposed a waste material classification system based on the Convolutional Neutral Network model to classify waste into different types. Another work [11] presented a binary classification of waste into plastic and non-plastic using a Convolutional Neural Network. In [12], a hierarchical deep learning approach was proposed for waste classification in food trays. The authors of [13] presented a waste segregator system in which machine learning tools were applied for waste material identification and multi-class classification.

All found papers on classification with machine learning algorithms are based on waste images. However, no papers concerning waste-related text classification were found.

One of the growing global problems is ensuring adequate waste management. As part of this, information management is significant. There is a necessity for a fast reaction to the observed problems. In the analyzed system, waste-related complaints such as failure to collect waste according to the schedule are reported and categorized by residents. However, reliable and fast response to residents' complaints about waste management services performance cannot be fully satisfied. This is mainly connected with the occurrence of long delays in responses to the reported complaints due to incorrect categorization of them. Following this, it is justifiable to introduce an automatic classification of residents' complaints.

The aim of this research is to check the possibility of applying Machine Learning algorithms to polish waste-related complaints classification. Unlabeled dataset, polish language, and specific waste vocabulary are the most severe difficulties in the considered case. The article is structured as follows. In Sect. 2, the complaint reporting system is described. Section 3 defines the considered problem. In Sect. 4, the main steps of the complaints classification model are presented. Section 5 shows obtained results of classification with the use of 10 selected classifiers. Section 6 gives comments on results and indicates the directions of future research.

2 Complaint Reporting System

In Wroclaw, one of the biggest Polish cities, residents can send their waste-related complaints to the company responsible for the proper waste management (waste collection, city cleaning, etc.). There can be separated three types of reporting users in the system:

I. Residents – people, who in most cases, do not have expert knowledge in waste management. They are ordinary residents and can report their complaints through the phone, e-mail, or the website.
II. Administrators – a separate group of residents with its own category in the reporting system to demand a bulky waste container or a green waste container. This type of users is treated as experts.
III. Company employees – people with expert knowledge in waste management who are employed in the company and receive telephone or e-mail complaint reports.

When reporting complaint via the website, there are seven categories to choose from: waste collection, winter-connected problems, waste containers/bags, waste segregation, city cleaning, administrator zone and other. The complaint report recorded in the system through the website is later analyzed by the company's employees and forwarded to the appropriate subcontractor. Incorrect addressing of the complaint by a resident may lead to a longer response time for the company. This is mainly connected with the necessity of sending messages between individual departments in the company regarding the analyzed report (Fig. 1).

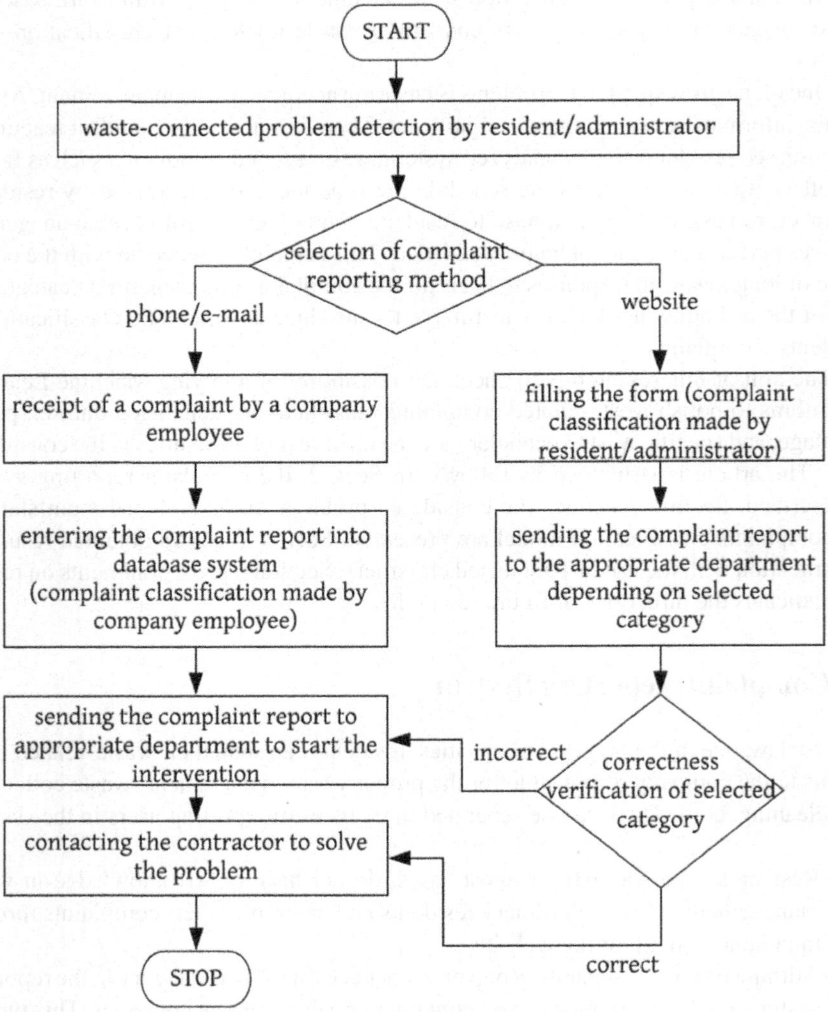

Fig. 1. Scheme of the complaint reporting system in the analyzed municipal waste company

From August 2017 to April 2019, the company received about 24,539 complaint reports (about 1,169 reports per month). The largest share of complaints according to the defined categories regards waste collection (32%), and then administrator zone (24%), container/waste bags (21%), and city cleaning (14%). The least numerous categories include others(7%), segregation(1%) and winter-connected problems (1%).

Each complaint report is stored in the database and mainly contains the date, the address, category (chosen by the resident or assigned by the employee), subject, and description. The description contains the information necessary to identify the problem and perform appropriate actions. Therefore, on its basis, it is possible to identify to which category it should be classified. The description contains an average of 38 words, with a standard deviation of 31. The maximum number of detected words per complaint is 667.

Based on the company's employees' information and their own examination of complaints reports (Table 1), it has been noticed that residents' classification is characterized by an unacceptable number of errors. Due to a lack of expert knowledge in waste management, only this one group of reporting users has an unacceptable classification accuracy. The consequence of this is an extended response time to the observed problems.

Table 1. The preliminary analysis of complaint reports (for randomly selected 2400 complaint reports)

Type of reporting users	Number of all reported complaints	Number of reports classified correctly	Classification accuracy	Percentage of reports per given reporting group
Residents	1,043	880	84.37%	44%
Company's employees	947	893	94.30%	39%
Administrators	410	402	98.05%	17%
Sum:	2,400	2,175		
Weighted average:			90.63%	

3 Problem Statement

The requirement of reliable and fast responses for residents' complaints cannot be fully satisfied. The wrong categorization of complaints leads to long delays in responses. There is a necessity to automate the classification process in the reporting system.

Automatic assignment of complaints to defined categories is a multi-class text classification problem. In this work, we address this problem in the area of waste management and apply selected Machine Learning algorithms.

Difficulties connected with the considered case are as follows:

- the specificity of the Polish language
  The Polish language has a complex morphology same as other flexional languages. For this reason, Natural Language Processing operations are more complicated than for English [14].

- waste-related vocabulary
  In the area of waste management, the vocabulary is characterized by a small number of unique words characteristic only for specific, defined waste-related complaint categories.
- unlabeled dataset
  Due to the observed and examined incorrect classification of residents' complaints, an analyzed dataset has to be considered an unlabeled dataset. Consequently, a group of experts has to carry out the labeling process to enable supervised learning.

4 Automatic Classification of Complaint Reports

Text classification using Machine Learning requires the following steps: data pre-processing, feature extraction, choosing classifier, classifier training, classifier testing, evaluation. The procedure was carried out using the Python software.

In the first step concerning data pre-processing, test complaints, complaints without description, and Polish characters were removed. Dataset consisting of 24,539 was reduced to 23,466. As mentioned in Sect. 3, due to the observed classification errors, the analyzed dataset requires labeling. Limited time and the number of experts led to the decision to use 2,400 complaint reports in this research. 2,000 of them were chosen randomly for training and testing purposes and the remaining 400 complaints (validation sample) covering the first three weeks of March 2019 (newest reports in the reporting system) for external validation (Fig. 2).

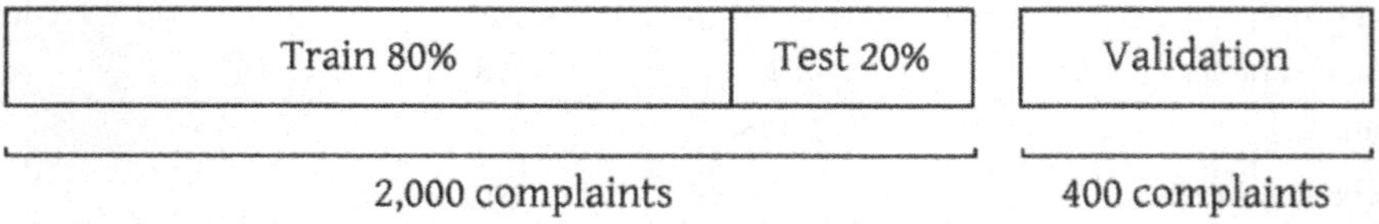

Fig. 2. Division of the selected data sample

A bag of words with term frequency-inverse document frequency (TF-IDF) was applied as a part of feature extraction. Bag of words technique allowed to perceive complaints through the word occurrence. According to this approach, tokenization and word occurrence counting was performed. The weight of the individual words was determined by introducing TF-IDF. Thanks to this, the most frequently occurring words in each complaint receive a smaller weight than the unique words that determine membership in a given class [15].

With extracted features, it was possible to apply selected classifiers. It was decided to examine the performance of ten classifiers:

- RF - Random Forest,
- KNN - Nearest Neighbors,
- MNB - Multinomial Naïve Bayes,
- DT – Decision Tree,

- BNB – Bernoulli Naïve Bayes,
- AB - AdaBoost,
- LR - Logistic Regression,
- SVC1 - Support Vector Classifier with Linear Kernel,
- SVC2 - Support Vector Classifier with Gaussian Kernel,
- SVC3 - Support Vector Classifier with Sigmoid Kernel.

Their detailed description can be found in [16, 17].

Regardless of the classifier used, it is necessary to carry out the learning and testing process. Supervised learning requires providing for the classifier the labeled dataset, which is divided into the part intended for the learning process and the part intended for the testing process. The testing process is carried out to check the classifier's operation on previously unseen data, i.e., not being part of the training set. There are two main approaches for training and testing: train/test split and k-fold cross-validation. In the case of train/test split, which was implemented, it is necessary to determine the division's proportion into learning and testing parts. The most commonly used ratio is 80/20, where 80% dataset is used as the training set, and the remaining 20% is the test set.

To assess and select the best classifiers, we chose few metrics commonly used in Text Classification for multi-class categorization [18]. Accuracy (Acc), Precision (Prec), F1 score, and Recall were selected.

5 Results

Based on the classification described in Sect. 4, the results presented in Table 2 and Table 3 were obtained.

Table 2. Classification results for the train/test data sample

Algorithm	Acc.	Prec.	Recall	F1 Score
RF	90.30%	88.86%	90.45%	89.18%
KNN	84.75%	86.19%	84.75%	84.96%
MNB	82.50%	82.25%	82.50%	80.18%
DT	83.35%	82.75%	83.10%	83.10%
BNB	82.00%	81.66%	82.00%	78.92%
AB	65.00%	59.12%	65.00%	59.75%
LR	91.75%	87.89%	91.75%	89.74%
SVC1	**92.75%**	**91.69%**	**92.75%**	**92.18%**
SVC2	89.25%	86.01%	89.25%	87.29%
SVC3	**92.75%**	**91.63%**	**92.75%**	**92.12%**

The assessment procedure's base value is the accuracy of residents' classification, which amounts to 84.37%. Six out of ten analyzed algorithms allowed for a better

Table 3. Classification results for the validation data sample

Algorithm	Acc.	Prec.	Recall	F1 Score
RF	**84.33%**	84.39%	84.28%	83.35%
KNN	78.00%	83.60%	78.00%	79.73%
MNB	74.50%	77.63%	74.50%	73.92%
DT	73.65%	74.97%	73.40%	73.70%
BNB	69.25%	76.17%	69.25%	70.43%
AB	40.75%	27.21%	40.75%	30.67%
LR	83.25%	**86.02%**	83.25%	81.61%
SVC1	**86.00%**	85.39%	**86.00%**	**85.50%**
SVC2	82.00%	**85.67%**	82.00%	79.83%
SVC3	**85.75%**	84.85%	**85.75%**	**85.13%**

classification than for the classification made by residents. Both the SVC1 and SVC3 algorithms allowed us to obtain the best variant: 92.75% and for improvement by 8.38% compared to the residents' accuracy (3.64% for the whole system).

For the validation sample, only two algorithms obtained a better result than the classification made by residents. These were SVC1 and SVC3 algorithms. Best result: 86% achieved the SVC1 algorithm and enabled an improvement of 1.63% (0.69% for the entire system).

The obtained results show that the SVC1 and SVC3 algorithms can successfully replace the current classification procedure, resulting in an improvement in the accuracy of the classification made by residents.

There are also presented the results of Precision, Recall and F1 score ratio calculation. All of the obtained metrics are on the acceptable level.

6 Discussion

Based on the obtained results, it can be concluded that the automatic classification based on ML algorithms allows to successfully replace the currently used classification procedure in the complaint reporting system.

The use of the ML-based classification will allow, on the one hand, to simplify the complaint reporting process, and on the other, to reduce the number of incorrectly classified complaints, which may result in a shorter response time. To determine the best classifier for the analyzed case, four basic measures were analyzed – Accuracy, Precision, Recall, and F1 score. The obtained results indicated that the best classifiers for the analyzed multi-label class problem are SVC algorithms.

Following this, from a practical point of view, the obtained results confirm the usefulness of the use of automatic classification for text-based datasets in the form of complaint reports related to the subject of waste management (causing difficulties in classification due to the small number of unique words within the classes) and characterized by the presence of Polish language (problematic in classification operations). However, it should be noted that the methods used are based on those used for English. As an initial step, we wanted to test the applicability of basic, widely used methods to our dataset. Newer methods are planned to be implemented in the future to improve the obtained results.

The waste management system is susceptible to seasonal changes and external factors. Following this, further research analysis should be focused on investigating their influence on classification process accuracy. This will allow for a better understanding of the relationships occurring in the system and for the development of possible changes in the presented approach in order to achieve an even higher accuracy level.

References

1. Kannangara, M., Dua, R., Ahmadi, L., Bensebaa, F.: Modeling and prediction of regional municipal solid waste generation and diversion in Canada using machine learning approaches. Waste Manag. **74**, 3–15 (2018)
2. Abbasi, M., El Hanandeh, A.: Forecasting municipal solid waste generation using artificial intelligence modelling approaches. Waste Manag. **56**, 13–22 (2016)
3. Kontokosta, C.E., Hong, B., Johnson, N.E., Starobin, D.: Using machine learning and small area estimation to predict building-level municipal solid waste generation in cities. Comput. Environ. Urban Syst. **70**, 151–162 (2018)
4. Rutqvist, D., Kleyko, D., Blomstedt, F.: An automated machine learning approach for smart waste management systems. IEEE Trans. Ind. Inf. **16**(1), 384–392 (2020)
5. Ferrer, J., Alba, E.: BIN-CT: Urban waste collection based on predicting the container fill level. BioSystems **186**, 103962 (2019)
6. Khoa, T.A., Phuc, C.H., Lam, P.D., Mai, L., Nhu, B., Trong, N.M., Thi, N., Phuong, H., Dung, N., Van, T.-Y.N., Nguyen, H.N., Ngoc, D., Duc, M.: Waste management system using IoT-based machine learning in University. Wirel. Commun. Mob. Comput. **2020**, 1–13 (2020)
7. Solano Meza, J.K., Orjuela Yepes, D., Rodrigo-Ilarri, J., Cassiraga, E.: Predictive analysis of urban waste generation for the city of Bogotá, Colombia, through the implementation of decision trees-based machine learning, support vector machines and artificial neural networks. Heliyon **5**(11), e02810 (2019)
8. Niska, H., Serkkola, A.: Data analytics approach to create waste generation profiles for waste management and collection. Waste Manag. **77**, 477–485 (2018)
9. Vu, D.D., Kaddoum, G.: A waste city management system for smart cities applications. In: 2017 Advances in Wireless and Optical Communications, RTUWO 2017, pp. 225–229 (2017)
10. Adedeji, O., Wang, Z.: Intelligent waste classification system using deep learning convolutional neural network. Procedia Manuf. **35**, 607–612 (2019)
11. Tarun, K., Sreelakshmi, K., Peeyush, K.P.: Segregation of plastic and non-plastic waste using convolutional neural network. In: IOP Conference Series: Materials Science and Engineering, vol. 561, no. 1, p. 012113 (2019)
12. Sousa, J., Rebelo, A., Cardoso, J.S.: Automation of Waste Sorting with Deep Learning. In: Proceedings - 15th Workshop of Computer Vision, WVC 2019, pp. 43–48 (2019).
13. John, N.E., Sreelakshmi, R., Menon, S.R., Santhosh, V.: Artificial neural network based intelligent waste segregator. Int. J. Sci. Eng. Res. **10**(4), 367–370 (2019)
14. Walkowiak, T., Malak, P.: Polish texts topic classification evaluation. In: ICAART 2018 - Proceedings of the 10th International Conference on Agents and Artificial Intelligence, vol. 2, pp. 515–522 (2018)

15. Dzisevic, R., Sesok, D.: Text classification using different feature extraction approaches. In: 2019 Open Conference of Electrical, Electronic and Information Sciences, eStream 2019 – Proceedings, pp. 1–4 (2019)
16. Kadhim, A.I.: Survey on supervised machine learning techniques for automatic text classification. Artif. Intell. Rev. **52**(1), 273–292 (2019)
17. Kowsari, K., Meimandi, K.J., Heidarysafa, M., Mendu, S., Barnes, L., Brown, D.: Text classification algorithms: a survey. Information (Switzerland) **10**(4), 1–68 (2019)
18. Hossin, M., Sulaiman, M.N.: A Review on evaluation metrics for data classification evaluations. Int. J. Data Min. Knowl. Manag Process **5**(2), 01–11 (2015)

Migration of Unit Tests of C# Programs

Anna Derezińska(✉) and Sofia Krutko

Institute of Computer Science, Warsaw University of Technology, Nowowiejska 15/19, 00-665 Warsaw, Poland

A.Derezinska@ii.pw.edu.pl

Abstract. Maintenance of a project with unit tests could require moving from one test platform to another. It can be established by an automated test re-generation or by transformation of the test code. The latter is recommended in case of a high quality test set. Preservation of the intrinsic knowledge introduced by the test developers could also be of high importance. NUnit and MSTest of Visual Studio are among the most popular unit test frameworks for C# programs. The ability of test transformation from NUnit to MSTest has been investigated. Transformation rules were implemented in a prototype tool integrated with Visual Studio, but only a subset of possible constructions have their straightforward equivalents. Experiments confirmed the potential benefits of the approach, but also limitations of the target tests.

Keywords: Unit test migration · Test maintenance · C# · NUnit · MS Test · Legacy code

1 Introduction

Creation of unit tests is an important task in program development [1, 2]. Unit tests are regarded as a specification notion, e.g., in test-driven development [3], agile approaches, and other methodologies. Maintenance of long-living software requires keeping up with changes of a production code and modifications of an environment, which also refer to unit tests associated with the code.

Tests of C# programs could be prepared using many platforms. NUnit [4], developed similarly to JUnit for Java, has been rewritten as a specific .NET solution since its 3rd version. This free tool has still been widely used and counted as the best unit testing tool in 2020 according to the Software Testing Help service [5]. Testing in the .NET environment [6] has also been supported by MS Visual Studio Testing Framework [7], which includes MSTest to prepare and run unit-like tests. Tests using both platforms are often applied in real-word programs, while NUnit has a longer history and offers more capabilities. The testing support of MS VS has been steadily extended and MSTest benefits from the tight integration with the development environment.

As no definite unit test leader for C# programs exists, there could be different reasons for migrating tests from one platform to another, such as: client requests, code reuse in another project, organizational changes, compatibility requirements for code integration or outsourcing, reducing of the number of platforms to be supported in an enterprise, etc. In general, there are two basic approaches to cope with this problem:

W. Zamojski et al. (Eds.): DepCoS-RELCOMEX 2021, AISC 1389, pp. 53–62, 2021.
https://doi.org/10.1007/978-3-030-76773-0_6

1. Automatic generation of test cases for the target test platform; the tests are based on the given application code.
2. Transformation of test cases from one test platform to another.

Both approaches have their pros and contra. Automatic tests could meet different criteria and show the high ability to detect faults [8]. On the other hand, generated tests could not take into account specific domain or logical constrains. If a project is accompanied with a set of high-quality tests or if manually tests contain a valuable specification and expert experience, there could be worthwhile to transform the existing tests into the tests of a target platform.

In this paper, we deal with unit test migration from NUnit to MSTest. The main contributions of the work are: (i) preparation of the transformation rules taking into account all NUnit attributes and possible assertion structures, (ii) implementation of a prototype tool integrated with MS VS, (iii) experimental evaluation of the approach.

The paper is organized as follows. The next Section describes the background and related work. Basic issues of the test migration are reported in Sect. 3. Tool support and experiments are discussed in Sect. 4. Finally, Sect. 5 concludes the paper.

2 Background and Related Work

Preparation and maintenance of unit tests is a labor-intensive activity. Therefore, a lot of research has been done on their automation. Test generation has achieved promising advances [8]. However, maintaining generated tests can take more time, and using test generation tools could result in creating more tests than manually [9].

Software maintenance refers not only to a production code but also to various kinds of its tests. Test cases can be treated in different ways. In disposable testing, tests are not maintained but substituted by automatically generated tests from the code [10]. Comparison of the latter and former tests is recommended to inspect the tests, reveal some changes, and throw away unnecessary tests. Broken tests that do not compile can be treated by a test repair procedure. Analyzing of a data-flow graph has been applied to rebuild tests that preserve the intent of the original tests [11].

However, testing activities cannot only be restricted to automation [12]. The most important thing in testing is thinking, analyzing, and creating good effective tests. Hence, having such tests we want to keep them going in the future software revisions.

Not all kinds of tests are equally worthwhile to be maintained. Automatically generated tests are often successful in fulfilling different coverage conditions or detecting given classes of faults, but could be not very realistic and hardly to read and comprehend [9]. Initial high-quality tests from test-driven development could be treated as a program specification, but sometimes TDD could result in poorly maintainable tests.

Maintenance of tests could be indispensable in legacy systems. Although some authors treat legacy code as a code without tests that need to be supplemented [13], in general, legacy systems could have long live cycles, a degraded structure, lack of documentation, but implement an important business logic. The experiments showed that the lack of unit tests created at the beginning provides to a hardly testable code. Efficiency of manual tests could be complemented with automatically generated tests, giving the best results for the hybrid solutions. Therefore, it could be profitable to retain existing tests, but the decision should be based on a cost analysis [14]. Addressing 10–20 year live cycles, as many government and military programs live, there should be decided whether to replace or maintain different types of test platforms.

Most of the work discussed above has been performed for Java programs. A set of guidelines for preparing readable and maintainable unit tests and working with legacy code in C# can be found in [15]. Building of integration tests for.NET platform has been reported in [16]. This approach is also based on the code analysis provided by Roslyn [17], but the ideas have not been realized yet. However, this direction could be combined with the test migration support to enhance the test creation.

Necessity of test migration can be motivated by different factors. Tests are run many times, need to be looked at and modified. Therefore, except of the efficiency in detecting errors, also readability or performance are the important obligations [2]. We could be interested in the preservation of the good features of a test set.

The problem of test migration between mobile applications has been addressed in [18]. GUI-based test cases written using the Espresso testing framework are analyzed and categorized according to different GUI elements. The tests of an Android application are adopted to another one that shares a common functionality.

To the best of our knowledge, migration of C# unit tests and automation of such a process has not been reported so far.

3 Migration of NUnit Tests into MS Test

Translation of unit tests requires modification of attributes, test code, and assertions. Directives (e.g., *using*) will also be converted accordingly.

3.1 Attributes

Attributes, a kind of metadata, are used in unit tests for specification of test cases and description of the environment behavior during test execution. NUnit uses more than 40 attributes to manage assemblies, classes, and test methods. Attributes associated with their parameters are used to control threads, execution time, value of test parameters, and other test features.

Translation of tests with all NUnit attributes has been analyzed. Only part of these attributes have their equivalents in MSTest. Functionality of another part of the attributes could be fully or partially realized using other MSTest mechanisms. The remaining attributes have no support in the structures of MSTest.

Detailed translation rules for all attributes of NUnit have been proposed. These rules could be clustered into translation schemata that refer to the attribute processing and modification of the corresponding test code. Six translation schemata have been identified (Table 1). The attributes of a schema are listed in the last column.

An attribute can be substituted by another one, removed, or converted into a comment. If an attribute is translated, its name is changed accordingly, and attribute parameters are modified, or additional attributes are added if necessary. In some cases, the equivalent attribute can encompass a broader or a narrower domain, or additional constraints have

Table 1. Attribute translation schemata.

Schema	Description	Attributes
TE-NM	Attribute is Equivalently Translated and test code Not Modified	Category, Description –for method, Ignore, LevelOfParallelism, MaxTime, NonParallelizable, Paralleizable – for assembly, Property, Range, Sequential, SetUp, TearDown, Test, TestCaseSource, TestFixture, TestOf, Values
TE-M	Attribute is Equivalently Translated and test code Modified	Author, OneTimeSetUp, OneTimeTearDown, TestCase, TestFixtureSetup, TestFixtureTeardown
TP-NM	Attribute is Partially Translated and test code Not Modified	Combinatorial, Explicit, Pairwise
R-NM	Attribute is Removed and test code Not Modified	Apartment, Description – for class or assembly, NonTestAssembly, Paralleizable – for method or class, Repeat, Retry
R-M	Attribute is Removed and test code Modified	DefaultFloatingPointTolerance, SetCulture, SetUICulture
C-C	Attribute and test Code are converted into Comments	Culture, Theory (Datapoint, DatapointSource), Order, Platform, Random, RequiresThread, SetUpFixture, SingleThreaded, TestFixtureSource, Theory, ValueSource

to be met. There could also be some related attributes. In these cases a sub-dependent attribute can be handled accordingly to the main attribute. They are shown in parentheses in the list of attributes in (Table 1).

An attribute could also be removed from the target test, usually when there is no equivalent attribute and the removal has no influence on the test result.

Sometimes, the functionality of the attribute cannot be reflected, and the attribute has an impact on the test result. In those cases, the attribute and the test will be not deleted but converted into a comment. Such commented tests will be omitted in the target test suite or could be manually adopted towards the desired notation.

Apart from attributes, the test code could be changed. The main modifications refer to the addition of a piece of code to the method or to a class under concern.

3.2 Assertions

There are two approaches to specifying assertions in NUnit: Classical Model and Constraint Model.

In the classical model, separate methods are used to verify different constrains concerning a unit under test and its behavior. The methods are collected in the following classes: *Assert*, *StringAssert*, *CollectionAssert*, *FileAssert*, and *DirectoryAssert*. Many commonly used methods have their equivalents in MSTest (Table 2). These methods will be translated and their parameters adjusted accordingly, e.g., swapped parameters in the methods from the StringAssert class. The remaining methods are not straightforwardly translatable and would be converted into comments.

Table 2. Methods from the NUnit classical model that have equivalents in MSTest.

Class	Methods
Assert	True, False, Null, NotNull, AreEqual, AreNotEqual, AreSame, AreNotSame, Pass, Fail, Ignore, Inconclusive
StringAssert	StartsWith, EndsWith
CollectionAssert	AllItemsAreInstancesOfType, AllItemsAreNotNull, AllItemsAreUnique, AreEqual, AreEquivalent, AreNotEquivalent, Contains, DoesNotContain, IsSubsetOf, IsNotSubsetOf, IsSupersetOf, IsNotSupersetOf

Constraint model is an alternative approach to express conditions to be verified. All assertions are specified with one method *Assert.That*. The first parameter of the method takes some data to be verified, and the second parameter represents the assertion logic. For example, the following assertion checks whether the given string variable includes the requested text.

```
Assert.That(someString, Is.EqualTo("Hello"));
```

The constraint model is recommended to be used in NUnit. The *Assert.That* method can be overloaded in a hundred ways, therefore, many assertions can be specified. However, MSTest supports only a classical model of assertions. Therefore, only a subset of assertions of the constraint model will be translated into their equivalents (Table 3). The remaining assertions will be commented.

Table 3. A subset of constraint model assertions (NUnit) and their equivalents (MSTest).

NUnit	MSTest
Assert.That([…], Is.EqualTo([…]))	Assert.AreEqual([…],[…])
Assert.That([…], Is.Not.EqualTo([…]))	Assert.AreNotEqual([…],[…])
Assert.That([…], Is.Null)	Assert.IsNulll([…])
Assert.That([…], Is.Not.Null)	Assert.IsNotNulll([…])
Assert.That([…], Is.True)	Assert.IsTue([…])
Assert.That([…], Is.False)	Assert.IsFalse([…])

A simple example illustrates the translation of attributes of TE-NM and TE-M schemata.

NUnit test method (before translation):

```
[TestCase(12, 3, ExpectedResult = 4)]
[Category("TestCaseCategory")]
public int DivideTest(int n, ind d)
{   return n/d; }
```

MSTest test method (after translation) with an assertion that has been added:

```
[DatRow(12, 3, 4)]
[TestCategory("TestCaseCategory")]
[TestMethod]
public void DivideTest(int n, ind d, int expectedResult)
{   Assert.AreEqual(n / d, expectedResult); }
```

4 Experimental Evaluation of Test Migration

We describe now experiments on the automated migration of unit tests of C# programs from the NUnit (version 3) to MSTest platform (version 2).

4.1 Test Translator – Automated Support for Test Migration

Experiments on the test migration from NUnit to MSTest have been supported by a prototype Test Translator [19]. The tool is integrated with the MS Visual Studio, as its extension. The translation engine uses the .NET Compiler Platform (Roslyn) [17] to handle the code in the form of an Abstract Syntax Tree.

Processing of the original unit tests consists of two phases: analysis and transformation. During the analysis phase, the code is scanned and expected modifications are recognized. At first, modifications are specified that refer to the whole set of unit tests. They could correspond to attributes of the whole assembly, *using* directives, etc. Then, all test classes are analyzed and their transformations collected, e.g., modifications of class attributes, modification of class declaration. Finally, the test methods are examined and their transformations identified, which could modify the method attributes, parameters, and code.

During the transformation phase, all previously specified modifications are performed and the target tests are generated.

4.2 Subject Programs

The experiments have been conducted on three programs of different origin and goals. The first one, *Benchmark*, was a simple program with a comprehensive test set that was developed to cover all attributes of NUnit and a variety of assertion structures. The second subject was an open source code, a part of *OpenRA* [20] game engine with tests prepared and published by their developers [21]. The third example, *IntroToNUnit* [22], was a loan handling program. It was aimed at learning testing with NUnit, in particular, using assertions in the constraint model [23]. The basic complexity metrics of the programs are given in Table 4, where:

- *LSC, Lines of Source Code* – the number of source code lines including comments and blank lines.
- *LEC, Lines of Executable Code* – the number of operations in the executable code.
- *NC, Number of Classes* – the number of classes in a project or assembly calculated as the object number of the *ClassDeclarationSyntax* type.

Table 4. Characteristics of the subject programs.

Programs	LSC	LEC	NC
1) Benchmark	15	1	1
2) OpenRA (classes of Game and Common modules)	4802	1354	35
3) IntroToNUnit	270	60	8

4.3 Results of Test Migration

While using TestTranslator, the sets of tests accompanied with the subject programs have been translated from NUnit into MSTest. The programs were tested with both variants of the test sets and the corresponding code coverage checked. Software metrics of the tests are given in Table 5. The consecutive rows correspond to subjects 1)-3) and their tests before (NUnit) and after (MSTest) translation. Apart from the metrics specified in the previous section, the following values have been provided:

- *NM, Number of Methods* - the number of method declarations in a project or assembly calculated as the object number of the *MethodDeclarationSyntax* type.
- *NA, Number of Assertions* – the number of assertions used in a project or assembly, calculated as the object number of the *InvocationExpressionSyntax* type included in the classes with an assertion token.
- *NTC, Number of Test Cases* – the sum of the number of nonparametric test methods plus the number of input parameter sets for the parametric test methods.
- *CA, Code Coverage* – the percentage of code lines covered by the unit tests during a program execution.

The metrics change is summarized in Table 6. It shows the relation of a metric after translation to its value before translation, in percent. A result of 100% denotes no change in the values, more than 100% stands for the increase, and less than 100% for the decrease of the corresponding metrics.

Table 5. Comparison of the software metrics of the tests before and after translation.

Test platform	LSC	LEC	NC	NM	NA	NTC	CA
1) NUnit	629	189	18	63	101	90	60.0%
1) MSTest	649	127	15	48	57	51	60.0%
2) NUnit	1330	672	19	63	164	55	54.2%
2) MSTest	1396	601	19	63	103	55	47.1%
3) NUnit	431	118	7	32	39	34	55.8%
3) MSTest	431	61	7	28	12	23	32.6%

Table 6. Change of test characteristics after test translation [in %].

Program	LSC	LEC	NC	NM	NA	NTC	CA
1) Benchmark	103.2	67.2	83.3	76.2	56.4	56.7	100
2) OpenRA	105	89.4	100	100	62.8	100	87
3) IntroToNUnit	100	51.7	100	87.5	30.8	35.3	58.4

4.4 Discussion of Results

The subject programs have different characteristics and various goals, hence their test results are diverse.

The Benchmark program (1) was intended to use all NUnit elements that could be modified during the test translation. It included attributes from all translation schemata as well as assertions written in the classic but also in the constraint model. In result, 3 classes, 15 methods, and 44 assertions were not converted into the corresponding target units. Hence, the number of code lines executed in the tests (LEC) was lowered. The number of source code lines (LSC) was not lowered, as all not translated units remained as comments. The remaining test parts were transformed according to the given rules, and the outcome of all translated tests was the same as the outcome of the original tests. The number of test cases lowered, but as some tests referred to the same code units, the coverage measures obtained for both test sets were identical.

Results of the Benchmark test translation have been used for verification of the approach and the translation rules in particular. Though, it was intentionally an artificial program and its results could not be generalized to usual application programs.

The modules and tests of OpenRA (2) have been prepared for purposes of the application development, independently of the experiment under concern. It can be observed that all test classes and test methods have been translated and the number of test cases remained unchanged (100% change of NC, NM, and NTC).

Unit tests of OpenRA consisted mainly of test methods with a single attribute - *TestCase*. This attribute could be used for passing argument values to a parametrized test method. However, in this program, the test methods were not parametrized. The *TestCase* attribute served as an indicator of a test method and was associated with an argument with a test name. Therefore, after translation, this attribute was converted into two attributes: *TestMethod* - to point at the test method and *TestProperty* which stores a test name. This is also a main cause of the rise in the number of code lines (LSC) after the test translation.

In OpenRA tests, a variety of assertions from the classical and constraint models have been used. Most of them were translated, but about 60 had no equivalents in MSTest (drop in NA) and were commented. Therefore, these tests could have required additional effort to get the complete original tests.

A special case concerned six assertions of the constraint model that changed test outcomes and caused the tests to fail. There were assertions structured as `Assert.That([..], Is.EqualTo([..]))` that were used for comparing collections. They were directly transformed into assertions of the form `Assert.AreEgual.` This assertion checked whether its arguments are equal, while the tests needed comparison of the collection content, and therefore, application of the `CollectionAssert.AreEqual` assertions. This modification was not anticipated before, and would have been handled automatically only after a successful identification of the cases during the test code analysis.

Unit tests of the third program, IntroToNUnit, were intentionally designed to present various kinds of parameterized NUnit tests and were mainly based on the constraint model of assertions. They used many mechanisms that are not implemented in MSTest. Therefore, the number of executable code lines lowered significantly. The number of target assertions also decreased considerably.

5 Conclusions

Automating of a test migration can assist the test maintenance, especially for long-living systems. This problem has been considered in the context of C# programs with original NUnit tests and target MSTest. A set of translations was developed, taking into account attributes and assertions that could encounter in a testing code.

Application of the tool support gave promising results, while all test classes, test methods, and test cases of a part of the production application OpenRA were successfully transformed. About 87% of the statements covered by the original tests remained also covered by the target ones. However, the lack of many assertion mechanisms in the target notation of unit tests caused missing almost 40% of the assertions.

The translation results highly depend on the kind of tests used in an original program. In the case of tests based on the assertion constraint model, the straightforward translation not always could be applicable. Consequently, in an application of such kind, only about 30% of assertions and 35% of test cases were transformed and 58% of the code lines stayed covered by tests.

Although the number of subject programs was too small to generalize the outcomes, we could observe the benefits of the migration automation and the drawbacks due to non-transformed or partially transformed test cases. The approach could be further enhanced with a more comprehended analysis of the test code. It would allow translating more kinds of assertions. Furthermore, the approach could be aimed at different platforms, such as xUnit.net; and/or combined with the automated test generation, which could supplement missing test cases according to the given criteria.

References

1. Daka, E., Fraser, G.: A survey on unit testing practices and problems. In: IEEE 25th International Symposium on Software Reliability Engineering, pp. 201–211. IEEE Computer Society (2014). https://doi.org/10.1109/ISSRE.2014.11
2. Fields, J.: Working effectively with unit tests. Leanpub (2014)
3. Beck, K.: Test Driven Development: by Example. Addison-Wesley Professional, Boston (2002)
4. NUnit. https://nunit.org/, Accessed 28 Dec 2020
5. Most popular Unit testing tools in (2020). https://www.softwaretestinghelp.com/unit-testing-tools/, Accessed 28 Dec 2020
6. Ritchie, S.: Pro.NET best practices. Apress, New York (2011)
7. MS Test V2 framework. https://github.com/microsoft/testfx, Accessed 28 Dec 2020
8. Ramler, R., Klammer, C., Buchgeher, G.: Applying automated test case generation in industry: a retrospective. In: International Conference on Software Testing, Verification and Validation Workshops, pp. 364–369. IEEE (2018). https://doi.org/10.1109/ICSTW.2018.00074
9. Shamshiri, S., Rojas, J.M., Galeotti, J.P., Walinshaw, N., Fraser, G.: How do automatically generated unit tests influence software maintenance? In: 11th International Conference on Software Testing, Verification and Validation, pp. 250–261. IEEE Computer Society (2018). https://doi.org/10.1109/ICST.2018.00033
10. Shamshiri, S., Campos, J., Fraser, G., McMinn, P.: Disposable testing: avoiding maintenance of generated unit tests by throwing them away. In: 39th International Conference on Software Engineering, pp. 207–209. IEEE/ACM (2017). https://doi.org/10.1109/ICSE-C.2017.100
11. Li, X., d'Amorim, M., Orso, A.: Intent-preserving test repair. In: 12th IEEE Conference on Software Testing, Validation and Verification (ICST), Xi'an, China, pp. 217–227. IEEE Computer Society (2019). https://doi.org/10.1109/ICST.2019.00030
12. Roman, A.: Thinking-driven testing. Springer, Cham (2018). https://doi.org/10.1007/978-3-319-73195-7
13. Ramler, R., Winkler, D., Schmidt, M.: Random test case generation and manual unit testing: substitute or complement in retrofitting tests for legacy code? In: 38th Euromicro Conference on Software Engineering and Advanced Applications, Cesme, Izmir, pp. 286–293, IEEE Computer Society (2012). https://doi.org/10.1109/SEAA.2012.42
14. Kent, J., Dewey, M.: Legacy test systems—Replace or maintain. In: 2016 Autotestcon, Anaheim, CA, 2016, pp. 1–5. IEEE (2016). https://doi.org/10.1109/AUTEST.2016.7589645
15. Osherove, R.: The art of unit testing with examples in C#, 2nd edn. Manning Publications Co., Shelter Island (2014)
16. Saadatmand, M.: Towards automating integration testing of .NET applications using Roslyn. In: International Conference on Software Quality, Reliability and Security (Companion Volume), pp. 573–574. IEEE Computer Society (2017). https://doi.org/10.1109/QRS-C.2017.99
17. Roslyn .NET compiler. https://github.com/dotnet/roslyn, Accessed 28 Dec 2020
18. Behrang, F., Orso, A.: Test migration between mobile apps with similar functionality. In: 34th International Conference on Automated Software Engineering (ASE), San Diego, CA, USA, 2019, pp. 54–65. IEEE/ACM (2019). https://doi.org/10.1109/ASE.2019.00016
19. Krutko, S.: Translation of NUnit tests to MSTest in C# programs. Bachelor Thesis, Warsaw University of Technology (2020). (in polish)
20. OpenRA – real-time game engine. https://www.openra.net/, Accessed 03 Jan 2021
21. OpenRA code. https://github.com/OpenRA/OpenRA, Accessed 28 Mar 2020
22. Introduction to .NET Testing with NUnit3. https://www.pluralsight.com/courses/nunit-3-dotnet-testing-introduction, Accessed 03 Jan 2021
23. IntroToNUnit. https://github.com/irmoralesb/IntroToNUnit, Accessed 22 Apr 2020

Synchronization and Scheduling of Tasks in Fault-Tolerant Computer Systems with Graceful Degradation

Mieczyslaw Drabowski(✉)

Department of Theoretical Electrical Engineering and Computing Science, Cracow University of Technology, Warszawska 24, 31-155 Kraków, Poland
drabowski@pk.edu.pl

Abstract. The paper presents a deterministic operational model for fault-tolerant computer systems and selected examples in executed such systems. This model can specify general problems and specific instances of problems about the behaviors, and structures of such systems. For these problems it is possible then to conduct research about computational complexity and, as a consequence, develop algorithms for solving these problems. This model define of optimized problems and for engineering practice an optimal or suboptimal solution is needed, and for these latter, preferably with a known or at least estimated distance from the optimal solution. These problems are computational generally NP-complete (precisely, their decision versions), so in practical application utility procedures usually calculate suboptimal but polynomial solutions. Such procedures are primarily an area of research interest because of practical motivations. The paper includes also specific example computer system with self-testing and self-repairs. Presented the example is the system with diagnostic tasks and as a consequence is he fault-tolerant and characterized by graceful degradation. A strategy for tasks scheduling in this system, both for functional tasks (external, which come to the system) and inside testing was proposed.

Keywords: Scheduling · Synchronization · Allocation · Dependable · Fault-tolerant · Graceful degradation

1 Introduction

Dependability is a feature of the system that reflects the user's degree of trust in the system. It manifests itself the continuity of functioning of the hardware and installed programs.

To achieve high system dependability it is necessary to:

1. During the design, construction and production of components and system integration - avoiding mistakes.
2. Before starting to use the system - troubleshooting.
3. While the system is running - limiting the damage caused by failures.

W. Zamojski et al. (Eds.): DepCoS-RELCOMEX 2021, AISC 1389, pp. 63–73, 2021.
https://doi.org/10.1007/978-3-030-76773-0_7

4. Introduction to internal control system and use of redundant modules - ensuring fault tolerance and graceful degradation capability.

Additional costs: synthesis and designing, production, implementation, maintenance and service of systems with the above dependability may drastically increase all the costs of such a system.

Elimination of the occurrence of damage to a component of a computer system is in general impossible, so it is justified to analyze the possibility of making the system capable of surviving damage to its components.

A higher level of dependability can be achieved at the expense of system efficiency. Dependability is a feature of the system more important than efficiency. The inefficiency of the system can be compensated for, but it is difficult to regain its dependability.

Dependability is a system concept that integrates hardware and software. Both the hardware can break and the software can crash.

2 The Model for the Problem of Computer Fault Tolerant System Synthesis

The substantive basis point for constructing our approach to the issues of task synchronization and their scheduling in fault-tolerant systems is the general deterministic theory of task scheduling. The theory may serve as a methodological basis for fault tolerant multiprocessor systems design. Accordingly, decomposition of the general task scheduling model is suggested, adequate to the problems of fault tolerant systems design.

Deterministic scheduling problems are described by three objects [1, 2]: the computing environment R comprising processor set P and other resources A, where $(R = P \cup A)$, task set T, and the optimality criterions.

Let $P = \{P_1, P_2, \ldots, P_m\}$ denote the set of processors. Two classes of processors can be distinguished: Parallel processors (general, CPU, Central Processor Unit) and Specialized processors (dedicated, embedded, ASIC, Application Specific Integrated Circuit). In the case of parallel processors each processor can execute any task. Parallel processors are divided into three classes: identical processors where all processors execute all tasks with the same speed, uniform processors where different processing speeds independent of the tasks, and unrelated processors for which execution speed depends on the processor and on the tasks. Specialized processors are dedicated devices performing different functions. For example, we often say that dedicated processors such as arithmetic, graphic, signal processors, vector etc. In certain situations even identical processors can behave like specialized one.

Apart from the processors there can be also a set $A = \{A_1, A_2, \ldots, A_s\}$ of additional resources, each available in A_i units $(i = 1, 2, \ldots, s)$. Computer memory, shared variables, storage, input/output devices are examples of these resources. Tasks may be acquire resources, and thus exclude concurrent execution of other tasks.

The second constituent of model for scheduling problems is the task subset. First, we explain the relation between certain concepts in parallel processing and operating systems, and the notations used in scheduling models. An application (or a program) ca be executed (at least potentially) by many processors working concurrently. A task is the basic unit in scheduling theory. The relation between the task, application and thread or stream depends on the schedule model. We consider task set *T*. A model should take into account the tasks, which may be either preemptable or nonpreemptable. Tasks are preemptable when each task can be interrupted and restarted later without incurring additional costs. In such a case the schedules are called to be preemptive. Otherwise, tasks are nonpreemptable and schedules no preemptive. Preemptability of tasks in our approach cannot be a feature of the searched schedule – as in the task scheduling model so far. The schedule contains all assigned tasks with individual attributes: preemptive, non-preemptive.

General the set of tasks is divided into k subsets [3, 4] $T^1 = \{T_1^1, T_2^1, \ldots, T_{n1}^1\}$, $T^2 = \{T_1^2, T_2^2, \ldots, T_{n2}^2\}, \ldots, T^k = \{T_1^k, T_2^k, \ldots, T_{nk}^k\}$, where $n = n1 + n2 \ldots + nk$. Each task T_i^1, $i = 1, 2, \ldots, n$, requires exactly one of the processors for its processing and its processing time is equal to t_i^1. Similarly, each task T_i^q, where $1 < q < k$, required q arbitrary processors simultaneously for its processing during a period of time which length is equal to t_i^q. In particular, we will consider the T_i^1 and T_i^2 tasks, which are single-processor and dual-processor tasks with execution times t_i^1 and t_i^2, respectively.

Tasks can be dependent or independent. The dependency is usually represented by digraph with nodes representing tasks, and directed edges representing precedence constraints [5]. The dependency relation between T_i and T_j is denoted by $T_i \prec T_j$.

A schedule is an assignment of tasks to processors and resources in time satisfying the following requirements.

- Each task is assigned to all resources needed for its execution.
- For each task pair $T_i \prec T_j$, task T_i is completed before T_j starts.
- Nonpreemptable tasks are not interrupted, and preemptable tasks are interrupted a limited number of times.

The following criteria of optimality are usually considered: costs of system implementation, and his dependability, i.e. reliability, availability, safety, security, fault tolerant and graceful degradation. In addition, system operating speed and energy consumption are considered [6, 7]. Such criteria are also taken into account in particular for computer systems.

Let us assume for example that fault tolerant system resources include universal parallel processors and specialized processors [8, 9]. As for tasks, we assume one-processor tasks used for modeling usable preemptable/nonpreemptable and dependent tasks, and two-processor tasks, for which we assume are time and resource non-preemptable. Two-processor tasks model the system testing tasks (e.g. one processor checks the other). Testing tasks may be dependent on the defined time moments of readiness to perform and to complete assigned tasks [10, 11]. Two-processor tasks (multiprocessor) and their schedules may realize a defined strategy of testing a fault tolerant system [12]. Modeling of tests with an arbitrator is possible with of three-processors tasks (or more processors-based) - e.g. one processor is tested by two others and a common negative test result disqualifies the tested processor.

3 Example of Computer System with Features Fault Tolerant and Graceful Degradation

We will consider the example of the computer system, heterogeneous, included hardware and software. The system will be implemented in the multiprocessors structure with features fault tolerant, self-testing and self-repair realized during the normal operation of the system [13, 14].

3.1 System Specification

The system – Fig. 1 – is consist of eight parallel and identical processors connected in a self -testing structure with testing tasks: T_{ij}, $i \neq j$, $i, j = \{1, 2, 3, 4, 5, 6, 7, 8\}$. Test task T_{ij} performs a test by processor "*I*" and the test concerns processor "*j*" [15].

For example, the following tasks are assigned to the P1 processor for P1 processor: $T_{12}, T_{21}, T_{13}, T_{31}, T_{14}, T_{41}, T_{15}, T_{51}, T_{16}, T_{61}, T_{17}, T_{71}, T_{18}, T_{81}$. The same is true for other processors. Each processor is connected to every other processor and can be tested and can also test, every other processor. The execution times for dual-processor testing tasks are – we assume - 1 unit each.

This is a classic model of the structure of a multiprocessor and dependable system. This structure allows hardware redundancy at the processors level and you can use it in it a strategy for monitoring system operation - the use of Multiple Modular Redundancy - for checking processor actions [16].

The hardware potential of this system is control and used by its operating system. Figure 2 shows a Gantt chart depicting the system schedule showing a certain strategy for self-testing of processors.

A Gantt chart is a convenient method of presenting schedules graphically. In these charts time is represented on the horizontal axis while the processors are placed along

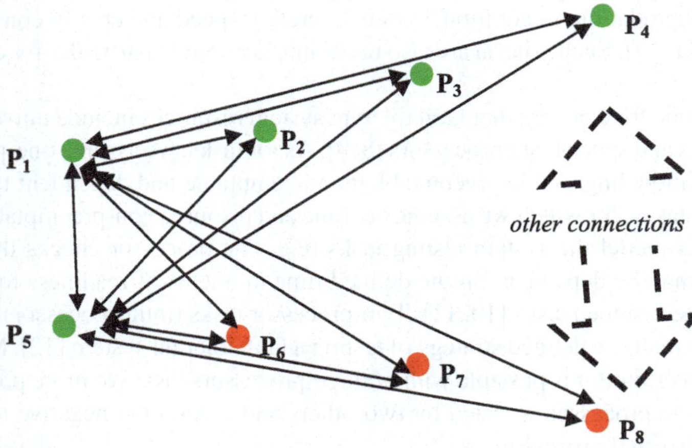

Fig. 1. Eight-processor system in a fully interconnected structure; five processors in action (P1, P2, P3, P4, P5) and three redundant processors (P6, P7, P8)

time	*1*	*2*	*3*	*4*	*5*	*6*	*7*	*8*	*9*	*10*	*11*	*12*	*13*	**14**
P1	T_{12}				T_{13}			T_{14}		T_{15}	T_{21}			
P2	T_{12}	T_{23}				T_{24}			T_{25}		T_{21}	T_{32}		
P3		T_{23}	T_{34}		T_{13}		T_{35}					T_{32}	T_{43}	
P4			T_{34}	T_{45}		T_{24}		T_{14}					T_{43}	T_{54}
P5				T_{45}			T_{35}		T_{25}	T_{15}				T_{54}

Fig. 2. A five-processor system in a fully connected structure

the vertical axis. In this chart, in the schedule there are presented the allocation the tasks (or their parts) to the processors.

The system is of course intended for performing utility tasks [15] – different applications. Let exemplary utility tasks be the following set of divisible, dependent tasks with specific execution times: $t_0= 3$, $t_1= 2$, $t_2= 2$, $t_3= 1$, $t_4= 3$, $t_5= 3$, $t_6= 4$, $t_7= 3$, $t_8= 1$, $t_9= 1$ - (Fig. 3).

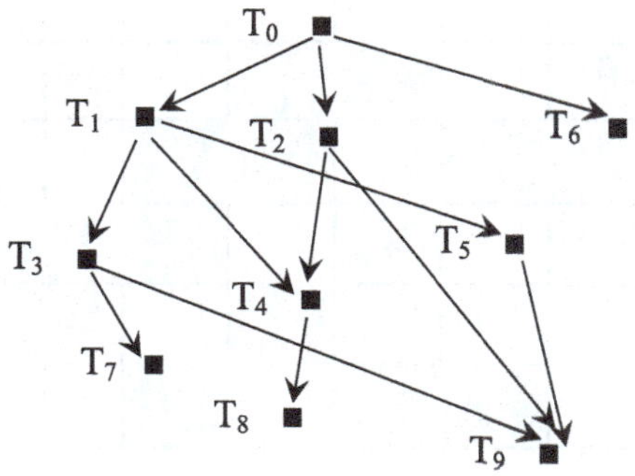

Fig. 3. A set of utility (external) tasks – digraph of precedence constraints

The operating system allocates processors testing tasks according to a certain strategy. As already mentioned, it is that testing of subsequent processors is adopted and the notation T_{ij} means that the task T tests the processor *j-th* and is performed on the processor *i-th*.

After checking the entire cycle - for five processors this cycle will take ten quanta of time - testing is resumed, then the processor being tested and testing is changed, now the processor *i* is tested by *j*. After two testing cycles, the testing order is repeated etc. Figure 4 shows the allocation of the T_0 utility task to the system.

time	*1*	*2*	*3*	*4*	*5*	*6*	*7*	*8*	*9*	*10*	*11*	*12*	*13*	14
P1	T_{12}				T_{13}			T_{14}		T_{15}	T_{21}			
P2	T_{12}	T_{23}				T_{24}			T_{25}		T_{21}	T_{32}		
P3	T_0	T_{23}	T_{34}		T_{13}		T_{35}					T_{32}	T_{43}	
P4		T_0	T_{34}	T_{45}		T_{24}		T_{14}					T_{43}	T_{54}
P5			T_0	T_{45}			T_{35}		T_{25}	T_{15}				T_{54}

Fig. 4. The allocation of the T_0 utility task to the system in time *(1, 3]* for processors

Further, the operating system assigns subsequent utility tasks to perform by processors, e.g. an example allocation of utility tasks T_1, T_2, T_6 at time $t = 4$ is shown in Fig. 5.

The allocation of other and all utility tasks will take place within a maximum of *10 units* - Fig. 6.

The system has completed all utility tasks and at the same time completed tasks that check the operation of the system. If the processor fails, it is possible to automatically reconfigure the system and continue its operation.

time	*1*	*2*	*3*	*4*	*5*	*6*	*7*	*8*	*9*	*10*	*11*	*12*	*13*	14
P1	T_{12}			T_1	T_{13}			T_{14}		T_{15}	T_{21}			
P2	T_{12}	T_{23}		T_2		T_{24}			T_{25}		T_{21}	T_{32}		
P3	T_0	T_{23}	T_{34}	T_6	T_{13}		T_{35}					T_{32}	T_{43}	
P4		T_0	T_{34}	T_{45}		T_{24}		T_{14}					T_{43}	T_{54}
P5			T_0	T_{45}			T_{35}		T_{25}	T_{15}				T_{54}

Fig. 5. The allocation of the T_1, T_2 and T_6 utility tasks to the system in time (4] for processors

For example, suppose that at $t = 6$, the T_{24} test detected a processor malfunction. A processor test running on a *P2* processor was testing a *P4* processor and a positive result (that it detected a malfunction) of the test does not mean that the *P4* processor is invalid. It could happen that the *P2* processor is not functional correct and pointed to the incorrectly functioning *P4* processor, which was working properly. The T_{24} test is therefore only an indication that one of these processors (*P2 or P4*) is malfunctioning. The performance of the T_{24} test at $t = 6$ is shown in Fig. 7.

Using a pool of redundant processors, the system switches ("on the fly") processors *P2* to *P6* and *P4* to *P7* while continuing to perform operational (external) tasks. All

time	*1*	*2*	*3*	*4*	*5*	*6*	*7*	*8*	*9*	*10*	*11*	*12*	*13*	**14**
P1	T_{12}			T_1	T_{13}	T_3	T_4	T_{14}	T_4	T_{15}	T_{21}			
P2	T_{12}	T_{23}		T_2	T_1	T_{24}	T_5	T_4	T_{25}	T_7	T_{21}	T_{32}		
P3	T_0	T_{23}	T_{34}	T_6	T_{13}	T_5	T_{35}	T_5	T_7	T_8		T_{32}	T_{43}	
P4		T_0	T_{34}	T_{45}	T_2	T_{24}	T_6	T_{14}	T_9	T_9			T_{43}	T_{54}
P5			T_0	T_{45}	T_6	T_6	T_{35}	T_7	T_{25}	T_{15}				T_{54}

Fig. 6. The allocation of all utility tasks

"suspect" processors are isolated [17, 18]. There is no waste of time performing tasks in the system – Fig. 8. Suspicious processors may still be tested and their diagnosis may indicating exactly which processor is faulty. This diagnosis can be carried out, for example, later, with less load on the system with the implementation of utility tasks.

time	*1*	*2*	*3*	*4*	*5*	*6*	*7*	*8*	*9*	*10*	*11*	*12*	*13*	**14**
P1	T_{12}			T_1	T_{13}	T_3		T_{14}		T_{15}	T_{21}			
P2	T_{12}	T_{23}		T_2	T_1	T_{24}			T_{25}		T_{21}	T_{32}		
P3	T_0	T_{23}	T_{34}	T_6	T_{13}	T_5	T_{35}					T_{32}	T_{43}	
P4		T_0	T_{34}	T_{45}	T_2	T_{24}		T_{14}					T_{43}	T_{54}
P5			T_0	T_{45}	T_6	T_6	T_{35}		T_{25}	T_{15}				T_{54}

Fig. 7. The performance of the T_{24} test at $t = 6$

time	*1*	*2*	*3*	*4*	*5*	*6*	*7*	*8*	*9*	*10*	*11*	*12*	*13*	**14**
P1	T_{12}			T_1	T_{13}	T_3	T_4	T_{17}	T_4	T_{15}	T_{61}			
P6	T_{12}	T_{23}		T_2	T_1	T_{24}	T_5	T_4	T_{65}	T_7	T_{61}	T_{36}		
P3	T_0	T_{23}	T_{34}	T_6	T_{13}	T_5	T_{35}	T_5	T_7	T_8		T_{36}	T_{47}	
P7		T_0	T_{34}	T_{45}	T_2	T_{24}	T_6	T_{17}	T_9	T_9			T_{47}	T_{57}
P5			T_0	T_{45}	T_6	T_6	T_{35}	T_7	T_{65}	T_{15}				T_{57}

Fig. 8. Switching of processors: P2 to P6 and P4 to P7

If there was no pool of redundant processors in the system or this pool was exhausted (after all processors were connected from it), it is possible to run through the system the so-called graceful degradation of the system [19]. The behavior of the system for this case is presented when there are 5 processors in the system, and the test task indicated a malfunction of the *P2* and *P4* processors pair (without the possibility of connecting the processors from the redundant pool) on Fig. 9, Fig. 10 and Fig. 11.

Beginning with $t = 7$, utility tasks are not performed on processors that are "suspicious", i.e. *P2* and *P4*. At $t = 7$, the T_{12} test is performed, checking the *P2* processor by the test performed on the *P1* processor, and the *P4* processor is state "suspended" – Fig. 9.

At time $t = 8$, test T_{14} is performed, the *P2* processor is "suspended". The T_4 utility task can be performed on a tested and functional *P3* processor – Fig. 10.

Additional tests T_{12} and T_{14} enable unambiguous indication of a damaged processor – e.g. let it be a *P4* processor, this processor will be isolated in the system – the system may drive it to a high impedance state.

time	*1*	*2*	*3*	*4*	*5*	*6*	*7*	*8*	*9*	*10*	*11*	*12*	*13*	14
P1	T_{12}			T_1	T_{13}	T_3	T_{12}	T_{14}		T_{15}	T_{21}			
P2	T_{12}	T_{23}		T_2	T_1	T_{24}	T_{12}		T_{25}		T_{21}	T_{32}		
P3	T_0	T_{23}	T_{34}	T_6	T_{13}	T_5	T_{35}					T_{32}	T_{43}	
P4		T_0	T_{34}	T_{45}	T_2	T_{24}	X	T_{14}					T_{43}	T_{54}
P5			T_0	T_{45}	T_6	T_6	T_{35}		T_{25}	T_{15}				T_{54}

Fig. 9. The system at $t = 7$: test T_{12} is run and processor *P4* is suspended, at this time no utility task is performed

time	*1*	*2*	*3*	*4*	*5*	*6*	*7*	*8*	*9*	*10*	*11*	*12*	*13*	14
P1	T_{12}			T_1	T_{13}	T_3	T_{12}	T_{14}		T_{15}	T_{21}			
P2	T_{12}	T_{23}		T_2	T_1	T_{24}	T_{12}	X	T_{25}		T_{21}	T_{32}		
P3	T_0	T_{23}	T_{34}	T_6	T_{13}	T_5	T_{35}	T_4				T_{32}	T_{43}	
P4		T_0	T_{34}	T_{45}	T_2	T_{24}	X	T_{14}					T_{43}	T_{54}
P5			T_0	T_{45}	T_6	T_6	T_{35}	T_5	T_{25}	T_{15}				T_{54}

Fig. 10. The system at $t = 8$: test T_{14} is run and processor *P2* is suspended, at this time the utility task T_4 is performed, allocated to processor *P3*

In the presented emergency situation, the system will perform utility tasks in *12* quantum of time.

The system has completed all utility tasks and at the same time completed tasks that check the operation of the system. Failure of the *P4* processor did not cause the entire system to freeze. There has been a system automatic dynamic process reallocation - process migration from the damaged processor and isolation of the damaged processor. The system continues to operate, but with limited capabilities. There has been a graceful degradation of the system from a five-processor system to a four-processor system – Fig. 11.

time	*1*	*2*	*3*	*4*	*5*	*6*	*7*	*8*	*9*	*10*	*11*	*12*	*13*	14
P1	T_{12}			T_1	T_{13}	T_3	T_{12}	T_{14}	T_6	T_{15}	T_{21}	T_7		
P2	T_{12}	T_{23}		T_2	T_1	T_{24}	T_{12}	X	T_{25}	T_4	T_{21}	T_{32}		
P3	T_0	T_{23}	T_{34}	T_6	T_{13}	T_5	T_{35}	T_4	T_5	T_7	T_8	T_{32}	T_{43}	
P4		T_0	T_{34}	T_{45}	T_2	T_{24}	X	T_{14}	X	X	X		X	X
P5			T_0	T_{45}	T_6	T_6	T_{35}	T_5	T_{25}	T_{15}	T_9			T_{54}

Fig. 11. The system has completed all utility tasks, damaged processor *P4* is isolated, the four-processor system is still working properly

4 Conclusions

The model and the example presented in this paper are an attempt of formal description of problems and illustration of the action, which are characteristic for fault-tolerant systems and particularly for scheduling of tasks occurring in these systems. A special feature of such systems is the redundant structure. When designing the system, it is necessary to calculate what system resources are needed (such as the processing power of processors, memory address space, or the characteristics of bus connections and interfaces) to fully implement all utility tasks and then apply appropriate redundancy.

For the proposed system's relevance to real systems [20], one should take into account the processes of communication between resources and tasks, preventing resource conflicts, as well as extend the available resources sets, for example by programmable and configurable structures implemented by hardware.

A characteristic feature of the behavior of fault-tolerant systems is self-testing by the use for example of dual-processor tasks and self-diagnosis proposed in this article, which in turn leads to the system obtaining fault tolerance and ensuring his a graceful degradation.

The article presents selected methods of ensuring resistance to damage and self-testing. The redundant system is particularly important for real-time systems, therefore the schedule should take into account the optimization criterion of scheduling tasks ahead of schedule [21]. The above issues are now studied.

References

1. Błażewicz, J., Ecker, K., Plateau, B., Trystram, D.: Handbook on Parallel and Distributed Processing. Springer-Verlag, Heidelberg (2000)
2. Blazewicz, J., Ecker, K., Pesch, E., Schmidt, G., Sterna, M., Weglarz, J.: Handbook on Scheduling: From Theory to Practice. Springer, Cham (2019)
3. Błazewicz, J., Drabowski, M., Węglarz, J.: Scheduling multiprocessor tasks to minimize schedule length. IEEE Trans. Comput. **C-35**(5), 389–393 (1986)
4. Garey, M., Johnson, D.: Computers and Intractability: A Guide to the Theory of NP-Completeness. Freeman, San Francisco (1979)
5. Lee, C.Y.: Machine scheduling with availably constraints. In Leung, J.Y.T. (ed.) Handbook of Scheduling, pp. 22.1–22.13. CRC Press, Boca Raton (2004)
6. Yhang, Z., Dick, R., Chakrabarty, A.: Energy-aware deterministic fault tolerance in distributed real-time embedded systems. In: 41st Proceedings Design Automation Conference, Anaheim, California, pp. 550–555 (2004)
7. Schmitz, M.T., Al-Hashimi, B.M., Eles, P.: Energy-efficient mapping and scheduling for DVS enabled distributed embedded systems. In: Proceedings of the Design Automation and Test in Europe Conference, pp. 514–521 (2002)
8. Dick, R.P., Jha, N.K.: MOGAC: a multiobjective genetic algorithm for the cosynthesis of hardware-software embedded systems. In: Proceedings of the International Conference on Computer Aided Design, pp. 522–529 (1997)
9. Drabowski, M.: Boltzmann tournaments in evolutionary algorithm for CAD of complex systems with higher degree of dependability. In: Wojciech, Z., Jacek, M., Jarosław, S., Tomasz, W., Janusz, K (eds.) Advances in Intelligent Systems and Computing, Theory and Engineering of Complex Systems and Dependability: Proceedings of the Tenth International Conference on Dependability and Complex Systems DepCos-RELCOMEX- 2015, vol. 365, pp. 141–152, Springer, Heidelberg (2015)
10. Drabowski, M., Wantuch, E.: Coherent concurrent task scheduling and resources assignment in dependable computer systems design. Int. J. Reliab. Qual. Safety Eng. **13**(1), 15–24 (2006)
11. Elburi, A., Azizi, N., Zolfaghri, S.: A comparative study of a new heuristic based on adaptive memory programming and simulated annealing: the case of job shop scheduling. Eur. J. Oper. Res. **177**, 1894–1910 (2007)
12. Ziegenbein, D., Richter, K., Ernst, R., Thiele, L., Teich, J.: SPI – a system model for heterogeneously specified embedded systems. IEEE Trans. VLSI Syst. **10**(4), 379–389 (2002)
13. Drabowski, M.: Modification of concurrent design of hardware and software for embedded systems – a synergistic approach. In: Grzech, A., Świątek, J., Wilimowska, Z., Borzemski, L. (eds.), Information Systems Architecture and Technology: Proceedings of 37th International Conference on Information Systems Architecture and Technology – ISAT 2016, vol. 522, pp. 3–13. Springer, Heidelberg (2017)
14. Gajski, D.: Principles of Digital Design. Prentice Hall, Upper Saddle River (1997)
15. Graphs STG. https://www.kasahara.elec.waseda.ac.jp/schedule/
16. Pricopi, M., Mitra, T.: Task scheduling on adaptive multi-core. IEEE Trans. Comput. **C-59**, 167–173 (2014)
17. Drabowski, M.: Concurrent, coherent design of hardware and software embedded systems with higher degree of reliability and fault tolerant. In: Borzemski, L., Świątek, J., Wilimowska, Z., (eds.), Information Systems Architecture and Technology: proceedings of 39th International Conference on Information Systems Architecture and Technology – ISAT 2018, vol. 852, pp. 7–18, Springer, Heidelberg (2019)
18. Montgomery, J., Fayad, C., Petrovic, S.: Solution representation for job shop scheduling problems in ant colony optimization. LNCS **4150**, 484–491 (2006)

19. Drabowski, M., Kiełkowicz, K.: A hybrid genetic algorithm for hardware–software synthesis of heterogeneous parallel embedded systems. In: Świątek, J., Borzemski, L., Wilimowska, Z. (eds.), Information Systems Architecture and Technology: Proceedings of 38th International Conference on Information Systems Architecture and Technology – ISAT 2017, vol. 656, pp. 331–343, Springer, Heidelberg (2018)
20. Yen, T.Y., Wolf, W.H.: Performance estimation for real-time distributed embedded systems. IEEE Trans. Parallel Distrib. Syst. **9**(11), 1125–1136 (1998)
21. Agraval, T.K., Sahu, A., Ghose, M., Sharma, R.: Scheduling chained multiprocessor tasks onto large multiprocessor system. Computing **99**(10), 1007–1028 (2017)

Comparison of Selected Algorithms of Traffic Modelling and Prediction in Smart City - Rzeszów

Paweł Dymora(✉) and Mirosław Mazurek

Faculty of Electrical and Computer Engineering, Rzeszów University of Technology, al. Powstańców Warszawy 12, 35-959 Rzeszów, Poland
{Pawel.Dymora,Miroslaw.Mazurek}@prz.edu.pl

Abstract. In recent years, the smart city concept has become a very actual topic of many scientific publications and the implementation of many technical and IT solutions. The smart city is a city that uses information and communication technologies to increase the interactivity and efficiency of urban infrastructure and its components, as well as to raise the awareness of its inhabitants. The article presents Smart City's concepts and the prospects for their development in the context of the growing population of cities in the world. Nowadays cities are forced to combine various aspects in order to provide a high quality of life, comfort and a friendly environment. These aspects include areas related to the economy, environment, management and mobility. Using ARIMA - Autoregressive Integrated Moving Average and Exponential Smoothing algorithms, forecasts of growth of selected parameters from current solutions were carried out. In particular, the analysis and prediction of traffic were concentrated on the example of a selected city of Rzeszów in Poland, which has been a leader in the implementation of Smart City services for many years. Based on historical data, the correctness of predictions was assessed. Moreover, the directions and possibilities of development of smart cities as well as production organization in Industry 4.0 were determined.

Keywords: Open data · ARIMA · ETS · Data mining · Smart city · Industry 4.0

1 Introduction

Today, highly developed countries are witnessing a very rapid increase in digitization and the practical use of new solutions and technologies in many aspects of everyday life. This situation is also visible in production, industrial and monitoring solutions, as well as in the public sector, which allows for simple and automatic control, management and supervision. Urbanization is still on the rise. Today there are more than 7.5 billion people in the world, half of whom live in urban areas. In Poland, cities constitute about 7% of the whole country's area. In the world, this percentage amounts to 2% [1–3]. This makes it a priority to provide residents with services that save time, energy, financial resources and increase comfort and quality of life. More and more important are city

W. Zamojski et al. (Eds.): DepCoS-RELCOMEX 2021, AISC 1389, pp. 74–85, 2021.
https://doi.org/10.1007/978-3-030-76773-0_8

management methods, practical use of new solutions, and their use to minimize technological progress's negative effects. Such technologically advanced cities are referred to as "Smart City" or "Intelligent City" and take advantage of the latest developments, including the Internet of Things.

This concept has been developed in literature over the last several years, and it has been developed in the field of literature. The term is relatively complex due to the fact that it covers many areas and aspects. It focuses on the integration of existing and new technologies. During this time, many country and city rankings have been created for applications in selected areas. These include areas related to smart economy, environment, people, management, mobility, and quality of life. Each of these areas can be divided into smaller ones, and specific actions towards implementing the Smart City concept can be distinguished from them. Modern technologies connect all these areas, the development of which is currently very fast and at the same time allows to achieve specific goals to a large extent [4]. The ISO 37120 standard was developed in May 2014, which defines and establishes a methodology for a set of indicators to control and measure the performance of urban services and quality of life [5]. This standard was amended in 2018. In the context of such technological solutions and aspects, case studies have been used as a research method.

Smart City, or intelligent city, is defined as an urban area that uses different types of electronic sensors to collect data to provide information that is used to manage different types of resources efficiently. It facilitates many activities that have been more time-consuming or complicated so far. The term Smart City appeared at the beginning of the 21st century, although the first mention of the smart city appeared in literature in 1997 [4–6]. They mainly concerned solutions based on the World Wide Web soon after its creation. Due to the wide range of technologies that have been implemented under the concept of a smart city, it is difficult to define this term precisely. According to the IEEE, a smart city combines technology, government and society to enable the integration of areas such as smart cities, smart economy, smart mobility, smart environment, smart population, smart living conditions, smart management [6].

In the Smart City concept, many areas can be distinguished depending on the analysis's level of detail [7]. One of the most popular models is the one developed by the Technical University of Vienna. It distinguishes the 6 most important areas for cities that plan or implements the Smart City strategy. This method takes into account heterogeneity within groups and maintains metric information and achieves high sensitivity to changes. Based on these rankings, which consider a number of different criteria, it is possible to compare several different cities in a number of ways objectively. Without this approach, the results of comparisons could be unreliable, erroneous, and focused on a distinctive solution rather than on sum and overall assessment. These are the following areas and criteria [6, 7]:

- *Smart economy,* where we can distinguish such factors as innovative spirit, entrepreneurship, the image of the city, efficiency, the labor market, international integration;
- *Smart mobility*, where the main components are local transport system, availability (between) national, ICT infrastructure, durability of the transport system.

- *Smart environment,* with air quality (no pollution), environmental awareness and sustainable management of resources.
- *Smart population*, where the most important role is played by education, lifelong learning, ethnic plurality and openness.
- *Smart living conditions*, where one can distinguish such factors as cultural and recreational facilities, health conditions, individual security, quality of the housing, educational facilities, tourist attractiveness, social cohesion.
- *Smart governance* focused on political awareness, public and social services, and efficient and transparent administration.

2 Rzeszow Smart City and ISO Standard and Certification

In 2014, the ISO 37120 standard "Sustainable Social Development - Indicators of Urban Services and Quality of Life" was developed and updated in 2018. This standard also has an equivalent in Polish. It contains a set of indicators that can be used to measure and monitor the level of development by cities. These indicators have been classified into 17 thematic groups, which in total contain 100 indicators: 46 have a basic character (obligatory to obtain certification) and 54 - an additional character (optional or obligatory in case of applying for certification of a higher level) [5].

Certification under this standard is carried out, among others, by the World of Council on City Data (WCCD) organization with its registered office in Canada, which grants an international certificate or, for example, by the Polish Committee for Standardization, which grants a certificate at the national level [8]. Currently, 4 Polish cities have such certification: Lublin, Gdańsk, Gdynia and Kielce [9–12, 20]. There are more than 100 cities in the world holding such a certificate. Certification on the international level is the cost of approx. USD 7.5 thousand [13]. The level of the certificate depends on the number of indicators reported by the city:

- *Aspirational* – 30–40 basic indicators.
- *Brown* – 46–59 indicators (46 basic and 0–13 additional indicators).
- *Silver* – 60–75 indicators (46 basic and 14–29 additional indicators).
- Gold – 76–90 indicators (46 basic and 30–44 additional indicators).
- *Platinum* – 91–100 indices (46 basic and 45–54 additional indicators).

The certificate level depends on the CIMI value, i.e., the calculated final indicator through the function based on the weighted model of aggregation of partial indicators [14, 15]. The requirement of obtaining the certificate is to go through the entire certification path and present reliable data and measurement results confirming individual indicators' value.

The Municipality of Rzeszów is a place friendly to people, encouraging them to settle down, build families, and improve their quality of life. The center takes care of economic, social and environmental development. The best example of the colossal development of this area is the systematically developing aviation, chemical and IT industries, as well as trade and services. The academic and scientific sphere of the city also plays an important role [4]. Additionally, when analyzing the ranking data on the smart-cities.eu website from the point of view of Polish cities of medium size, one can notice that their development in a given aspect is very similar. For example, comparing Rzeszów, Bydgoszcz and Kielce, one can only see small fluctuations in specific areas (Fig. 1).

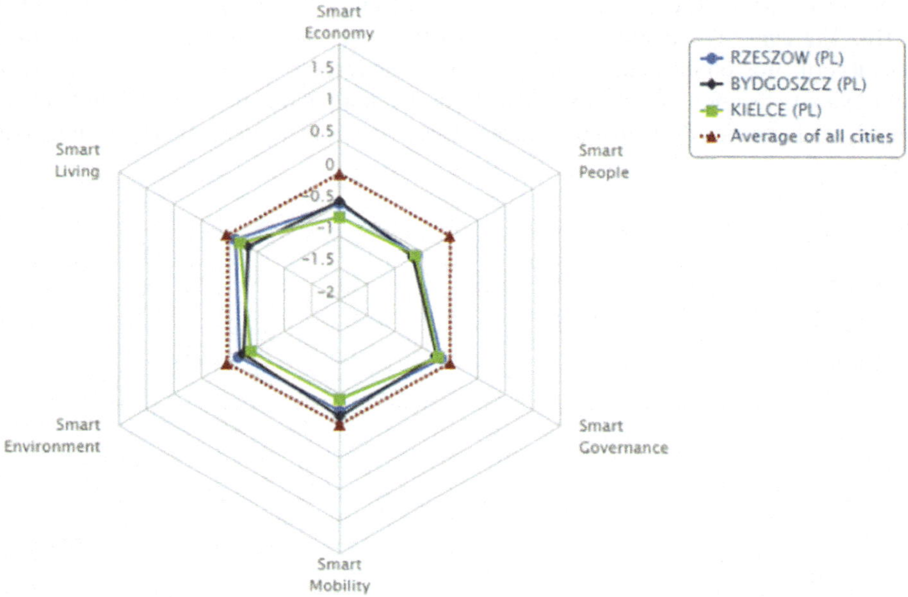

Fig. 1. Comparison of city profiles.

In Poland, however, it is challenging to find unified development strategies in this direction. Most of today's implementations are still single solutions, but they are related to Smart City. In this way, such solutions as intelligent traffic control, lighting, city monitoring, systems monitoring the number of available parking spaces, continually growing number of services allowing for non-cash or contactless payments, e-services, or local loyalty programs are available. Such solutions are already widespread, but they cover only a fraction of the aspects and solve only a part of the problems. The same applies to the issue of ecology. Legal issues and the growing awareness of society force the use of solutions limiting the level of emitted pollutants, solutions monitoring air quality or appropriate standards [5, 7].

According to the Philips report on the challenges and benefits of Smart City in Poland, residents have slightly different goals and priorities in relation to the investment [16]. In the ranking https://www.smart-cities.eu for European cities of 100–500 thousand and 0.3–1 million inhabitants, one can also find Polish cities, which, depending on the selected criteria, occupy higher or lower positions in the ranking. Nevertheless, they do not differ significantly from the European average and do not indicate gross errors or deficiencies in certain aspects. On the other hand, the ranking can be used to determine the direction in which the city should take steps and in which more development is needed. The ranking also allows determining the general direction of cities' development and solutions belonging to the Smart City concept [4, 17].

3 Traffic Modelling – Methodology

For the purposes of the article, the following data were collected from inductive loop readings installed in the surface and used to monitor and manage the traffic on the roads of Rzeszów. The data covered the period from 1 January 2018 to 31 December 2018. The analysis included data from 11 entry and exit roads, with daily intervals. The information on the completeness of data at the level of 97.88% was analyzed. The approximate location of inductive loops from the first point is presented on the map in Fig. 2. Each point on the map indicates the place of traffic measurement in both directions [4].

Fig. 2. Induction loop deployment.

In the years 2008–2014, Rzeszów Intelligent Transport System was created as part of the EU project. It consists of several elements. The main one is the area traffic control system. It includes 53 intersections, which are equipped with traffic lights. They are equipped with a combination of 68 PTZ cameras, which send images from the streets to the control center. By default, the system works to keep the green light on for as long as possible on the busiest routes. In addition, there are several emergency modes, which in the event of a collision, for example, allow eliminating obstacles. The system is also responsible for counting vehicles and assessing traffic volume through induction loops..

The collected data were imported into a MySQL database. Prior to analysis, the data was preprocessed. This operation consisted of grouping and aggregating daily data into monthly and annual statements.

To forecast traffic, ARIMA model - Autoregressive Integrated Moving Average was used, which was implemented in a script created in Python [18]. To verify the correctness of forecasts, Microsoft Excel used functions based on ETS - Exponential Smoothing algorithm and additionally implemented ETS model in a script created in R language.

3.1 ARIMA Model

This algorithm was created by the Russian scientist Vladimir Levenshtein, who discovered this equation in 1965. It is used to calculate the editing distance, which is why it is often called the Edit Distance algorithm. Levenshtein's distance is a metric sequence to measure the difference between two sequences. Informally, Levenshtein's distance between two words is the minimum number of edits of single characters (i.e., inserts, deletions, or substitutions) required to replace one word with another. Levenshtein's distance can also be referred to as the edit distance, although it can also mean a larger family of distance indicators. It is closely related to the alignment of sequences in pairs [3, 12–14].

The Levenshtein distance (Eq. 1) between two strings, *a* and *b* (of length $|a|$ and $|b|$ respectively), is given by $lev_{a,b}(|a|, |b|)$ where [3, 9–11]:

$$lev_{a,b}(i,j) = \begin{cases} \max(i,j) & \textit{if } \min(i,j) = 0, \\ \min\begin{cases} lev_{a,b}(i-1,j)+1 \\ lev_{a,b}(i,j-1)+1 \\ lev_{a,b}(i-1,j-1)+1_{(a_i \neq b_j)} \end{cases} & \textit{otherwise}. \end{cases} \tag{1}$$

where $1_{(a_i \neq b_j)}$ is the indicator function equal to 0 when $a_i \neq b_j$ and equal to 1 otherwise, and $lev\ a, b(i, j)$ is the distance between the first *i* characters of *a* and the first *j* characters of *b*. The first element in the minimum corresponds to deletion (from *a* to *b*), the second to insertion, and the third to match or mismatch, depending on whether the respective symbols are the same [12–14].

3.2 Jaro Winkler Algorithm

ARIMA model, i.e. autoregressive integrated model of moving mean, consists of 3 elements: autoregression (AR), degree of integration (differentiation) (I) and moving mean (MA) [18, 19]. Autoregression is a process in which the searched value is determined on the basis of previous values. The degree of integration presents the value of the difference between the successive values. The moving mean is a process similar to autoregression, but it focuses on the values of current and previous value disturbances. These elements form the ARIMA (*p, d, q*), model, where *p, d, q* is the value of autoregression, the degree of integration and the value of moving averages, respectively. The model is based on the adjustment of observed values and it aims to reduce the difference between the values produced in the model and the observed values as much as possible.

ARIMA is used in forecasts based on seasonal and non-seasonal data. It is based on several assumptions, mainly on data that do not contain anomalies, fixed model parameters, or continuous time series that do not change. It is one of the most reliable and effective ways of forecasting time series. In some cases, the name ARIMA and Box-Jenkins models are mentioned as synonyms.

Given the time series of X_t data, where t is an integer and X_t are real numbers, the model gives (Eq. 2):

$$(1 - \sum_{i=1}^{p'} \alpha_i L^i) X_t = (1 + \sum_{i=1}^{q} \theta_i L^i) \varepsilon_t \tag{2}$$

where L is the delay operator, α_i are the parameters of the autoregressive part of the model, θi are the parameters of the moving average and error conditions ε_t. Error terms are generally assumed to be independent, identically distributed sampling variables from the normal distribution with zero mean [18, 19].

An ARIMA(p,d,q) process expresses the polynomial factorization property with $p = p' - d$. The model gives (Eq. 3):

$$(1 - \sum_{i=1}^{p} \phi_i L^i)(1 - L)^d X_t = (1 + \sum_{i=1}^{q} \theta_i L^i) \varepsilon_t \tag{3}$$

and therefore can be considered a special case of the ARMA process $(p + d, q)$ having an autoregressive polynomial with the elements d of the unit. (For this reason, no ARIMA model with $d > 0$ is stationary).

3.3 ETS Model

The ETS model is based on the exponential smoothing algorithm [19, 20]. This algorithm is a method of forecasting time series for one-dimensional data, which can be extended to handle data with a systematic trend or seasonal component. It calculates or predicts subsequent values based on already existing (historical) values. A predicted value is a continuation of historical values on a specific date, which should be a continuation of a time series. This function, however, requires a timeline to be organized with a constant step between the successive points. For example, it could be a monthly or annual timeline with values on each month or year's first day.

The ETS model consists primarily of three values: the date or value for which the forecast is made, the values of historical data and the range of dates in the form of a time series. Optionally, it is possible to define seasonality (for example, with hourly, monthly or annual interval) or a parameter that determines the completeness of data (helpful in case of incomplete data sets). The syntax of this function is as follows: FORECAST.ETS (target date, values, time axis, [seasonality], [data completeness], [aggregation]). The exponential smoothing method can be considered equivalent and alternative to the popular ARIMA method for forecasting time series. The ETS algorithm is also commonly known as the Holt-Wints algorithm, after the scientists who described the features of this model.

The simplest form of exponential smoothing can be written as (Eq. 4):

$$s_t = \alpha \cdot x_{t-1} + (1 - \alpha) \cdot s_{t-1} = s_{t-1} + \alpha \cdot (x_{t-1} - s_{t-1}) \tag{4}$$

where α is the smoothing factor and $0 < \alpha < 1$. In other words, smoothed statistics s_t' is a simple weighted average of the current observation x_t and the previous smoothed statistics s_{t-1}. The concept of smoothing factor used here in relation to α is something confusing because higher values of α actually reduce the level of smoothing, and in a limiting case for $\alpha = 1$ output series is just an ongoing observation. Simple exponential smoothing is easy to apply and provides smoothed statistics as soon as two observations are available.

A values close to one have a smaller smoothing effect and attach more weight to recent data changes, while α values closer to zero have a larger smoothing effect and are less sensitive to recent changes. There is no formally correct procedure for choosing α. Sometimes statistics judgment is used to choose the right factor. Alternatively, a statistical technique can be used to optimize the α value. We can use the least-squares method to determine the value of α for which the sum of $(s_t - x_{t+1})^2$ is minimized [19, 20].

4 Implementation of the ARIMA Model in Orange 3 Environment

Orange 3 is a program for the visualization of open-source data and machine learning. It contains a set of data mining tools. It has a visual programming interface for exploratory analysis and interactive data visualization and can be used as a Python library. Orange 3 components are called widgets and include simple data visualization, subset selection and preprocessing, as well as the empirical assessment of learning algorithms and predictive modelling. Visual programming is carried out using an interface where workflows are

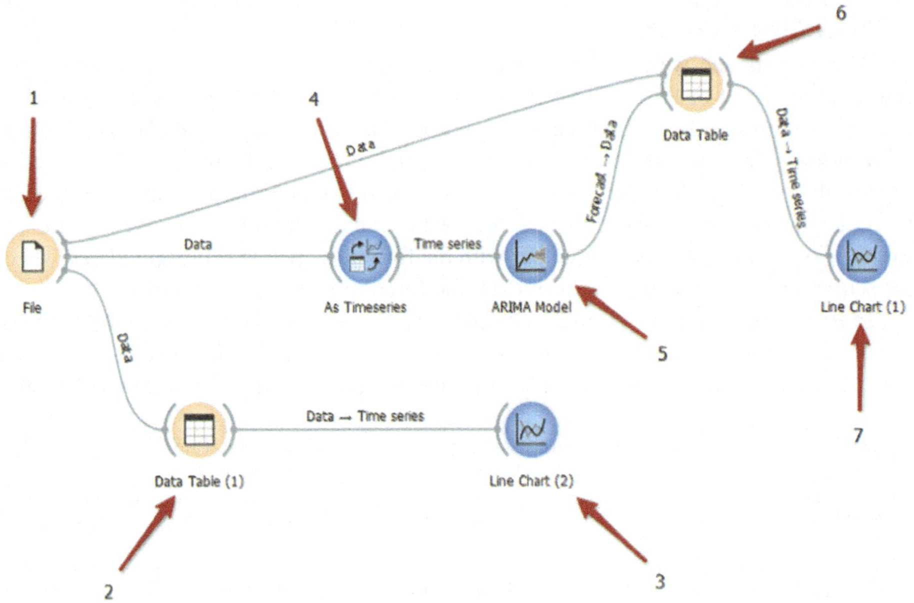

Fig. 3. Implementation of the forecasting model in the Orange 3 program.

created by combining predefined or user-designed widgets. In advanced and complex projects, users can use Orange 3 as a Python library to manipulate data and change widgets [21]. The analysis uses the Time Series widget, which allows the analysis of time series. A model was created in the program, the final form of which is shown in Fig. 3. The complete model used to forecast time series from the source data. In order to carry out the forecast, certain data were imported at the beginning (number 1). Next, a table of imported data (2) and a graph were created on the basis of the original source data (3). The source data were converted into time series (4) depending on the detected interval (days, months or years). Then, on the basis of data in the form of time series, forecasts of subsequent values were made on the basis of the ARIMA model (5). These data were transferred to table (6), from which graphs were created, which additionally contained lines of confidence intervals (7).

5 Results

On the basis of a selected fragment of data, as test data, the model has verified whether it meets the assumed objective. Generated data were compared with real data. This model, with noticeably high accuracy, was able to predict the next values, also taking into account the seasonality of the data. In this case, it was the number of vehicles entering Rzeszów in 2018. This is shown in the graph in Fig. 4 generated after the script was launched. A better match could be achieved with data from a more extended period of time. Similar results can also be obtained by creating a script in R.

In order to compare the results obtained in Orange, analogous forecasting in Microsoft Excel was performed using the ETS model [20]. For the analysis, available formulas from the statistical library were used: REGLINX.ETS and REGLINX.ETS.CONFINT.

The second one was used to determine the confidence intervals, both upper and lower, at the level of 95%. These intervals mean that 95% of future points should be located at a distance from the function forecast result, which the REGLINX determined ETS formula. Determination of confidence intervals allows for better and more effective determination of the accuracy of the predicted model. Obtaining a smaller confidence interval means that the forecast is more certain and accurate for a given point. An example of the analyzed table together with the formulas created is presented below. For subsequent values from the series of data, the forecast is made in an analogous way.

In Excel, the ETS model is available in two versions: AAA and AAN. In the analysis, the AAA version was used for data containing seasonality. All of the analyzed data sets were data from at least one year daily or monthly. Therefore this version was chosen.

Forecast (Warsaw Street Entry)	Lower confidence limit (Warsaw Street Entry)	Upper limit of confidence (Warsaw Street Entry)
REGLINX.ETS (A14;$B$2:$B$13;$A$2:$A$13;1;1)	C14REGLINX.ETS.CONFINT (A14;$B$2:$B$13;$A$2:$A$13;0,95;1;1)	C14 + REGLINX.ETS.CONFINT (A14;$B$2:$B$13;$A$2:$A$13;0,95;1;1)

Similarly, as in the case of the ARIMA model, prediction using the ETS model was also checked and compared with the actual data. In this case, it was also data on the number of vehicles entering Rzeszów in 2018, so that it was possible to compare the

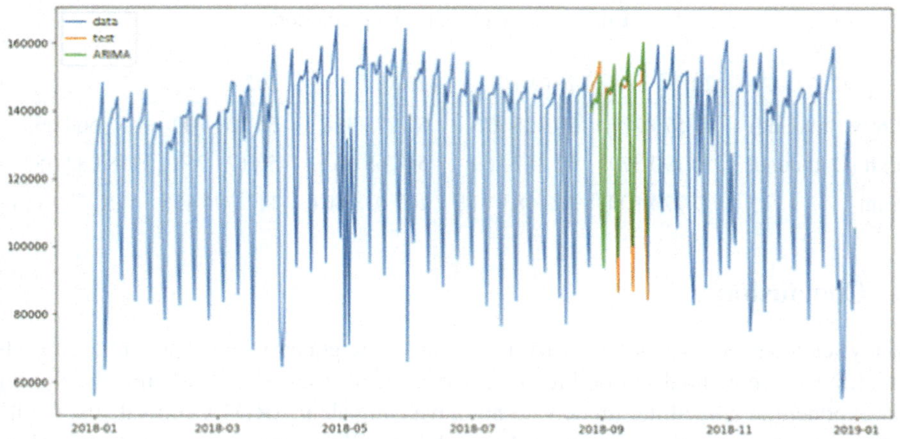

Fig. 4. Prediction with ARIMA model.

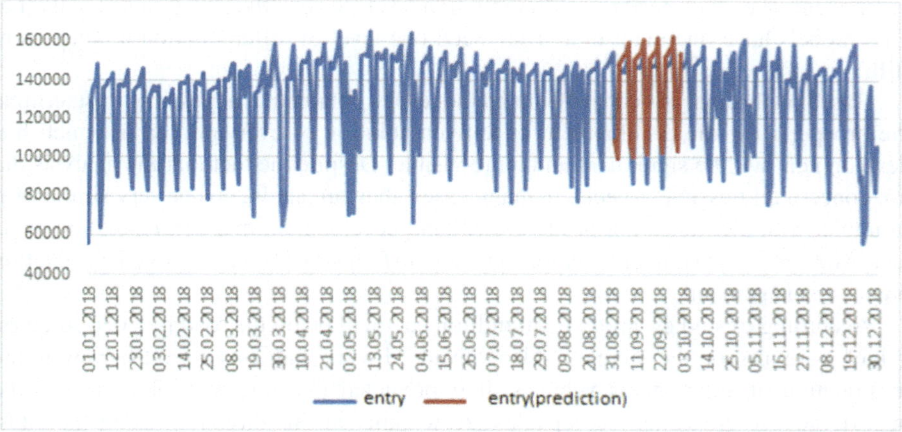

Fig. 5. Prediction with ETS model.

results with those obtained using the ARIMA model. The results are presented in Fig. 5. In this case, the forecast was also mainly in line with actual values. This makes it possible to obtain reliable and reliable predictive analyses. The prerequisite for a better match is a longer time frame analysis. The changes potentially appearing to be anomalous could be a normal increase or decrease with a seasonal frequency.

A detailed analysis of the data shows that the results are very close to each other and have a low confidence level. The prediction return of the number of vehicles in the future years is shown in Table 1.

Table 1. Sum of tours and prediction.

	2018	2019	2020	2021	2022
Entry (Inbound)	47 579 721	50 325 975	52 773 338	55 220 700	57 668 063
Exit (Outbound)	50 190 379	52 357 169	54 265 463	56 173 758	58 082 052
Sum	97 770 100	102 683 144	107 038 801	111 394 458	115 750 114

6 Conclusion

Many scientists believe that ARIMA models are more general than exponential smoothing. Linear, exponential smoothing models are special cases of ARIMA models. Nonlinear exponential smoothing models do not have equivalent ARIMA equivalents. On the other hand, there are also many ARIMA models that do not have exponential smoothing equivalents. In particular, all ETS models are non-stationary and only some ARIMA models are stationary. ETS models with or without seasonal trend, or both have two components (i.e., they need two differentiation levels to keep them stationary). All other ETS models have one main unit. They need one level of differentiation to keep them still.

Based on the analysis carried out and the results obtained, it is possible to determine the prospects for the development of a given solution or even an area in which this development will be significant and more visible. Despite the fact that the analysis did not concern all possible aspects, in many cases, they are analogous or very similar, for example, with the indication of the growth and development of a given solution or its lack. Moreover, such implementation may refer to Industry 4.0 and intelligent production process management.

According to studies, around 250 million cars will be on the road by 2020. Electric vehicles are also on the increase. Also, advanced work and tests are underway on the introduction of autonomous vehicles. It is predicted that this will take place in the next decade. A great many companies are working in this direction, using the latest technologies. Testing solutions is the most important and, at the same time, the longest phase of production. It is necessary to eliminate or minimize as much as possible the possibility of error occurrence. This leads to the time-consuming analysis of sensor readings and analysis of existing problems.

Acknowledgments. We are thankful to the graduate student Lukasz Marcola of Rzeszów University of Technology, for supporting us in the collection of useful information.

Funding. This work is financed by the Minister of Science and Higher Education of the Republic of Poland within the "Regional Initiative of Excellence" program for years 2019–2022. Project number 027/RID/2018/19, the amount granted 11 999 900 PLN.

References

1. https://stat.gov.pl/files/gfx/portalinformacyjny/pl/defaultaktualnosci/5468/7/13/1/powierzchnia_i_ludnosc_w_przekroju_terytorialnym_w_2016_r.pdf

2. https://www.eea.europa.eu/pl/articles/obszary-miejskie-2014-od-przestrzeni-miejskich-do-ekosystemow-miejskich
3. https://data.worldbank.org/indicator/SP.POP.TOTL
4. Dymora, P., Mazurek, M., Kowal, B.: Open data – an introduction to the issue. In: ITM Web Conference Computing in Science and Technology (CST 2018), vol. 21 (2018). https://doi.org/10.1051/itmconf/20182100017
5. https://www.iso.org/standard/62436.html
6. https://sites.ieee.org/isc2-2018/files/2018/05/IEEE-Smart-Cities.pdf
7. https://www.smart-cities.eu/index.php?cid=2&ver=4
8. https://www.pkn.pl/sites/default/files/sites/default/files/imce/files/Program%20certyfikacji%20PN-ISO%2037120.pdf
9. https://www.portalsamorzadowy.pl/komunikacja-spoleczna/lublin-dolaczyl-do-miast-z-certyfikatem-smart-city,122037.html
10. https://www.gdansk.pl/wiadomosci/co-laczy-gdansk-z-londynem-paryzem-i-berlinem-jakosc-zycia,a,88959
11. https://www.gdynia.pl/mieszkaniec/numer-biezacy,2790/gdynia-z-miedzynarodowym-certyfikatem-iso-37120-w-liczbach,515399
12. https://idea.kielce.eu/wccd_strona_glowna.html
13. https://www.surrey.ca/bylawsandcouncillibrary/CR_2015-R227.pdf
14. https://media.iese.edu/research/pdfs/ST-0471-E.pdf
15. https://media.iese.edu/research/pdfs/ST-0335-E.pdf
16. https://www.lighting.philips.pl/systemy/tematy/raport-cyfrowe-miasta
17. Śniegocki, A., Buchholtz, S., Bukowski, M.: Big and open data in Europe: A growth engine or a missed opportunity? Demos EUROPA WISE Inst. **1**(1), 1–116 (2014)
18. https://support.predictivesolutions.pl/ekspress/download/EB54/mat/model_ARIMA_istotnosc_parametrow.pdf
19. https://otexts.com/fpp2/arima-ets.html
20. https://freerangestats.info/blog/2016/11/27/ets-friends
21. https://orange3-timeseries.readthedocs.io/en/latest/widgets/arima.html

Dependability Analysis Using Temporal Fault Trees and Monte Carlo Simulation

Ernest Edifor[1(✉)], Neil Gordon[2], and Martin Walker[3]

[1] Manchester Metropolitan University, Manchester M15 6BH, UK
e.edifor@mmu.ac.uk
[2] University of Hull, Hull HU6 7RX, UK
[3] Dymodian Systems Ltd., Hull HU1 1TJ, UK

Abstract. The safety and reliability of high-consequence systems is an issue of utmost importance to engineers because such systems can have catastrophic effects if they fail. Fault Tree Analysis (FTA) is a well-known probabilistic technique for assessing the reliability of safety-critical systems. Standard FTA approaches are primarily static analysis techniques and as such cannot effectively model systems with dynamic behaviours, such as those with standby components or multiple modes of operation. There have been several efforts to address this limitation, one of which is Pandora, a temporal fault tree approach. Pandora uses three temporal gates—Priority-AND, Simultaneous-AND, and Priority-OR—to model the effects of sequences of events. Hitherto, Pandora was unable to perform a holistic evaluation of a full system that is repairable, taking account of useful system operating environment variables (such as time of operation, flow rate, etc.) or system data such as repair state and preventive maintenance. This paper aims to address these limitations. Algorithms to evaluate different system configurations have been generated and techniques for modelling and analyzing different system data in a simulation platform have been proposed. This paper extends the capabilities of Pandora so that it is capable of analyzing a modern system that features different failure modes, has diverse component failure distributions, considers the system's operation environment data, and models different system configurations. The outcome of such analysis enables analysts to understand the operation and dynamics of a system holistically and aids in the implementation of appropriate risk mitigating strategies.

Keywords: Temporal fault tree analysis · Dependability · Monte Carlo simulation

1 Introduction

Safety-critical systems are high consequence systems that can have detrimental effects on the environment or human life if they should fail. Fault Tree Analysis (FTA) [1] is one of the Probabilistic Risk Assessment (PRA) techniques used to evaluate the reliability of systems. FTA is typically used to identify the combinations of component faults (basic events) that can lead to the occurrence of a system failure (top event).

W. Zamojski et al. (Eds.): DepCoS-RELCOMEX 2021, AISC 1389, pp. 86–96, 2021.
https://doi.org/10.1007/978-3-030-76773-0_9

Fault trees are based on Boolean logic and most commonly use OR gates and AND gates to connect events. Fault tree analyses can be performed both qualitatively (logically) and quantitatively (probabilistically). The logical analysis involves using Boolean algebra to derive minimal cut sets (MCSs). An MCS is the smallest logical combination of basic events that will trigger the occurrence of a top event. Qualitative analysis helps to identify critical and non-critical components of the system, as well as providing a better understanding of the causes of system failures. Quantitative analysis involves the mathematical evaluation of the top event probability and the determination of importance measures (components contribute to the top event occurrence). The quantitative analysis gives analysts an indication of how likely a system is to fail, as well as the contribution of each component to the system failure.

Despite its widespread use, traditional FTA has some drawbacks. Various efforts have been made to extend FTA. One relatively recent solution is Pandora [2], a temporal fault tree technique. It maintains the structure of FTA but extends its laws and semantics with the introduction of three temporal gates – Priority-AND, Simultaneous-AND, and Priority-OR. These new gates enable the representation and analysis of sequences of events in fault trees.

Pandora's logical analysis capabilities allow it to eliminate redundancies and contradictions in the fault tree to produce a set of minimal cut sequences (MCSQs), which are analogous to MCS but which retain information about the order of events. Pandora is capable of performing a comprehensive qualitative analysis. Analytical and simulation approaches have been proposed to allow Pandora to perform quantitative analysis. However, these techniques are restricted to at least one of the following: unable to perform full quantitative analysis, limited to an exponential distribution, limited to non-repairable events, unable to capture system environment data (such as flow-rate), limited to producing only reliability data.

In this paper, a Monte Carlo based solution is presented to enable the full quantitative analysis of Pandora. This solution includes the definition of a simulation procedure for modelling and evaluating different types and combinations of MCSQs. An alternative solution is also prescribed to allow the dependability analysis of Pandora to take place by using an established simulation software platform based on Monte Carlo simulation. The second solution involves the creation and modelling of temporal gates and behaviours on the simulation platform. The outcome of these solutions produces different quantitative data such as reliability, availability, mean-time-to-failure, and criticality measures. These enable the quantitative analysis of modern systems that feature different component failure distributions, have repair and maintenance regimes, feature different system configurations, and are responsive to changes in their operating environment.

The remaining sections of this paper are arranged as follows: Sect. 2 describes temporal gates and the mathematical models and evaluations of these gates using analytical and simulation approaches. Section 3 contains new Monte Carlo simulation algorithms and modelling techniques for evaluating different types of MCSQs in temporal fault trees. Section 4 shows the application of the proposed solutions to a hypothetical autonomous underwater vehicle. Concluding remarks are given in Sect. 5.

2 Related Work

Due to the popularity and long history of FTA, many research efforts have focused on expanding FTA to enhance its capabilities and encompass a greater range of systems. One of the best-known techniques is the Dynamic Fault Tree (DFT) technique, an extension of static FTA that was proposed to tackle computer-based systems in which the outcomes are affected by the order of occurrence of events [3]. Like FTA, DFT has seen several further developments, extensions, and enhancements for both qualitative and quantitative analysis [4, 5], but it remains primarily a quantitative analysis technique. There are other approaches, such as the Temporal Fault Tree (TFT) technique by Palshikar [6]. A temporal fault tree is one in which temporal dependencies between events can be specified. Palshikar's extension of the classical FT introduces new operators and it is intended primarily to aid in the post-failure diagnosis of log data. Kabir [7] provides a critical review and evaluation of various FTA techniques.

2.1 Pandora

Pandora [2] is a more recent technique for analyzing temporal fault trees. It introduces three new gates: the Priority-AND (PAND), Simultaneous-AND (SAND), and Priority-OR (POR). The PAND (Priority-AND) is true if an input event occurs strictly before another input event in the left-to-right order (leftmost first). It only becomes true when its last input event occurs. The PAND is represented by the symbol '<', thus '$A < B$' indicates that A should have occurred before B.

The POR (Priority-OR) is true only if its first input event occurs before any other input event, but unlike the PAND, those other input events are not strictly required to occur. Thus whereas the PAND is a type of conjunction, the POR is a type of disjunction and represents a type of trigger condition. In Pandora, the symbol '|' represents POR; '$A|B$' means 'A POR B', and is true if either A occurs and B does not, or both occur, with A occurring before B.

The SAND (Simultaneous-AND) gate is used to represent the situation where all input events occur at the same time. '&' is used to indicate a SAND gate: '$A\&B$' means 'A SAND B', and is true if A and B occur simultaneously. Throughout this paper, Boolean AND is represented with the symbol '•' and Boolean OR with the symbol '+'.

2.2 Quantitative Analysis Using Analytical Approaches

In terms of quantitative analysis, techniques exist for the probabilistic evaluation of the Boolean gates (AND and OR) [1], PAND gate [7], and POR gate [8]. Nearly simultaneous events (that is, events that occur together within a small interval of time) can be evaluated using the pSAND according to a technique presented in [9]. Unless otherwise stated, it is assumed that E is a set of n basic events such that $E = \{X_1, X_2, \ldots X_{n-1}, X_n\}$ with corresponding failure rates $\lambda_1, \lambda_2, \ldots \lambda_{n-1}, \lambda_n$, where $0 \leq \lambda_i \leq 1$ respectively. Equations for the AND and OR gates can be found in [1]. Equations 1 to 3 are analytical formulae for the PAND, POR, and pSAND; it is assumed that all events have exponential failure distribution.

Henceforth, any reference to an analytical solution of PAND refers to the solution proposed by Fussell et al. [10] in (1), where $a_0 = 0$ and $a_m = -\sum_{j=1}^{m} \lambda_i$ for $m > 0$.

$$P\{X_n < X_{n-1} < \ldots < X_2 < X_1\}(\mathrm{t}) = \prod_{i=1}^{n} \lambda_i \sum_{k=0}^{n} \left[\frac{e^{(a_k t)}}{\prod_{\substack{j=0 \\ j \neq k}}^{n} (a_k - a_j)} \right] \tag{1}$$

The mathematical expression for calculating the probability of MCSQs with two or more POR gates at a time t is explained in [8]. The formula is shown in (2).

$$P\{X_1 | X_2 | \ldots | X_{n-1} | X_n\}(t) = \frac{\lambda_1 \left(1 - \left(e^{-(\sum_{i=1}^{n} \lambda_i)t}\right)\right)}{\sum_{i=1}^{n} \lambda_i} \tag{2}$$

Equation (3) is the formula for evaluating the pSAND gate for an MCSQs with two or more POR gates given that the interval between the occurrences of all events is d.

$$P\{X_1 \&_d X_2 \&_d \ldots \&_d X_n\}(t_0, t_1) = \sum_{i=1}^{n} \left(\left(1 - e^{\lambda_i t_0}\right) \bullet \left(\prod_{\substack{j=1 \\ j \neq i}}^{n} \left(1 - e^{\lambda_j (t_1 - t_0)}\right) \right) \right) \tag{3}$$

2.3 Quantitative Analysis Using Simulation

A simulation is a means of learning something about the real world by replicating a scenario using a model. Simulations are used in situations where real-world scenarios are financially costly, could be dangerous, are overly complicated to design, or are too time-consuming to implement. Monte Carlo (MC) simulation is a popular simulation technique used in various fields such as chemistry, engineering, medicine, games, finance, and telecommunications. MC begins with modelling the system under study. Once this is done, the model is simulated or 'run' by generating random numbers for the model variables to create a unique 'instance' of the model. The system variables are generated several times, called trials, to create several instances of the model. These instances are examined for some common predetermined property, which eventually determines the behaviour of the model.

Simulation algorithms have been developed [11] for the PAND, POR and pSAND gates. There are techniques [11] for evaluating MCSQs with multiple events but one type of operator, MCSQs with multiple events and multiple different operators, and MCSQs with different events, operators and failure distributions. Unfortunately, these techniques struggle to holistically analyze real-world systems with all of the following features: allows for repairs, maintenance, and replacements, considers other system data apart from failure data and capture different system configurations. In this paper, a full quantitative analysis of temporal fault trees which overcome these limitations will be presented using algorithms for simulation and a technique for modelling temporal fault trees in the Goldsim software.

3 Dependability Analysis

For an MCSQ with input events $X_1, X_2, \ldots, X_n$, the input events could be basic or intermediate. Regardless of the type of input events an MCSQ has, an intermediate event can be considered an input event depending on the investigators level of abstraction. Given that an input event X_i is basic, $P(X_i)\{t\} = F(X_i)\{t\}$, where the function P is the probability of failure and F is the cumulative distribution function (CDF). However, if an input event X_i is an intermediate event, $P(X_i)\{t\}$ will have to be evaluated based on the individual sub-input events of X_i. To evaluate X PAND Y, where X has an exponentially distributed failure rate of λ and Y has Weibull failure distribution with α and β representing the scale and shape parameters respectively, the simulation condition can be generated as $R_A <= F_E(\lambda_A)\{t\}$ && $R_B <= F_W(\alpha_B, \beta_B)\{t\}$ && $TTF_E(\lambda_A, R_A)\{t\} < TTF_W(\alpha_B, \beta_B, R_B)\{t\}$. Where F_E and F_W are the CDFs of the events A and B respectively and TTF_E and TTF_W are the time-to-failure (TTF) of A and B respectively.

3.1 Evaluating All MCSQs

The top-event evaluation of a temporal fault tree with only exponentially distributed component failure modes is straightforward but a simulation approach is relatively slower to compute and produces estimates, although it is not restricted to any particular failure distribution. Combining both analytical and simulation approaches to harnesses the strengths of both techniques has been proven to be very useful [12]. Algorithm 1 is proposed for the top event evaluation of a fault tree with n MCSQs. *X[n]* represents the *n*th MCSQ, *static* represents events with no dynamic gates and *FTA* is a function that evaluates Boolean expressions [1]. *ANA* is a function that evaluates non-static MCSQs with components that have only exponentially distributed failure modes using their analytical Eqs. (1), (2), and (3). *SIM* is a function that evaluates MCSQs with any combination of different failure distributions [11]. *EP* is a function that uses the Esary-Proschan formula in [13] to calculate the top event probability.

Algorithm 1. Evaluation of top-event of a dynamic system.

```
Require: X
Z ← 0
for i = 1 to n do
    if (X[i] is static) then
        Z ← Z + FTA (X[i])
    else
        if (X[i] is exp) then
            Z ← Z + ANA (X[i])
        else //not exponential distribution
            Z ← Z + SIM (X[i])
        end if
    end if
end for
return EP (Z)
```

3.2 Modelling in Goldsim

To ensure that temporal fault trees can capture repair and maintenance data and different system operating environment data [14] such as flow-rate, time of operation, etc., the Goldsim software [15] will be used. Goldsim is a software that allows the modelling and probabilistic analysis of complex systems based on Monte Carlo simulation. It has elements for representing static AND and OR Boolean gates. However, it has no pre-constructed elements for the PAND, POR, or pSAND gates and it cannot perform qualitative analysis. Once a qualitative analysis has been performed with Pandora, it can be modelled in Goldsim.

Dependability analysis in Goldsim can be performed in one of two ways: using Fault Tree Analysis (FTA) or using Requirement Tree Analysis (RTA). Figure 1 is an example of an FTA model of the PAND gate in Goldsim. The fuel cell is the primary power source so it is initially turned on while the diesel secondary source is turned off. The sensor needs to be turned on when the system starts so that it can activate the diesel engine if the fuel cell engine fails. When the fuel cell stops operating, the sensor activates the diesel system; when the fuel cell resumes operation, the diesel system is turned off. RTA is based on a set of conditions necessary for a system or component to succeed. RTA in Goldsim is rather simple to implement. For the entire dual-fuel engine to work, the input command must be issued, the engine and the fuel-cell or diesel sub-systems must be working. RTA can be used without the manual construction and modelling of temporal gates.

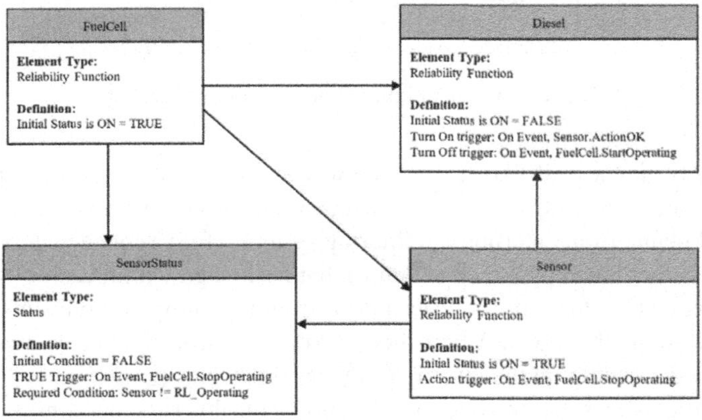

Fig. 1. Model of a hybrid fuel system in Goldsim.

It is well known that, unlike analytical approaches that produce exact results, simulation approaches produce approximated results. Unfortunately, there are no analytical approaches that address all the limitations stated earlier. The Goldsim simulation software uses state-of-the-art sampling algorithms to improve the accuracy of its results. Safety engineering seldom relies on a single analysis method. The results of a Monte Carlo simulation would be valuable as a guide, particularly early in the development process; some critical decisions are unlikely to be made based on simulation alone.

4 Case Study

To illustrate the above techniques, the authors have designed a hypothetical Autonomous Underwater Vehicle (AUV) to help solve one of the big concerns of the modern world – to collect microplastics from the ocean. The setup of the vehicle is such that it forms part of a collection of similar robots – referred to as a "shoal"– which is deployed from a mother sea vessel. The shoal functions as a large organism via artificial intelligence. A control centre on-board the vessel is kitted to control and monitor the shoal as a unit, or each seabot. For example, the shoal can be controlled to create a formation, to move to a different location, or to move to the mother vessel for safekeeping before a storm strikes. Each robotic AUV unit (called a seabot) functions autonomously but is capable of communicating with other seabots. Figure 2 is a simplified version of a seabot.

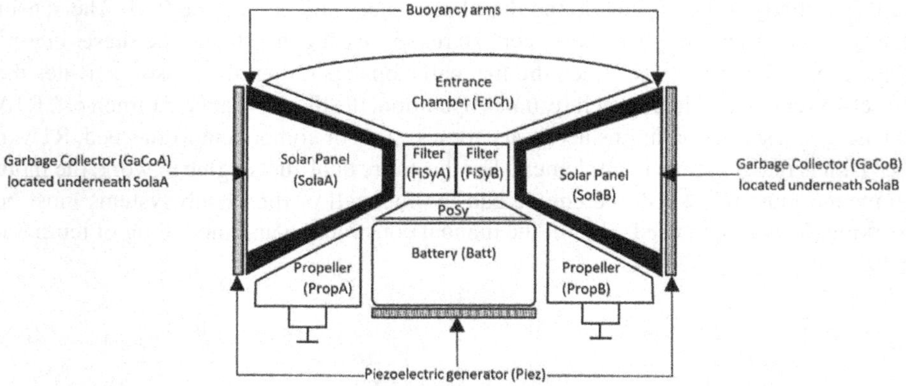

Fig. 2. An abstract model of the proposed seabot.

The seabot has a dish-like feature, known as the entrance chamber (EnCh) that collects seawater. The EnCh is designed with various dome-like features to prevent sea animals and plants from entering two filtering systems (FiSy) connected to it. The flow of seawater is facilitated by a pump (Pump) that is situated within each FiSy. The FiSy also contains a valve that allows water into it from the pump and prevents water from going back through the EnCh. A flow meter (FoMe) controls the rate of water the pump should allow depending on the reading of the level detector (LeDe).

The microplastic isolation system (MiSy) is responsible for separating the microplastics from the seawater within the FiSy and depositing the residue into either of the garbage collectors (GaCo) situated under the solar panels (SoPa). Each GaCo stores the microplastics as long as its level detector (GaDe) does not read full. When both GaCos are full, the seabot moves to the mother vessel for its GaCos to be replaced with empty ones. The GaDe are also able to detect the situation where its GaCo bursts. In the event where a GaCo bursts, the seabot will signal some of the seabots closest to it to assist in collecting the microplastics it is releasing back into the sea. The seabot will then move to the mother vessel for the defective GaCo to be replaced.

There are two onboard propellers or thrusters (Prop) that move the seabot backwards or forward and two buoyancy arms (Bouy) that are responsible for the floating and sinking of the seabot. Collision sensors are fixed on all sides of the seabot to allow it to avoid collision with other objects during navigation. All operations of the seabot are powered by a stack of batteries (Batt). During the day, the seabot operates afloat and harnesses solar power (Sola) to charge the Batts. During underwater movements, cilia boards located on the sides of the seabot agitate piezoelectric actuators (Piez) to generate electricity to charge Batts. In the event where the Batt fails, the seabot is powered by either the SoPa and/or the Piez if they can provide enough energy required. There are various controllers responsible for fault diagnosis (FaDi), filtering system (FiSy), communication system (CoSy), power management system (PoSy), and navigation system (NaSy).

To perform quantitative analysis, the data in Table 1 are assumed. From the table, λ represents the hazard rate. α and β represent the scale and shape parameters of a Weibull distribution respectively, and μ and δ are the mean and standard deviation of a lognormal distribution respectively. (F) and (R) stand for failure and repair data respectively. A full qualitative analysis and dependability analysis of the entire seabot system is outside the scope of this paper. This paper will consider only the operations necessary for collecting the microplastics from the EnCh and storing them in the GaCo. After a qualitative analysis using techniques in [2], the temporal fault tree can be represented by the following CSQ expression.

$$Top - event = EnCh + (FiSyA.FiSyB) + (GaCoA \& GaCoB) + (Batt < Sola)$$
$$(Batt|Piez).(Sola|Piez) + \left[PoSy|Sola\right] + \left[PoSy|Batt\right] + \left[PoSy|Piez\right]$$

Table 1. System operating data (in days or per day).

Entity	Failure type	λ (F)	α (F)	β (F)	λ (R)	μ (R)	δ (R)
Seabot	General system failure	2.33E-5	–	–	2.0	–	–
Seabot	Preventive maintenance	2.74E-3	–	–	–	0.5	0.0
EnCh	Blocked/covered	1.37E-3	–	–	–	0.8	0.2
FiSyA	Internal failure	2.74E-3	–	–	–	1	0.5
FiSyA	Valve stuck closed	–	480	1.5	–	1	0.5
FiSyB	Internal failure	2.74E-3	–	–	–	1	0.5
FiSyB	Valve stuck closed	–	480	1.5	–	1	0.5
Batt	Internal failure	1.10E-3	–	–	–	1	0.25
Piez	Internal failure	1.34E-3	–	–	–	1	0.5
Sola	Internal failure	1.83E-3	–	–	–	1	0.75
PoSy	Internal failure	–	635	2	0.5	–	–
GaCoA	Valve stuck closed	–	522	1	–	0.5	0.25
GaCoA	Replacement	–	–	–	0.1	–	–
GaCoB	Valve stuck closed	–	522	1	–	0.5	0.25
GacoB	Replacement	–	–	–	0.1	–	–

It is assumed that the solar panels are turned on and off between 09.00 and 17.00 from January to March, 08.00 to 18.00 from April to June, 06.00 to 20.00 from July to September and 07.00 to 17.00 on the other months. The garbage collector has a maximum capacity of 100 m^3 and a rate of addition with an exponential distribution mean of 1.5 m^3/day. When both garbage collectors are full, the seabot moves to the mother shipping vessel for replacements. Annual preventive maintenance is scheduled for each seabot on the first of June. Using the proposed algorithms, the case study was modelled and run using Monte Carlo simulation. The simulation was computed for a system lifetime from 1 to 50 days with 100000 iterations per day and a time-step of 1 h. Only failure data was used in the simulation. The result of the simulation is displayed in Fig. 3.

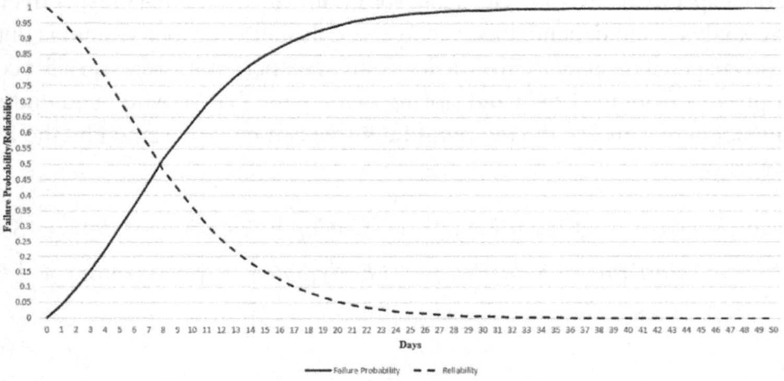

Fig. 3. Result of Monte Carlo simulation with only failure data.

It is clear that the probability of the top-event occurring increases with an increasing system lifetime. By the 50th day, it is expected that the top-event should have occurred – that is, the system should have failed. Modelling the same failure data, including the repair and maintenance data and system operating data in Goldsim using 5000 iterations over 50 days with a time-step of 1 h produces the result in Fig. 4.

A lower number of iterations is used in Fig. 4 because a sampling technique (Latin Hypercube Sampling) is implemented. Given repair, maintenance, and failure data, the mean system reliability stays over 90% after the 50th day. Such results are expected because if components are maintained and repaired, the entire system fails far less. Other quantitative measures can be extracted from the Goldsim simulation. In Table 2, *FP*, *Rel.*, *TF*, and *TR* are the failure probability, reliability, mean-time-to-failure (MTTF), and mean-time-to-repair (MTTR) respectively. *FP*, *Rel.*, *MTTF* and *MTTR* retain their original definitions from [1]. *IA* is the inherent availability and *OA* is the operational availability [15].

The exhaustive analysis of these results is outside the scope of this paper. The mean reliability of the seabot over the entire duration of the simulation is 0.932 with 0.9261 and 0.9379 as the 5% and 95% confidence bounds respectively. The filtering systems are the least reliable components of the system; they are less reliable than the entrance chamber. The ratio of the contributions of each of the filtering systems to that of the

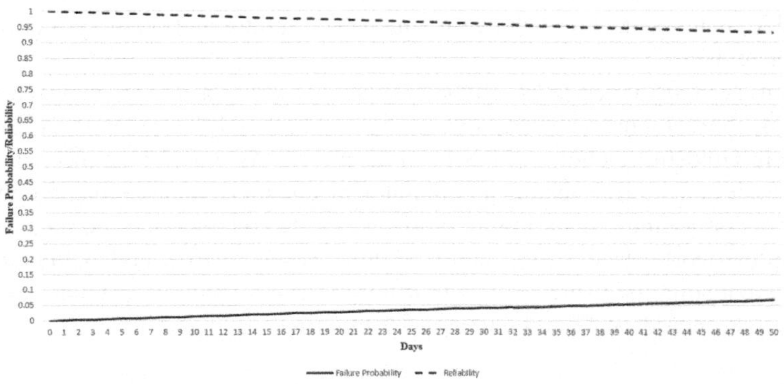

Fig. 4. Result of Monte Carlo simulation with all system data.

Table 2. Seabot reliability analysis.

Entity	FP	Rel	IA	OA	TF (days)	TR (days)
FiSyA	0.151	0.849	0.997	0.986	260	0.977
GaCoB	0.091	0.909	0.999	0.988	491	0.518
GaCoA	0.09	0.91	0.999	0.988	479	0.507
Seabot	0.068	0.932	0.989	0.989	696	0.791
Piez	0.064	0.936	0.999	0.988	746	0.986
Ench	0.057	0.943	0.999	0.989	724	0.782
Batt	0.053	0.947	0.999	0.988	904	1.011
Sola	0.038	0.962	0.999	0.423	1096	1
PoSy	0	1	1	0.002	–	–

entrance chamber leading to the seabot failure is 1:118. However, the entrance chamber fails less frequently with an MTTF of 724 days and takes a shorter time to repair with an MTTR 0.782. Meaning, even though the entrance chamber is more reliable than the filtering systems, it is a single point of failure and contributes more to the system failure so it needs to be improved if a system improvement is desired.

5 Conclusion

Two techniques have been proposed for the evaluation of temporal fault trees using simulation approaches. In the first approach, an algorithm (limited to non-repairable events) has been developed for evaluating all MCSQs in the temporal fault tree. The second approach, using Goldsim, allows for the modelling of dynamic systems featuring some temporal behaviours. This technique applies to almost any system with repairable, replaceable, maintenance, preventive-maintenance, failure and other system operating environment data.

An autonomous underwater vehicle system case study has been analyzed with the proposed techniques and a comparative analysis of both only failure data and all system data (using Goldsim) techniques has been discussed. As would be expected, the proposed technique featuring repairs and maintenance is more reliable than the technique to considers only failure data. Future works will focus on applying the proposed techniques to real-world systems; this could be done by aggregating and synthesizing data from different sourcing using various fourth industrial revolution technologies such as Big Data, the Internet of Things and cloud computing.

References

1. Vesely, W.E., et al.: Fault Tree Handbook with Aerospace Applications. NASA Office of Safety and Mission Assurance (2002)
2. Walker, M.: Pandora: A Logic for the Qualitative Analysis of Temporal Fault Trees. University of Hull (2009)
3. Dugan, J.B., Bavuso, S.J., Boyd, M.: Dynamic fault-tree models for fault-tolerant computer systems. IEEE Trans. Reliab. **41**, 363–377 (1992)
4. Tang, Z., Dugan, J.B.: Minimal cut set/sequence generation for dynamic fault trees. In: Annual Symposium Reliability Maint, 2004 – RAMS, pp. 207–213 (2004)
5. Merle, G., Roussel, J., Lesage, J.: Improving the efficiency of dynamic fault tree analysis by considering gate fdep as static. Reliab. Risk, 1–7 (2010)
6. Palshikar, G.K.: Temporal fault trees. Inf. Softw. Technol. **44**, 137–150 (2002)
7. Kabir, S.: An overview of fault tree analysis and its application in model-based dependability analysis. Expert Syst. Appl. **77**, 114–135 (2017)
8. Edifor, E., Walker, M., Gordon, N.: Quantification of priority-OR gates in temporal fault trees. In: Lecture Notes in Computer Science 7612 LNCS, 99–110 (2012).
9. Edifor, E., Walker, M., Gordon, N.: Quantification of simultaneous-AND gates in temporal Fault Trees. Adv. Intell. Syst. Comput. 224, 141–151 (2013)
10. Fussell, J.B., Aber, E.F., Rahl, R.G.: On the quantitative analysis of priority-AND failure logic. IEEE Trans. Reliab. **R-25**(5), 324–326 (1976)
11. Edifor, E.E.: Quantitative analysis of dynamic safety-critical systems using temporal fault trees. University of Hull (2014)
12. Herrera, F., Sander, I.: Combining analytical and simulation-based design space exploration for time-critical systems. In: IEEE Specification & Design Languages, pp. 1–8 (2013.
13. Esary, D., Proschan, F.: Coherent structures with non-identical components. Technometrics **5**, 191–209 (1963)
14. Hong, Y., Zhang, M., Meeker, W.Q.: Big data and reliability applications: the complexity dimension. J. Qual. Technol. **50**, 135–149 (2018)
15. Goldsim: Goldsim. A Dynamic Simulation Approach to Reliability Modeling and Risk Assessment Using GoldSim (2020). https://media.goldsim.com/Documents/WhitePapers/GoldSim_Reliability_and_PRA.pdf, Accessed 23 Jan 2021

Hybrid Parallel Programming in High Performance Computing Cluster

Alexander Fedulov[1], Anastasiya Fedulova[2], and Yaroslav Fedulov[1(✉)]

[1] National Research University "Moscow Power Engineering Institute" (Branch) in Smolensk, Energetichesky Proyezd 1, Smolensk 2014013, Russia

[2] National Research University "Moscow Power Engineering Institute", Krasnokazarmennaya 14, Moscow 111250, Russia

Abstract. This article provides a brief overview of approaches to calculating the complexity function for sequential programs. To determine the complexity of parallel programs, an approach based on operation analysis was used. The features of the parallelization technologies for sequential programs OpenMP and MPI are described. The main hardware and software factors affecting the parallel programs execution speed on computing cluster nodes are presented. The research of the impact on performance of the ratio of the computing and exchange operations number in the program is the main focus of the work. To implement the research, test parallel OpenMP and MPI programs were developed, in which the total number of operations and the ratio between computational and exchange operations are set. A high performance computing cluster consisting of several nodes was used as a hardware and software platform. The experimental studies have confirmed the hypothesis about more efficient operation of the OpenMP program in comparison with MPI on one node of the computing cluster for programs of a certain class, characterized by a significant share of exchange operations. The efficiency of a parallel program hybrid model in multi-node systems with heterogeneous memory is shown using OpenMP in shared memory subsystem, and MPI in a distributed memory subsystem.

Keywords: Complexity of algorithms · Types of operations · Parallelization technologies · High performance computing cluster

1 Introduction

At present, high performance parallel systems of various architectures and configurations are used to solve a number of applied and scientific problems that require a large amount of computing resources. One of the most accessible solutions in terms of price and obtaining computing capacity are high performance computing (HPC) clusters, consisting of one or more nodes with multi-core processors, where each node usually represents a parallel symmetric multiprocessing (SMP) computing system with shared memory. The cluster nodes are connected by a communication environment with high throughput and operate in parallel to solve one set task, forming a high performance system with distributed memory [3, 14].

W. Zamojski et al. (Eds.): DepCoS-RELCOMEX 2021, AISC 1389, pp. 97–105, 2021.
https://doi.org/10.1007/978-3-030-76773-0_10

One of the main problems in systems of this class is writing an effective parallel program that solves the given task using the maximum available number of processor cores on the cluster nodes in order to obtain the corresponding performance gain in relation to the sequential version of the algorithm [5, 6].

Among the factors affecting the program execution efficiency in parallel HPC cluster systems, two classes can be distinguished: hardware and software [15].

The hardware factors include the following: performance of processors and their number; communication environment bandwidth; delay (latency) of sending messages in systems with distributed memory; the speed of data exchange between processor cores through shared memory; amount of Cache memory, frequency and number of processor cores on cluster nodes.

A distinctive feature of hardware-class factors affecting the efficiency of parallel programs execution on HPC clusters is the complexity of their operational change.

The software factors include: parallel programming technology; operating system capabilities; compiler with the ability to parallelize sequential programs; characteristics and features of algorithms and programs.

Effective use of these factors will improve the performance of parallel programs [7].

In this paper, the main attention is paid to program factors. The primary purpose is to study the cooperative use possibilities of parallelization technologies OpenMP and MPI, considering the characteristics and features of algorithms and programs to improve the performance of parallel programs on HPC clusters.

2 Approaches to Determining the Complexity and Efficiency of Algorithms and Programs

To evaluate the execution effectiveness of sequential algorithms and programs, various approaches are used to determine their complexity [8]. Complexity is usually understood as the cost of computing system resources, the total number of operations performed, and program execution time.

In work [12] algorithms that differ in the form of the complexity functions are considered. Three types of algorithms have been identified: quantitative-dependent, parametric-dependent, and quantitative-parametric. In quantitative-dependent algorithms, the complexity function depends only on the dimension of the input, and not on the values supplied to the input. In parametric-dependent algorithms, the complexity is calculated from the input values. In quantitative-parametric algorithms, the complexity depends both on the dimension of the input data and on their values. Algorithms of the first type are most suitable for use in parallel programs.

Another approach to determining the complexity function of an algorithm is asymptotic analysis [6]. This approach makes it possible to determine the rate of growth of labor intensity with an increase in the amount of data using special indicators of estimates. In asymptotic analysis, the algorithm complexity is usually referring to operating costs or costs of computing system resources.

Often the most relevant is the evaluation of the program execution time. This assessment is more critical to specific hardware and software factors, for example, significant differences between the formal algorithm writing instructions and the real processor system; dependence of the execution time of the same type of commands on operands values and data types; the impact of compiler and its settings on the effectiveness of the final program implementation.

One of the approaches to the evaluation of the algorithm time complexity is the operation analysis [9], which allows calculating the expected execution time as the sum of the execution times of all operations. For complex algorithms and programs, the use of operation analysis explicitly is not always convenient. Easier to use is the Gibson method [8], which applies the certain type operations frequency and, taking into account their execution time, allows to calculate the total execution time of the algorithm.

To evaluate the performance of parallel algorithms and programs, the following main indicators are used: average degree of parallelism, parallel algorithm acceleration, of a parallel algorithm efficiency [10].

An alternative approach to the analysis of the efficiency and complexity of the programs is the use of profilers. There are a large number of profiler programs [15–17] that calculate the execution time of each program function. The profiler is a useful tool for analyzing the execution time of algorithms and programs [13].

The main purpose of this paper is to analyze the cooperative use of OpenMP and MPI technologies in hybrid parallel programming [11]. The ratio of the exchange and computational operations number in the program is proposed as the main software factor affecting parallel program performance. This is justified by the fact that they take the most time in solving scientific, technical and applied problems, and also use various interfaces of the computing system, which differ significantly in throughput.

The following indicators are introduced as parameters used to analyze the execution time of parallel programs: D_c – share of computational operations in the program relative to the total number of operations, D_{ex} – share of exchange operations in the program relative to the total number of operations.

Indicators D_c and D_{ex} are calculated as follows:

$$D_c = \frac{K_c}{K_{ex} + K_c}, \quad D_{ex} = \frac{K_{ex}}{K_c + K_{ex}}, \tag{1}$$

where K_c – number of computational operations, K_{ex} – number of exchange operations, $K_c + K_{ex}$ – total number of operations, $D_o, \ D_v \ \in \ ]0, 1[$.

Operations of other types are not considered in the analysis.

3 Hardware and Software Platform

The research was conducted on a high-performance computing cluster [2], the architecture of which is shown in Fig. 1. In its structure, the HPC cluster has two computing nodes, the control node, a communication environment, memory, input-output devices. Compute nodes have central processing units (CPU) and graphics processing units (GPU). For the experiments, the CPU part of the cluster nodes was used.

The internal network of HPCC is built on the basis of a 16-port commutator with a bandwidth of 1 Gbit/s, and the inner exchange between ports is carried out at the speed of 40 Gbit/s.

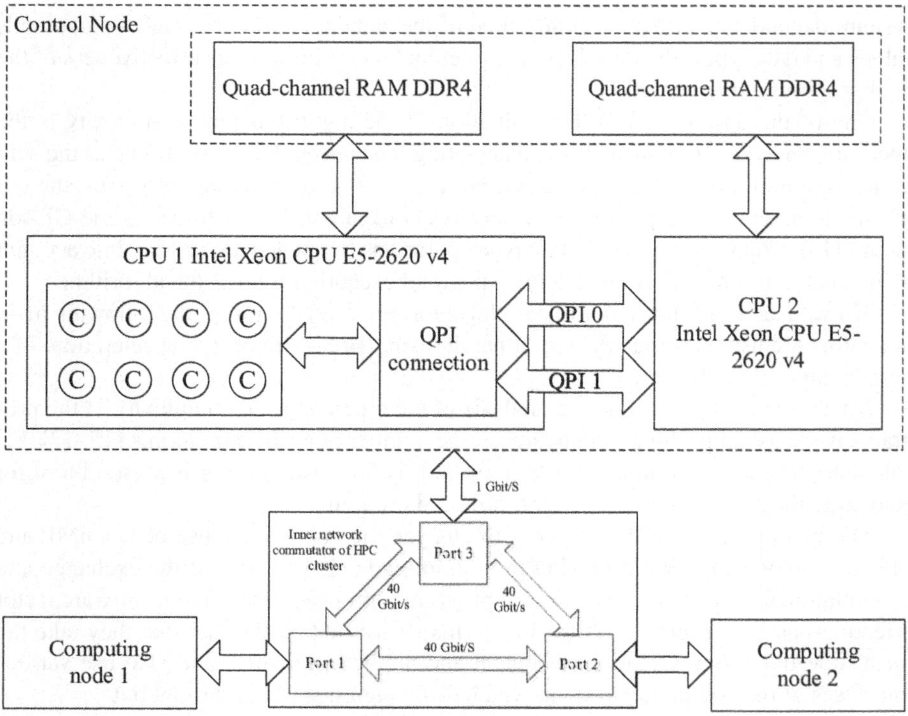

Fig. 1. The architecture of HPC cluster.

The control node includes interface components, two 8-core processors connected to DDR4 RAM, each core can process 2 threads in parallel. Data exchange between the two processors is performed 2 QPI bus connections. The bandwidth of one processor-memory channel is 17064 MB/s. Considering the possibility of four channels parallel operation, the maximum speed of processors-memory exchange is 68.3 GB/s.

The architecture of computing nodes 1 and 2 is similar to the structure of the control node, but the main difference is two 10-core processors, and the bandwidth of one processor-memory channel is 14933 MB/s. Considering the possibility of four channels parallel operation, the maximum speed of processors-memory exchange is 59,7 GB/s. The cluster nodes implement a non-uniform access architecture in which each processor has its own local memory with fast access, but slower access to the memory of another processor via 2 channel QPI connection.

The software environment for research is a set of software installed on each cluster node: operational system CentOS Linux release 7; compilers gcc 6.4.0 and mpiexec 1.10.7; MPI implementation – OpenMPI 1.10.0; OpenMP implementation – OpenMP 3.1. Computing resources are managed by the SLURM resource manager version 15.08.

4 OpenMP and MPI Features on the Architecture of a Computing Cluster

The OpenMP software environment is designed for parallelization of programs in computing systems with shared memory [1]. In this case, memory is at the same time a medium for the information exchange between threads. This means that the fastest node interface CPU – RAM is involved in exchange operations. MPI is designed to work on distributed memory systems. The exchange of information between processes (including working on one node) is performed using external interfaces, the performance of which is noticeably lower.

The Fig. 2 shows a diagram of the cooperative work of MPI and OpenMP on three nodes of the HPC cluster.

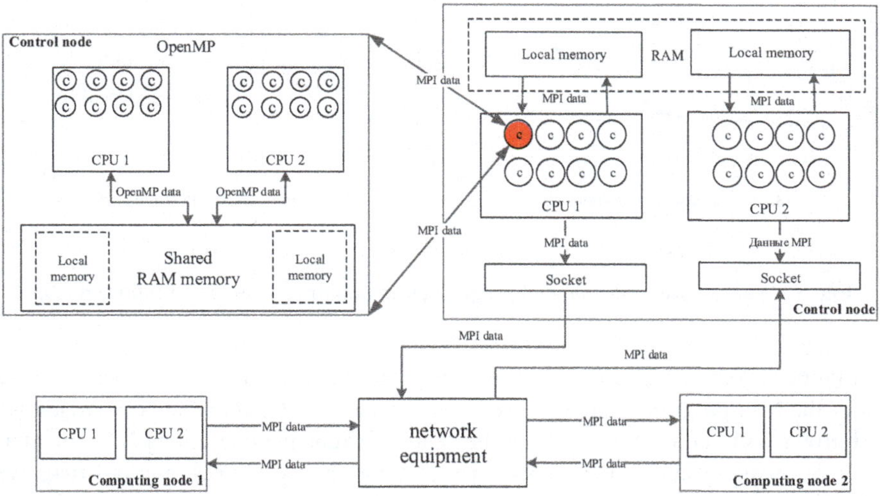

Fig. 2. The operation diagram of the hybrid parallel program MPI + OpenMP on three nodes.

In the diagram of a hybrid parallel program presented in Fig. 2 data transfer between nodes is conducted using MPI, and on one node a separate MPI process running on the core can launch several parallel threads of the OpenMP program, using free processors cores, using the advantages of fast exchange with shared RAM.

5 Results of Experimental Studies

For experimental verification of the hypothesis about more efficient operation of OpenMP in comparison with MPI for a certain class programs on single HPC cluster node, test programs were developed that simulate computational and exchange operations with the ability to change their total number and the ratio between them [17].

The share of other types of operations in test programs is insignificant. To control the correct functioning of the test programs, a parallel calculation of a definite integral by the method of rectangles was implemented [4]. Also, the test programs provided measuring of their execution time [18]. After the measurements were made, graphs of the dependences of the program execution time on various parameters were built.

Figure 3(a) shows the dependence of the parallel program execution time on the control node depending on the ratio of exchange and computational operations number with a constant total of 250 million operations.

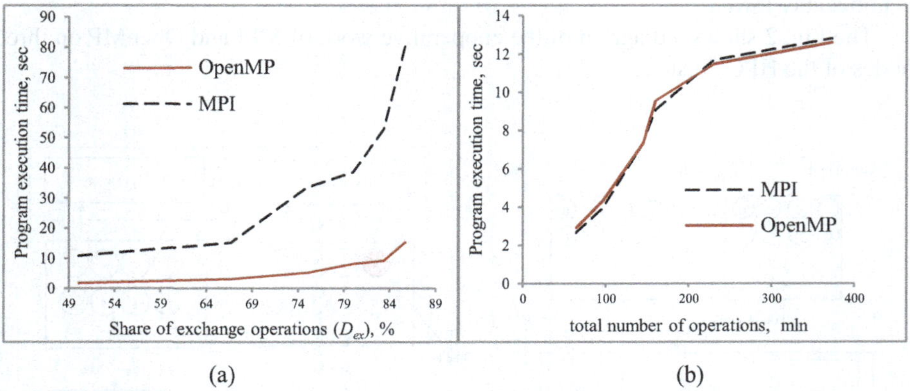

Fig. 3. Dependence of the parallel program execution time on the control node on D_{ex}

Figure 3 (a) shows that the execution time of the OpenMP program is always less than that of the MPI program. The increase in performance of OpenMP in comparison with MPI grows with the magnification in the share of exchange operations D_{ex}, calculated as (1). So, with a share equal to 85%, the acceleration of OpenMP in comparison with MPI on one cluster node reaches 5.3 times.

Figure 3(b) shows the dependence of the program execution time on the control node on the total number of operations calculated as (1) when $D_{ex} \to 0$ and $D_c \to 1$. As seen from the figure, the execution times of OpenMP and MPI programs are approximately equal.

Figure 4 shows the experimental results when executing programs on three nodes with various total number of operations and different combinations of exchange and computational operations shares in the case of applying only MPI or MPI in conjunction with OpenMP (inside the node) according to the diagram of a hybrid parallel program presented in Fig. 2.

Figure 4(a) shows the dependence of the program execution time on the total number of operations while the share of computing operations is insignificant $D_{ex} \to 0$ and $D_c \to 1$. As can be seen from the figure, the execution times of hybrid MPI + OpenMP and pure MPI programs are approximately equal. Figure 4(b) reflects the time results of the program execution at $D_{ex} = D_c = 0.5$. The faster operation of hybrid parallel program compared to pure MPI can be detected.

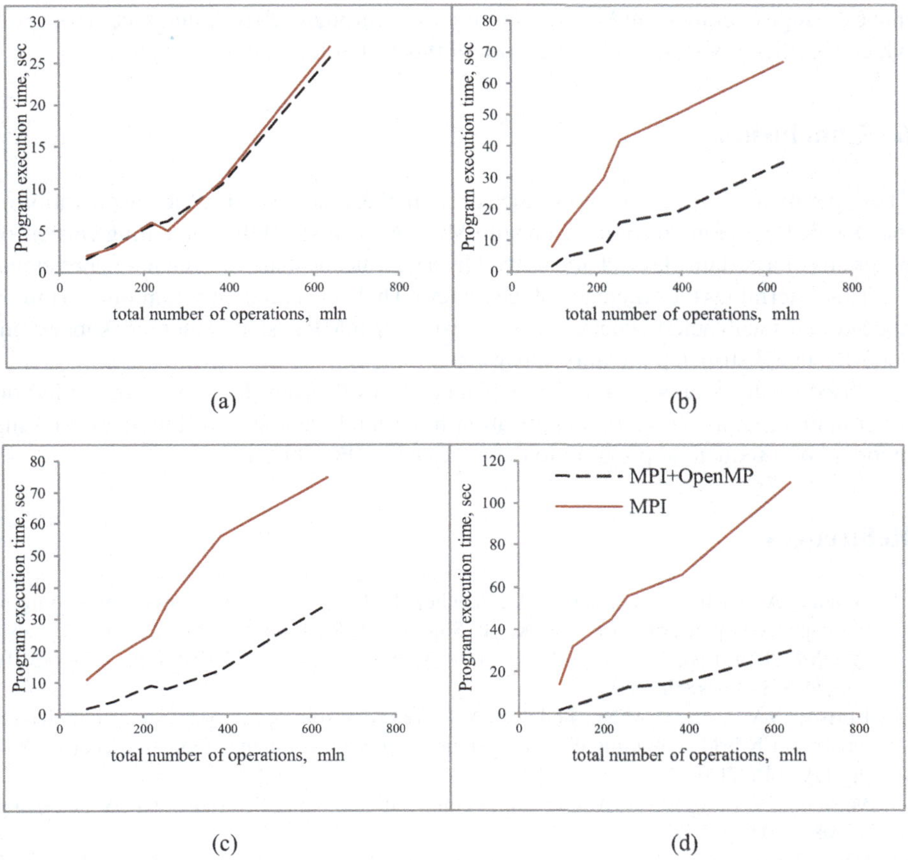

Fig. 4. Executing a program on three nodes using MPI and MPI in conjunction with OpenMP. (a) $D_{ex} \to 0$ and $D_c \to 1$ (b) $D_{ex} = D_c = 0.5$ (c) $D_{ex} = 0.7$ (d) $D_{ex} \to 1$ and $D_c \to 0$.

Figure 4(c) reflects the time results of the program execution at $D_{ex} = 0.7$. The time gap between versions of parallel programs begins to grow as the share of exchange operations increases.

Figure 4 (d) shows the results for $D_{ex} \to 1$ and $D_c \to 0$. In this case, the difference in performance between hybrid MPI + OpenMP and pure MPI parallel programs is maximum and grows with an increase in the total number of operations.

In general, according to the research results, it can be concluded that exchange operations both on a single and on several HPC cluster nodes in MPI are processed slower, which becomes more noticeable with an increase of the exchange operations share. Computational operations are performed using parallelization technologies in almost the same way.

With a small total number of operations, the difference in the execution time of OpenMP and MPI exchange operations is small, but with an increase in the total number of transfer operations, the time increases by more than 5 times, this is due to the fact that exchange operations in MPI technology work through network interfaces (sockets), which significantly slows down the work of this technology on the cluster.

6 Conclusion

The experimental studies have confirmed the hypothesis about more efficient operation of the OpenMP program in comparison with MPI on one node of the computing cluster for programs of a certain class, characterized by a significant share of exchange operations.

This determines the efficiency of the reviewed hybrid parallel programming in multi-node systems with heterogeneous memory using OpenMP in shared memory subsystem, and MPI in a distributed memory subsystem.

Based on the results obtained, it is planned to further develop the described hybrid programming approach and its application in an intelligent system for supporting the mapping of a sequential program to the nodes of a HPC cluster.

References

1. Álvarez, Á., Ugarte, Í., Fernández, V., Sánchez, P.: OpenMP dynamic device offloading in heterogeneous platforms. In: Fan, X., de Supinski, B.R., Sinnen, O., Giacaman, N. (eds.) IWOMP 2019. LNCS, vol. 11718, pp. 109–122. Springer, Cham (2019). https://doi.org/10.1007/978-3-030-28596-8_8
2. Borisov, V.V., Zernov, M.M., Fedulov, A.S., Yakushevskii, K.A.: Analysis of the characteristics of hybrid computing cluster. Systems of control, communication and security. №4, pp. 129–146 (2016)
3. Voevodin, V.V., Voevodin, V.V.: Parallel computations, Sankt-Peterburg: BKhV-Peterburg, p. 608 (2004)
4. Voitsitskaya, A.S., Fedulov, A.S.: Sovmestnoe ispol'zovanie tekhnologii rasparallelivaniya OpenMP i MPI na gibridnom vychislitel'nom klastere, Energetika, informatika, innovatsii, pp. 305–309 (2018)
5. Gergel', V.P.: Teoriya i praktika parallel'nykh vychislenii. Moskva: BINOM. Laboratoriya znanii, p. 424 (2007)
6. Kolpakov, A.A., Kropotov, Yu.A.: Increase of productivity of heterogeneous computer data processing systems, Moscow: Direkt-Media, p. 122 (2019). https://doi.org/10.23681/496776.
7. Kutepov, V.P., Fal'k, V.N.: Algoritmicheskie parallel'nye protsessy i ikh slozhnost'. Vestnik MEI, pp. 119–126. (2020). https://doi.org/10.24160/1993-6982-2017-4-117-128
8. Ovsyannikov, A.V., Pikman, Yu.A.: Algoritmy i struktury dannykh: uchebno-metodicheskii kompleks dlya spetsial'nosti. Minsk: BGU, p. 124 (2015)
9. Selivanova, I.A., Blinov, V.A.: Postroenie i analiz algoritmov obrabotki dannykh. Ekaterinburg: Ural, p. 109 (2015)
10. Starchenko, A.V., Bertsun, V.N.: Metody parallel'nykh vychislenii. Tomsk: Izd-vo Tom, p. 223 (2013)
11. Ul'yanov, M.V., Sheptunov, M.V.: Matematicheskaya logika i teoriya algoritmov, chast' 2: Teoriya algoritmov. Moskva: MGAPI, p. 80 (2003)
12. Ul'yanov, M.V.: Resursno-effektivnye komp'yuternye algoritmy. Razrabotka i analiz. Moskva: Nauka FIZMATLIT, 2007. – 376 s
13. Chaikovskii, D.S., Gulevich, N.A.: Analiz sovremennykh sredstv profilirovaniya parallel'nykh program. Informatsionnye tekhnologii (Vestnik SGTU), pp. 109–113 (2014)

14. Chamberlain, B.L., Callahan, D., Zima, H.P.: Parallel programmability and the chapel language. Int. J. High Perform. Comput. Appl. **21**(3), 291–312 (2007). https://doi.org/10.1177/1094342007078442
15. Brian Gough Foreword by Richard M. Stallman. An introduction to the GNU C and C++ Compilers, gcc and g++, which are part of the GNU Compiler Collection (GCC). https://www.linuxtopia.org/online_books/an_introduction_to_gcc/gccintro_79. Accessed 15 July 2020
16. Intel VTune Profiler. https://software.intel.com/en-us/vtune. Accessed 10 July 2020
17. Kubacki, M., Sosnowski, J.: Exploring operational profiles and anomalies in computer performance logs. Microprocess. Microsyst. **69**, 1–15 (2019). https://doi.org/10.1016/j.micpro.2019.05.007
18. Voitsitskaya, A., Fedulov, A., Fedulov, Y.: Resource managing method for parallel computing systems using fuzzy data preprocessing for input tasks parameters. In: Abraham, A., Kovalev, S., Tarassov, V., Snasel, V., Sukhanov, A. (eds.) IITI'18 2018. AISC, vol. 874, pp. 410–419. Springer, Cham (2019). https://doi.org/10.1007/978-3-030-01818-4_41

Computing of Blocks of Some Combinatorial Designs for Applications

Alexander Frolov(✉)

National Research University Moscow Power Engineering Institute, Krasnokazarmennaya 14, Moscow 111250, Russian Federation

Abstract. Combinatorial block designs are used in various fields of technology; recently, they have been implemented, for example, as models of computer systems and key distribution schemes. They are defined on the basis of some finite set as satisfying certain conditions the set of its subsets called blocks. In this chapter, we study combinatorial block designs whose order is a prime number or a prime power. The tasks sufficient for the practical use of such block designs are considered: determining the composition of a particular block by its number and determining the numbers of blocks containing given element. It is shown that for their solution, a memory volume linear in the design order is used. Solutions are given with a certain numbering of elements and blocks. The sets on which the designs are defined are initial nonnegative, the blocks are their subsets. Computations are performed by mapping this set into a vector space of a certain dimension over the prime field or its extention. Four algorithms are given: Algorithm 1 and Algorithm 2 for constructing of projective plane and dual projective plane and Algorithms 3 and 4 for constructing of separate blocks of those projective planes.

Keywords: Computer network · The preliminary key distribution · Finite field · Vector space · Combinatorial design · Projective plane · Linear complexity of used memory

1 Introduction

Some combinatorial designs are implemented as two- or three-level structural models of complex systems such as distributed computer systems, key redistribution systems, mobile connection systems, and agro industrial test stations and so on. Among others, the most popular are balanced incomplete block designs (BIBD) [1–4]. They are defined on some finite set $\boldsymbol{X}$ of cardinality v and consist of a set $\boldsymbol{B}$ of b proper subsets called blocks, each of which contains k elements that appear in r blocks, each a pair of different elements appears in λ blocks. They are abbreviated to (v, k, λ)-BIBD. The remaining parameters are restored by the abbreviated notation using two elementary relations $bk = vr$ and $r(k-1) = \lambda(v-1)$. Dual BIBD is defined on the set of blocks which subsets form so called duality blocks. Duality block corresponds set of blocks, containing given element. According to proposed it this chapter numerating duality rule elements as well as dual block numbers are the numbers $0, 1,\ldots, v-1$, block numbers are $0,\ldots,b$, duality

W. Zamojski et al. (Eds.): DepCoS-RELCOMEX 2021, AISC 1389, pp. 106–115, 2021.
https://doi.org/10.1007/978-3-030-76773-0_11

block consist of block numbers, containing its number. Here are some BIBDs used in computing technology and cryptography. A projective plane of order n is a BIBD-($n^2 + n + 1, n + 1, 1$). It is the projective geometry PG(2,n) as well. The projective geometry PG(3,n) is a BIBD-($n^3 + n^2 + n + 1, n^2 + n + 1, n + 1$).

The usage of combinatorial design in the keys pre-distribution systems was initiated by D. Stinson [5, 6] as an intermediate option between the most unsafety version of the common key of all participants and the most difficult to implement version of assigning different keys to different pairs of participants. A unique element (key) that is the number of given duality block is assigned to all blocks represented by their numbers in this duality block. So each block has different keys of the duality blocks to which belongs its number, i.e. block corresponds to network node, duality block consists of numbers of nodes containing key that number coincides with duality block number. Projective planes for key distribution in wireless sensor networks were implemented in [7, 8]. But their use was problematic due to the need to store a large number of keys in network nodes that exceed the square root of the network size. This is unacceptable in conditions of limited memory resources of wireless sensor networks. In this regard, transversal systems have found application, in which, however, the local network connectivity is violated, but the possibility of communication through one intermediary node remains [9].

The use of combinatorial designs for construction of computer systems is discussed, for example, in [10–12]. Attention is drawn, under a certain condition, the establishment of connections in a computer network in accordance with a combinatorial block design allows us to guarantee communication between any two computers using no more than one intermediary. It turned out that this condition precisely correspond to projective plane. The method of constructing projective planes proposed in [10–12] is combinatorial geometric and involves some enumeration and analysis of characteristic configurations.

A number of algebraic methods for constructing projective planes whose order n is a prime number or a prime power are known. Each uses a certain basic field F_n defined by a primitive polynomial of degree n and its root, the primitive element $x \in F_n$. One method uses a three-dimensional vector space over this field [2], and the other uses algebraic extension of degree three of the field F_n defined by a primitive polynomial of degree three over this field and the corresponding primitive element [1, 2]. Algebraic algorithms for constructing of afore mentioned transversal block designs are available in [9].

Remark that description of a separate element needs a few bytes, but the descriptions of the desired block designs occupy a large amount of memory. So the representation volume of the original set is quadratic with respect to the order n of the projective plane, and the representation volume of the blocks is cubic. Volumes representations of these sets of projective geometry are to order more. But it is clear that, this data are not used in aggregate at the same time, their use is distributed: blocks are associated with various subsystems. Therefore, for their practical application, it is enough to be able to solve two basic problems: determine the composition of a particular block by its number with a certain ordering and determine the numbers of blocks containing given element. This chapter shows that they can be solved using an amount of memory linear with respect to the order of the block design. At the same time, the possibility of parallel computing of block design is demonstrated. The next Sect. 2 is devoted to algebraic formalization of the representation of the studied combinatorial block designs, it provides sequential algorithms for constructing projective and dual projective planes (Algorithm 1 and Algorithm 2) and two algorithm for constructing separate blocks of this designs (Algorithm 3 and Algorithm 4). An analogous approach to the distributed construction of blocks and dual blocks of linear and quadratic transversal combinatorial block diagrams is given in [13, 14].

2 Algebraic Methods for Constructing Projective Planes

2.1 Projective Planes

Let F_n be a finite field of order $n = p^k$, x be its primitive element, and $F(n^3)$ be its algebraic extension of degree 3, considered as a vector space V of dimension 3, and also as a field, generated by a primitive polynomial of degree 3 over the field F_n. The projective space of dimension 2 is defined as the set of normalized elements of the set V*. Its elements are its subspaces of zero dimension, or point. Pairs of different points form the basis of the subspaces of dimension one i.e. lines, or blocks. There are $N_2 = (n^3 - 1)/(n - 1) = n^2 + n + 1$ points and the same number of lines. Sets of points and lines form the projective geometry $PG(2,n)$ of the dimension 2. It is called the projective plane of order n and is denoted by $PP(n)$. When studying combinatorial block designs, it is customary to call points of the projective plane as elements, and lines as blocks. We will use the following methods for numbering elements of vector spaces of dimension $s + 1$ and projective spaces of dimension s [13, 14]. The number $\varphi(\mathrm{e_s}, \mathrm{e_{s-1}},\ldots,\mathrm{e_1}, \mathrm{e_0})$ of the element $(\mathrm{e_s}, \mathrm{e_{s-1}},\ldots,\mathrm{e_1}, \mathrm{e_0}) \in F(n^s)$, $s \le 3$ is defined as the numerical equivalent of the set of indices (discrete logarithms at base n) of its elements in n-digit number system: $\varphi(\mathrm{e_s}, \mathrm{e_{s-1}},\ldots,\mathrm{e_1}, \mathrm{e_0}) = n^s \mathrm{ind}\ \mathrm{e_s} + \ldots + n\ \mathrm{ind}\ \mathrm{e_1} + \mathrm{ind}\ \mathrm{e_0}$. The number $\psi(1, \mathrm{e_{s-1}}, \mathrm{e_1}, \mathrm{e_0})$ of the element of the projective space of dimension $s - 1$, is defined by the formulae

$$\psi(1) = 0,$$
$$\psi(1, \mathrm{e_0}) = \varphi(e_0) + 1 = 1 + \mathrm{ind}\ \mathrm{e_0} + 1,$$
$$(1, \mathrm{e_1}, \mathrm{e_0}) = \varphi(\mathrm{e_1},\mathrm{e_0}) + n + 1 = n\ \mathrm{ind}\ \mathrm{e_1} + \mathrm{ind}\ \mathrm{e_0} + n + 1.$$

The above numbering $\varphi{:}F(n^s) \rightarrow \{0, 1,\ldots, n^s - 1\}$ and $\psi{:}\{V'\} \rightarrow \{0,1,\ldots, (n^s - 1)/(n - 1) - 1\}$, where V' is the set of normalized polynomials from the vector space $F(n^s)$ are bijections and are reversible:

$\varphi^{-1}(M) = 1$, if $\mathrm{M} = 0$,

$\varphi^{-1}(M) = (M_\mathrm{s}, M_{s-1},\ldots, M_1, M_0)$, where $(M_\mathrm{s}, M_{s-1},\ldots, M_1, M_0)$ is the set of expansion coefficients of the number M in powers of n otherwise;

$\psi^{-1}(\mathrm{M'}) = 1$, if $\mathrm{M'} = 0$,

$\psi^{-1}(\mathrm{M'}) = (1, M_{s-1},\ldots, M_1, M_0)$, where $(M_{s-1},\ldots, M_1, M_0) = \varphi^{-1}(M_{\mathrm{s}-1} - \mathrm{M'}_{\mathrm{s}-1})$, $M'_0 = 1$, $M'_1 = 1 + n$, otherwise.

In this chapter, we restrict ourselves to computing in vector space V without using the operations of multiplication and exponentiation in the field $F(n^3)$. Therefore, the consideration of projective planes of Singer [1] will be constrained by short remark in Subsect. 2.3. Our algebraic approach differs from that discussed in [2] using algebraic basis's of one-dimensional subspaces instead of these subspaces themselves, which allows us to use memory economically.

2.2 The Algebraic Method to Construct the Projective Plane Using basis's of Their Blocks

We take as a set X elements of BIBD-$(n^2 + n + 1, n + 1, 1)$ the set of non-negative integers $\{0, 1,\ldots, n^2 + n\}$ and, using the function φ'^{-1}, we map it into the projective space of dimension 2 consists of normalized polynomials of degree at most two over the field F_n. For the set $\boldsymbol{B}$ of blocks, we take the set of lines generated by linearly independent pairs of elements. Let the algebraic images of the first elements of these pairs be normalized polynomials 1 or $1 + xe$ of degree at most one. Let polynomials be listed with n multiples repetitions in order of non-decreasing their numbers $\psi(1)$ and $\psi(1, e)$:

$$1(1)^n(10)^n(1x)^n\left(1x^2\right)^n\ldots\left(1x^{n-2}\right)^n(11)^n.$$

The second elements form a list

$$10\left(100,\ 1x0,\ 1x^20,\ldots,\ 1x^{n-2}0,\ 110\right)\left(100,\ 10x,\ 10x^2,\ldots,\ 10x^{n-2},\ 101\right)^n.$$

Above (list)^n denote n times repetitions of polynomial or series of polynomials.

It is easy to see that pairs of polynomials corresponding to one another in arrangement in these lists are linearly independent. They form the algebraic bases of lines. Numerical images of line elements form blocks. In abbreviated form, blocks will be represented by numerical images of two elements of line bases instead of all $n + 1$ numeric images. With the indicated numbers $\psi(\mathrm{e}^{(1)})$ and $\psi(\mathrm{e}^{(2)})$ of the first $\mathrm{e}^{(1)}$ and second $\mathrm{e}^{(2)}$ elements of the basis, the serial number $\mathrm{N}(\psi(\mathrm{e}^{(1)}, \psi(\mathrm{e}^{(2)}))$ of the block $(\psi(\mathrm{e}^{(1)}, \psi(\mathrm{e}^{(2)})$ generated by this basis is determined [13, 14]:

$$N(\psi(e^{(1)}), \psi(e^{(2)})) = \begin{cases} (\psi(e^{(2)}) - 1)/n \text{ if } \psi(e^{(1)}) = 0, \\ (\psi(e^{(1)}) - 1)n + \psi(e^{2}), \text{if } \psi(e^{(1)}) > 0. \end{cases} \tag{1}$$

The algebraic closure $<e^{(1)}, e^{(2)}>$ of the basis is calculated by adding linear combinations $e^{(2)} + x^t e^{(1)}$, $t = 1,\ldots, \mathrm{n} - 1$ to it (since all elements of the block are normalized polynomials). In this case, the corresponding block is also computed $(\psi(e^{(1)}), \psi(e^{(2)}), \psi(e^{(2)} + x^1 e^{(1)}),\ldots, \psi(e^{(2)} + x^{n-1} e^{(1)})\ \psi\ (e^2 + x^{n-1} e^1))$.

Example 1. Let us represent the inverse images of the basis's of blocks images of the projective plane (21, 5, 1) with a list of pairs of elements ordered by increasing block numbers:

((0, 1), (0, 5), (0, 9), (0, 13), (0, 17), (1, 5), (1, 6), (1, 7), (1, 8), (2, 5), (2, 6), (2, 7), (2, 8), (3, 5), (3, 6), (3, 7), (3, 8), (4, 5), (4, 6), (4, 7), (4, 8)).

For example, we calculate the image of block B_{12} as the closure $< \varphi'^{-1}(2), \varphi'^{-1}(8) > = ((1, x), (1, 0, 1)) > = \{(1, x), (1, 0, 1), (1, 0, 1) + x \times (1, x)), (1, 0, 1) + (x + 1) \times (1, x)), (1, 0, 1) + 1 \times (1, x))\} = \{(1, x), (1, 0, 1), (1, x, x), (1, x + 1{,}0), (1, 1, x + 1)\}$.

In Table 1, the described method for constructing the projective plane is presented as Algorithm 1 using Python notation.

Next, we consider how to construct the dual projective plane ($\boldsymbol{X}^*$, $\boldsymbol{B}^*$), where $\boldsymbol{X}^*$ is the set of block numbers and $\boldsymbol{B}^*$ is the set of sets of block numbers containing a given element from the original set $\boldsymbol{X}$. According to the Algorithm 1 for constructing the projective plane ($\boldsymbol{X}$, $\boldsymbol{B}$), element 0 is present in blocks $B_0, B_1, B_2, \ldots, B_n$, containing elements 0, 1, 2,…, n, and elements i, $1 \le i \le n$ are included in block B_0 and in blocks B_{tn+i}, $t = 1, \ldots, n$. Now note that each of the remaining elements of the original set $\boldsymbol{X}$, is present in $n + 1$ blocks each time with one of the elements 0, 1,…, n. If its number does not exceed $2n$, then it is the second element of the block, the number of which is determined by its number j and the number of the first element by inversion of rule (1) [13, 14]:

$$\varphi'(e^{(2)}) = \begin{cases} jn + 1 \text{ if } \varphi'(e^{(1)}) = 0, \\ j - (\varphi'(e^{(1)}) - 1)n + \varphi'(e^{(2)}), \text{ if } \varphi'(e^{(1)}) > 0. \end{cases} \quad (2)$$

Table 1. Algorithm 1. Construction of a projective plane of order $n = p^k$, p is prime.

Input	Prime number p, positive number k, $f(X)$ is primitive polynomial of degree k over the field F_p, x is its root.
Output	The projective plane ($n^2 + n + 1$, $n + 1$,1) of order $n = p^k$: ($\boldsymbol{X}$, $\boldsymbol{B}$).

1	Create list $\boldsymbol{B}$ of m empty lists and include in them the first elements of the blocks:
	for i in range (0,n+1):
	    $\boldsymbol{B}$[i].append(0)
	for i in range (1,n+1):
	    for j in range(i+n+1,(i+1)*n +1):
	        $\boldsymbol{B}$[j].append(i)

2	Include the second elements of the blocks:
	for i in range (0,n+1):
	    $\boldsymbol{B}$[j].append(1+j*n)
	for i in range (1,n +1):
	    for j in range (0,n):
	        $\boldsymbol{B}$[i*n+j+1].append(n+j+1)

3	Add missing elements to blocks of list $\boldsymbol{B}$:
	for i in range (0,m):
	    for i in range (1, n-1):
	        $\boldsymbol{B}$[i].append(φ (x^i φ^{-1}($\boldsymbol{B}$[i][0])+ φ^{-1}($\boldsymbol{B}$[i][1])))

4	Return $\boldsymbol{B}$

Table 2. Algorithm 2. Construction of a dual projective plane of order $n = p^k$, p is prime.

Input	Prime number p, positive number k, $f(X)$ is primitive polynomial of degree k over the field F_p, x is its root.
Output	The dual projective plane $(n^2+n+1, n+1, 1)$ of order $n = p^k$: (X^*, B^*).

1	Create list B* of m empty lists and place in the first of them the sets of block numbers containing 0,1, ..., n-th elements of the original set:
	for i in range(0,n+1):
	  B*[0].append(i)
	for i in range(1,n+1):
	  B*[i].append(0)
	  for j in range (1,n+1):
	    B*[i].append(n*i+j)

2	Create the first elements of the remaining sets of block numbers:
	for i in range (1,n+1):
	  for j in range(i*n+1,(i+1)*n +1):
	    B*[j].append(i)

3	Replenish the sets of block numbers containing elements with numbers n+1, ..., 2n:
	for i in range (n+1,2*n+1):
	  for j in range (0,n):
	    B*[i].append(i+jn)

4	Replenish the sets of block numbers containing elements with numbers 2 n + 1, ...,m:
	def nextelement(i,j):
	    for t in range (1,n):
	      NN= φ(vectorsub($\varphi^{-1}(j)$,mulbyscalar(x^t $\varphi^{-1}(j)$)))
	      if NN<=2*n:
	        return n*(i-1)+NN
	for i in range (1,n+1):
	  for j in range (2*n+1,m):
	    Nxtel=nextelement(i,j)
	    B*[j].append(Nxtel)

5	Return [B*]

If its number exceeds $2n$, then it is 3, 4,... n or $n + 1$-th element of the block. With the known number of the first element by the rule that is inverse to the rule of its calculation by the first and second elements, it is easy to determine the second element of the block and then (by the first and second elements) determine the number of the block into which this element is included. In the Table 2 the described method is presented as Algorithm 2.

Above there is justified.

Proposition 1. *Algorithms 1 and 2 construct the projective and dual projective planes of order $n = p^k$ if n is a prime number or a prime power.*

2.3 Building a Block of a Projective Plane by Its Number and a Block of a Dual Projective Plane Containing This Element

As noted above, the first two elements of any block uniquely determine the number of this block according to rule (1). This rule is reversible: by the block number j, one can determine the numbers of its two first elements $n_1(j)$ and $n_2(j)$ [13, 14]:

$$n_1 = \begin{cases} 0 \text{ if } j \le n, \\ \left\lfloor \frac{j}{n} \right\rfloor \text{ if } j > n; \end{cases} \quad n_2 = \begin{cases} 1 + nj \text{ if } j \le n, \\ j - (n_1 - 1)n \text{ if } j > n. \end{cases} \tag{3}$$

The remaining elements of the block are calculated by the first two according to Algorithm 3 given in Table 3.

Table 3. Algorithm 3. Building a given block of the projective plane.

Input	Prime number p, positive number k, $f(X)$ is primitive polynomial of degree k over the field F_p, x is its root, and the number j of constructed block
Output	j-th block of the projective plane $(n^2 + n + 1, n + 1, 1)$ of order $n = p^k$.

1.	Initializing of block B and including of the first element in it:
	B=[]
	if*j*<*n*+1:
	  B.append(0)
	else:
	  B.append((*j*-1)/*n*)

2	Computing and including the second element:
	if*j*<=*n*:
	  B.append(*j***n*+1)
	else:
	  B.append((j-1)%*n*+*n*+1)

3	Computing and including of other elements of the block: for i in range (0,*n*+1): *B*.append(φ(vectorsum(mulbyscalar(x^i, φ^{-1}(*B*[0])), φ^{-1} (B[1]))))
4	Return *B*

By constructing a list of blocks according to Algorithm 1, element 0 is included in.

By constructing a list of blocks according to Algorithm 1, element 0 is included in blocks 0, 1,…, n, elements j, $0 < j < n + 1$ are contained in blocks 0, $jn + t$, t = 1,…,n. Any of the remaining elements is contained in blocks in which in addition to it there is one of the elements 0, 1,…, n. Using such a pair of elements, it is possible to determine the second element of the block, and according to the known first and second blocks determine the block number in the list according to rule (1). This entails Algorithm 4 for constructing a plurality of blocks containing a given element, given in Table 4.

Table 4. Algorithm 4. Building *j*-th block of the dual projective plane containing element *j*.

Input	Prime number p, positive number k, $f(X)$ is primitive polynomial of degree k over the field F_p, x is its root, and the number j of given element
Output	The block of the dual projective plane $(n^2+n+1, n+1,1)$ of order $n = p^k$, containing given element j.

```
def DualProjectivePlaneBlokCreation():
    def newelement(r,s):
        if r==0:
            for t in range(0,n):
                vsub=vectorsub(PsiInverse(s),[[0],[0],xPow(t)])
                newel=Psi(vsub)
                if newel%n==1:
                    res=newel//n
        else:
            for t in range(0,n):
                vsub=vectorsub(PsiInverse(s), mulbyscalar(PsiInverse(r),xPow(t)))
                newel=Psi(vsub)
                if Psi(vsub)<=2*n:
                    res=(r-1)*n+newel
        return res
   global n,s
   D=[]
   if s<=n:
      D.append(0)
      for t in range (1,n+1):
         D.append(s*n+t)
   else:
      for r in range (0,n+1):
         nel=newelement(r,s)
         D.append(nel)
   return sorted(D)
```

Example 2. Implementing Algorithm 4 one can build block number 14 of the dual projective plane.

$$14.[1, 14, 23, 32, 41, 50, 59, 68, 77, 86]$$

that consists of the numbers of (91, 10, 1) projective plane blocks, containing element 14 (they are computed implementing algorithm 3 10 times):

01.[0, 10, 11, 12, 13, 14, 15, 16, 17, 18]
14.[1, 14, 23, 32, 41, 50, 59, 68, 77, 86]
23.[2, 14, 24, 30, 45, 49, 62, 65, 73, 88]
32.[3, 14, 21, 36, 40, 53, 56, 64, 79, 87]
41.[4, 14, 27, 31, 44, 47, 55, 70, 78, 84]
50.[5, 14, 22, 35, 38, 46, 61, 69, 75, 90]
59.[6, 14, 26, 29, 37, 52, 60, 66, 81, 85]
68.[7, 14, 20, 28, 43, 51, 57, 72, 76, 89]
77.[8, 14, 19, 34, 42, 48, 63, 67, 80, 83]
86.[9, 14, 25, 33, 39, 54, 58, 71, 74, 82]

Remark. In relation to the so-called Singer projective planes of order n that are constructed on the basis of difference sets, to construct a j-th block, it is enough to add consequently modulo n the number j to each element of the difference set, and to construct a block of the dual projective plane containing the numbers of blocks of the projective plane, possessing element j, it is enough to subtract consequently modulo n from this number j elements of the difference set [13,14].

3 Conclusion

This chapter describes algorithms for constructing individual blocks of projective plane using a memory volume that is linear with respect to projective plane order n. There is no need to create and store many elements used in blocks, because by default, it is assumed that it is composed of initial non-negative numbers. Algorithms for constructing a block only use the expansion of its number in the base number system n, the amount of discrete logarithm tables used in algorithms correspond the order of the block design, when calculating the block, an empty list is announced, replenished with no more than $n + 1$ elements. The presented basic algorithms can be used to solve derivative problems: finding a block containing given two elements or finding two blocks (blocks containing this element), as well as to build a complete set of blocks, if the allocated memory allows. To construct blocks of affine planes as residual block diagrams, it is sufficient to construct a block that is removed from the projective plane and then remove its elements from subsequent blocks to be constructed.

Acknowledgements. The presented algorithms were tested using the Sage [15] computer algebra system. This research was supported by the Russian Foundation for Basic Research, project 19-01-00294a.

References

1. Hall, M.J.R.: Combinatorial Theory, 2nd edn. Wiley, New York (1998)
2. Stinson, D.: Combinatorial Designs: Constructions and Analysis. Springer, Germany (2003)
3. Colbourn, C., Dinitz, J. (eds.): The CRC Handbook of Combinatorial Designs, 2nd edn. CRC Press, Boca Raton (2007)
4. Linder, C.C., Rodgen, C.A.: Design Theory. Taylor & Francis, Ltd., Boca Raton (2008)
5. Stinson, D.R., van Trung, T.: Some new results on key distribution patterns and broadcast encryption. Des. Codes Cryptography **14**, 261–279 (1998)
6. Stinson, D.R., van Trung, T., Wei, R.: Secure frameproof codes, key distribution patterns, group testing algorithms and related structures . J. Statist. Plan. Infer. **86**(2), 595–617 (May 2000)
7. Çamtepe, S., Yener, B.: Combinatorial design of key distribution mechanisms for wireless sensor networks. In: Samarati, P., Ryan, P., Gollmann, D., Molva, R. (eds.) ESORICS 2004. LNCS, vol. 3193, pp. 293–308. Springer, Heidelberg (2004). https://doi.org/10.1007/978-3-540-30108-0_18
8. Lee, J., Stinson, D.: A combinatorial approach to key pre-distribution for distributed sensor networks. In: IEEE Wireless Communications and Networking Conference WCNC 2005, vol. 2, pp. 1200–1205. IEEE Communications Society (2005)
9. Lee, J., Stinson, D.R.: On the construction of practical key predistribution schemes for distributed sensor networks using combinatorial designs. ACM Trans. Inf. Syst. Secur. **11**(2), 1–35 (2008). Article no. 5
10. Karavay, M.F., Parkhomenko, P.P., Podlazov, V.S.: Combinatorial methods for constructing bipartite uniform minimal quasi complete graphs symmetrical block designs. Autom. Remote Control **70**(2), 312–327 (2009)
11. Parkhomenko, P.P., Karavay, M.F.: Multiple combinatorial block diagrams. Autom. Remote Control **74**(6), 995–1003 (2013)
12. Parkhomenko, P.P.: Algorithmizing design of a class of combinatorial block diagrams. Autom. Remote Control **77**(7), 1216–1224 (2016). https://doi.org/10.1134/S0005117916070080
13. Frolov, A.B., Klyagin, A.O., Kochetova, N.P., Temnikov, D.Yu.: Raspredelennoye vychisleniye kombinatornykh blok-skhem. Problemy teoreticheskoy kibernetiki. Materialy zaochnogo seminara XIX mezhdunarodnoy konferentsii. Pod redaktsiyey YU. I. Zhuravleva. — Kazan', pp. 126–129 (2020). (in Russian)
14. Frolov, A.B., Kochetova, N.P., Klyagin, A.O., Temnikov, D.Y.: The algorithmic aspects of creating and using wireless sensor network key spaces based on combinatorial block diagrams. Bull. MPEI **2**, 108–118 (2021). (in Russian)
15. Sage. https://www.sagemath.org. Accessed 10 Jan 2020

A Stopwatch Automata-Based Approach to Schedulability Analysis of Real-Time Systems with Support for Fault Tolerance Techniques

Alevtina Glonina and Vasily Balashov(✉)

Department of Computational Mathematics and Cybernetics, Lomonosov Moscow State University, Leninskie Gory, MSU, 1, Bldg. 52, Room 764, 119991 Moscow, Russia
alevtina@lvk.cs.msu.su, hbd@cs.msu.su

Abstract. During design of a fault tolerant real-time computer system (RTCS), it is necessary to guarantee that real-time constraints on system operation are met (i.e. all jobs are executed within deadlines) despite the increase of workload due to use of fault tolerance techniques (FTT). Checking these constraints can be reduced to schedulability analysis of workload which is modified with respect to the used set of FTT. In this paper, we propose an approach to such analysis, based on simulation of RTCS operation in order to produce a time diagram. Simulation model is automatically constructed from the general RTCS operation model (corresponding to a class of systems) and the RTCS configuration description. A generalization of stopwatch automata networks is chosen as a formal base for RTCS modeling, allowing to prove correctness properties for the models. The approach is implemented as an open-source tool system. Results of experimental evaluation are presented, demonstrating applicability of the approach to checking real-time constraints for solutions of the reliability allocation problem, extending the applicability of an existing evolutionary algorithm for this problem to a larger class of systems.

Keywords: Real-time systems · Schedulability · Simulation · Stopwatch automata

1 Introduction

Modern complex technical systems (e.g. airplanes, power plants) are controlled by real-time computer systems (RTCS). A state-of-the-art RTCS is comprised by a set of standard computational modules connected by a data transfer network. The RTCS workload is a set of periodic tasks: for every period, an instance of a task (referred to as *job*) must be executed. RTCS must meet *real-time constraints*: all jobs must be executed within deadline intervals defined by tasks' periods. Data exchange between jobs of different tasks is performed by message passing. Data dependencies may exist between tasks with same period; in such case, the receiver job must wait for message arrival from the sender job. Every task is bound to a specific module. Execution of tasks bound to a module is controlled by the scheduler according to a scheduling scheme, typically a dynamic one.

W. Zamojski et al. (Eds.): DepCoS-RELCOMEX 2021, AISC 1389, pp. 116–125, 2021.
https://doi.org/10.1007/978-3-030-76773-0_12

In this paper, by *RTCS configuration* we mean the set of computational resources (set of modules, number and types of CPU cores on the modules), set and characteristics of the workload (including task periods and worst-case execution times, durations of message transfers through the network), binding of tasks to modules.

An important property of an RTCS is dependability, meant as probability of failure-free operation during a specified time interval [1]. RTCS dependability can be improved by use of fault tolerance techniques (FTT). These techniques allow prevention of RTCS failure in case of fault in some of RTCS components, by means of software and/or hardware redundancy: adding spare computational modules, using several independently developed versions of application tasks (AT), accompanied by service tasks (ST) such as voters and acceptance tests.

Use of hardware redundancy increases the necessary amount of RTCS computational resources, and development of several AT versions increases the work effort for RTCS software development; this leads to increase of RTCS cost, and such increase must be controlled. Moreover, during RTCS development it is necessary to check that the real-time constraints are met, with consideration of the set of FTT used in the system. Thus, the reliability allocation problem (RAP) arises, stated as follows. *Given:* RTCS workload (set and characteristics of the "basic" AT versions, as well as of "alternate" AT versions for use with FTT; set and characteristics of messages to be transferred between AT), constraints on computational resources (number and types of available modules), costs of resources; available types of FTT, costs of FTT use; constraint on the total system cost; *Find:* set of FTT which maximizes *dependability* of the RTCS while meeting *real-time constraints* and *constraint on the total system cost*. Besides the choice of a specific FTT (or no FTT) for every AT, the RAP solution must include binding of all application and service tasks, including those added due to FTT use, to the RTCS modules. The RAP statement may vary, e.g. to minimize the cost while providing minimum required dependability, but the general idea remains the same.

Quite a lot of papers are dedicated to RAP research (see overviews in [2, 3]), but only few [4–7] propose methods for RAP solving that check real-time constraints on RTCS operation. Of these papers, only in [4] use of different FTT types is supported *and* data dependencies between tasks are taken in account. Checking of real-time constraints is performed in [4] with several simplifying assumptions – in particular, it is assumed that every module executes a single AT (or a group of tasks corresponding to the same "FTT-enhanced" AT); all ATs are supposed to have the same period.

These assumptions significantly limit the class of RTCS to which the evolutionary algorithm for RAP solving from [4] is applicable. In this paper, we propose an approach to checking real-time constraints for RTCS with FTT, which removes these assumptions and extends the class of RTCS to which the algorithm from [4] is applicable. The proposed method is based on [8] and performs schedulability analysis using simulation models with verified correctness.

2 Fault Tolerance Techniques

A number of FTT are used to provide RTCS dependability [9, 10], for instance N-version programming (NVP), N self-checking programming (NSCP), Recovery block (RB).

Each of these FTT, applied to a specific AT, assumes execution of several independently developed versions of this AT (software redundancy, to tolerate software faults), as well as service tasks; tolerance to hardware faults is provided by use of additional computational modules (hardware redundancy). FTT of a specific TYPE, making a specific AT tolerant to X hardware faults and Y software faults, is denoted as TYPE/X/Y.

The RAP solving algorithm from [4] supports FTT NVP/0/1, NVP/1/1, and RB/1/1. In the present paper, we consider checking real-time constraints for these FTT, as well as for general FTT of NVP, RB and NSCP types, in which the original AT is replaced by a group of several AT versions and service tasks.

Use of NVP-type FTT for a specific AT involves N versions of this AT ($N \geq 3$, N is odd) and two ST: input data receiver and voter; the voter is combined with output data sender. Receiver ST receives all input messages addressed to the AT and dispatches them to versions of the AT. Every version of the AT is executed and sends the result to the voter, which selects the result produced by most of AT versions and sends this result to the output, as the original AT should do. In NVP/0/1, three AT versions are used, running on a single module. NVP/1/1 differs from NVP/0/1 by running AT versions on separate modules, i.e. three modules are used. Data dependency graph for AT versions and ST in case of NVP/0/1 or NVP/1/1 is shown in Fig. 1.

Use of NSCP-type FTT for a specific AT involves N versions of this AT ($N \geq 2$) and N+2 ST: input data receiver, N acceptance tests (possibly identical), and output data sender. Receiver ST receives all input messages addressed to the AT and dispatches them to versions of the AT. Every version of the AT is executed and sends the result to the corresponding acceptance test, which checks the result for correctness, and in case of success sends this result with pass mark to the sender (otherwise, a fail mark is sent); the sender takes the first received successful result and sends it to the output. With NSCP-type FTT, AT versions usually run concurrently on separate modules, or on one multi-core module. In NSCP/1/1, two modules and four AT versions are used, two versions for every module. Data dependency graph for AT versions and ST in case of NSCP/1/1 is shown in Fig. 1.

Use of RB-type FTT for a specific AT involves N versions of this AT ($N \geq 2$) and N+2 ST, with roles similar to those in NSCP. Main difference of RB from NSCP is that AT versions are executed sequentially (as a chain) and conditionally, so that the next AT version is started only in case the result of the previous version failed the acceptance test. There can be several such chains, each running on a separate module. In RB/1/1, two modules and two AT versions are used, both modules running the same AT chain. Data dependency graph for AT versions and ST in case of RB/1/1 is shown in Fig. 1. It should be noted that real-time constraints must be checked in the worst case, in which all tasks in all chains are executed.

A particular solution for the RAP problem described in Introduction corresponds to an RTCS configuration, which includes a set of modules with redundancy required by the set of FTT, a set of tasks with respect to the set of FTT (including service tasks and versions of application tasks, with necessary data dependencies); binding of tasks to modules is also specified. As real-time constraints are checked for the worst case, in which all versions of AT and all ST are executed (this is important for RB-type FTT), all information on FTT significant for checking these constraints is present in the

RTCS configuration in form of computational resources and workload description. Thus, checking real-time constraints for a RAP solution amounts to checking these constraints for the corresponding RTCS configuration in which the used FTT are taken in account but are not explicitly present. The method for real-time constraints checking must perform schedulability analysis (as several tasks, possibly with different periods, can be bound to the same module), as well as account for data dependencies between tasks (both initially present between different AT, and emerging with use of FTT).

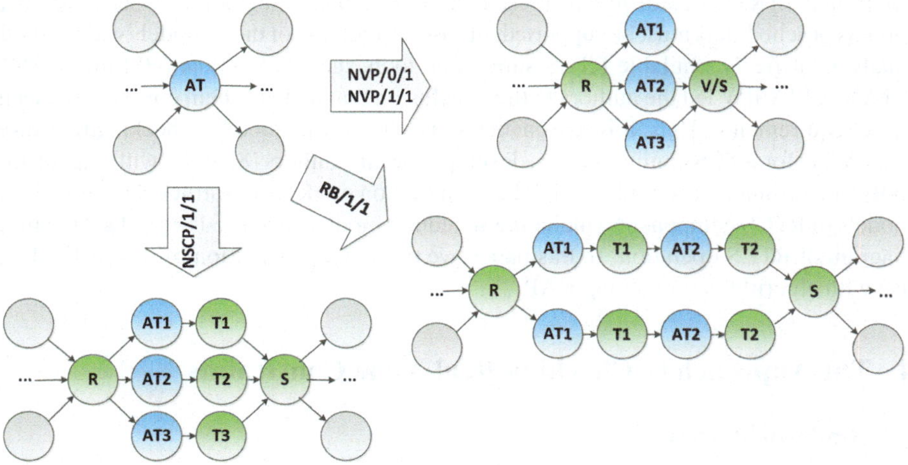

Fig. 1. Replacing an application task with a group of its versions and service tasks with introduction of FTT (AT: application task, V: voter, R: receiver, S: sender, T: acceptance test)

3 Related Work on Schedulability Analysis

In Sect. 2 we concluded that to check real-time constraints for an RTCS with FTT (e.g. obtained as a RAP solution), it is necessary to perform schedulability analysis for an RTCS configuration, the set of modules and the workload of which are specified with respect to the set of used FTT. During this analysis, both the scheduling scheme and data dependencies between tasks must be taken in account.

A group of analytical methods for schedulability analysis is known, based on classical approaches [11, 12]. These methods have a number of modern modifications [13–15]. A common disadvantage of these methods is significant over-estimation of tasks' worst case response times in case of task sets with data dependencies. This leads to false negative assessment of schedulability.

Another approach to schedulability analysis is construction and formal verification (e.g. by model checking) of RTCS model [16, 17]. This approach produces exact worst case response times of tasks, but severely limits the scale of analyzed systems (no more than several dozens of tasks), as its complexity is exponential by the number of tasks.

A more flexible and scalable approach to schedulability analysis for RTCS with a given configuration is construction of a time diagram for RTCS operation with job execution times equal to worst-case execution times (WCET) of tasks [18, 19]. Time diagram is constructed by running a simulation model of RTCS and contains intervals of job execution on CPU cores. With time diagram available, it is possible to check real-time constraints by direct comparison of job finish times to their deadlines.

RTCS simulation model must account for all aspects of RTCS operation significant for checking real-time constraints (e.g. task scheduling schemes, task preemption, data dependencies). Correctness of the model must be verified, as it is impossible during RTCS design to experimentally compare behavior of the model to behavior of actual RTCS. Since modern RTCS use different task scheduling schemes, integration of user-developed models of schedulers must be supported, in case correctness of these models was verified. Analysis of freely available RTCS simulation tools such as Cheddar, HSSim, MAST, DYANA, MASIW led the authors to the conclusion that none of them completely meets these requirements. Thus, it is reasonable to use the new method for checking real-time constraints for RTCS configurations, developed by the authors [8] along with supporting tools. According to this method, an RTCS simulation model is constructed from verified models of RTCS components (including scheduler models) and used to produce the time diagram of RTCS operation. In this paper, we present an adaptation of this method for use with algorithms for solving RAP.

4 The Approach to Checking Real-Time Constraints

4.1 Problem Statement

Let us introduce informally the problem of checking real-time constraints for an RTCS configuration; formal statement of this problem is given in [8]. Checking these constraints for a RAP solution, i.e. for an RTCS with a specific set of FTT, is reduced to this problem.

Given the RTCS configuration: set of computational modules, workload (set of tasks, set of messages), binding of tasks to modules; for every module – number and types of CPU cores, as well as the scheduling scheme (e.g. fixed priority preemptive); for every task – its period, WCETs for different CPU core types, priority unique within a module; for every message – sender task, receiver task, maximum durations of transfer through module's memory and through the network.

Construct the time diagram of RTCS operation and *check* the time diagram for meeting the real-time constraints (these constraints are formulated below in terms of time diagram).

While checking the real-time constraints, RTCS operation is analyzed on the scheduling interval, duration of which equals to the least common multiple of all tasks' periods. For every task T_i with period p_i, on the scheduling interval of duration L a job set $W_i = \{w_{ij}\}_{j=1}^{L/p_i}$ is defined.

Event in the RTCS is a tuple $\langle EType, Src, t \rangle$, where *EType* is the event type (*EX* – starting or resuming job execution; *PR* – job preemption; *FIN* – finishing job execution); $Src \in \bigcup_i W_i$ – event source job; $t \in \overline{1, L}$ – event time. The time is considered discrete, time unit can be chosen equal to RTCS CPU clock cycle.

Time diagram of RTCS operation is a set of events during the (real or simulated) RTCS operation. In practice during RTCS design it is assumed that execution time of every job is fixed and equals to WCET of corresponding task, message transfer times are fixed and equal to maximum possible ones, and all schedulers operate deterministically. Under these assumptions, a time diagram unambiguously corresponds to an RTCS configuration. RTCS simulation model is used for construction of this unambiguously defined time diagram.

Execution intervals for a job w_{ij} are intervals between the events $\langle w_{ij}, EX, t_1\rangle$ and $\langle w_{ij}, PR, t_2\rangle$, or between the events $\langle w_{ij}, EX, t_1\rangle$ and $\langle w_{ij}, FIN, t_2\rangle$, in the time diagram. Such intervals must not contain other events of types EX, PR,FIN with source job w_{ij}.

Real-time constraints are formulated in terms of time diagram as follows: for every job, total duration of its execution intervals equals duration of this job (i.e. WCET of corresponding task). If this condition is not met for some job, it indicates that execution of this job was terminated before completion, solely due to reaching the job's deadline. In other words, the job execution was not completed before its deadline. To check this condition for a specific RTCS configuration, it is necessary to construct time diagram of RTCS operation with this configuration. Since every task is bound to some module, it is correct to formulate the real-time constraints as such condition only if all CPU cores in a module have the same type, i.e. for every core of the same module a task has the same WCET. Real-time computer systems with modular structure, which are known to the authors, meet this assumption.

4.2 General Networks of Stopwatch Automata

Networks of stopwatch automata [20] were chosen as the basic formalism for modeling RTCS operation, aimed at time diagram construction via simulation. A stopwatch automaton (farther referred to as automaton) is a finite state machine with integer variables and special timer variables. A network of such automata is a set of cooperating automata, interacting via shared variables and synchronization through channels. *Parametrized automaton* is an automaton, expressions in which (e.g. for variable assignment, for transition conditions) include integer parameters, besides variables and numeric constants. By assigning values to all parameters, an *instance* of a parametrized automaton is created, which is a "normal" automaton.

To construct a general (i.e. abstract) model of RTCS operation, with support for modeling different (in particular, user-developed) schedulers, and to prove the correctness of this model, we propose a new abstraction level of automata networks: *general networks of stopwatch automata*. Main concepts of this abstraction level are introduced below.

Set of variables and synchronization actions, by which the automaton interacts with other automata in a network, is called *interface* of the automaton. *Base automata type* is described only by parameters and interface; no locations or transitions are specified for it. A parametrized automaton *implements* a base automata type if the interface and the set of parameters of the parametrized automaton match those of the base automata type. A set of base automata types is called *general automata network*, set of parametrized automata – *parametrized automata network*, network of instances of parametrized automata – *instance of automata network*.

These concepts can be explained by the example of scheduler modeling. A scheduler model, abstract from the specific scheduling scheme used in the module, is a base automata type. A model of a fixed priority preemptive scheduler is a parametrized automaton. A model of a scheduler operating in a specific module of an RTCS with a given configuration (including the set of tasks to schedule) is an instance of automaton, i.e. a "normal" automaton without parameters.

4.3 General Model of Real-Time System Operation

In [8] a general model for operation of a time-partitioned RTCS with modular structure was proposed, in form of a general automata network. In the present paper the RAP problem is considered for RTCS without time partitioning, thus an RTCS is modeled using following base automata types from [8]: base type `T`, modeling a task (with respect to its interaction with a scheduler); base type `TS`, modeling a scheduler operating in an RTCS module; base type `L`, modeling a virtual link (as means for message transfer with limited duration).

General model of RTCS operation specifies the interface for every base automata type comprising it. For instance, automata of base type `L` send signals through $\texttt{send}_{\texttt{ij}}$ and $\texttt{receive}_{\texttt{ij}}$ channels, where channel (*i,j*) corresponds to the *j*-th task of the *i*-th module. Structure of the general RTCS operation model is shown in Fig. 2; arrows represent directed channels, shared variables are not shown for brevity.

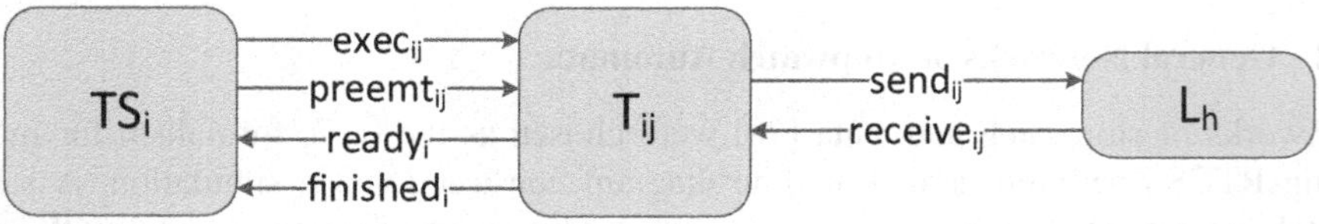

Fig. 2. Base automata types comprising the general model of RTCS operation

Only uniprocessor dynamic schedulers are modeled in [8], thus at present our capabilities for modeling RTCS with FTT are limited to systems in which every module contains only one CPU core (e.g. the central computer system of MC-21 airplane), or systems with multicore modules in which every task is bound to a specific CPU core.

4.4 Constructing a Simulation Model for Schedulability Analysis

Parametrized automata network implementing the general automata network described in Sect. 4.3, is a parametrized model of RTCS operation. Such model proposed in this paper contains following parametrized automata developed by the authors: task model, virtual link model, and models of three schedulers (fixed priority preemptive, fixed priority non-preemptive, preemptive EDF). Parameter values specify the set of computational resources, set and characteristics of the workload, binding of workload to computational resources.

According to values of parameters, defined by RTCS configuration, an instance of parametrized automata network (i.e. an instance of RTCS model) is constructed: channels and variables for automata interaction are created, for every RTCS component a parametrized automaton of corresponding type is created, and values are assigned to its parameters (e.g. period is set for a task model). An instance of RTCS model unambiguously corresponds to an RTCS configuration by construction. The result of running the RTCS model instance is the time diagram of RTCS operation. Running of the model, i.e. simulation of RTCS operation, is supported by stopwatch automata runtime library developed by the authors.

In [21] a set of correctness requirements was formulated for models of scheduler, task, virtual link, and RTCS as a whole. Using formal properties of general network of stopwatch automata, the authors proved that the developed parametrized models and the RTCS model constructed from them meet these requirements. E.g. for a fixed-priority preemptive scheduler model we proved that it always selects for execution a ready job of the task with highest priority.

5 Case Study

The proposed approach to checking real-time constraints for configurations of RTCS with FTT was implemented as an open-source tool system [22]. An experimental evaluation was performed to assess applicability of the approach to RTCS of realistic scale. We analyzed dependency of simulation time (RTCS model construction and execution time) on RTCS configuration scale and share of application tasks with FTT. The experiments were performed on AMD Ryzen 5 2600 3.4 GHz CPU, using one core.

In the first group or experiments, for an RTCS configuration with realistic scale (10 modules, around 150 ATs, more than 160 messages [18]), the share of AT with FTT was gradually increased (from 0% to 100% with 10% step); following FTT were used in equal numbers: NVP/0/1, NVP/1/1, RB/1/1. In the second group of experiments, for an RTCS configuration with 50% of AT using FTT, such quantitative characteristics were consistently increased as number of AT, messages and modules. Results of experiments are shown in Fig. 3.

RTCS considered in [18] is an avionics system with architecture described in patent RU2014115662A. According to this architecture, modules are connected by a switched network with support for virtual links, providing predictable message transfer times. The architecture is scalable, as it supports adding new computational modules grouped in crates by up to five; this allows using FTT that require hardware redundancy. Every crate has its own network switch module, which connects computational modules of the crate to each other and to other crates.

The experiments demonstrated quadratic dependency of simulation time on variation of selected characteristics of RTCS configuration. We can conclude that the developed tool system can be used with algorithms for RAP solving that enumerate hundreds of solutions (or thousands, if solutions are analyzed concurrently on a multicore CPU), including the evolutionary algorithm from [4].

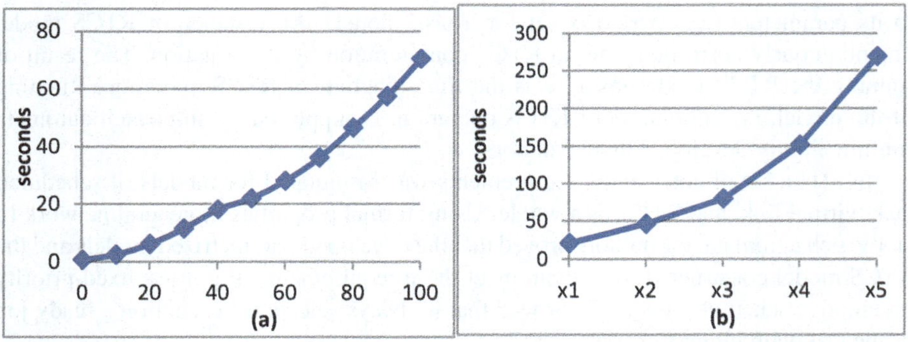

Fig. 3. Dependency of simulation time on share of AT with FTT (a), on configuration scale (b)

6 Conclusion

In this paper, we proposed an approach to checking real-time constraints for configurations of RTCS with fault tolerance techniques, e.g. obtained as solutions of the reliability allocation problem. The approach is based on automatic construction of RTCS simulation model, and its running to obtain a time diagram of RTCS operation, taking in account the scheduling scheme(s) used in the RTCS modules. Choice of generalized stopwatch automata networks allowed formal proof of correctness for the RTCS models. The proposed approach significantly extends applicability of RAP solution algorithm from [4] and can be used with other RAP solution algorithms.

Future research is aimed at development of multiprocessor dynamic scheduler models and proof of their correctness, in order to extend the supported class of RTCS. Support for time-partitioned fault tolerant RTCS is another possible goal.

Acknowledgements. This work is partially supported by the Russian Foundation for Basic Research under grant №19-07-00614.

References

1. Laprie, J.C., Coste, A.: Dependability: a unifying concept for reliable computing. In: Proceedings of the 12th Fault Tolerant Computing Symposium, pp. 18–21 (1982)
2. Kuo, W., Wan, R.: Recent advances in optimal reliability allocation. IEEE Trans. Syst. Man Cybern. - Part A Syst. Hum. **2**(37), 143–156 (2007)
3. Coit, D.W., Zio, E.: The evolution of system reliability optimization. Reliability Eng. Syst. Saf. **192**, 106259 (2019)
4. Volkanov, D., et al.: Simulation modeling based method for choosing an effective set of fault tolerance mechanisms for real-time avionics systems. Progress Flight Dyn. GNC Avionics **6**, 487–500 (2013)
5. Jha, P.C., et al.: Optimal component selection of COTS based software system under consensus recovery block scheme incorporating execution time. Int. J. Reliability, Qual. Saf. Eng. **3**(17), 209–222 (2010)

6. Attiya, G., Hamam, Y.: Reliability oriented task allocation in heterogeneous distributed computing systems. In: 9th International Symposium on Computers And Communications Proceedings, pp. 68–73. IEEE, Alexandria, Egypt (2004)
7. Kang, Q., He, H., Wei. J.: An effective iterated greedy algorithm for reliability-oriented task allocation in distributed computing systems. J. Parallel Distrib. Comput. **8**(73), 1106–1115 (2013)
8. Glonina, A.B., Bahmurov, A.G.: Stopwatch automata-based model for efficient schedulability analysis of modular computer systems. In: Malyshkin, V. (ed) PACT 2017, LNCS, vol. 10421, pp. 289–300. Springer, Cham (2017)
9. Laprie, J.C., et al.: Definition and analysis of hardware and software-fault-tolerant architectures. IEEE Comput. **23**(7), 39–51 (1990)
10. Wattanapongsakorn, N., Levitan, S.: Reliability optimization models for fault-tolerant distributed systems. In: International Symposium on Product Quality and Integrity Proceedings, pp. 193–199. IEEE, Philadelphia, USA (2001)
11. Liu, C.L., Layland, J.W.: Scheduling algorithms for multiprogramming in a hard-real-time environment. J. ACM **20**(1), 46–61 (1973)
12. Audsley, N., et al.: Applying new scheduling theory to static priority preemptive scheduling. Softw. Eng. J. **8**(5), 284–292 (1993)
13. Kim, J., et al.: A novel analytical method for worst case response time estimation of distributed embedded systems. In: 50th Annual Design Automation Conference Proceedings, pp. 1–10. ACM, New York (2017)
14. Palencia, J.C., et al.: Response-time analysis in hierarchically-scheduled time-partitioned distributed systems. IEEE Trans. Parallel Distrib. Syst. **28**(7), 2017–2030 (2017)
15. Amurrio, A.: Response-time analysis of multipath flows in hierarchically-scheduled time-partitioned distributed real-time systems. IEEE Access **8**, 196700–196711 (2020)
16. Han, P., et al.: Schedulability analysis of distributed multicore avionics systems with UPPAAL. J. Aerosp. Inf. Syst. **16**(11), 473–499 (2019)
17. André, É., et al.: Parametric schedulability analysis of a launcher flight control system under reactivity constraints. In: 19th International Conference on Application of Concurrency to System Design Proceedings, pp. 13–22. IEEE, Aachen, Germany (2010)
18. Balashov, V.V., Balakhanov, V.A., Kostenko, V.A.: Scheduling of computational tasks in switched network-based IMA systems. In: OPTI'2014 International Conference Proceedings, pp. 1001–1014. NTUA, Athens, Greece (2014)
19. Cheramy, M., et al.: Simulation of real-time scheduling with various execution time models. In: 9th International Symposium on Industrial Embedded Systems, pp. 1–4. IEEE, Pise, Italy (2014)
20. Cassez, F., Larsen, K.: The impressive power of stopwatches. In: Palamidessi, C. (ed) CONCUR 2000, LNCS, vol. 1877, pp. 138–152. Springer, Berlin (2000)
21. Glonina, A.B., Balashov, V.V.: On the correctness of real-time modular computer systems modeling with stopwatch automata networks. Automatic Control Comput. Sci. **7**(52), 817–827 (2018)
22. https://github.com/AlevtinaGlonina/MCSSim. Accessed 27 Feb 2021

Fractional Order Derivative Mechanism to Extract Biometric Features

Zbigniew Gomolka[1(✉)], Boguslaw Twarog[1], and Ewa Zeslawska[2]

[1] College of Natural Sciences, University of Rzeszow, Pigonia Street 1, 35-959 Rzeszow, Poland
{zgomolka,btwarog}@ur.edu.pl

[2] Department of Applied Information, University of Information Technology and Management in Rzeszow, Sucharskiego Street 2, 35-225 Rzeszow, Poland
ezeslawska@wsiz.edu.pl

Abstract. Currently growing market of mobile devices demands algorithms for biometric identification, which require low computational complexity and reliability. The paper presents the use of a fractional derivative to construct a feature vector. Using the IriShield scanner, a database of iris photos obtained in the near infrared was generated. The iris area was segmented and normalized using the Daugman algorithm. Iris feature vectors were obtained by weaving a standardized iris image with a kernel of a gradient edge detector using derivatives of fractional orders according to the Riemann-Liouville definition. For the efficiency assessment of the proposed approach, the comparison of results obtained with a Log-Gabor filter, which is a classic method of quadrant coding in the Daugman algorithm were performed. A statistical analysis of the effectiveness of the proposed method was carried out using the Hamming distance measurement to compare the generated iris codes. The conducted experiments allowed to determine the optimal order of derivative and size of the fractional derivative kernel in order to increase the discrepancy of feature vectors data. The use of fractional derivative convolution mechanism show the great potential for distinguishing iris. Proposed approach might be applied in programming tools for iris recognition.

Keywords: Fractional derivatives · Biometric recognition · Grünwald-Letnikov · IriShield

1 Introduction

Nowadays, it is becoming increasingly popular to use the fractional derivative mechanism in many areas. These include the description of diffusion processes [1], reverse thermal conduction issues [2], viscoelastic material dynamics modeling [3] and image recognition [4]. In image processing, the mechanism is used to define convolution masks in edge detection [5], denoising [4, 6] or even to estimate the optical flow [7]. This is due to the inherent property of the weave operation as a form of derivative estimation. An important area of application is the process of biometric identification of persons on the basis of detailed analysis of biometrics, among which efficient methods of extracting their characteristics are sought. The iris pattern (apart from color) is an epigenetic trait,

W. Zamojski et al. (Eds.): DepCoS-RELCOMEX 2021, AISC 1389, pp. 126–135, 2021.
https://doi.org/10.1007/978-3-030-76773-0_13

i.e. it does not depend on the human DNA code sequence and allows for high biometric entropy, which avoids identification clashes. The most popular and most commonly used iris recognition algorithm was proposed by John Daugman in his 1994 patent [8]. It was one of the first automatic identification systems based on the iris pattern. The idea of the principle of operation comes down to capturing the image of the eye, its segmentation and normalization [9–11]. It uses weaving of a Gabor filter with a normalized iris image in the frequency domain for the extraction of characteristics [12, 18–20]. The purpose of this study is to explore new methods of encoding iris feature vectors which would be computationally efficient and fast. The main research goal of this work is to conduct experiments that will show how effective the formation of the vector feature of the iris is using the Grunvald-Letnikov fractional order derivative. In addition, the computational complexities of the classical and proposed feature encryption methods will be compared. The experiments will show whether conducting the convolution in polar coordinates of the iris image will allow to obtain acceptable recognition accuracy with less computational complexity.

2 Research Methods

There are many definitions and methods of estimation of a fractional derivative [13–17]. A discreet fractional derivative as defined by Grünwald-Letnikov, uses the difference coefficients of the reverse discrete function:

$$ {}_{t_0}D_t^{(\alpha)} f(t) = \lim_{\substack{h \to 0 \\ t - t_0 = hk}} \left[\frac{1}{h^\alpha} \sum_{i=0}^{k} a_i^{(\alpha)} f(t - hi) \right] \tag{1} $$

where

$$ a_i^{(n)} = \begin{cases} 1, & i = 0 \\ (-1)^i \frac{n(n-1)(n-2)\ldots(n-i+1)}{i!}, & i = 1, 2, 3, \ldots \end{cases} \tag{2} $$

The Caputo definition uses the gamma function:

$$ {}_{t_0}D_t^{(\alpha)} f(t) = \frac{1}{\Gamma(n-\alpha)} \int_{t_0}^{t} (t-\tau)^{n-\alpha-1} f^{(n)}(\tau) d\tau \tag{3} $$

where $1 < \alpha < n; n \in \mathbb{Z}$.

Riemann-Liouville's fractional derivative is calculated as the n integer derivative of a non-integer order integral $(n - \alpha)$:

$$ {}_{t_0}D_t^{\alpha} f(x) = D^n \, {}_{t_0}I_t^{n-\alpha} f(x) \tag{4} $$

$$ D_{a+}^{\alpha} f(x) = \frac{1}{\Gamma(n-\alpha)} \frac{d^n}{dx^n} \int_a^x \frac{f(t)dt}{(x-t)^{\alpha-n+1}} \tag{5} $$

Using fractional order derivatives for the extraction of iris characteristics the use of gradient edge detector kernel was presented. The method is based on a standardized image of the iris of the eye obtained by means of the Daugman algorithm. The study used a database of eye images scanned using the IriShield biometric scanner with automatic image capture after iris detection. The base record consists of 3 images of the left eye and 3 images of the right eye in JPG 640 × 480 pixels format on an 8-bit grey scale, conforming to the ISO 19794-6 standard.

Segmentation and Normalization

The collected images were segmented and normalized using the Daugman algorithm. The image was exposed to the Canny edge detector and circle Hough Transform (CHT) to determine iris and pupil areas. A linear Hough transform was used to locate the eyelids that could obscure part of the iris, while the noise in the form of light reflections were located by means of brightness thresholding. Normalized iris images were obtained by developing the image to polar coordinates, with a resolution of 40px (radial) by 240px (angular), which were later used to create characteristic vectors. Figure 1 shows a diagram of the Daugman algorithm with visualization of the data received in each of the above steps.

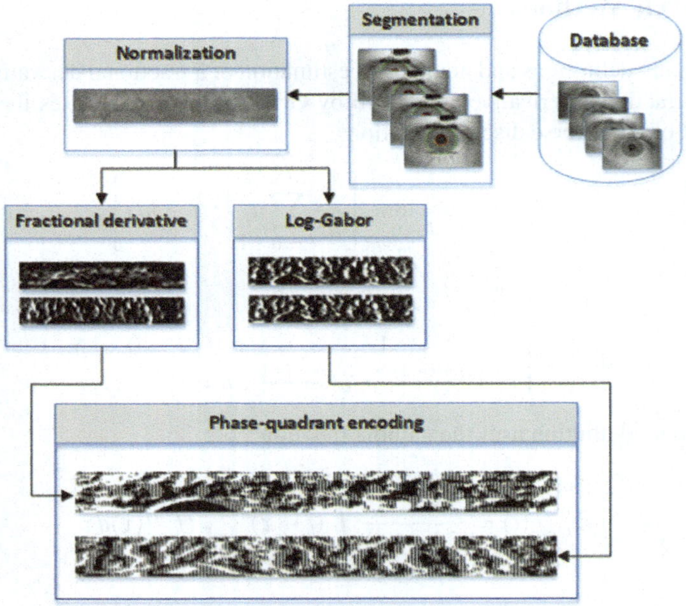

Fig. 1. Diagram of the Daugman algorithm used in the identification process.

Extraction of Characteristics

In the classic version of the Daugman algorithm, a Gabor filter is used to extract the iris characteristics, while in this work its logarithmic version is used as a reference for determining the effectiveness of the proposed method. A one-dimensional Log-Gabor filter in the frequency domain is described by the following equation:

$$G(f) = \exp\left(-(\log(f/f_0))^2 \Big/ 2(\log(\sigma/f_0))^2\right) \tag{6}$$

where: f – signal frequency, f_0 – basic filter frequency (for which the strongest response is obtained), σ – standard deviation of the Gaussian curve, responsible for the width of the filter.

In image analysis it is useful to keep the same filter shape for different base frequencies. This is achieved by requiring that the ratio of standard deviation of the Gaussian function to the basic filter frequency be constant $\sigma / f_0 = const$. The filter used in (6) is described by the equation in the frequency domain, so it is applied not to the image function but to its Fourier transform. The proper characteristic vector is obtained by inverse Fourier transform of the filter and image function Fourier transform weave. This operation is performed for each line of the image.

Fractional Derivative Filter

Starting from Riemann-Liouville's definition and using the properties of the weave operation, you can obtain a kernel of a gradient edge detector approximating the discrete fractional order derivative α:

$$\Theta_x(x_{i,j}, y_{i,j}) = \frac{\alpha \cdot x_{i,j}}{\Gamma(1-\alpha)} \left(x_{i,j}^2 + y_{i,j}^2\right)^{-\frac{\alpha}{2}-1} \tag{7}$$

$$\Theta_y(x_{i,j}, y_{i,j}) = \frac{\alpha \cdot y_{i,j}}{\Gamma(1-\alpha)} \left(x_{i,j}^2 + y_{i,j}^2\right)^{-\frac{\alpha}{2}-1} \tag{8}$$

where $0 < \alpha < 1$.

An example of a kernel mask of size 5×9 approximating the 0.2 x-axis derivative is shown in Fig. 2. A correctly adopted scale in red represents the maximum values of the mask and a blue one represents the minimum values of the mask.

0.025	0.031	0.035	0.029	0.000	-0.029	-0.035	-0.031	-0.025
0.030	0.041	0.058	0.080	0.000	-0.080	-0.058	-0.041	-0.030
0.033	0.046	0.075	0.172	0.000	-0.172	-0.075	-0.046	-0.033
0.030	0.041	0.058	0.080	0.000	-0.080	-0.058	-0.041	-0.030
0.025	0.031	0.035	0.029	0.000	-0.029	-0.035	-0.031	-0.025

Fig. 2. Kernel coefficients approximating the 0.2 x-axis derivative.

The horizontal gradient kernel is a transposition of the vertical gradient kernel. The extraction of iris characteristics (approximation of the directional derivative) is carried out by weaving the above kernels with the image function $f(x, y)$. Since two-dimensional kernels were used, to avoid anomalies occurring at the boundaries of the image function of the normalized iris, in the areas above and below the image, the average value of the image brightness was applied to convolution. The left and right ends are connected to form a kind of side surface of the cylinder. Both vertical and horizontal kernels are used to extract the characteristics. It has been assumed that the result of the weave will be recorded in the array of complex numbers, where the real part will contain values of vertical gradients and the imaginary part of horizontal ones. This allows you to record information about the vertical distribution of characteristics, as opposed to the one-dimensional Log-Gabor filter. This also makes it easier to further synthesize the binary code of the iris, which in this case is the same for both methods.

QuadrantCoding
The Daugman algorithm uses quadrant phase demodulation in coding. Depending on the position on the complex plane, two bits are assigned to each cell of the array of extracted characteristics. Since the results of operation of Log-Gabor methods and a fractional derivative are arrays of complex numbers with dimensions of a standardized image, it is possible to apply the same method in both cases. The result is a binary iris code, twice as long as the standard input image.

Hamming Distance
The Hamming distance (HD) was used to compare iris codes, which is a measure of the difference between two strings of equal length. It is expressed as the ratio of the number of different bits (at the same positions) to the number of bits being compared:

$$HD = \frac{\sum_{i=1}^{n} \mathrm{XOR}(a_i, b_i)}{n} \tag{9}$$

where: a_i – the bit of the first string being compared, b_i – the bit of the second string being compared, n – the number of bits being compared.

The algorithm for comparing iris codes takes into account the occurrence of noise, rejecting bits obscured by noise masks from the analysis. In order to avoid false rejection when comparing the same irises from different photos, bit shifts are used to find the best match. This allows you to correctly identify a person even if the orientation of the eye to the camera in two different attempts is different.

Optimization of Log Gabor filter parameters and fractional derivative
The following parameters affecting the effectiveness of both methods are distinguished:

- For the Log-Gabor filter:

– basic wavelength: $\lambda_0 = 1/f_0$,
– throughput parameter: σ/f_0.

- For the fractional derivative filter:

– derivative order: α,
– width of the convolution mask: $size(x)$.

Optimal parameters were sought through a series of simulation experiments for each possible combination of parameters from the tested ranges (see Table 1).

Table 1. Tested ranges of variable parameters of iris characteristics extraction.

Parameter	Log-Gabor		Fractional derivative	
	λ_0	$\alpha/0$	α	*size*(*x*)
min: step: max	14:2:30	0.25:0.25:0.75	0.1:0.1:0.9	5:2:9

To this end, iris codes were generated for each combination of variable coding parameters and the Hamming distance between different irises was determined for each set of parameters. The following criterion was adopted for the evaluation of parameter sets:

$$S = \left(\overline{HD_i} - 3\sigma_i\right) - \left(\overline{HD_c} + 3\sigma_c\right) \tag{10}$$

where: S – evaluation score of a given set of parameters; $\overline{HD}$ – average value of the Hamming distance of comparisons of different (i) or the same (c) irises; σ – standard deviation of the Hamming distance distribution of comparisons of different (i) or the same (c) irises.

3 Results

On the basis of the simulation tests carried out in the initial phase, sets of parameters for Log-Gabor filters and a fractional derivative were selected, which allowed for maximum effectiveness of iris discrimination by the system (see Table 2). The obtained results of evaluation for each parameter set are presented in Tables 3 and 4 (the highest evaluation score values are in bold).

Table 2. Optimal values for filter parameters.

Parameter	Log-Gabor		Fractional derivative	
	λ_0	$\sigma/0$	α	$size(x)$
Optimal value	16	0.5	0.2	9
Evaluation score	0.1000		0.0383	

Using optimal sets of filter parameters, weaves were carried out on individual collections of irises. Next, histograms of the distribution of Hamming distance were created for both methods (see Fig. 3).

It can be seen that the above distributions do not overlap. Taking into account that the parameter S is positive, and it assumes the distance of three standard deviations, the probability of correct recognition is over 99.7% if the Hamming distance limit is appropriately selected. On this basis, it can be concluded that the proposed method using a fractional order derivative mechanism effectively differentiates the iris. In addition, in the case of the Log-Gabor filter, the separation of Hamming distances is much greater than in the case of a method using a fractional derivative. The width of the distribution of the client's results (comparisons of the same irises from different photos) slightly differs from the one obtained by the fractional derivative method. However, we are observing a significant shift in the Hamming distances of differing irises towards higher values. Their distribution is also narrower. Figure 4 shows the obtained codes of two different irises using the fractional derivative method. Their Hamming Distance $HD(Eye_{412}, Eye_{222}) = 0.4557$.

Table 3. Evaluation of Log-Gabor filter parameter sets.

		Throughput parameter: $\alpha/0$		
		0.25	0.5	0.75
λ_0	14	−0.1126	0.0939	0.0674
	16	−0.1630	**0.1000**	0.0718
	18	−0.2074	0.0995	0.0762
	20	−0.2463	0.0950	0.0752
	22	−0.2824	0.0854	0.0660
	24	−0.3147	0.0733	0.0604
	26	−0.3444	0,0573	0.0596
	28	−0.3715	0.0373	0.0590
	30	−0.3972	0.0103	0.0561

Table 4. Results of the evaluation of the fractional order derivative filter parameter sets.

		Size(x)		
		5	7	9
α	0.1	0.0295	0.0374	0.0376
	0.2	0.0283	0.0365	**0.0383**
	0.3	0.0280	0.0363	0.0371
	0.4	0.0285	0.0355	0.0370
	0.5	0.0287	0.0350	0.0372
	0.6	0.0282	0.0344	0.0373
	0.7	0.0280	0.0334	0.0368
	0.8	0.0275	0.0326	0.0364
	0.9	0.0274	0.0320	0.0349

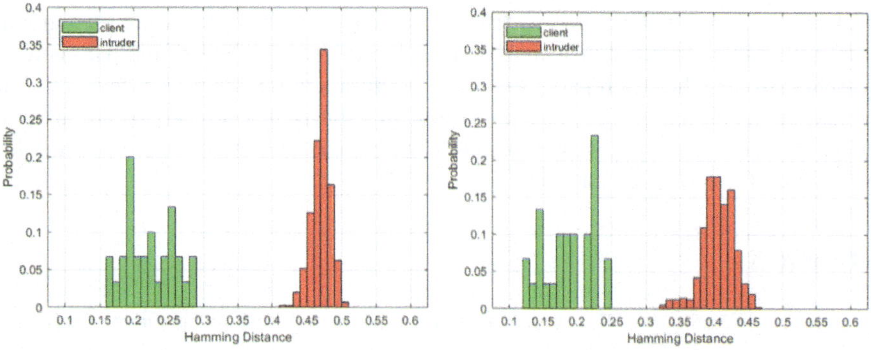

(a) (b)

Fig. 3. Distribution of Hamming distances of comparisons of the same (client) and different (intruder) irises coded using the Log-Gabor method (a) and the fractional derivative method (b).

Fig. 4. Iris codes of different eyes obtained using a fractional derivative.

4 Conclusion

In order to achieve the objectives of this work, a program was designed to code the vector of iris characteristics through a weave of a normalized image of the iris with a gradient edge detector kernel estimating a fractional derivative from the range]0, 1[as defined by Riemann-Liouville. In the experimental part of the work, a system was designed to optimize coding parameters based on the Hamming distance. For each set of parameters, within the assumed range of filter operation variability, the codes of all irises were created from the database of images prepared for the purposes of the work and the corresponding Hamming distances were calculated for each pair of obtained images. The analysis of the obtained results showed that for iris differentiation the most effective is a 5×9 weave kernel, estimating the 0.2 order derivative. Each configuration of the coding conditions with this method enables effective differentiation of the iris with appropriate selection of the acceptance limit (approx. 0.30 – 0.35 for different parameter configurations), as the distribution of Hamming distances is bifurcated between results for the same and different irises. Compared to the classical method using the Log-Gabor filter, the gradient edge detector, however, performs worse. In the method with a fractional derivative, a shift of Hamming distances towards lower values is observed, and their distribution when comparing different irises is wider. This method is less sensitive to changes in coding parameters because if the Log-Gabor filter parameters are incorrectly selected, a significant decrease in the ability to distinguish images is observed, while for a gradient kernel it is stable. This stability enables us to use different values of the parameters to different systems. This discrepancy between systems increase safety, as each system's iris database can't be used in other systems to gain unauthorized access. Also computational efficiency of using simple weave operation without Fourier transform is desirable in devices with limited specifications.

References

1. Sene, N., Adbelmalek, K.: Analysis of the fractional diffusion equations described by Atangana-Baleanu-Caputo fractional derivative. Chaos Solitons Fractals **127**, 158–164 (2019)
2. Brociek, R., Słota, D., Król, M., Maluta, G., Kwaśny, W.: Comparison of mathematical models with fractional derivative for the heat conduction inverse problem based on the measurements of temperaturę in porous aluminium. Int. J. Heat Mass Transfer **143**, 118440 (2019)
3. Shen, L.-J.: Fractional derivative models for viscoelastic materials at finite deformations. Int. J. Solids Struct. **190**, 226–237 (2019)
4. Ghandbari, B., Atangana, A.: A new application of fractional Atangana-Baleanu derivatives: designing ABC-fractional masks in image processing. Physica A **123516**, 6 (2019)
5. Amoako-Yirenkyi, P., Appati, J.K., Dontwi, I.K.: A new construction of a fractional derivative mask for image edge analysis based on Riemann-Liouville fractional derivative. Adv. Difference Equations **238** (2016)
6. Wang, Q., Ma, J., Yu, S., Tan, L.: Noise detection and image denoising based on fractional calculus. Chaos, Solitons and Fractals (2019)
7. Lavín-Delgado, J.E., Solís-Pérez, J.E., Gómez-Aguilar, J.F., Escobar-Jiménez, R.F.: Robust optical flow estimation involving exponential fractional-order derivatives. Optik – Int. J. Light Electron. Optics. **202**, 163642 (2020)
8. Daugman, J.: Biometric personal identification system based on iris analysis. United States Patent No. 5,291,560 (1994)
9. Ahmadi, N., Nilashi, M., Samad, S., Rashid, T.A., Ahmadi, H.: An intelligent method for iris recognition using supervised machine learning techniques. Optics Laser Technol. **120**, 105701 (2019)
10. Liu, X., Bai, Y., Luo, Y., Yang, Z., Liu, Y.: Iris recognition in visible spectrum based on multi-layer analogous convolution and collaborative representation. Pattern Recogn. Lett. **117**, 66–73 (2019)
11. Vyas, R., Kanumuri, T., Sheoran, G., Dubey, P.: Efficient iris recognition through curvelet transform and polynomial fitting. Optik – Int. J. Light Electron Optics **185**, 859–867 (2019)
12. Galdi, C., Dugelay, J.-L.: FIRE: fast Iris recognition on mobile phones by combining colour and texture features. Pattern Recogn. Lett. **91**, 44–51 (2017)
13. Gomolka, Z.: Neurons' transfer function modeling with the use of fractional derivative. In: Zamojski, W., Mazurkiewicz, J., Sugier, J., Walkowiak, T., Kacprzyk, J. (eds.) Contemporary Complex Systems and Their Dependability. DepCoS-RELCOMEX 2018. Advances in Intelligent Systems and Computing, vol. 761. Springer, Cham (2019)
14. Gomolka, Z.: Backpropagation algorithm with fractional derivatives. ITM Web Conf. **21**, 00004 (2018). https://doi.org/10.1051/itmconf/20182100004
15. Ortigueira, M., Machado, J.: Which Derivative?,Fractal and Fractional, number 1, 1, 3 (2017). ISSN: 2504–3110, https://doi.org/10.3390/fractalfract1010003
16. Ortigueira, M., Tenreiro M.J.: On the Properties of Some Operators Under the Perspective of Fractional System Theory. Communications in Nonlinear Science and Numerical Simulation. **82,** 105022 (2019). https://doi.org/10.1016/j.cnsns.2019.105022

17. Ortigueira, M., Machado, J.: Fractional derivatives: the perspective of system theory. Mathematics **7**, 150 (2019). https://doi.org/10.3390/math7020150
18. Gomolka, Z., Twarog, B., Zeslawska, E., Nykiel, A.: Biometric data fusion strategy for improved identity recognition. In: Zamojski W., Mazurkiewicz, J., Sugier, J., Walkowiak, T., Kacprzyk, J. (eds.) Theory and Applications of Dependable Computer Systems. DepCoS-RELCOMEX 2020. Advances in Intelligent Systems and Computing, vol.173. Springer, Cham. https://doi.org/10.1007/978-3-030-48256-5_24
19. Juneja, K., Rana, C.: Compression-robust and fuzzy-based feature-fusion model for optimizing the iris recognition. Wireless Pers Commun. **116**, 267–300 (2021). https://doi.org/10.1007/s11277-020-07714-3
20. Luo, Z., Gu, Q., Su, G., Zhu, Y., Bai, Z.: An adaptive face-Iris multimodal identification system based on quality assessment network. In: Lokoč, J., et al. (eds.) MultiMedia Modeling. MMM 2021. Lecture Notes in Computer Science, vol. 12572. Springer, Cham (2021). https://doi.org/10.1007/978-3-030-67832-6_8

Contract-Based Specification and Test Generation for Adaptive Systems

Bence Graics(✉), Vince Molnár, and István Majzik

Department of Measurement and Information Systems,
Budapest University of Technology and Economics, Budapest, Hungary
{graics,molnarv,majzik}@mit.bme.hu

Abstract. Control systems in railway, automotive or industrial robotic applications are generally tightly integrated into their environment to allow adapting to environmental changes. This paper proposes a contract-based specification and testing approach for adaptive systems based on the combination of a high-level scenario language (LSC variant) and an adaptive contract language (statechart extension). The scenario language supports high-level modeling constructs as well as configurable options for test generation. The adaptive contract language supports the flexible definition of scenario contract activation and deactivation based on environmental changes or interactions. Tests can be derived from adaptive contract descriptions using the combination of graph-traversal algorithms and integrated model checker back-ends. The applicability of the approach is demonstrated in the context of the Gamma framework.

Keywords: Adaptive systems · Contracts · Scenarios · Test generation

1 Introduction

Model-based systems engineering (MBSE) is becoming more prevalent in the design of adaptive systems, e.g., control systems in the automotive, railway and aerospace industries. Such systems are tightly integrated into their environment to facilitate the adaptation to external changes and the related changing of requirements. These systems often carry out critical tasks requiring thorough verification starting in the early development phases. Contract-based techniques can help system design by specifying high-level requirements, e.g., in the form of scenarios. They can also be the basis for verification, e.g., in test generation.

Many languages have been proposed for the contract-based description of system behavior with scenarios [1–4]. However, the contracts defined in these languages are generally static in the sense that they do not support the description of adaptive behavior, e.g., the reconfiguration (activation and deactivation)

This work was partially supported by the ÚNKP-20-3 New National Excellence Program of the Ministry for Innovation and Technology and by the National Research, Development and Innovation Fund of Hungary, financed under the [2019-2.1.1-EUREKA-2019-00001] funding scheme.

W. Zamojski et al. (Eds.): DepCoS-RELCOMEX 2021, AISC 1389, pp. 136–145, 2021.
https://doi.org/10.1007/978-3-030-76773-0_14

of system components and changing behavior in case of certain (external or internal) events. In addition, black-box test generation based on these models typically does not consider implementation constraints for system behavior like allowed latency. Therefore, approaches aiming to aid the contract-based specification and testing of adaptive systems should support *1)* high-level modeling languages for adaptive contract definition, and *2)* configuration options for test generation. We propose a solution in the context of the Gamma framework.

The Gamma Statechart Composition Framework [5] is an integrated modeling toolset for the semantically well-founded composition and analysis of statechart components with different semantic variations. At its core, it provides a composition language supporting hierarchical composition with precise semantics [6]. Gamma supports system-level formal verification and validation (V&V) by mapping models into formal models of various model checkers and back-annotating the results. The framework provides test generation functionalities as well as automated code generators for both statechart and composite models.

In this work, we introduce a contract-based specification and black-box test generation approach for adaptive systems in the context of Gamma. Contracts for adaptive behavior can be defined with a *scenario language* and an *adaptive contract language*, tailored to synchronous component execution semantics. *Tests can be derived* both from individual scenarios and adaptive contract descriptions. The main novelties of our approach are as follows. *1)* Scenarios can be defined in a configurable high-level *scenario language* tailored to test generation with various configuration options. *2)* A statechart-like *adaptive contract language* for the flexible description of contract (scenario) activation and deactivation triggered by events. The behavior of a system is specified by the set of active scenarios, which may change due to receiving environmental events (e.g., changes in context or communication), internal events (including error signals), or time delays. *3) Tests can be derived* from individual scenarios or adaptive contract descriptions using graph-traversal algorithms together with the integrated model checker back-ends. Test generation can be customized with coverage criteria in the adaptive contract description.

2 Contract-Based Specification and Test Generation

Our approach builds on a model transformation chain depicted in Fig. 1. The approach consists of six steps and builds on the following modeling languages.

- The **Gamma Scenario Language (GSCL)** is a configurable variation of LSC [2] (Live Sequence Chart) that supports the high-level description of system behavior in terms of input and output events (interactions).
- The **Gamma Statechart Language (GSL)** is a UML/SysML-inspired statechart [7] language supporting different semantic variants of statecharts.
- The **Gamma Adaptive Contract Language (GACL)** is a GSL extension supporting the definition of scenario contract activation and deactivation.
- The **Gamma Test Language (GTL)** is a high-level test language supporting the specification of execution traces for reactive systems.

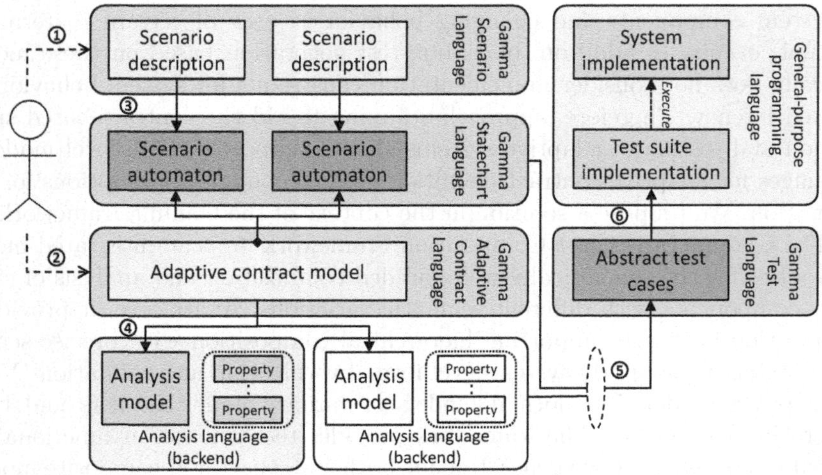

Fig. 1. The model transformation chains and modeling languages.

In Step 1, the user can specify contracts for the system in the form of GSCL scenarios. The language (see Sect. 3.1) has high-level constructs to model system behavior and configuration options for test generation.

In Step 2, the user can introduce adaptivity to the contract descriptions by creating an adaptive contract model in GACL, specifying the activation and deactivation of contracts upon certain events. The language (see Sect. 3.2) supports high-level constructs available in regular statecharts.

In Step 3, the scenarios are automatically transformed to automata modeled in GSL where execution paths represent the interaction sequences of scenarios. Since these automata are modeled as regular Gamma statecharts, all the functionalities of Gamma can be applied to them, which can further support the specification and verification of contracts, e.g., by deriving runtime monitors.

In Step 4, the adaptive contract model is transformed into input languages of integrated model checkers (currently UPPAAL [8] and Theta [9]). Properties are also generated depending on the specified coverage criteria for test generation (currently *state*, *transition*, *transition-pair* or *out-event* coverage).

In Step 5, abstract test cases are derived using model checking and graph-traversal techniques. Model checking is responsible for computing a path to the activation of certain scenario contracts. These contracts are then interpreted using graph-traversal algorithms to derive test inputs, delays and expected outputs in that state. Abstract test cases are the combination of the steps returned by the model checker with the steps returned by the graph-traversal algorithms.

Finally, in Step 6, abstract test cases are customized to concrete execution environments. Abstract test cases are mapped to sequences of concrete calls to provide test inputs, schedule system execution, and then retrieve and evaluate outputs. Currently, JUnit[1] is supported as a concrete execution environment.

[1] https://junit.org/.

```
1  component Controller
2
3  [@Permissive | @Strict]
4  @AllowedWaiting 1 .. 3
5  scenario atomicInteractions [
6   cold receives Control.toggle
7   hot sends Priority.toggle
8   hot delay (500 .. 550)
9   {
10   hot sends Priority.toggle
11   hot sends Secondary.toggle
12  }
13  negate cold sends Priority.toggle
14  optional {
15   cold sends Priority.toggle
16  }
17  [alternative | unordered | parallel] {
18   cold sends Priority.toggle
19  } {
20   cold sends Secondary.toggle
21   cold sends Tertiary.toggle
22  }
23  loop (2 .. 3) {
24   cold sends Priority.toggle
25  }
26 ]
```

Fig. 2. Model snippet demonstrating the features of GSCL.

3 Languages for Specifying Contracts and Adaptive Behavior

3.1 Scenario Language

GSCL is a Live Sequence Chart (LSC) [2] variant with restrictions and extensions to facilitate test generation. It can describe scenarios that specify *mandatory*, *forbidden* and *possible* executions of reactive systems in terms of interactions. The constructs of the language are demonstrated in Fig. 2.

GSCL scenarios support the specification of interactions for a *single lifeline* (the system, Line 1) communicating with its environment. Interactions are defined in a sequential order. Atomic interactions are *events* received (Line 6) or sent (Line 7) via a specific port, as well as *time delays* (Line 8). Hot and cold modalities distinguish between mandatory (hot) and optional (cold) atomic interactions. In order to support synchronous processing of batches of events (i.e., in execution turns), GSCL supports *interaction sets*, consisting of atomic interactions that need to occur at the same time/turn (Lines 9–12). In an interaction set, the direction and modality of the contained interactions must be the same. Atomic interactions and interaction sets can be *negated* to specify the absence of interactions in certain scenarios (Line 13).

GSCL supports *optional*, *alternative*, *unordered*, *parallel* and *loop* fragments (composite interactions) for complex behavior description. Each combined fragment type contains one or more interactions as operands. An *optional* fragment describes an interaction that may or may not occur (Lines 14–16). An *alternative* fragment describes a set of interactions out of which exactly one must occur (Lines 17 and 18–22). An *unordered* fragment describes a set of atomically handled (potentially composite) interactions that can occur in an arbitrary order (Lines 17 and 18–22). A *parallel* fragment describes a set of non-atomically handled interactions that can occur in parallel, that is, interleaving between the interactions can occur (Lines 17 and 18–22). A *loop* fragment describes the iteration of an interaction. The loop can have a *minimum* and a *maximum bound* or (if no bounds are specified) can be *infinite* (Lines 23–25).

Based on a GSCL scenario, an observed interaction sequence is *(i) inconclusive* if an atomic interaction violates (i.e., does not match) a *cold* interaction specification or does not cover the entire scenario; *(ii) invalid* if an atomic interaction violates a *hot* interaction specification, meaning that it contradicts the expected behavior; *(iii) valid* if it is neither *inconclusive* nor *invalid.*

GSCL supports test generation with configuration options for *1)* specifying constraints for system response and *2)* categorizing unspecified system behavior.

Option 1 provides solution for the problem of not knowing the exact latency between inputs and reactions. For example, Lines 6 and 7 specify that the system must respond to an incoming *toggle* event with a *toggle* event on the respective output ports. Depending on the implementation, this may take several execution turns. Therefore, GSCL introduces an annotation (Line 4) that specifies the accepted range of latency in terms of executions turns. If the response does not arrive within the specified interval, a violation occurs. Note that this kind of uncertainty does not apply to events sent to the system, because they do not depend on the system implementation.

Option 2 allows the different interpretation of unexpected events sent by the system during test execution. As we saw, after the reception of a *toggle* event (Line 6) the system may have "time" to respond, during which it may send unexpected events too. We can use annotations (Line 3) to decide whether this is permitted (*permissive* mode) by ignoring that event or treated as a violation of the interaction expecting the response (*strict* mode).

3.2 Adaptive Contract Language

GACL is a statechart language [7] extension to support the definition of adaptive contracts, that is, the activation and deactivation of static scenario contracts upon specific events. It builds on GSL and supports powerful constructs such as composite states, parallel regions, history states, variables, and complex transitions such as choice, fork and join. It also supports semantic variation points, e.g., conflict resolution between transitions of different hierarchy levels or the definition of priorities between transitions with the same source.

As a key feature, GACL supports linking a set of GSCL scenarios to states, therefore scenario management can also benefit from the high-level features of statecharts. During execution, the current state configuration indicates the active scenario set. When processing incoming events, the GACL statechart has priority over the active scenarios. When a state configuration is left, the scenarios linked to the left states get deactivated and the ones linked to the newly entered states get activated. As scenarios do not have history, the examination of behavior always starts at the beginning of the scenario.

We show an example adaptive contract model for a synchronous traffic light controller system from the tutorial[2] for Gamma in Fig. 3. In the system, there are two standard 3-phase *traffic lights* (*priority* and *secondary*), which loop

[2] https://github.com/ftsrg/gamma/tree/master/tutorial.

```
component Crossroads
@Strict @AllowedWaiting 1
scenario Init [
 {hot sends priority.red
  hot sends secondary.red}
 hot sends priority.green
]
```

```
component Crossroads
@Strict @AllowedWaiting 1
scenario Normal [
 loop { // Infinite loop
  hot sends priority.yellow
  hot delay (1000)
  {hot sends priority.red
   hot sends secondary.green}
  hot delay (2000)
  hot sends secondary.yellow
  hot delay (1000)
  {hot sends priority.green
   hot sends secondary.red}
  hot delay (2000)
 }
]
```

```
component Crossroads
@Strict @AllowedWaiting 1 .. 2
scenario Blinking [
 loop { // Infinite loop
  {hot sends priority.black
   hot sends secondary.black}
  hot delay (500)
  {hot sends priority.yellow
   hot sends secondary.yellow}
  hot delay (500)
 }
]
```

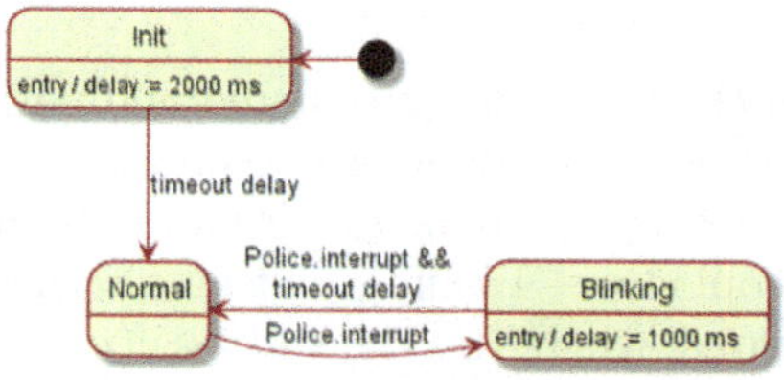

Fig. 3. Contracts for the Crossroads model.

through the red-green-yellow-red sequence (*normal mode*) scheduled by a *central controller* and have an *interrupted mode* where they blink in yellow.

The requirements for the system in the two modes are specified by the following GSCL models. The activation and deactivation of these scenarios are modeled by the GACL model. In the example, every scenario is linked to the GACL state with the same name.

According to the configuration options, at least one, at most two execution turns in the *Blinking* scenario, and exactly one execution turn in the other scenarios must be carried out between the reception of an interaction (time delay) and the response from the system's point of view (*@AllowedWaiting 1 [.. 2]*) and unexpected events must not be sent while waiting (*@Strict*). In addition, every scenario uses *hot* modality interactions, indicating that no difference is acceptable from the specified behavior in the system states. The *Init* scenario specifies the initialization of the system. After 2000 ms, the system starts its normal operation (*Normal*) with the red-green-yellow-red traffic light sequences in specified timed intervals (1000 ms and 2000 ms). Interrupted mode can be turned on and off (*Blinking*) by sending an *interrupt* signal, in which case we expect the system to send alternating black and yellow signals every 500 ms.

4 Deriving Test Cases from Adaptive Contract Models

4.1 Translating Scenarios to Automata

Scenario models are transformed into automata in two steps. First, as a preprocessing step (that will be replaced by a more sophisticated solution as future

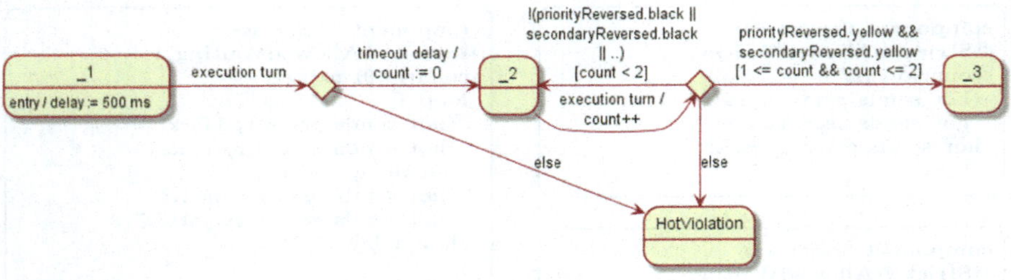

Fig. 4. The GSL automaton excerpt generated from the *Blinking* scenario.

work), every loop, unordered and parallel fragment is mapped to a combination of alternative fragments to create a *simplified scenario* model.

For a *loop* fragment with lower bound n and upper bound m, the loop operand is copied $n - 1$ times, and an alternative fragment with $m - n + 1$ operands is created, where the $1st, 2nd, \ldots, (m-n+1)th$ operand contains the loop operand $1, 2, \ldots, (m - n + 1)$ times. For *unordered* and *parallel* fragments, the permutations describing the possible interaction sequences are modeled using alternative fragments based on Heap's algorithm [10]. These simplified scenario models (with only optional and alternative combined fragments) are then transformed into automata capable of classifying interaction sequences as *inconclusive*, *invalid* and *valid*. Figure 4 shows an excerpt from the GSL automaton generated from the *Blinking* scenario.

Every automaton has three specific states: *cold violation*, *hot violation* and *accept* associated with the classification of an observed interaction sequence. An atomic interaction is mapped to a source state and a complex choice transition targeting *(i)* the source state of the next interaction if the specified behavior is matched or the accepting state if it is the final interaction (highest priority), *(ii)* a violation state based on the modality of the interaction if the specified behavior is not matched (lowest priority), and potentially *(iii)* the source state of the current interaction if it is a *sends* interaction, modeling different configuration options regarding the execution time (medium priority).

Construct *(iii)* relies on an integer variable (*count*) that counts the execution turns until an expected event arrives. The transition going back to the source state does not have a trigger (therefore triggered by execution) in *permissive* operation mode and has a complex trigger as the negated disjunction of every unexpected event that can be sent by the system in *strict* mode.

Alternative fragments are mapped to choice transitions that target the source states of the mapped interaction operands. Optional fragments are also mapped to complex choice transitions where one target is the source state of the mapped interaction operands, and the other one is the source state of the next interaction after that. Notice that per se, this mapping does not create a deterministic *monitor* to classify a sequence in one pass. It can be regarded as a non-deterministic finite automaton, but since we are about to generate test cases from its execution paths, it is enough that it can simulate the described behavior.

With the scenarios transformed, an adaptive contract model can be created using a GACL model, with states linked to a set of scenarios and transitions describing the activation/deactivation of such scenarios.

4.2 Deriving Test Cases from Adaptive Contract Models

Test derivation are performed in two steps: generating paths in the adaptive contract model with model checkers and traversing the scenario automata. Building on the verification features of Gamma, we use model checkers to get diagnostic traces describing the steps to get to different state configurations based on the specified coverage criteria. These automata then can be traversed to derive *positive* or *negative* tests. The traversal of automata can be carried out with a graph-traversing algorithm that considers transition triggers and basic relational operations for the single integer variable (*count*). For positive tests, *every path* (possible interaction sequence) from the initial state to the *accept* state is retrieved as the system implementation must conform to at least one of them. Note that there are a finite number of paths from the initial state due to the employed constructs (transition guards). For negative tests, paths from the initial state to the *hot violation* state are retrieved. Abstract test cases are created by combining the paths returned by the model checker with the paths returned by the graph-traversal algorithm. Abstract test cases can be easily mapped to JUnit tests using the high-level GTL constructs.

To summarize, we generate a test suite where each scenario activated during the traversal of the GACL model gets a test case, e.g., during state coverage, each scenario in each GACL state gets a test case. A test case consists of several test executions derived from that scenario, and its succeeds if any of them is successfully executed (for positive tests) or all of them fail (for negative tests).

4.3 Test Generation Example

We transformed the GACL model of Sect. 3.2 into UPPAAL automata and derived tests with *transition coverage* criterion. We generated 4 *positive tests*: one for *Init* with one execution path, one for *Normal* with one execution path, one for *Blinking* with 4 execution paths because both two *sends* interactions can happen either after 1 or 2 turns, and one for the *Normal* state again, but first going to *Blinking* (to cover all the transitions). By evaluating these tests on the *initial* design of the Crossroads system in the Gamma tutorial, we could find that it was *incorrect* as the combination of *1)* the interaction sequence leading to the *Normal* state through *Blinking* in the GACL model {*delay 2000 ms, schedule, Police.interrupt, schedule, delay 1000 ms, schedule, Police.interrupt, schedule*} and *2)* the interaction sequence obtained from the *Normal* scenario automaton {*priority.yellow, schedule, priority.red, secondary.green*} leads to an assertion error. The problem was caused by the fact that the controller was not properly stopped in the design when the lights are switched into interrupted mode. See the tutorial for an explanation of this problem, which was deliberately introduced for demonstration purposes.

5 Related Work

There are many approaches for the specification and verification of adaptive behavior. Adaptive behavior is generally specified using domain-specific modeling languages that support the capturing of adaptation policies in the form of event-condition-action (ECA) rules, similar to our proposed state-based approach [11,12]. In order to cope with scalability issues arising when large sets of rules are specified, optimization techniques have been introduced that capture high-level goals and facilitate the rule selection process [13,14]. However, these approaches focus on the specification, simulation and runtime verification of the adaptive behavior rather than test generation.

Model-based test generation approaches that target the testing of scenarios oftentimes rely on the combination of UML Sequence Diagrams and State Machine Diagrams [15], which describe typical scenarios of system use and the (independent) behavior of system components, respectively. [16] presents a test sequence generator (TeStor) algorithm that relies on sequence diagrams as an abstract or even incomplete specification of what the test should include. TeStor synthesizes more detailed sequence diagrams (conforming to the abstract scenarios) by recovering missing information from state diagrams. The approach in [17] aims to extract message sequences from sequence diagrams, which are complemented by initialization sequences for the participating objects derived from State Machine diagrams. [18] presents an approach aiming to improve sequence model based testing by considering the effect of each message on the states of the target object. They use sequence model interactions to select key transitions and states from a state machine. Similarly to ours, these approaches use the combination of sequence and state diagrams for test generation, but the motivation to use state diagrams is different: none of them considers adaptivity, i.e., the activation and deactivation of scenario contracts.

Generally, in our approach, we aim to integrate adaptive behavior modeling with test generation. To our best knowledge, there is no research focusing on test derivation based on adaptive contract models.

6 Conclusion and Future Work

In this paper, we proposed a contract-based specification and black-box test generation approach for adaptive systems. We introduced a scenario language and an adaptive contract language supporting contract activation and deactivation based on environmental events, error events or other internal events, which allows for environment-dependent adaptive behavior, including the reconfiguration of the expected behavior as a result of errors. We presented how tests can be generated from these adaptive contract descriptions using graph-traversal algorithms and model checking.

Subject to future work, we plan to improve the mapping of combined fragments of the scenario language and evaluate our approach in large scale industrial case studies. We also plan to investigate how these adaptive contracts can be employed in runtime verification.

References

1. Message Sequence Chart (MSC) – Annex B: Algebraic semantics of Message Sequence Charts. Standard, ITU-TS (1995). https://itu.int/rec/T-REC-Z.120
2. Damm, W., Harel, D.: LSCs: breathing life into message sequence charts. Formal Methods Syst. Des. **19**(1), 45–80 (2001)
3. Unified Modeling Language Version 2.5.1 (UMLv2.5.1) - Sequence Digrams. Standard, Object Management Group (OMG) (2017)
4. Harel, D., Maoz, S.: Assert and negate revisited: Modal semantics for UML sequence diagrams. Softw. Syst. Model. **7**(2), 237–252 (2008)
5. Molnár, V., Graics, B., Vörös, A., Majzik, I., Varró, D.: The Gamma Statechart Composition Framework: design, verification and code generation for component-based reactive systems. In: 40th International Conference on Software Engineering (ICSE), pp. 113–116. ACM, Gothenburg, Sweden (2018). https://doi.org/10.1145/3183440.3183489
6. Graics, B., Molnár, V., Vörös, A., Majzik, I., Varró, D.: Mixed-semantics composition of statecharts for the component-based design of reactive systems. Softw. Syst. Model. **19**(6), 1483–1517 (2020). https://doi.org/10.1007/s10270-020-00806-5
7. Harel, D.: Statecharts: a visual formalism for complex systems. Sci. Comput. Program. **8**(3), 231–274 (1987). https://doi.org/10.1016/0167-6423(87)90035-9
8. Behrmann, G., David, A., Larsen, K.G., Håkansson, J., Pettersson, P., Yi, W., Hendriks, M.: Uppaal **4.0**, (2006)
9. Tóth, T., Hajdu, A., Vörös, A., Micskei, Z., Majzik, I.: Theta: a framework for abstraction refinement-based model checking. In: Stewart, D., Weissenbacher, G. (eds.) Proceedings of the 17th Conference on Formal Methods in Computer-Aided Design, pp. 176–179 (2017). https://doi.org/10.23919/FMCAD.2017.8102257
10. Heap, B.R.: Permutations by Interchanges. Comput. J. **6**(3), 293–298 (1963). https://doi.org/10.1093/comjnl/6.3.293
11. Keeney, J., Cahill, V.: Chisel: a policy-driven, context-aware, dynamic adaptation framework. In: Proceedings POLICY 2003. In: IEEE 4th International Workshop on Policies for Distributed Systems and Networks, pp. 3–14 (2003)
12. David, P.C., Ledoux, T.: Safe dynamic reconfigurations of fractal architectures with FScript. In: Proceeding of Fractal CBSE Workshop, ECOOP, vol. 6 (2006)
13. Kephart, J.O., Das, R.: Achieving self-management via utility functions. IEEE Internet Comput. **11**(1), 40–48 (2007). https://doi.org/10.1109/MIC.2007.2
14. Fleurey, F., Solberg, A.: A domain specific modeling language supporting specification, simulation and execution of dynamic adaptive systems. In: Schürr, A., Selic, B. (eds.) Model Driven Engineering Languages and Systems, pp. 606–621. Springer, Heidelberg (2009)
15. Shirole, M., Kumar, R.: UML behavioral model based test case generation: a survey. ACM SIGSOFT Softw. Eng. Notes **38**(4), 1–13 (2013)
16. Pelliccione, P., Muccini, H., Bucchiarone, A., Facchini, F.: TeStor: Deriving test sequences from model-based specifications. In: Heineman, G.T., Crnkovic, I., Schmidt, H.W., Stafford, J.A., Szyperski, C., Wallnau, K. (eds.) Component-Based Software Engineering, pp. 267–282. Springer, Heidelberg (2005)
17. Sokenou, D.: Generating test sequences from UML sequence diagrams and state diagrams. INFORMATIK 2006–Informatik für Menschen–Band 2, Beiträge der 36. Jahrestagung der Gesellschaft für Informatik eV (GI) (2006)
18. Bandyopadhyay, A., Ghosh, S.: Test input generation using UML sequence and state machines models. In: 2009 International Conference on Software Testing Verification and Validation, pp. 121–130 (2009). https://doi.org/10.1109/ICST.2009.23

New Loss Function for Multiclass, Single-Label Classification

Krzysztof Halawa(✉)

Wroclaw University of Science and Technology, 27 Wyb. Wyspiańskiego Street, 50-370 Wrocław, Poland
krzysztof.halawa@pwr.edu.pl

Abstract. Deep neural networks can perform complex transformations for classification and automatic feature extraction. Their training can be time consuming and require a large number of numerical calculations. Therefore, it is important to choose the good initial learning settings. Results depend, inter alia, on a loss function. The paper proposes a new loss function for multiclass, single-label classification. Experiments were conducted with convolutional neural networks trained on several popular data sets. Tests with multilayer perceptron were also carried out. The obtained results indicate that the proposed loss may be a good alternative to the categorical cross-entropy.

Keywords: Deep learning · Convolutional neural network · Loss function

1 Introduction

Deep neural networks can perform complex nonlinear data transformations that are much more complicated than operations made by shallow architectures. Deep networks achieve the best results in many applications, including, but not limited to, image analysis, natural language processing, speech synthesis and recognition. Convolutional neural networks (CNNs) can contain many layers, e.g. Xception, Inception-V3, and Resnet151v2 have more than one hundred layers and tens of millions of parameters [1, 2]. Some networks are trained on huge dataset, for instance Resnet was learned on the ImageNet database which contains over 14 million images. Efficient graphics processing unit (GPUs) or special designed tensor processor units (TPUs) are used for training because of the large number of numerical operations that have to be performed in each epoch. Due to time-consuming and tedious learning, a sensible selection of loss function, layers and hyperparameters is important, as it is not always possible to afford multiple training attempts. Among the many deep architectures the following networks are commonly used: autoencoders, CNNs, Long short-term memories (LSTM), gated recurrent units (GRU). Autoencoder and CNN are feedforward networks, while LSTM and GRU are recurrent networks. These networks can handle regression and classification problems. CNNs are often applied for image classification. The type of task determines both the choice of the loss function and the activation function in the last layer. Their typical choice for a few common problems [2, 3] is shown in Table 1.

W. Zamojski et al. (Eds.): DepCoS-RELCOMEX 2021, AISC 1389, pp. 146–155, 2021.
https://doi.org/10.1007/978-3-030-76773-0_15

Table 1. The common choice of last-layer activation and loss function

Task type	Loss function	Last layer activation
Binary classification	Binary crossentropy	sigmoid
Multiclass, single-label classification	Categorical crossentropy	softmax
Multiclass, multilabel classification	Binary crossentropy	sigmoid
Regression	Mean Squared Error (MSE)	Linear f(x) = x

The properties of the softmax function make them particularly well suited to multiclass classification. This function ensures the normalization of the last layer outputs. The values of all outputs sum up to one and are in the range [0,1]. As a result, these values are often treated similar to class probability. The output of the j-th neuron with the softmax activation can be written as

$$y_j = \frac{e^{\mathbf{w}_j^T \mathbf{x}}}{\sum_{k=1}^{K} e^{\mathbf{w}_k^T \mathbf{x}}}, \tag{1}$$

where $\mathbf{w}_{\mathrm{j}}$ denotes the weight vector of the jth neuron, K is the number of classes, $\mathbf{x}$ is the vector with outputs of the previous layer.

Cross-entropy is a measure from the field of information theory. The categorical cross-entropy loss for online training is given by

$$E_{CCE} = -\sum_{k=1}^{K} t_k log(y_k), \tag{2}$$

where t_{k} denotes desired value of k-th output, values of $t_1,\ldots,t_k$ are determined by one-hot encoding, hence they equals 1 or 0.

For mini-batch training the average across N elements is minimized

$$E_{CCE2} = -\frac{1}{N}\sum_{i=1}^{N}\sum_{k=1}^{K} t_{i,k} log\left(y_{i,k}\right), \tag{3}$$

where N is the batch size.

This paper proposes a novel loss for multiclass, single-label classification. The new loss function is presented in Sect. 2. The conducted experiments are described in Sect. 3. Several popular datasets where used for training CNNs to classify images. Tests were also carried out with multilayer perceptrons (MLPs) without the use of images. The summary is at the end of the paper.

2 Proposed Loss Function

The paper assumes that target classes are indicated with one-hot encoding. The relationship between the value of $\mathrm{E_{CCE}}$ and $\mathrm{y_k}$ for $t_{\mathrm{k}} = 1$ is depicted in Fig. 1. From this graph it is easy to notice that the cross-entropy loss function strongly penalizes cases in which target classes are indicated as very unlikely. As the value of $\mathrm{y_k}$ decreases, the value of

E_{CCE} increases very rapidly. It can also be seen that there is only a minor penalty for non-zero network outputs if they are not related to the target class (this small penalty results indirectly from the property that the sum of the all outputs is equal to 1 for softmax). In order to overcome this drawback, a new target function is proposed. It is given by

$$E_{CCEMSE} = \frac{1}{N}\sum_{i=1}^{N}\sum_{k=1}^{K}\left[-t_{i,k}log\left(y_{i,k}+\varepsilon\right)+\beta\frac{1}{K}\left(t_{i,k}-y_{i,k}\right)^2\right], \quad (4)$$

where β is a positive real number, ε is added to avoid a sharp increase in the loss function to infinity when $y_{i,k} = 0$. This function is a combination of the categorical cross entropy and the mean squared error (MSE). The value of β determines the impact of MSE. Thanks to K in the denominator, the same beta value may be used regardless of the number of classes.

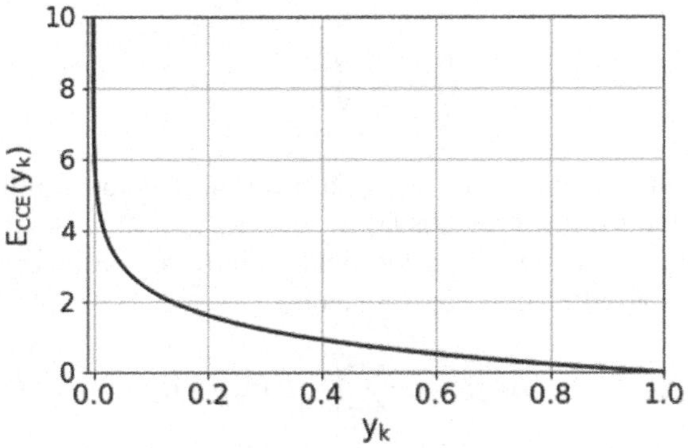

Fig. 1. The relationship between the value of categorical cross-entropy (2) and y_k for $t_k = 1$ (training online i.e. $N = 1$)

In all conducted experiments it was assumed that $\beta = 10$ and $\varepsilon = 10E{-}10$.

3 Conducted Experiments

All tests were performed on the author's personal computer. The networks were programmed and trained in python3 using open-source library Keras [1] from TensorFlow 2.4.1. The operating system was Ubuntu 20.04. In order to speed up training, computations were performed on Nvidia RTX3070 GPU.

The experiments were conducted with the use of six datasets listed in Table 2. The MNIST [4] database consists 70,000 greyscale images of handwritten digits. All images have a small resolution of 28 × 28 pixels. The database is split into two parts. There are 60,000 examples in the training set and 10,000 images in the test set. Fashion-MNIST shares the same image size and structure of training and testing splits [5]. This dataset is comprised of images of fashion products from 10 categories. Fashion-MNIST is considered to be more challenging classification problem than regular MNIST. CIFAR10 consists of 50,000 32 × 32 colour photos of objects belonging to 10 classes shown in Table 3. CIFAR20 and CIFAR100 are like the CIFAR10, except they have more classes. Sample images from CIFAR10 and MNIST are shown in Fig. 2. Main information on all data sets is presented in Table 2. The first five datasets in Table 2 were used with convolutional neural networks (CNNs), which were trained to classify images. Moreover, multilayer perceptrons (MLPs) were trained with the Forest Covertype dataset [12] to check if the proposed function could be used for tasks other than image classification. The instances in this dataset correspond to 30 × 30 m patches of forest in the United States. The neural network was trained to recognize the dominant tree species from 54 features described in [12] and [13].

Table 2. Datasets used in experiments

Dataset	Train set	Test set	Classes	Input shape
MNIST	60,000	10,000	10	28 × 28
Fashion-MNIST	60,000	10,000	10	28 × 28
CIFAR10	50,000	10,000	10	3 × 32 × 32
CIFAR 20	50,000	10,000	20	3 × 32 × 32
CIFAR100	50,000	10,000	100	3 × 32 × 32
Forest Covertypes	387,342	193,670	8	54

All neural networks were trained using mini-batches. The performed experiments required training a total of 120 neural networks, because for each dataset, the results of 10 networks were averaged for the two compared loss functions. Batch size of 500 was used to achieve a fairly short training time. The structure of the convolutional network for MNIST and MNIST FASHION is shown in Fig. 3. CNNs for CIFAR10 had more layers, because it performs a much more difficult task than handwritten digit recognition. The network architecture for CIFAR10 is shown in Fig. 4 and Table 6. CNNs for CIFAR20 and CIFAR100 had the same architecture except for the number of neurons in the last layer. The simplest network was the multilayer perceptron that was used to classify tree species. Its layers are presented in Table 5. Forest type detection is carried out on the basis of a variety of features without the use of any images. All networks had softmax activations in the last layer, while in the remaining layers was maxrelu. The total number of trainable parameters (i.e. weights) for MNIST was 952,478. Network for CIFAR10 had 2,076,818 parameters. The multilayer perceptrons had only 67,158 weights. Tables 4, 5 and 6 show dropout rates and the number of weights in each layer. All neural networks were trained using the stochastic gradient descent algorithm with momentum [9, 10]. The learning rate was equal to 0.005. Momentum was set to 0.9 according to some hints from [9]. Tests were carried out for $\beta = 10$. The initial weight values were pseudo-numbers generated using the method proposed in [11]. One-fifth of the learning sets were used only to observe whether there is overfitting that could lead to incorrect conclusions. The number of epochs was chosen to avoid overfitting that would make it impossible to reliably compare obtained results. For CIFAR10, CIFAR20, and

CIFAR100, the number of epochs was 150. MLP training lasted 50 epochs. For other datasets, the number of epochs was 70.

After the training of each network was completed, the accuracy on the test set was determined. The accuracy is defined as the ratio of number of correctly predicted label to the total number of samples. The test sets contained data that was not used for training. The obtained results are shown in Table 7. Training times are presented in Table 8.

Table 3. Classes in CIFAR10 and MNIST FASHION

Class number	CIFAR 10	MNIST FASHION
1	Airplanes	T-shirts / tops
2	Automobiles	Trouser
3	Birds	Pullovers
4	Cats	Dresses
5	Deer,	Coats
6	Dogs	Sandals
7	Frogs	Shirts
8	Horses	Sneakers
9	Ships	Bags
10	Rucks	Ankle boots

Table 4. The number of parameters and dropout rates of CNNs for MNIST and MNIST-FASHION

Layer (type)	Output shape	Trainable parameters
Conv2D	26, 26, 32	320
Conv2D	24, 24, 32	9248
MaxPooling	12, 12, 32	0
Dropout (dropout rate = 0.2)	12, 12, 32	0
Flatten	4608	0
Dense	200	5538816
Dense	100	1049600
Dropout (dropout rate = 0.4)	100	0
Dense	10	10250

From the comparison of the results from Table 7, it can be noticed that with the proposed function better results were achieved for the vast majority of datasets. Only for the CIFAR100 set, worse results were obtained.

a)

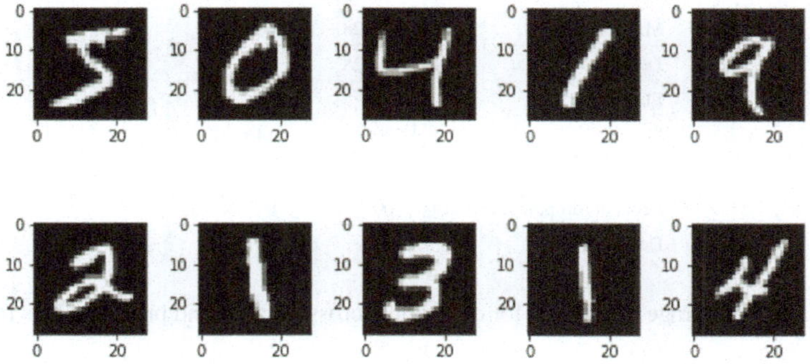

b)

Fig. 2. Sample images from a) CIFAR10 b) MNIST

Table 5. The number of parameters and dropout rates of MLP for Forest Covertype

Layer (type)	Output shape	Trainable parameters
Dense	200	11000
Dense	150	30150
Dropout	150	0
Dense	100	15100
Dense	100	10100
Dropout	100	0
Dense	8	808

Table 6. The number of parameters and dropout rates of CNNs for CIFAR10

Layer (type)	Output shape	Trainable parameters
Conv2D	32, 32, 32	896
Conv2D	32, 32, 32	9248
MaxPooling	16, 16, 32	0
Dropout (dropout rate = 0.2)	13, 13, 32	0
Conv2D	16, 16, 64	18496
Conv2D	16, 16, 64	36928
MaxPooling	8, 8, 64	0
Dropout (dropout rate = 0.2)	8, 8, 64	0
Conv2D	8, 8, 128	73856
Conv2D	8, 8, 128	147584
MaxPooling	4,4,128	0
Dropout	4,4,128	0
Flatten	2048	0
Dense	700	1434300
Dense	500	350500
Dropout (dropout rate = 0.4)	500	0
Dense	10	5010

Table 7. Average accuracies for categorical cross-entropy and proposed loss function

Dataset	Average accuracy for categorical cross-entropy	Average accuracy for the proposed loss function
MNIST	0.9912	0.9918
Fashion-MNIST	0.9170	0.9205
CIFAR10	0.8171	0.8210
CIFAR20	0.6235	0.6242
CIFAR100	0.4994	0.4973
Forest Cover	0.6895	0.6927

Table. 8 Average training time in seconds

Dataset	Average training time for categorical cross-entropy	Average training time for the proposed loss function
MNIST	89.11	91.56
Fashion-MNIST	89.56	91.78
CIFAR10	393.41	397.45
CIFAR20	393.73	397.48
CIFAR100	393.71	397.44
Forest Cover	199.72	208.42

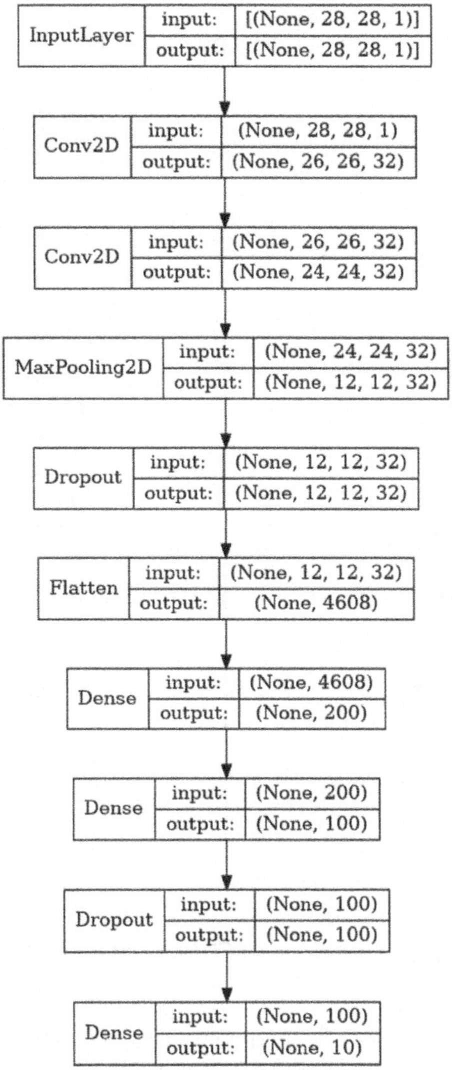

Fig. 3. Architecture of CNNs used for MNIST and MNIST_FASHION

4 Summary

The commonly used categorical cross-entropy unevenly accounts errors for false positive and false negative. The proposed loss function seems to be a potential alternative. It can be used with modern network architectures such as CNNs. In the conducted experiments, the proposed method allowed to achieve slightly better results for five datasets. Only for CIFAR100 a deterioration was obtained. The proposed function can be applied not only to CNNs recognizing images. Better results were also obtained for identification of forest types on the basis of various features without using any images. This classification

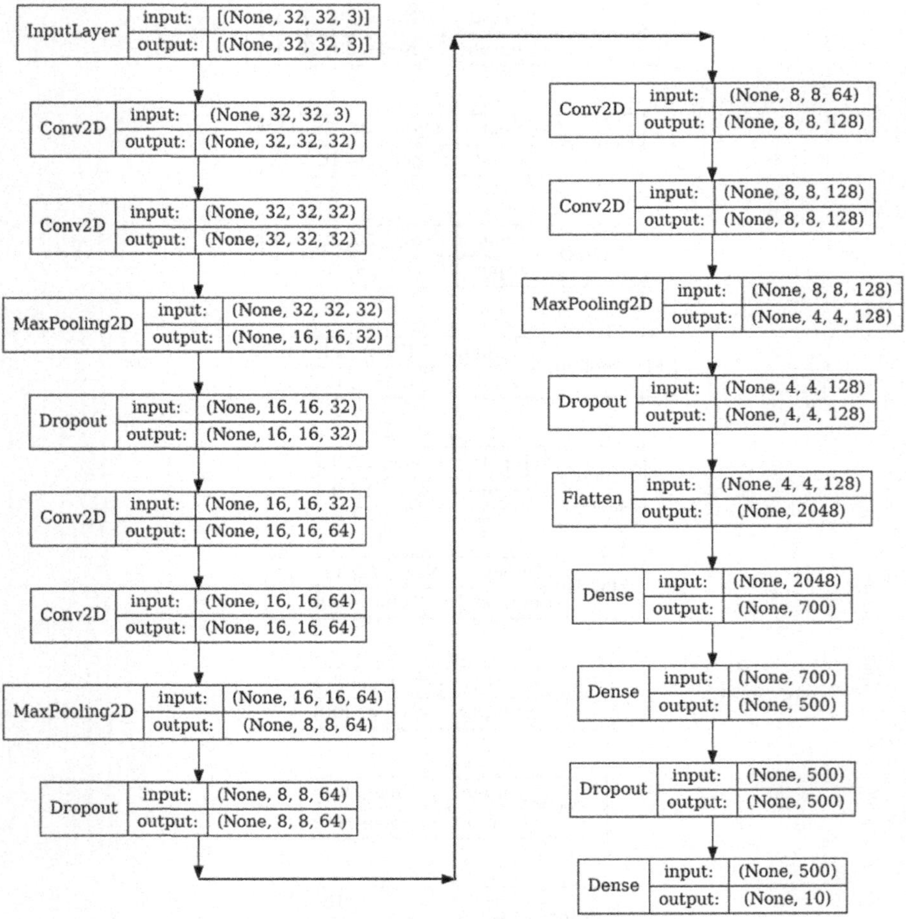

Fig. 4. Architecture of CNNs used for CIFAR10

was made using MLPs. Probably the proposed function can be applied to classification problems not only with CNNs and MLPs. The effect of increasing β was also tested, but the best results were obtained with $\beta = 10$. Due to the limitation of the article length, outcomes for other β values are not shown. The proposed target function slightly extends the training time. The larger the network, the smaller the time difference. This may be due to the method of numerical computation performed by the GPU, and in particular to the influence of the time needed for data transfer between GPU memory and tensor cores. Communication with the CPU can also have an significant impact. The training time of a small multilayer perceptron was the most extended. Due to the large number of elements of the Forest Covertype dataset, training smaller networks on this set took more than twice as long as learning larger networks on the MNIST and MNIST FASHION sets.

When implementing part related to computing the logarithm in (4), it is important to add a small value ε in order to prevent the loss from growing very rapidly.

References

1. Khan, A., Sohail, A., Zahoora, U., Qureshi, A.S.: A survey of the recent architectures of deep convolutional neural networks. Artif. Intell. Rev. **53**(8), 5455–5516 (2020)
2. Francois, C.: Deep learning with Python. Manning Publications, Shelter Island NY (2017)
3. Géron, A.: Hands-on Machine Learning with Scikit-Learn, Keras, and TensorFlow: Concepts, Tools, and Techniques to Build Intelligent Systems. O'Reilly Media, Sebastopol (2019)
4. LeCun, Y., Cortes, C.: MNIST database. https://yann.lecun.com/exdb/mnist/. Accessed 07 Jan 2021
5. Xiao, H., Rasul, K., Vollgraf, R.: Fashion-mnist: a novel image dataset for benchmarking machine learning algorithms. arXiv preprint arXiv:1708.077472017
6. Krizhevsky, A.: CIFAR Dataset, https://www.cs.toronto.edu/~kriz/cifar.html. Accessed 07 Jan 2021
7. Krizhevsky, A.: Learning Multiple Layers of Features from Tiny Images (2009). https://www.cs.toronto.edu/~kriz/learning-features-2009-TR.pdf. Accessed 07 Jan 2021
8. Bengio, Y.: Practical recommendations for gradient-based training of deep architectures. In: Neural Networks: Tricks of the Trade, pp. 437–478. Springer, Heidelberg (2012)
9. Bengio, Y., Goodfellow, I., Courville, A.: Deep Learning, vol. 1, MIT Press (2017)
10. Yazan, E., Talu, M.F.: Comparison of the stochastic gradient descent based optimization techniques. International Artificial Intelligence and Data Processing Symposium (IDAP), pp. 1–5. IEEE (2017)
11. He, K., Zhang, X., Ren, S., Sun, J.: Delving deep into rectifiers: surpassing human-level performance on imagenet classification. In: Proceedings of the IEEE International Conference on Computer Vision, pp. 1026–1034 (2015)
12. UCI Machine learning repository, Covertype Dataset. https://archive.ics.uci.edu/ml/datasets/covertype. Accessed 09 Mar 2021
13. Blackard, J.A., Dean, D.J.: Comparative accuracies of artificial neural networks and discriminant analysis in predicting forest cover types from cartographic variables. Comput. Electron. Agriculture **24**(3), 131–151 (1999)

Network Risk Assessment Based on Attack Graphs

Damian Hermanowski and Rafał Piotrowski(✉)

Military Communication Institute, Street Warszawska 22A, 05-130 Zegrze, Poland
{d.hermanowski,r.piotrowski}@wil.waw.pl

Abstract. The paper discusses the problem of computer networks' security evaluation. It focuses on attack graph based approach. The proposed method is based on MulVAL reasoning engine that identifies possible attack paths leading from an attacker to pointed assets of the assessed IT network. These paths create an attack graph used for attack probability calculation. The method takes advantage of information from vulnerability scanners and topology snapshot. A typical enterprise network has been examined and attack graph based security evaluation- presented. The case study probability calculations have been provided including possible remediation. Benefits and limitations of proposed method have been discussed.

Keywords: Computer network · Network security · Cybersecurity · Cyberattack · Attack graph · Risk assessment · Intrusion detection · Vulnerability

1 Introduction

Nowadays every business is dependent on an IT infrastructure. All modern management and control solutions require reliable and secure IT infrastructure, which supports business processes. Security attributes: Confidentiality, Integrity and Availability should reflect appropriate level of security to avoid any theft of data (e.g. loss off enterprise secret), or their unauthorized change. The simplest way to check if the system is susceptible to cyber-attack is to use security scanners that look for security gaps and check if the system has vulnerabilities that can be exploited by a potential attacker. Software and hardware vendors offer patches and updates for identified vulnerabilities, but developing and applying them is a time consuming process. For the newest, so called *0-day vulnerabilities*, attackers prepare *0-day exploits* enabling them to take control over the target system elements, but the remedy (patch) is usually not available for some period of time. In specific systems patching is not always possible because of procedural limitations or verification of system's behavior after patching process completion (e.g. in production systems) is required. The above restrictions raise many doubts and questions: Which vulnerabilities should an organization respond to first? Which response is the most effective (including cost-effect ratio) and should be undertaken first? Which of vulnerabilities cause small risk for an organization's business and can remain open? Foregoing questions are common and exemplify everyday company's issues. The risk of attack assessment and its impact on the business processes as well as the eventual

W. Zamojski et al. (Eds.): DepCoS-RELCOMEX 2021, AISC 1389, pp. 156–167, 2021.
https://doi.org/10.1007/978-3-030-76773-0_16

benefit to the attacker seem to be crucial for the organization. Total risk calculations, which take into account detected vulnerabilities and interrelationship between systems elements, are the most desired information for the company's decision makers.

There are several approaches for assessing the security of a computer system available for security analysts. The most popular are: matching security controls' checklist, vulnerability scanning, penetration testing or security audit. Vulnerability scanning is a semi-automated method delivering a list of vulnerabilities for each host. Some of them are described in detail and reference global security databases.

According to [1, 2] promising IT system risk assessment methods are those taking the advantage of attack graphs. They can analyze identified vulnerabilities in the context of host and network configuration to find possible techniques used by an attacker to compromise network elements. A logical attack graph reflects the relationship between vulnerabilities, potential attacker privileges and configuration settings. Moreover, analysis of an attack graph helps the analyst to understand where the system's weak points are and assists in deciding which safety measures will be the most effective and adequate to the present status of evaluated system.

This paper presents results of research using *MulVAL (Multi-host, multistage Vulnerability Analysis)* tool as reasoning engine, delivering an attack graph used further on for total risk calculation for the examined system. This tool is publicly available as the result of a research project carried out at the Kansas State University, USA [3]. Our previous work [4] related to attack graphs and risk assessment described in details the implemented processing chain. It was based on the following pillars:

1. it is necessary to select attack targets reflecting the critical assets in the examined system,
2. it focuses on the most severe vulnerabilities resulting in code execution and this way enables modelling the worst case scenario,
3. risk calculation algorithm takes into consideration parameters describing validity of hosts for business processes,
4. risk assessment considers topological context and communications limitations among hosts.
5. networking topological model from *MulVAL* improved (by authors) includes multi-homed hosts, subnets, routers and expressive firewall policies,
6. extended logical attack graph generated by the modified *MulVAL* provides the analyst the possibility to choose multiple possible ways of attack counteraction, which cut off logical attack paths within the model (e.g. by executing real – life actions: software update, vulnerability patching, change of network configuration, switching off a service).

This paper demonstrates results of risk assessment for a typical network enterprise topology including several subnetworks. To present benefits of our approach two locations of an attacker are considered: the first case when an attacker is located in the Internet (outside of the enterprise), and the second case: when an attacker is located in a local network (inside the enterprise).

The remainder of the paper is as follows. Section 2 includes basics of the risk assessment. Proposed Attack Graph Based Risk Assessment (AGRA) method is described in

Sect. 3. Results of our research for typical network topology are presented in Sect. 4. Section 5 summarises benefits and weaknesses of the method. At the end conclusions are formulated.

2 Basics of Risk Assessment

2.1 Definitions

Risk definitions can be divided into three groups [5] relaying on:

1. events description, consequences and uncertainties,
2. probabilities originated form relative frequencies,
3. subject matter expert knowledge-based probabilities.

ISO 31000 [6] defines risk as the effect of objectives' uncertainty. In [7] Security Risk is defined as *"the likelihood of a communication and information system's inherent vulnerability being exploited by the threats, leading to the system being compromised"*. Regardless of the definition, the principal purpose of the risk assessment is to understand that the future performance of a system may be disturbed by an attacker exploiting internal vulnerabilities of the system.

Generally the risk R_i may be expressed as a product of consequences/losses L_i and their likelihood $p(L_i)$

$$R_i = p(L_i)L_i \tag{1}$$

The process of risk assessment includes the following steps: risk identification, risk analysis and risk evaluation.

2.2 Risk Assessment Methods

Commonly risk assessment bases usually on high-level methods [5] (e.g. CRAMM, EBIOS, ISAMM, ISO 27002, ISO 27001, ISO 27005, IT-Grundschutz, MEHARI, OCTAVE, NIST SP 800–30, AS/NZS4360, CORAS, MAGERIT and ETSI TVRA) and a number of specific techniques (Attack Graphs, Bayesian Networks etc.). Risk assessment methods can be classified as qualitative, semi-quantitative or quantitative.

Qualitative methods define consequence, probability and three levels of risk: "*high*", "*medium*" and "*low*". Mathematical product of consequence and its probability express resultant level of risk against qualitative criteria. This method usually relies on subject matter expert's experience and received results may be different depending on individual risk auditor's assessment.

Quantitative methods identify vulnerabilities, estimate real values for consequences and their probabilities, and deliver calculated values of the level of risk for a specified network element. A complete quantitative analysis requires usually lots of detailed information regarding the system examined as well as dependencies between its constituent network elements. Insufficient information, lack of data about the analysed system or its activity (change of state) can make the assessment impossible. Quantitative approach is less dependent on expert's experience, and ensures repeatability of the results for the same input data.

Topological context, in terms of reachability between network nodes, often requires a manual analysis or a rough approximation to be effectively incorporated in either qualitative or quantitative risk assessment process.

3 Attack Graph Based Risk Assessment

Attack graphs present how a sequence of attack steps can potentially enable an adversary to take control over a network asset. To enable building an attack graph some initial requirements must be fulfilled: specific network connections between hosts should be possible and one or more vulnerabilities must exist.

The method of risk assessment on the basis of attack graph assumes that a logical attack path between the attacker location and appointed target of an attack does exist. Given an attack graph, probability calculations are performed for each network element. Then system level risk values are estimated. The method tries to automate vulnerability assessment process by combining relations between vulnerabilities and network architecture properties of a system being assessed.

Basically, attack graph illustrates the opportunity to exploit vulnerabilities combined among many hosts in the examined network. Proposed method is based on the following set of parameters:

a) Topology of the examined network (hosts, services, permitted connections),
b) List of detected vulnerabilities on the host/service described by Common Vulnerabilities and Exposures (CVE) [8] identifier and
c) additional parameters described in related databases e.g. Common Vulnerability Scoring System (CVSS) [9] and Common Weakness Enumeration (CWE) [10].

This article describes quantitative approach to risk assessment, where the influence of the human factor on results is negligible. Topology describes all hosts in the examined network including OS/services/applications, permitted connections (as a *HACL* facts) among them (described by subnetworks) and restrictions enforced by firewalls' rules. Topology must reflect all connection limitations in the real network, because redundant or non-existing links may result in false positive paths in the attack graph. Moreover, all vulnerabilities discovered by scanners (e.g. Nessus, OpenVas) on hosts and services in the examined network should be validated beforehand automatically or by the analyst. Similarly to additional topology links, each additional false positive vulnerability entered as an input (as *VulnExist* fact) may cause generation of additional paths in an attack graph.

Prior to the attack graph generation the attacker location and attack target(s) must be determined. Targets may be defined as hosts belonging to critical infrastructure of the enterprise network, because usually attackers are focused on their goals. For example: stealing data, modifying a database, denying the service, which respectively drive to loss of confidentiality, integrity, availability. Setting the attack targets by a wildcard enables the analyst to check which hosts are prone to cyber-attack for the given attacker location. However this is a less relevant case than setting attack targets according to the importance of the asset.

The abovementioned data delivered to *MulVAL* enables to generate a logical attack graph. Calculation of the likelihood of reaching the indicated target node by an attacker bases on the method described by Homer et al. in [2]. This algorithm is well documented, mature and fairly, formally proven to calculate cumulative values of the likelihood of reaching the designated target node of an attack graph by an attacker. Moreover component metrics (linked to each graph node) are input values for computing the cumulative likelihood.

Individual *i* – host can be breached using many attack paths resulting in gaining different privileges on that host. Thus, there may be multiple nodes in a graph linked to the same host. Calculating a cumulative value of the probability of breaching i – host $p(L_i)$ has to include all cumulative values of all nodes related to that host (X_i).

Straightforward approach to aggregate these values uses one of the Homer's formulas [2]:

$$p(L_i) = 1 - \phi\left(\overline{X_i}\right) \tag{2}$$

because:

$$\phi\left(\overline{X_i}\right) + \phi(X_i) = 1$$

where:

$p(L_i)$ - is the probability of breaching the host – *i* or, in the other words, probability that the loss *Li* will occur.
$\phi\left(\overline{X_i}\right)$ - is cumulative value of the probability that all X_i nodes are false (i.e. the probability that none of the attack steps represented by all X_i nodes can be successfully taken).
X_i - is a set of nodes enabling gaining a privilege on the single host – *i*.

In a typical *MulVAL* rule set scenario X_i nodes are chosen as those that have experienced privilege escalation through code execution, what is assumed the most severe exploitation consequence.

Component metrics are basically weights used for computing cumulative probability. Essentially, these node weights are Access Complexity (AC) metrics, which are part of *Common Vulnerability Scoring System* [11] (CVSSv2 vector) linked to relevant vulnerabilities. AC metric expresses the complexity of an individual attack required to exploit the vulnerability once an attacker has gained access to the target system.

The Homer's method is quite intricate. It bases on a wide variety mathematical assumptions, recognises and overcomes problems in calculations for different types of attack graph nodes. Primarily, it relies on common aspects of probability theory such as conditional probabilities and *d*-separation widely used in Bayesian nets. The abovementioned aspects and exhaustive explanation of Homer's method are described in [2]. Simplified examples of probability calculations for an attack graph are presented in the following case study section. For the sake of simplification we focused only on probability computation, because risk for host is a mathematical product of probability and loss – see formula (1). The value of loss product is usually defined by the user and can be omitted.

Table 1. Vulnerable hosts in the exemplary network.

Subnet	Host	OS	Vulnerable software
DMZ	*Old-server*	Windows 2008	OS not updated for 6 years
	www1	Red Hat	Tomcat
Users	*Infector*	Windows 8	OS missing half year of updates (EternalBlue vulnerability)
Privileged	*Dev-host*	Windows 10	
Servers	*Testbed*	Windows 2016	OS missing 1 year of updates (Jenkins)
	DB	Debian	OS missing 2 years of updates (MySQL, phpMyAdmin, Bash)
	Old-DB	Debian	
Guests WLAN	*Guests*	Windows 10	N/A

4 Case Study

The section presents two-step approach to evaluate risk of compromising most valuable asset in the exemplary network. The first evaluation step assumes an attacker in the Internet, the second one – an attack coming from an internal network. For the test cases we use the exemplary network comprised of 7 hosts in 5 subnets (see Fig. 1).

Each host has a vulnerable network service running. These security gaps enable an adversary remote code execution (RCE) (see Table 1). The network configuration policy is defined as in Table 2 (e.g. every host has an access to the Internet). Only *www1* server can be reached from the Internet. One of the core servers (server *DB*) was designated in our scenario as an attack target (marked in red, dashed rectangle in Fig. 1a)). Topological description of network characteristics, vulnerabilities, attacker's location and attack target have been expressed in Datalog [12] facts which are an input to MulVAL for the attack logical graph generation. On a network level *MulVAL* semantics abstracts the allowed connections between hosts/services with *HACL* (Host Access Control List) facts (e.g. routing and firewall rules).

These facts are used within the logical reasoning process to reason about possible network connections (multi - step) among hosts throughout the defined subnets. Besides exploited vulnerabilities, user permissions and other featured semantics, in fact, quite a big part of an attack graph, describes possible network interconnections. Resulting logical graph has been evaluated with Homer's algorithm and likelihood of an attack for each host was calculated. Final values are shown in Fig. 1a) (in blue rectangles). The AGRA evaluation procedure is as follows:

- 1st iteration of the risk assessment assumes an adversary actions to be taken from the Internet (Case 1).
- Based on the identified attack graph, remedy actions are estimated and applied to make generation of an attack graph to the target DB server impossible.
- 2nd risk assessment places an attacker within network's perimeter – in the guests subnet (Case 2).

Table 2. Exemplary network's ACL (Accesss Control List) policy.

<table>
<tr><th>From\To</th><th>Old-server</th><th>www1</th><th>Infector</th><th>Dev-host</th><th>Testbed</th><th>DB</th><th>Old-DB</th><th>Guests</th><th>Internet</th></tr>
<tr><td>Old-server</td><td colspan="2" rowspan="8">X</td><td rowspan="2"></td><td rowspan="2"></td><td rowspan="2"></td><td rowspan="2"></td><td>X</td><td rowspan="2"></td><td rowspan="9">X</td></tr>
<tr><td>www1</td><td></td></tr>
<tr><td>Infector</td><td>X</td><td>X</td><td></td><td></td><td></td><td></td></tr>
<tr><td>Dev-host</td><td></td><td>X</td><td>X</td><td></td><td></td><td></td></tr>
<tr><td>Testbed</td><td rowspan="3"></td><td rowspan="3"></td><td colspan="3" rowspan="3">X</td><td></td></tr>
<tr><td>DB</td><td></td></tr>
<tr><td>Old-DB</td><td></td></tr>
<tr><td>Guests</td><td></td><td></td><td></td><td></td><td></td><td>X</td></tr>
<tr><td>Internet</td><td></td><td>X*</td><td></td><td></td><td></td><td></td><td></td><td></td></tr>
</table>

4.1 Case 1. Attacker Located in the Internet

This case should be simulated at the beginning of the network security analysis when potential attack vector from the Internet is possible. The *Infector* machine is operated by the user "*grazyna*" marked as an incompetent one (see Fig. 1a)) and thus can be compromised easily as a result of client side exploitation (e.g. a phishing attack). Then the attack may be carried out through the vulnerable *old-server* and the old-DB. Cumulative likelihood calculated using Homer's algorithm for this machine is 0,13. The other possible attack path goes through the *dev-host* used by a privileged user (a developer). The *testbed* machine can be compromised either by abusing *devop's* credentials or by exploiting server's vulnerability – the cumulative likelihood equals 0,16. The final step in the attack graph is breaching the DB server (marked in red as a target of an attack – the most valuable asset) with likelihood 0,10. Graph evaluations show that attack steps through the *testbed* machine are more likely to be taken by an adversary than exploiting the *old-DB*. Figure 2 shows a tiny part of the logical attack graph, which is represented in the topological view as the single attack step between the *infector* and the *PhishServer* (Fig. 1a)).

The logical attack graph consists of three types of nodes:

a) LEAFs (facts defining analysed network/system; in the rectangles),
b) ANDs (representing an attack step/progression; in the ovals) and
c) ORs (representing an obtained privilege or consequence of an attack step; in the rhombuses).

AND/OR nodes conform to logical operators of conjunction/disjunction requiring all/any of its preconditions (in-neighbours) to be met. The attack path consists of actions to be taken by an attacker to pass step by step through the infrastructure and reach the target node.

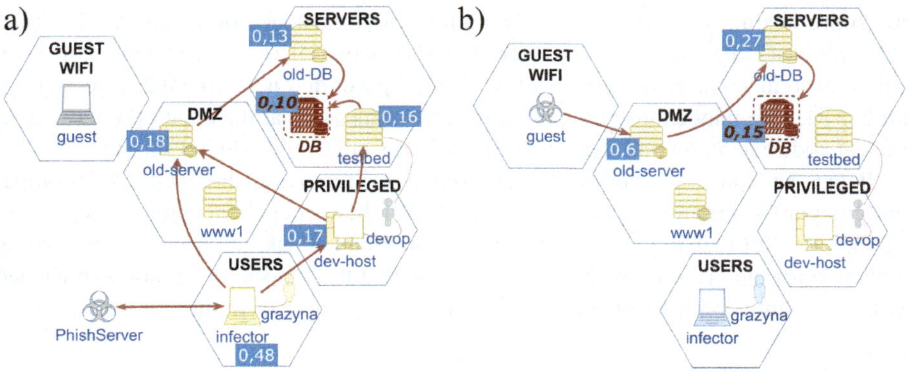

Fig. 1. Scored possible attack paths originating from the a) the Internet b) internal subnetwork.

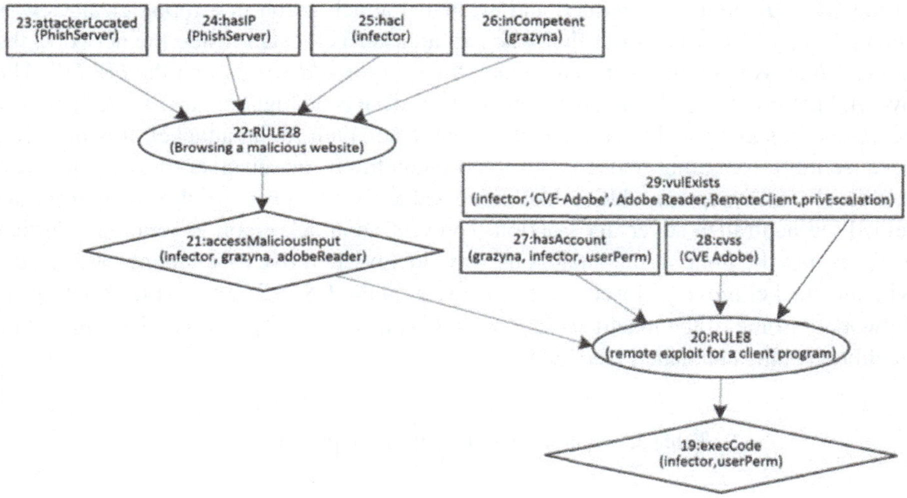

Fig. 2. Part of the logical attack graph for the Case 1.

It is clear that proper remedy actions should "break the attack path" making it impossible to perform one of the steps (the most desirably – the closest to the source node). Host-level actions may include patching, disabling a network service or shutting down a machine.

One may prefer limitation of network access to vulnerable service or network configuration refinement. Undertaken remedy actions may impair availability of a network asset which cannot be acceptable in some cases (from the company's business point of view). One may use likelihood evaluation results as suggestions for repair actions. Nevertheless, optimizing the remediation decisions is out of scope of this paper. Breaking an attack path is literally preventing the logical engine from deriving particular, subsequent attack steps. Within the AGRA itself, the easiest way to achieve this goal is to remove parts of the input data (facts) enabling logical derivation by *MulVAL* engine. Patching procedure is simply removing an information about existence of a vulnerability (fact '29:*vulExists*(...)') – see Fig. 2. Network-based remediation (e.g. firewall blocking rule) can be introduced as removing a fact describing permitted network ACL (the fact'25: *hacl*(...)'). One may choose to apply remediation measure at the edge of a network to prevent code execution on infector host ('19:*ExecCode*(...)' node in Fig. 2).

It can be done by training the *grazyna* user (removing the fact: '26:incompetent(grazyna)') or removing vulnerability in the AdobeReader client application (removing the fact: '29:*vulExists*(...)'). After remediation, an attack graph is no longer being generated for the given input data, which means that the DB server cannot be breached (within the limits of *MulVAL*'s attack model) by an attacker from the Internet.

4.2 Case 2. Attack from the Internal Network

Based on the topology refined with remediation as in Case 1, we placed an adversary in the *GUEST* subnet. Generated and evaluated attack graph in a topological view is shown in Fig. 1b). It turns out that when an attacker is located inside the network, the target *DB* server is still open to an attack through the *old-server* and the *old-DB*. The overall likelihood of breaching the target is 0,15, what is far higher than in Case 1, mainly because of a shorter and thus less demanding attack path for an attacker. It is intuitive, because more foregoing nodes express more conditions an adversary has to meet and this results in lower cumulative likelihood value. Next step is to choose appropriate remedy to neutralize adversary's actions (prevent an attack graph generation). Further analysis may be done in the analogous way: involve iterative assessments and repairs with an attacker moving closer to the target (e.g. in the *USERS* and *PRIVILEGED* users networks). Some of the machines (e.g. *www1*, *Infector* in the 2nd Case) have no attack likelihood value calculated (see Table 3).

Table 3. Vulnerable hosts in the exemplary network.

Host	*Old-server*	*www1*	*Infector*	*Dev-host*	*Testbed*	*Old-DB*	*DB*
Case 1	0,18	-	0,48	0,17	0,16	0,13	0,1
Case 2	0,6	-	-	-	-	0,27	0,15

One might deduce that the aforementioned hosts are safe from attacks, which is a false belief. When the attack target is defined all hosts listed on attack paths leading to the target node have some likelihood values. These values decrease along the path from the attacker to the target node, because it requires intruder to meet additional conditions and perform more attack steps (e.g. time and extra skills to exploit vulnerability on the next node). The absence of a host on the attack graph (e.g. *www1*) means that it could not be used as a pivot in any of attack steps to achieve the ultimate goal. Simply hosts not used to achieve the defined target are excluded from building an attack graph and computing likelihood. On the other hand setting the *www1* machine as the attack target (instead of the *old-DB* machine) could result in generating an attack graph and evaluating the risk.

When an attack graph cannot be generated for a given target, it cannot be compromised (within the limits of *MulVAL*'s attack model) and no likelihood nor risk evaluation can be performed.

4.3 Discussion

The first AGRA evaluation case presented traditional approach where the analysis assumes threat coming from the outside of the network's perimeter. The second case rejects this naive assumption by placing an attacker in the rather untrusted GUEST subnet, adjacent to the subject network.

Simulating attacks from internal subnets (Case 2) reflects the MITRE's "assume breach" philosophy [13] justified by the fact that there is always a way for an attacker to gain an initial foothold by:

a) A social-driven attack dedicated to prone users.
b) Utilizing *0-day* exploits.
c) Making use of an acquired internal user (an insider threat modelling).

Successive placing an attacker closer to the target simulates more detrimental scenarios and more demanding defensive decisions. It enables much deeper security analysis, e.g. by answering the question: From which internal hosts it is possible to harm the target network asset? Determining attack-enabling hosts and conditions is crucial for addressing security gaps and hardening the network.

It is worth noticing that even hosts having serious weaknesses like RCE may not be included in deriving attack paths and be excluded from risk assessment if they do not allow lateral movement towards the target (set in the analysis) or are unreachable from the attacker's location (see Table 3). Yet, they still might be considered attack targets (e.g. the *www1* server). In both cases attack paths have been presented in a topological aspect on purpose. The logical attack graph built in Case 1 consists of ~100 nodes and is hardly readable for a human. Actionable and usable presentation of knowledge emerging from large logical attack graph is noted a challenge.

One of the benefits of logical graphs is a detailed, explicit description of the attack steps. Having explicitly given conditions enabling particular attack steps, it is natural to perform some post-processing on generated attack-graph to decide on the remediations to follow. Typical actions to be taken in order to mitigate security vulnerabilities on-host include: software patching, reconfiguration or turning off a service or host when applicable. Other remedy actions may involve changing network configuration: e.g. applying firewall blocking rules, adding IPS rules or setting up redundant assets for load balancing.

Attack Graph Based Risk Assessment enables quantitative assessment of the network (likelihood of attack). There is possible a two-stage assessment:

a) An attack graph has been generated by *MulVAL* – it means that there exists probability $p_a > 0$ where an attacker may reach selected assets (target) form the starting point;

b) *MulVAL* was not able to generate attack graph – it means that probability of an attack from the starting point to the target is equal to zero $p_a = 0$ (the assessment does not take into account *0-day* vulnerabilities).

It is possible that a vulnerable host (vulnerability scanner discovered at least one or more vulnerabilities) does not take part in logical attack graph. It may take place when *MulVAL* cannot find any path through this host to the target.

The proposed assessment method enables analysis of large computer networks and ensures repeatability of the calculated likelihood (human factor is minimized). The method enables simulating insider threats by changing the location of an attacker (Case 2). This approach gives more information about possible attack paths from selected hosts inside the examined network. The method may support a network administrator in the security assessment or a pentester in the operation planning. The set of vulnerabilities used for an attack graph generation should realistically represent the evaluated system. Exploitation of false positive data creates additional paths in the attack graph and may increase total risk related to the examined system. In an opposite case, when not all severe vulnerabilities are detected by the scanner, it results in reduction of possible attack paths in the graph. To achieve reliable result of the risk assessment, it is necessary to find all known severe vulnerabilities on hosts and their services. A detailed vulnerability scan is time consuming, but it guaranties that the attack graph takes into account all possible (and realistic) attack paths and that the ultimate risk assessment is credible. Moreover, intrusion detection and prevention systems should be turned off during the vulnerability scanning process to minimize problems with false negative results.

5 Conclusions

Analysis of attack graphs helps to understand where the system's weaknesses lie and to decide which security measures will be effective and the most adequate to the evaluated system. It enables to discover possible ways used by an attacker to compromise an enterprise network by analysis of the hosts and the network configuration.

The AGRA method where *MulVAL* framework is used for graph generation was described in detail in [4]. This method utilizes a quantitative analysis estimates of attack probabilities (Homer algorithm [2]), and produces potential attack likelihood values for specific network element, which takes part in attack graph structure. Quantitative assessment approach requires detailed information about the examined system (actual vulnerabilities on hosts and all possible communication paths among hosts). Moreover our AGRA implementation offers a simple functionality of modelling network level remediations based on logical attack graph to reduce identified risk (Case 2). Networking model for *MulVAL* improved by authors, includes multi-homed hosts, subnets, routers and expressive firewall policies enables more precise network remediations. Results of attack likelihood calculations for a typical enterprise network topology showed that different attacker locations have great influence on results and enable simulation of potential *0-day* vulnerabilities (by changing of attacker location) and insider threats. Iterative moving of an adversary deep inside the network, evaluating attacks and applying remediations is a fruitful way to determine and address unobvious, harmful security gaps.

References

1. Wang, L., Islam, T., Long, T., Singhal, A., Jajodia, S.: An attack graph-based probabilistic security metric. In: Data and Applications Security XXII. DBSec. LNCS, vol. 5094. LNCS, pp. 283–296. Springer, Heidelberg (2008). https://doi.org/10.1007/978-3-540-70567-3_22
2. Homer, J., Zhang, S., Ou, X., Schmidt, D., Du, Y.: Aggregating vulnerability metrics in enterprise networks using attack graphs. J. Comput. Secur. **21**(4), 561–597 (2013). https://doi.org/10.3233/JCS-130475
3. Ou, X., Govindavajhala, S., Appel, A.W.: MulVAL: a logic-based network security analyzer. In: Proceedings of 14th Conference USENIX Security Symposium,vol. 14, p. 8 (2005)
4. Hermanowski, D., Piotrowski, R.: Proactive risk assessment based on attack graphs. An element of the risk management process on system, enterprise and national level. In: IEEE International Conference on Data Science and Systems DSS-2018, Exeter, Great Britain, https://doi.org/10.1109/HPCC/SmartCity/DSS.2018.00237
5. Frank, M.S., Konrad, W.: Comparative Study and Roadmap DRA, Comparative study and roadmap for the development of the dynamic risk assessment function. Technical report 2012/SPW007956/03
6. ISO 31000 - Risk management. https://www.iso.org/iso-31000-risk-management.html. Accessed 15 Dec 2020
7. North Atlantic Treaty Organization: Management Directive on CIS Security (2005). https://www.nbu.cz/download/pravni-predpisy---nato/AC_35-D_2005-REV3.pdf. Accessed 15 Dec 2020
8. Common Vulnerability Enumeration. https://cve.mitre.org Accessed 15 Dec 2020
9. Common Vulnerability Scoring System v3.0: Specification Document (2015). https://www.first.org/cvss/v3.0/specification-document. Accessed 15 Dec 2020
10. Common Weakness Enumeration - About CWE (2020). https://cwe.mitre.org/. Accessed 15 Dec 2020
11. Common Vulnerability Scoring System v3.0: User Guide. https://www.first.org/cvss/v3.0/cvss-v30-user_guide_v1.6.pdf. Accessed 15 Dec 2020.
12. Huang, S., Green, T., Loo, B.: Datalog and Emerging applications: An Interactive Tutorial. http://www.cs.ucdavis.edu/~green/papers/sigmod906t-huang.pdf. Accessed 15 Dec 2020. https://doi.org/10.1145/1989323.1989456
13. Strom, B.E., et al.: MITRE ATT&CK™: Design and Philosophy (2018). https://www.mitre.org/sites/default/files/publications/pr-18-0944-11-mitre-attack-design-and-philosophy.pdf. Accessed 15 Dec 2020

Cost Results of Block Inspection Policy with Imperfect Testing in Multi-unit System

Anna Jodejko-Pietruczuk(✉)

Wrocław University of Science and Technology, Wybrzeże Wyspiańskiego 27, 50-370 Wrocław, Poland

anna.jodejko@pwr.edu.pl

Abstract. The maintenance practice confirms that sometimes it is more profitable to apply a less precise but cheaper testing method to a system if there is more than one available. This paper investigates the profitability of inspections with various accuracies and their combination in a common inspection-based maintenance policy of a multi-unit system. The main novelties of the paper concern the profitability evaluation of applying imperfect inspections in the maintenance of a multi-unit system and the cost comparison of the one- and two-level inspection policies in a multi-unit system. For this research, Monte Carlo simulation model has been used. The obtained results confirm that perfect and expensive testing methods usually give worse cost results in a *k-out-of-m* system than other programs based on cheaper and less accurate methods. It has been shown that in many cases the two-level inspection policy is more economical than the one-level one.

Keywords: Predictive maintenance · Imperfect inspection · Multi-unit system

1 Introduction

One of the most critical requirements of successfully applying a condition-based maintenance strategy is the existence of testing methods that can identify a state of a system. However, when executing an inspection (the testing process resulting in a system state diagnosis) one must consider that a diagnosis may be wrong due to the limited inspection accuracy. It may be caused by measurement errors, a low quality of algorithms of data analysing and reasoning, inspector/measurer experience, and limited knowledge of the relations between tested technical parameters of a system and its reliability or functional state [1]. There can be one of the two (e.g. [2]) or two inspection error types considered (e.g. [3, 4]): a false positive or a false negative error. A false positive mistake occurs when an inspection finds that the system is defective when in fact it is in a good condition. A false negative occurs when the inspection says the system is good, when in fact it is defective [5]. The main parameters that reflect the accuracy of an inspection inference are the probability of correct diagnosis and the probabilities of false positive and/or false negative errors. The most habitual assumption is that the inspection accuracy is constant in time (e.g. [5, 6]) however, there are also models where the error probability is a function of time (e.g. [7]). Berrade et al. [6] emphasize that the probability of inspection

W. Zamojski et al. (Eds.): DepCoS-RELCOMEX 2021, AISC 1389, pp. 168–178, 2021.
https://doi.org/10.1007/978-3-030-76773-0_17

faults (of any type) should not exceed the value 0,2 and Geara et al. [8] point out that the cost of inspection is usually an increasing function of its accuracy. This assumption, frequently reflecting maintenance practice, is an important factor for the findings of this paper.

2 Consideration of Inspection Accuracy in System Maintenance

There are a lot of publications that take an imperfect inspection into consideration, however a great majority of them assume that the inspection accuracy is a given parameter of a maintained system. It makes the inspection accuracy a constraint of a maintenance policy and other parameters (mainly – the inspection period) are decision variables and are optimized (e.g. [6, 9]).

There is very limited number of papers considering inspection accuracy as a decision variable together with other parameters of a maintenance policy. Jing et al. [10] analyse a single-component system where three levels of imperfect inspection, differing in their cost and accuracy, may be applied. In [2], Wang et al. consider the policy of periodic inspection in a single-component system where two kinds of inspections, which are different in their possibility of defect revealing, are carried out with different frequencies. Wang [11] considers a single-element system whose state is controlled periodically by the measurement of its degradation signal. If the signal value is in a defined range, an additional manual, perfect inspection is executed to check real state of the system. In [12], Hao et al. also analyse two maintenance policies applied to a single-component system. The first policy assumes imperfect, cheap testing, the second – imperfect, cheap inspections carried out only below some level of a system degradation signal. After exceeding that level, only perfect inspections are implemented. They conclude "that the policy with imperfect inspections can be better than the classical one, and that the proposed policy with a two-stage inspection scheme always leads to the minimum long run maintenance cost rate" [12]. Berrade et al. [3] consider two periodic, imperfect inspection maintenance policies, applied in a single-component system. They analyse the best parameters of a policy with one-level inspection and one with two-level inspection. In [13], Parmigiani analyses a two-level inspection policy applied in a single-element system. In conclusion the author emphasises that a two-level inspection policy can be fruitful for some range of policy parameters in comparison with a one-level policy. The problem of imperfect inspection is even less considered when the maintenance of a multi-unit system is modelled. In [1], Liu et al. model the maintenance of a system composed of subsystems and elements. The authors develop the Bayes-based formula to estimate system reliability at the time of every new inspection. In [14], Seyedhosseini et al. present the two-components system in which inspection may be imperfect. The aim of that paper is to find the optimal inspection interval which minimizes the total expected cost of maintenance operations. In [15], the authors investigate Block Inspection Policy performance for multi-unit systems with imperfect inspection. They define the principal relations between the system performance under the policy and parameters of the policy.

The maintenance practices, as well as the conclusions drawn from the literature survey, confirm that sometimes it is more profitable to apply a less precise but cheaper testing method if there is more than one available. Sometimes the best cost or reliability results can be obtained by using a combination of inspections with different accuracy in one maintenance program. Thus, this paper investigates the profitability of inspections with various accuracies and their combination in the inspection-based maintenance policy in a multi-unit system. The main novelties of this paper are:

- the profitability evaluation of applying imperfect inspection in the maintenance of a multi-unit system,
- the comparison of the one- and two-level inspection policies in a multi-unit system with respect to their cost-effectiveness.

In the third section, the main assumptions of the simulation model are indicated, the model of the maintenance process is presented, and in the next sections the sensitivity analysis of the model is shown. The paper ends with a summary and recommendations for the direction of future research.

3 Simulation Model of System Operation with One- and Two-Level Inspection Maintenance Policy

3.1 Modelling Assumptions

- The system in consideration is a multi-unit system working in the *k-out-of-m* reliability structure. The configuration is a special case of a complex system that requires (for the overall system to be functional) at least k out of the total of m components must be working.
- All elements are identical, they work and fail independently of each other. Their failure is a two-stage process consistent with the delay-time (DT) concept. It means that before a real element failure comes, there is a period when a signal (so called defect) of approaching failure may be observed in the element [16, 17].
- The backup components fail during their standby regime with the same failure rate as the operating elements. They are also inspected during the Block Inspections moments.
- The implemented maintenance policy is the Block Inspection (BI). This means that the inspection services are carried out in the whole system according to a fixed calendar time, regardless of whether the component being inspected was just renewed or not. Inspections are carried out at regular time intervals of T.
- A defect may be diagnosed only by inspection while a system failure (caused by failures of $m - k + 1$ elements) is self-announcing. After discovering the defect in an element, it is perfectly renewed. After a system failure, the system is stopped until the failed elements are perfectly renewed.
- There are a few ($n = 1,2,3,\ldots,N$) control methods (testing levels) available. They differ in accuracy and cost. The accuracy of the n-th control method is defined by the constant probability of a false positive (α_n) and a false negative (β_n) error of diagnosis. The probabilities are independent of operation time or of delay-time of an element. Every inspection method, apart from the N-th, is imperfect and may give false positive as well as false negative conclusions. The accuracy and the unit cost of a higher-level test are greater than those of a lower-level one. The inferring of the N-th testing level is perfect and the inspection is the most expensive.

- There are two maintenance policies that may be applied to the system. The one-level inspection policy means that the same, n-th level control method is used every time when the system is inspected. The two-level inspection policy means that the system is controlled using the n-th imperfect method ($n \neq N$) and, if the inspection diagnoses a defect, the additional control of the N-th level is executed to confirm the result. The inspection programme is fixed once for the whole lifetime of the system.
- The maintenance services (inspections, replacements) of the system generate the cost of elements' replacement, system failures, and the inspection (dependent on the testing method). Durations of all maintenance services (inspections, corrective, and predictive replacements) are not considered in the model because it is assumed that they do not generate any additional cost except those described above.

3.2 Simulation Model

To achieve the defined goals of the paper, the Monte Carlo simulation model was written in the GNU Octave program. The general stages of the simulation algorithm are depicted in Fig. 1. The algorithm reflects the operation and maintenance processes in the system in which the two-level inspection policy is applied. If there is no stage 9th of the algorithm, it models the operation and maintenance for the one-level inspection policy.

Table 1 presents the set of model input parameters together with the range of tested values while model examining. The first two rows present cumulative distribution functions (Cdf.) of random period until the defect arrival in every system element and delay-time period of an element. The values of m and k (rows 3, 4) define the reliability structure of the tested system. Composition of their values gives 3 cases for the five-elements system (1-*out-of*-5, 3-*out-of*-5, 5-*out-of*-5) which were examined, and 3 cases for the twenty-elements system. They show the influence of a chosen maintenance policy upon the cost results of systems working in various reliability structures. The variables T, $\propto_n$, β_n were tested in the ranges of values given in the last column of the Table 1 (*min*: *increment*: *max*). Additional explanation to the last row of the Table 1 (variable d) should also be given. It presents the cost factor which defines the relation between the n-th inspection accuracy and the single element inspection cost. When the N-th testing method is applied to the system, its unit cost is the highest due to the fact that this test is 100% accurate ($\propto_n = \beta_n = 0$): $c_{iN} = d$. When the d factor equals 0, neither inspection (of any level) generates any costs. However, the application of the N-th testing method eliminating false positive errors of inspection shows the cost effect of the two-level inspection policy resulting from reliability outcomes of the system operation.

The unit cost of the operation and maintenance process of the modelled system was calculated according to Eqs. 1–3.

$$C_{ni} = \frac{\left[(m-k+1)\cdot NO_{SF}(OT) + \sum_J NO_{PR}\right]\cdot c_e}{OT} + \frac{NO_{SF}(OT)\cdot c_c}{OT} + \frac{CI_{ni}(OT)}{OT} \tag{1}$$

where: C_{ni} – the cost ratio of the operation and maintenance process of the system in which the i-th maintenance policy (i = 1,2) is applied using the n-th testing method, OT – the length of simulated operation period, $NO_{SF}(OT)$ – the number of system

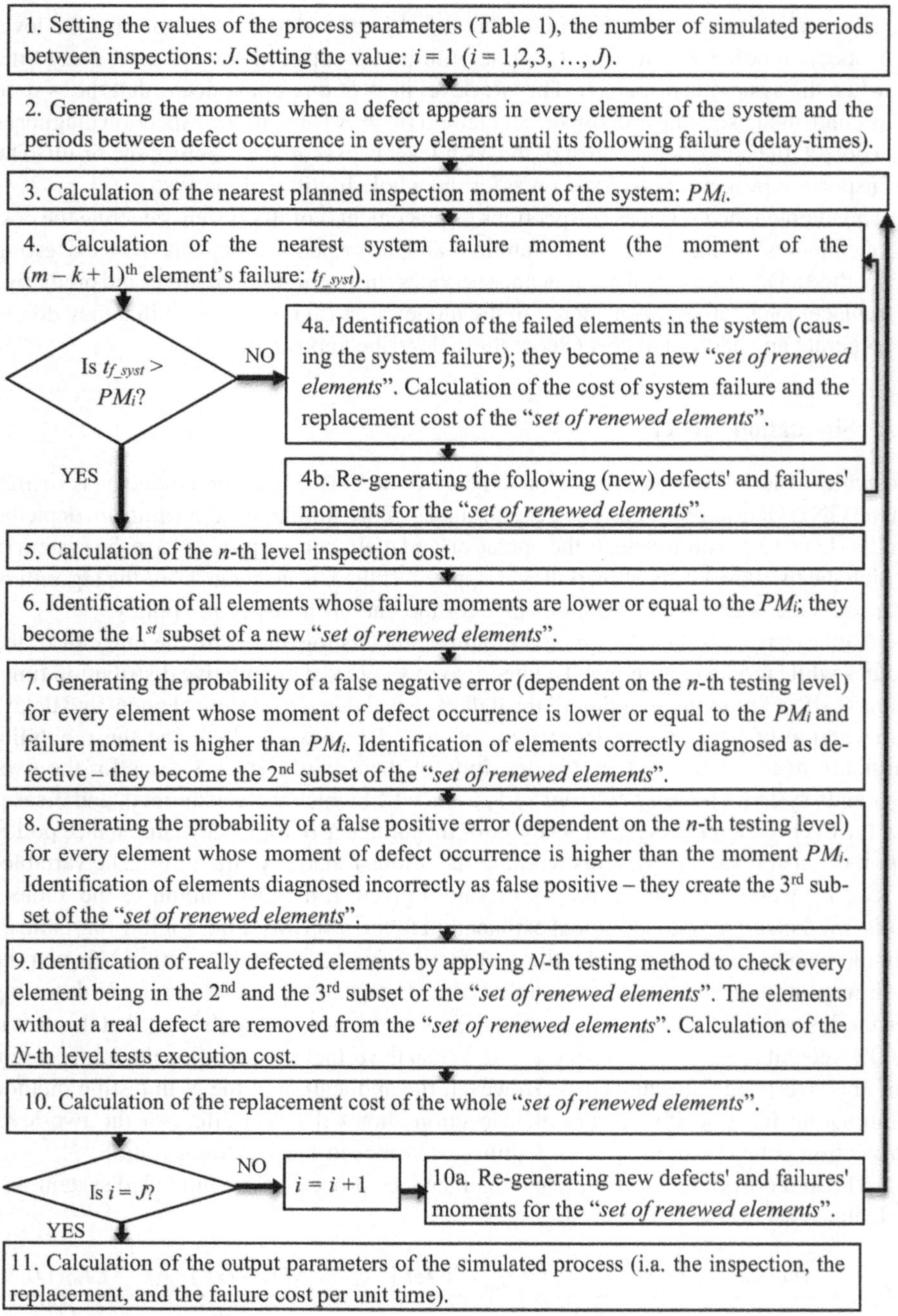

Fig. 1. The simulation algorithm of the operation and maintenance processes

Table 1. The input variables and parameters of the simulation model

Notation	Description	Basic value/Tested variants
$F_U(t)$	Cdf. of element's period to defect arrival	$F_U(t) = 1 - e^{-\left(\frac{t}{70}\right)^{3,5}}$
$F_H(t)$	Cdf. of element's delay time (since defect appearance until failure arrival)	$F_H(t) = 1 - e^{-\left(\frac{t}{30}\right)^{3,5}}$
m	number of elements working in the system	5/20
k	minimal number of up-stated elements for having system in an operational state	{1; 3; 5}/{1; 10; 20}
T	length of the period between inspections	10: 5: 110
αn	probability of false positive error of the n-th testing level	0: 0 05: 0,5
βn	probability of false negative error of the n-th testing level	0: 0,05: 0,5
c_e	the cost of a single element replacement	100
c_c	the unit cost of a system failure	10 000
c_{in}	the cost of single element control using the n-th testing level	$(1 - \alpha_n) \bullet (1 - \beta_n) \bullet d$
d	the factor of inspection cost increase when a higher testing level is applied in the system	{0; 0,5; 1; 2; 3; 4; 5; 6; 7; 8; 9; 10}

failures in OT, J – number of simulated inspection periods in OT, NO_{PR} – the number of preventively replaced elements in every out of J inspection periods, CI_{ni} – the total cost of inspections carried out in the system in OT if the i-th maintenance policy ($i = 1$, 2) is applied using the n-th testing method.

$$CI_{n1}(OT) = J \cdot m \cdot c_{in} \tag{2}$$

$$CI_{n2}(OT) = J \cdot m \cdot c_{in} + \sum\nolimits_J NO_{PP} \cdot c_{iN} \tag{3}$$

where: NO_{PP} – the number of "*potentially defected elements*" tested additionally with the N-th testing level in every out of J inspection periods.

The first component of the Eq. 1 represents the total cost of elements' replacement in the system, the second – the system failures cost, while the third – the total cost of inspection if the i-th maintenance policy ($i = 1, 2$) is applied in the system.

4 Sensitivity Results of the Simulation Model

The simulation model presented in the previous section of this paper was investigated with respect to the input variables presented in Table 1. The chosen cost and reliability results are shown in Figs. 2, 3 and 4. Due to the limited volume of this paper only a few of the obtained results are shown.

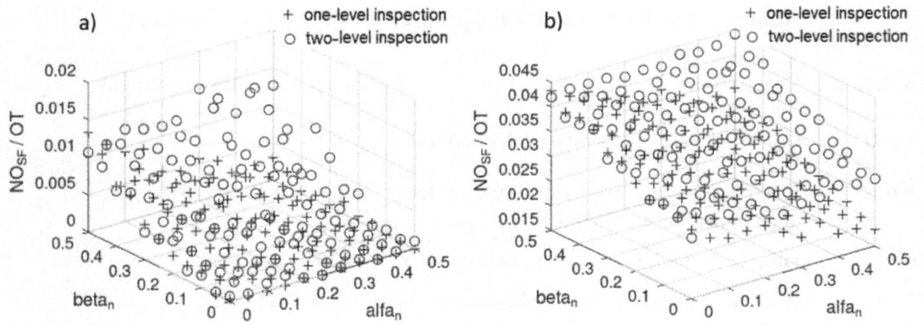

Fig. 2. The average number of system failures per a unit time in the 5-*out-of*-5 system for $T = 10$ (a) and $T = 50$ (b).

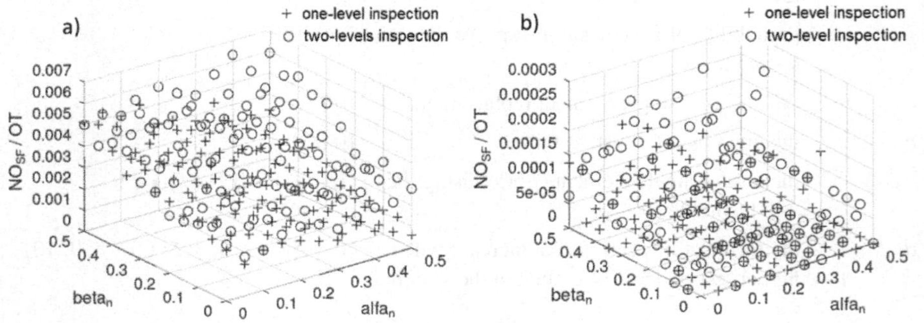

Fig. 3. The average number of system failures per a unit time in the 3-*out-of*-5 (a) and 1--*out-of*-5 (b) systems for T = 50.

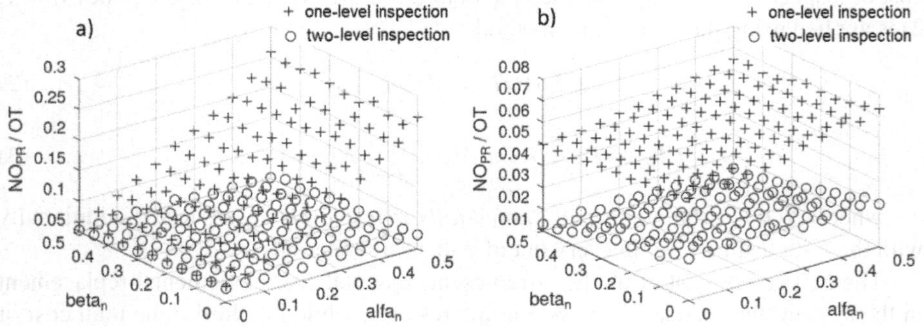

Fig. 4. The average number of preventively replaced elements per unit time in the 1-*out-of*-5 elements for $T = 10$ (a) and $T = 50$ (b).

Figures 2, 3 and 4 show the average number of the most important maintenance events which occur during the operation and maintenance of the modelled system. When analysing the number of system failures per simulated time unit, it can be seen that in the *m-out-of-m* system (Fig. 2a, 2b) the two-level maintenance policy yields more failures than the one-level policy. It is especially explicit when the period between planned inspections is long (Fig. 2b). It can even be noticed there that the increasing probability of false positive error (α_n) causes the drop of the number of system failures if the one-level policy is used. The same effect can be observed in the 3-*out-of*-5 and 1-*out-of*-5 systems (Fig. 3a, 3b). Such result is the effect of an incorrect length of the period between inspections. If the period is too long, all false positive mistakes decrease the number of unexpected system failures by excessive elements replacing (without real defects while their controlling). In the system that is the least sensitive to a single element failure (Fig. 3b), the mentioned effect can be seen only for the cases when the probability of false negative and false positive inspection errors is very high ($\alpha_n \geq 0, 2$, $\beta_n \geq 0{,}4$). The conclusion is that the inspection accuracy should be optimized jointly with the period between inspections because they have a synergy effect on each other. Figure 4 presents the number of elements preventively replaced for both considered policy variants and for the two periods between inspections. The presented examples reflect quite well the results obtained for other tested cases – the number of replaced elements for the one-level inspection policy is never lower than for the two-level case. Growing probability of false positive mistake (α_n) increases the number of preventively replaced elements, while the higher probability of false negative error (β_n) – decreases their number.

5 Comparison of One- and Two-Level Inspection Policies in Multi-unit System

In this section, the cost comparison of the one- and two-level inspection policies is presented. There are also conclusions on the profitability of applying imperfect inspection in the maintenance of a multi-unit system presented.

When considering the results presented in the previous section of this paper, initially it seems that the one-level inspection policy gives better results. However, a more detailed analysis of the best-found cost ratios shows that this relation is not true. Figure 5a depicts the proportion of the examined cases for which a given kind of policy is better if applied in a multi-component system. Figure 5b compares the policies' predominance for various values of the inspection cost factor "*d*".

Based on Fig. 5 the first conclusions can be drawn. Generally, the two-level inspection policy is more cost-effective than the one-level one, especially when the system is not the *m-out-of-m* system. If a system has the *m-out-of-m* structure, then the one-level policy is usually better. On the other hand, the effectiveness of both considered policies strongly depends on the cost factor *d* (Fig. 5b). Nevertheless, all afore presented results confirm the fact that there is a complex dependency between all input parameters of the model and the cost results of the maintenance process in a multi-unit system. Figures 6 and 7 present the probabilities of inspection errors and the length of the periods between inspections for which the modelled system gets minimum costs ratio (Eq. 1).

The probabilities of false positive and false negative inspection errors depicted in Fig. 6 represent various testing methods available when controlling the modelled system. The best cost results are obtained for nonperfect methods, regardless of the kind of the applied maintenance policy. It proves that the inspection precision should be carefully chosen if there is more than one testing method available in a system. When investigating

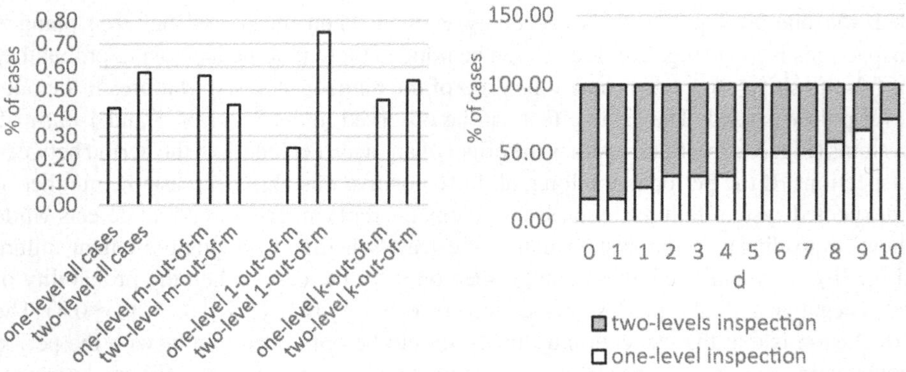

Fig. 5. The proportion of the cases when the one- or the two-level inspection policy is better when applied to a system with a given reliability structure (a) and for various values of the inspection cost factor "*d*" independently of system structure (b).

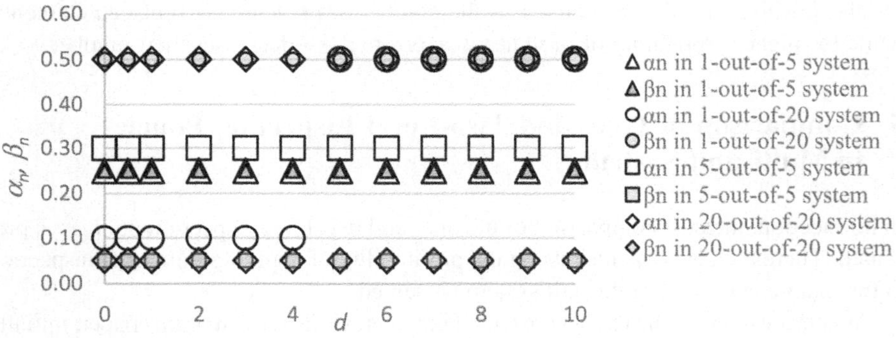

Fig. 6. The probability of inspection errors for which the system gets the minimum costs ratio

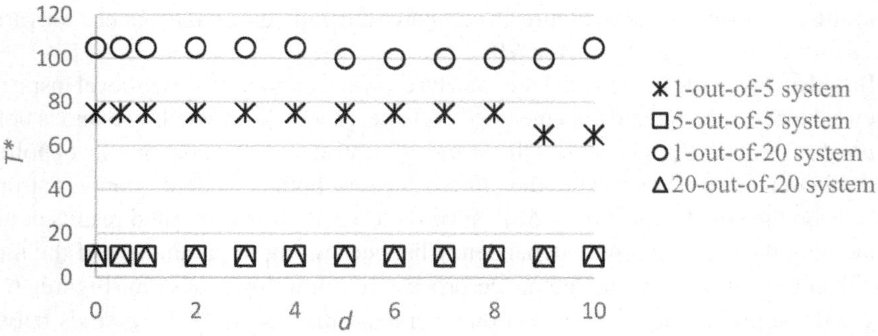

Fig. 7. The length of the period between inspections for which the modelled system gets the minimum costs ratio

the relations between the reliability structure of a multi-unit system (given by the k parameter), the period between inspections (T) and the probabilities of mistakes when inspecting the system (α_n, β_n), following rules can be defined.

- In the 1-*out-of*-20 system which is the least sensitive for a single element's failure, it is profitable to use cheaper but very imperfect inspection methods. When the cost factor d is low, the unit inspection cost c_{in} is also low and it is worth to utilize controlling methods with the lower probability of false positive error ($d = <0{,}4> \; => \alpha_n =$ 0, 1 in Fig. 6). When the difference in cost of n-th and N-th test level is higher ($d \geq$ 5), it is more profitable to use less precise methods ($\alpha_n = 0{,}5$) as a standard and to pay for diagnosis confirmation by the N-th inspection method. It is also cheaper then to reduce the length of the period between inspections (for $d \geq 5$ in the Fig. 7).
- In all tested 1-*out-of-m* systems (Fig. 6) it is profitable to allow for false negative mistakes. As it appears, rare errors of this type do not have any great influence on system failures due to the redundancy built in the structure of the system. For the highest examined values of factor d, it is rather cheaper to shorten the period between inspections than to change the testing method (Fig. 7).
- When investigating the error probabilities (Fig. 6) of the most economic results obtained in *m-out-of-m* systems (which are very sensitive to a single element's failure), it appears that the best effects are achieved if the false negative probability of inspection is low (but not the lowest possible). The probability of the false positive error should be, respectively, higher. Such inspections are cheaper than applying more accurate control methods and may reduce the system failure probability.
- The period between inspections in the *m-out-of-m* systems should be as short as possible (Fig. 7) because it is better to test the state of the system element a few times while operating than to test it less often with higher precision.

6 Conclusions

A proper design of a predictive maintenance policy in a multi-unit system is a complex problem. There are a lot of interdependencies between structural, technical, and organizational factors having influence on a system operation process. Some of these factors are difficult or impossible to shape by the decision-maker, some of them are decision variables. The most common approach observable in the literature is to treat the inspection accuracy as a given parameter of a given maintenance process and to optimize the policy mainly by the period between inspections steering. However, in practice, it is widespread to use a few control methods with various inferring precisions in one maintenance program. It makes the inspection accuracy a decision variable of such program. This paper is the first attempt to assess the profitability of using inspections with different precisions in one maintenance policy applied to a multi-unit system. The results confirm the conclusions from some papers [12] that it may be better to test the state of a system with cheaper but less accurate inspection methods than to use more expensive, more exact methods. Additionally, it has been shown that in many cases the two-level inspection policy is more economical than the one-level one. The planning problem of predictive maintenance policies joining various controlling methods is complex, difficult for analytical modelling and optimization, especially when it regards multi-unit systems. It requires a lot of further research to develop effective optimization algorithms for deliberate shaping decision variables of the maintenance policy to get the expected reliability, cost and safety results.

References

1. Liu, Y., Chen, C.-J.: Dynamic reliability assessment for nonrepairable multistate systems by aggregating multilevel imperfect inspection data. IEEE Trans. Reliab. **66**(2), 281–297 (2017)
2. Wang, W., Zhao, F., Peng, R.: A preventive maintenance model with a two-level inspection policy based on a three-stage failure process. Reliab. Eng. Syst. Saf. **121**, 207–220 (2014)
3. Berrade, M.D., Cavalcante, C.A.V., Scarf, P.A.: Maintenance scheduling of a protection system subject to imperfect inspection and replacement. Eur. J. Oper. Res. **218**, 716–725 (2012)
4. Ge, E., Li, Q., Zhang, G.: Condition-based maintenance policy under imperfect inspection using Monte-Carlo simulation. Appl. Mech. Mater. **201–202**, 955–958 (2012)
5. Berrade, M.D., Scarf, P.A., Cavalcante, C.A.V., Dwight, R.A.: Imperfect inspection and replacement of a system with a defective state: a cost and reliability analysis. Reliab. Eng. Syst. Saf. **120**, 80–87 (2013)
6. Berrade, M.D., Scarf, P.A., Cavalcante, C.A.V.: A study of postponed replacement in a delay time model, Reliability Engineering and System Safety, 168, 70–79 (2017)
7. Zhang, F., Shen, J., Ma, Y.: Optimal maintenance policy considering imperfect repairs and non-constant probabilities of inspection errors. Reliab. Eng. Syst. Saf. **193**, 1–12 (2020)
8. Geara, C., Faddoul, R., Chateauneuf, A., Raphael, W.: Hybrid inspection-monitoring approach for optimal maintenance planning. Struct. Infrastruct. Eng. **6**(11), 1551–1561 (2020)
9. Yang, L., Ye, Z., Lee, C., Yang, S., Peng, R.: A two-phase preventive maintenance policy considering imperfect repair and postponed replacement. Eur. J. Oper. Res. **274**, 966–977 (2019)
10. Cai, J., Zuo, H.F., Xu, Y.-M.: Maintenance cost analysis under different inspection levels for aircraft structure. In: 2010 Prognostics & Health Management Conference, PHM2010 Macau, pp. 1–7, January 2010
11. Wang, W., Wang, H.: Preventive replacement for systems with condition monitoring and additional manual inspections. Eur. J. Oper. Res. **247**, 459–471 (2015)
12. Hao, s., Yang, J., Berenguer, C.: Condition-based maintenance with imperfect inspections for continuous degradation processes. Appl. Math. Modelling **86**, 311–334 (2020)
13. Parmigiani, G.: Optimial inspection and replacement policies with age-dependent failures and fallible tests. J. Oper. Res. Soc. **44**(11), 1105–1114 (1993)
14. SeyedHosseini, S.M.N, Moakedi, H., Shahanaghi, K.: Imperfect inspection optimization for a two-component system subject to hidden and two-stage revealed failures over a finite time horizon. Reliability Eng. Syst. Saf. **174**, 141–156 (2018)
15. Jodejko-Pietruczuk, A., Werbińska-Wojciechowska, S.: A Delay-Time Model with Imperfect Inspections for Multi-Unit Systems/Model Opóźnień Czasowych Z Nieperfekcyjną Diagnozą Stanu Systemu Dla Systemów Wieloelementowyc. J. KONBiN, 0, 157–172 (2012)
16. Wang, W., Banjevic, D.: Ergodicity of forward times of the renewal process in a block-based inspection model using the delay time concept. Reliab. Eng. Sys. Saf. **100**, 1–7 (2012)
17. Werbińska-Wojciechowska, S.: Technical System Maintenance, Delay-Time-Based Modelling, Springer, Cham (2019)

Reliability Assessment of Multi-cascade Redundant Systems Considering Failures of Intermodular and Bridge Communications

Vyacheslav Kharchenko[1,3], Andriy Kovalenko[2,3](✉), Eugene Ruchkov[1,3], and Ievgen Babeshko[1,3]

[1] National Aerospace University "KhAI", Kharkiv, Ukraine
{v.kharchenko,e.babeshko}@csn.khai.edu
[2] Kharkiv National University of Radio Electronics, Kharkiv, Ukraine
andriy.kovalenko@nure.ua
[3] RPC Radiy, Kropyvnytskyi, Ukraine
rev@radiy.com

Abstract. The paper is devoted to research of reliability aspects of complex safety-critical systems (CSs). Nowadays such CSs are used in variety of applications, including safety systems for nuclear power plants (NPPs), power grids industrial systems and others. In the paper, models of reliability for CSs are developed and studied, taking into account intermodular and bridge communications. The following problems are resolved: a set of CSs is created based on majority voting according to the "2oo3" logic and redundancy according to the "1oo2" logic considering intermodular and bridge communications; 2) reliability block diagrams (RBDs), analytical models of CS reliability with (multi)cascade redundancy of "2oo3" and "1oo2" principles are developed (including RBDs of reactor trip systems considering communications); 3) the models are studied, the dependencies for failure-free operation probabilities of various CSs, failure rates of intermodular and bridge communications and voting elements are determined; 4) recommendations for the selection of CS types are stated.

Keywords: Reactor trip systems · Redundant structures · Reliability · Communications

1 Introduction

Redundancy is used by modern industry in many areas, including power generation, mechanical engineering, aerospace and chemical industries, processing and more. Reliability assessment of redundant structures is a complex problem, solution of which is impossible without prior formalization and presentation of the primary model. The most common type of models for reliability assessment are Reliability Block Diagrams.

To date, many authors have described possible approaches and solutions for majority-voting systems, redundant systems and different types of voting logic, ranging from classical monographs [1] to modern works [2, 3].

W. Zamojski et al. (Eds.): DepCoS-RELCOMEX 2021, AISC 1389, pp. 179–188, 2021.
https://doi.org/10.1007/978-3-030-76773-0_18

Most authors do not take into account the reliability of communications, which can lead to overestimation of reliability. Traditionally, these tools are assumed to be absolutely reliable, but for many industrial systems this assumption is quite optimistic given the complexity of cable communications [4, 5], as well as the large number of additional connections between structures (cabinets, individual modules…), which are located inside the enterprise.

In [6–8], the impact of communication failure on the failure of majority-voting systems operating according to "2oo3" and "2oo4" logic was studied. However, different options for redundancy of structures, firstly, were not systematized, and, secondly, were not studied with different options of voting elements inclusion. Considering practical application, redundant structures have more potential in terms of means of communications, which in their turn can be redundant in different ways [9].

In [10] several "1oo2" and "2oo3" configurations have been modeled by Markov chains so as to understand safety parameter's contribution on safety. In [11] hardware architecture that can be applied in critical systems is analyzed.

The goal of the paper is to systematize the redundant structures of complex systems, as well as to develop and to study CS operating according to "2oo3" and "1oo2" logic taking into account various communications.

The paper is structured in the following way. Section 2 introduces reliability block diagrams of NPP Reactor Trip System (RTS) considering communications. Section 3 contains results of reliability block diagrams development for multi-cascade redundant systems. Section 4 is devoted to development of the models of reliability and, finally, Sect. 5 represents real results for such models application.

2 Reliability Block Diagrams of Reactor Trip Systems Considering Communications

One of the most important modern Instrumentation and Control (I&C) systems that are resistant to failures are RTSs. These are the most critical I&C systems that perform the functions of guaranteed shutdown of the reactor in various situations.

Figure 1 represents a functional diagram for one channel of NPP Reactor Trip System. The channel consists of AIM (Analog Input Module) to process input data from sensors, LM (Logic Module) to execute program logic, SFM (Signal Forming Module) to form signals and SDM (Signal Distribution Module) to distribute signals.

Reliability and functional safety of the RTS is provided by redundancy (up to 6 or even more channels) and diversity (the presence of two independent protection systems). Figure 2 depicts RBD of a single channel for the main RTS.

Using an approach based on decomposition of the sequential process of signal transmission along the entire path into separate failures related to data processing (P_D) and data transmission (P_C), it is possible to graphically represent a reliability model in one layer for the entire sensor-actuator path (Fig. 2a) and for the same layer within the boundaries of RTS equipment (intermodular communications, Fig. 2b).

Then, for the same RTS, but already in three layers (channels), a detailed analysis and quantification of RTS performance is required to ensure that safety requirements are met.

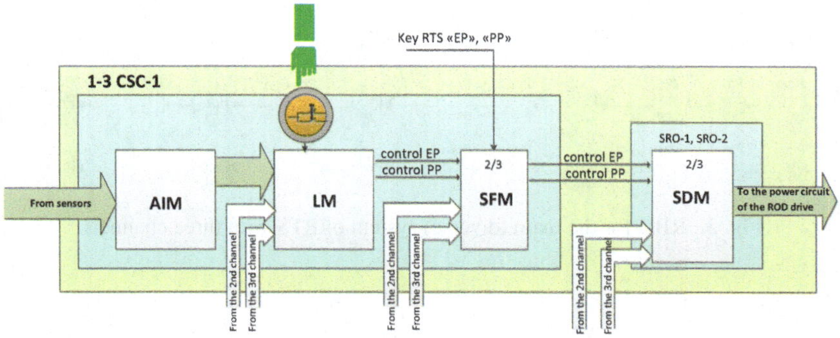

Fig. 1. Functional diagram for one channel of NPP Reactor Trip System.

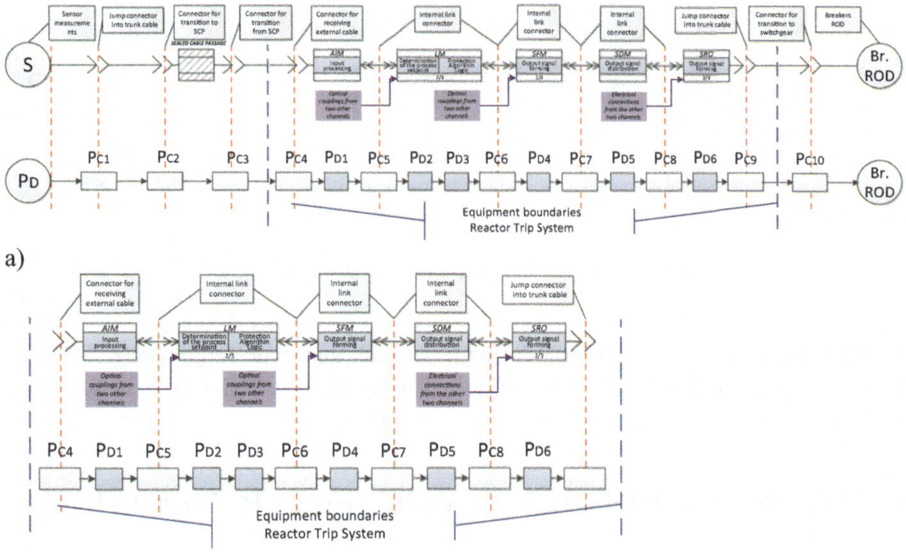

Fig. 2. RBD of a single channel for the main RTS: a) For the entire path of control signals; b) Transmission within the boundaries of RTS equipment; P_{Di}, P_{Cj} – failures probabilities of processing and communication components.

Figure 3 represents the RBD for the main (diverse) system of RTS with three channels. For three channels, RBD becomes a complex model with network and node redundancy. To assess the reliability and functional safety, taking into account the means of communication, it is necessary to analyze the real RTS circuits from sensors to actuators with different connection options.

Taking into account that two independent RTSs operate in parallel in order to manage a single reactor breaker actuator (the main one and diverse one, Fig. 4), the construction of which is similar, it is possible to create a model of their joint operation.

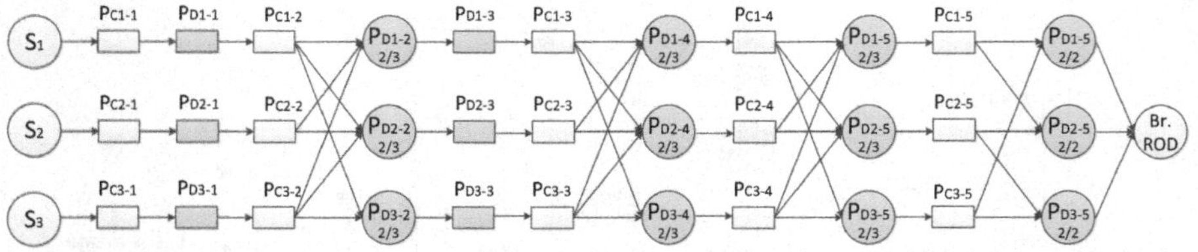

Fig. 3. RBD for the main (diverse) system of RTS with three channels.

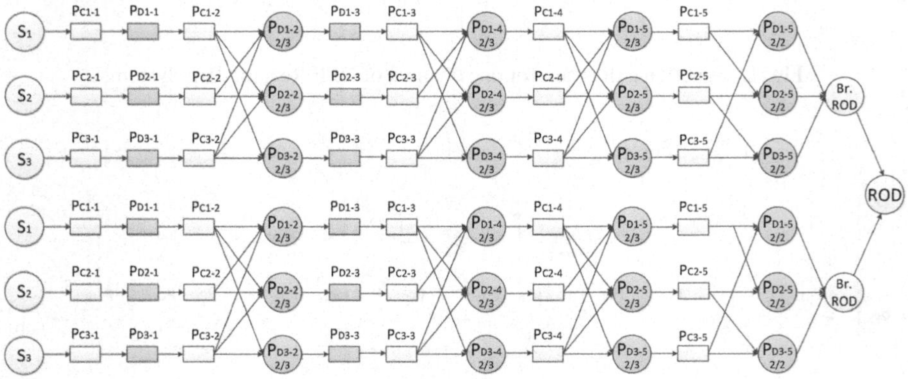

Fig. 4. Reliability block diagram for two-version RTS.

Such diagram is a complex model with network and node redundancy. To access reliability and functional safety it is required to analyze RTS tracks from sensors to actuators considering different communication options.

3 Reliability Block Diagrams of Multi-cascade Redundant Systems

Within the framework of the formulated problems, at the first stage there is a review of the redundant structures and their systematization. Consider the following structures, summarized in a form of Table 1 below. In appropriate figures (Fig. 5, 6, 7, 8, 9, 10, 11, 12, 13) the following notions are used: P_D – probability of failures related to data processing; P_C – probability of failures related to data transmission; $p23_1$, $p23_2$, $p12_1$ – probabilities of failures related to appropriate voting elements.

Table 1. Summary on developed structures.

Si	Type of redundant structure	Features of the structure	Figure no
S0	Sequential non-redundant structure	The simplest non-redundant structure that considers both data processing (D) and communication (C)	5
S1	Majority-voting structure with general nodal redundancy according to "2oo3" logic	A simple redundant structure that operates according to "2oo3" logic	6
S2	Majority-voting structure with general redundancy and "2oo3" cascade logic	Majority-voting structure with "2oo3" logic	7
S3	Redundant structure with "1oo2" general redundancy cascade logic and "2oo3" (network and node) voting	The structure is a special case of the previous structure and implements "1oo2" logic	8
S4	Majority-voting structure with separate nodal redundancy according to "2oo3" logic	Structure with separate redundancy according to "2oo3" logic	9
S5	Majority-voting structure with sectional division and voting according to "2oo3" logic	Structure with sectional division and voting according to "2oo3" logic	10
S6	Structure with separate voting according to "1oo2" and "2oo3" logic	A special case of the previous structure: there are also separate sections, but they perform separate voting according to "1oo2" and "2oo3" logic	11
S7	Bridge structure with "1oo2", "2oo3" cascade redundancy	A nonlinear option where connections are included in the majority structure	12
S8	Bridge structure with "2oo3" cascade network and node redundancy	An option of double (cascade) voting according to "2oo3" logic	13

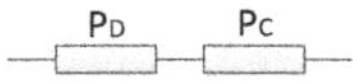

Fig. 5. Sequential non-redundant structure.

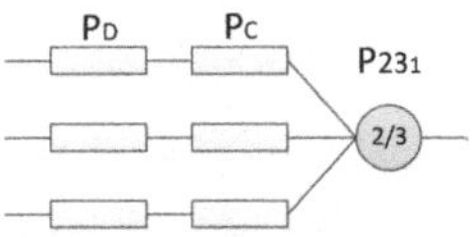

Fig. 6. Majority-voting structure with general nodal redundancy according to "2oo3" logic.

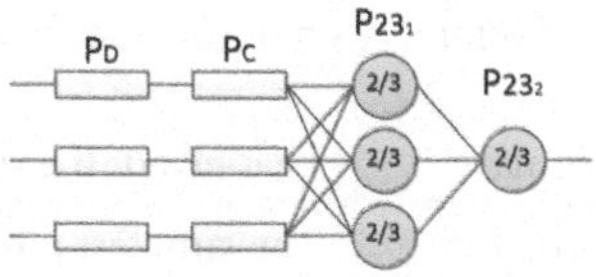

Fig. 7. Majority-voting structure with general redundancy and "2oo3" cascade logic.

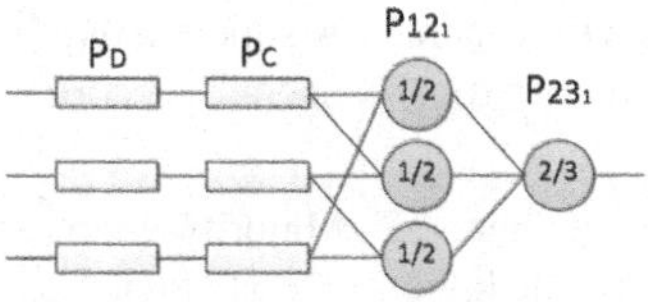

Fig. 8. Redundant structure with "1oo2" general redundancy cascade logic and "2oo3" (network and node) voting.

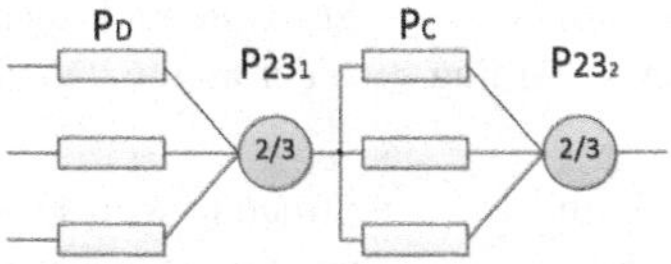

Fig. 9. Majority-voting structure with separate nodal redundancy according to "2oo3" logic.

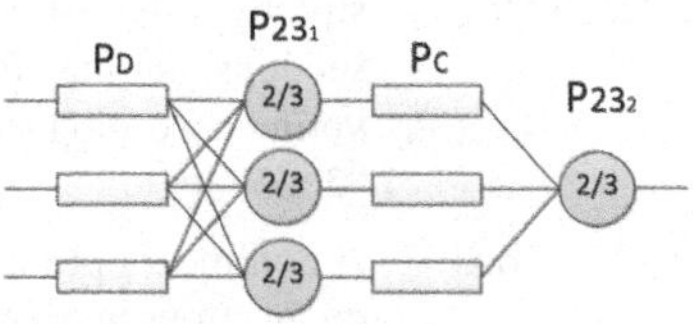

Fig. 10. Majority-voting structure with sectional division and voting according to "2oo3" logic.

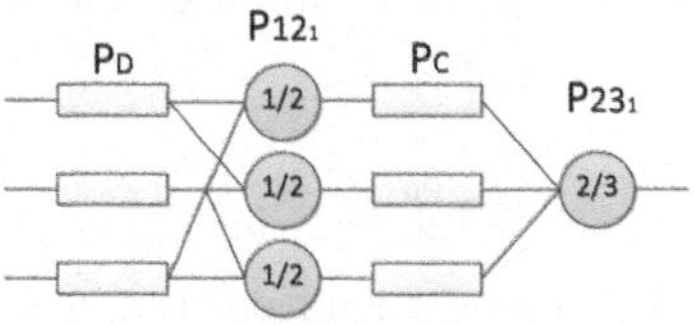

Fig. 11. Structure with separate voting according to "1oo2" and "2oo3" logic.

4 Reliability Models

This section presents the results of the development of analytical dependencies (1)–(8), which correspond to each of the structures S_0–S_8 from the previous section (Fig. 5, 6, 7, 8, 9, 10, 11, 12, 13).

$$P_{S_0} = p_D \cdot p_C \tag{1}$$

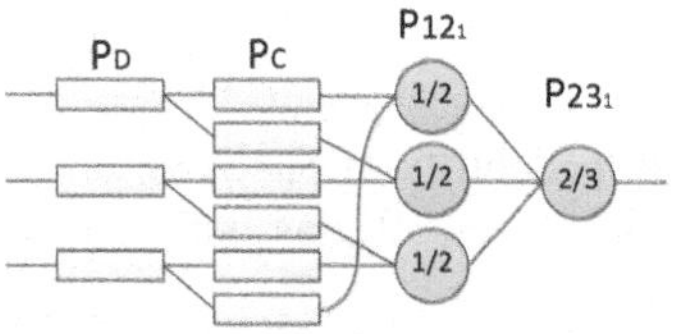

Fig. 12. Bridge structure with "1oo2", "2oo3" cascade redundancy.

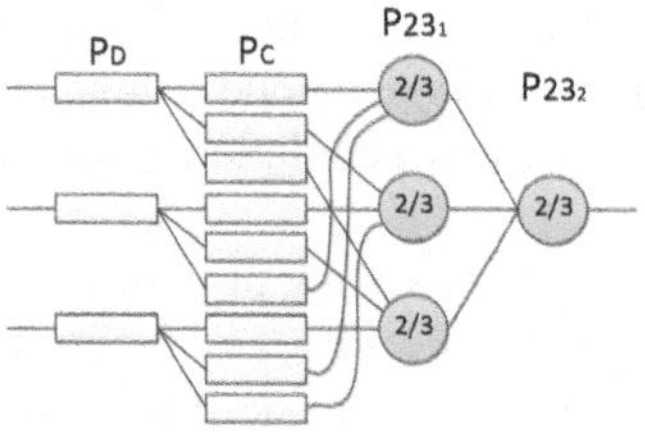

Fig. 13. Bridge structure with "2oo3" cascade network and node redundancy.

$$P_{S_1} = \left(3p_D^2p_C^2 - 2p_D^3p_C^3\right)p_{23_1} \tag{2}$$

$$P_{S_2} = \left(3p_D^2p_C^2 - 2p_D^3p_C^3\right) \cdot \left(3p_{23_2}^2 - 2p_{23_2}^3\right)p_{23_1} \tag{3}$$

$$P_{S_3} = \left\{\left[1 - (1 - p_Dp_C)^3\right]p_{12}^3 + \left(p_Dp_C + p_D^2p_C^2 - p_D^3p_C^3\right)3p_{12}^2(1 - p_{12})\right\}p_{23_1} \tag{4}$$

$$P_{S_4} = \left(3p_D^2 - 2p_D^3\right)p_{23_2} \cdot \left(3p_C^2 - 2p_C^3\right)p_{23_1} \tag{5}$$

$$P_{S_5} = \left(3p_D^2 - 2p_D^3\right) \cdot \left(3p_C^2p_{23_2}^2 - 2p_C^3p_{23_2}^3\right)p_{23_1} \tag{6}$$

$$P_{S_6} = \left\{\left[1 - (1 - p_D)^3\right]p_{12}^3p_C^3 + \left(p_D + p_D^2 - p_D^3\right) \cdot 3p_{12}^2p_C^2(1 - p_{12}p_C)\right\}p_{23_1} \tag{7}$$

$$P_{S_7} = \left[\begin{array}{c} p_{12}^3\left\{\begin{array}{c} p_D^3\sum_{i=0}^4 C_6^{6-i}p_C^{6-i}(1 - p_C)^i + 3p_D^2(1 - p_D)\times \\ \times\left[\sum_{i=0}^2 C_6^{6-i}p_C^{6-i}(1 - p_C)^i + \sum_{i=3}^4 C_4^{6-i}(1 - p_C)^i\right]+ \\ +3p_D(1 - p_D)^2p_C^2 \end{array}\right\}+ \\ +3p_{12}^2(1 - p_{12})\times \\ \times\left\{\begin{array}{c} p_D^3\left[\sum_{i=3}^2 C_6^{6-i}p_C^{6-i}(1 - p_C)^i + \sum_{i=3}^4 C_4^{6-i}p_C^{6-i}(1 - p_C)^i\right]+ \\ +3p_D^2(1 - p_D)\sum_{i=0}^1 C_3^{3-i}p_C^{3-i}(1 - p_C)^i + p_D(1 - p_D)^2p_C^2 \end{array}\right\} \end{array}\right] \cdot p_{23_1} \tag{8}$$

$$P_{S_8} = \left[\begin{array}{c} p_{23_2}^3\left\{\begin{array}{c} p_D^3\left[\begin{array}{c} \sum_{i=0}^3 C_9^{9-i}p_C^{9-i}(1 - p_C)^i + 3C_3^2C_3^2C_3^1p_C^5(1 - p_C)^4+ \\ +3C_3^2C_3^2p_C^4(1 - p_C)^5 \end{array}\right]+ \\ +3p_D^2(1 - p_D)\left[\sum_{i=0}^1 C_6^{6-i}p_C^{6-i}(1 - p_C)^i + C_3^2C_3^2p_C^4(1 - p_C)^2\right] \end{array}\right\}+ \\ +3p_{23_2}^2\left(1 - p_{23_2}\right)\left\{p_D^3\left[p_C^2 + C_6^1p_C^5(1 - p_C) + C_3^2p_C^4(1 - p_C)^2\right] + 3p_D^2(1 - p_D)p_C^4\right\} \end{array}\right] \cdot p_{23_1} \tag{9}$$

5 Research of Models

At the first stage, modeling of reliability of S_0–S_8 structures was carried out, in particular, dependencies of probabilities P_{S1}–P_{S8} as functions of p_D, p_C, p_{121}, p_{231} and p_{232} probabilities were investigated.

It can be noted that, in general, there is an increase in this probability with increasing in corresponding probability of individual components of the structure, and, at the same time, with complication (i.e. transition from the initial type of majority-voting structure with general nodal redundancy according to "2oo3" logic to the last type – bridge structure with "2oo3" cascade network and node redundancy) of the corresponding voting schemes in such structures. Such simulation results completely meet theoretical expectations.

At the second stage of research, modeling of failure models of S_0–S_8 structures was performed, at which the probabilities P_{S1}–P_{S8} as a function of time were chosen, as well as the corresponding failure intensities λ_D, λ_C, λ_{12}, λ_{231}, λ_{232} (1/h), where $p_z = \exp(-\lambda_z t)$.

The corresponding results for several sets of source data are shown in Fig. 13, 14, 15, 16. It should be noted that the results for all other intermediate sets of input data are not presented in this section due to their accurate change between the presented values.

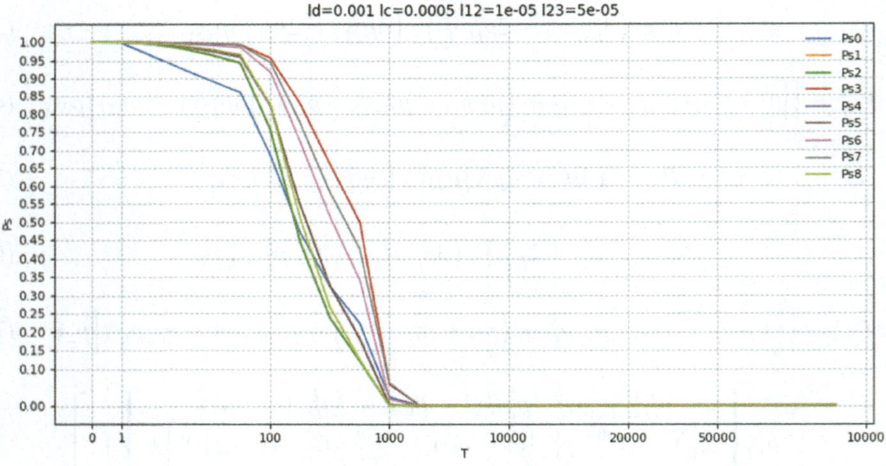

Fig. 14. The results for the 2nd set of source data.

Thus, it can be stated that the above time dependencies are quite predictable, based on a preliminary analysis of RBDs, which, in turn, confirm the results of the previous stage of research.

S_3, S_7 and S_6 structures have the greatest reliability. S_2 structure has the lowest reliability, while S_1, S_4, S_5 and S_8 structures have an average level of reliability with small differences. These findings persist and the level of difference between these groups of structures in reliability increases with decreasing reliability of information processing means.

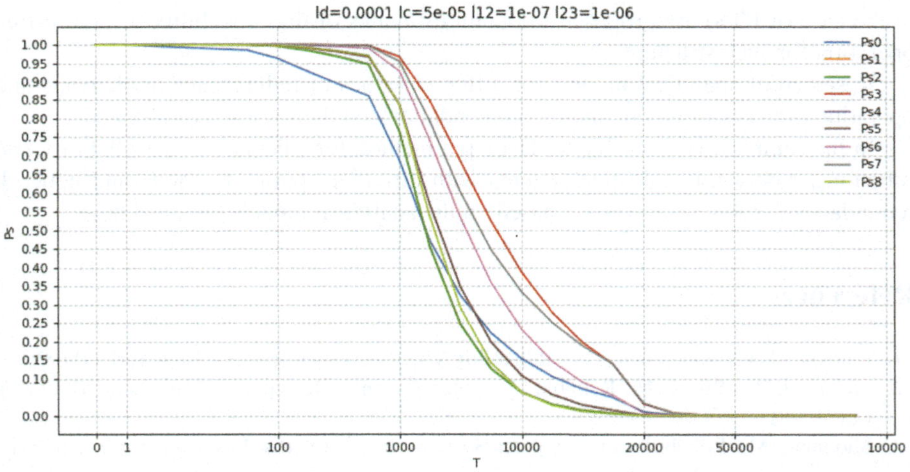

Fig. 15. The results for the 15th set of source data.

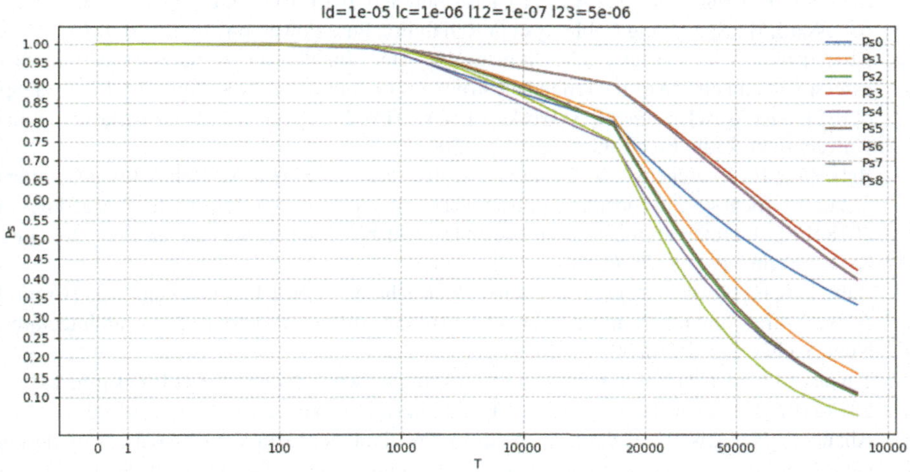

Fig. 16. The results for the 23rd set of source data.

The proposed analytical dependencies can be used to calculate the indicators of reliability and functional safety, as well as to formulate recommendations for the choice of redundant structures.

6 Conclusions

The paper introduces RBDs for a number of redundant structures with different variants of "2oo3" and "1oo2" voting cascades taking into account means of information processing and communications. Also, analytical models of reliability (functional safety in the case of RTS) are proposed to obtain estimates of the probability of failure-free operation for such structures.

Research of analytical models for different values of failure rates of elements and operation time is carried out and recommendations on their use are formulated.

In the future, it is advisable to develop a method for calculating the reliability and functional safety for multi-tier systems with different options for node and network redundancy ("2oo3" and "1oo2") based on the analytical models obtained here.

References

1. Polovko, A.M.: Fundamentals of Reliability Theory. Academic Press, New York (1968)
2. Siewiorek, D., Swarz, R.: Reliable Computer Systems: Design and Evaluation. AK Peters Ltd (1998)
3. Shooman, M.: Reliability of Computer Systems and Networks: Fault Tolerance, Analysis, and Design. John Wiley & Sons, Inc (2002)
4. Siirto, O., Vepsäläinen, J., Hämäläinen, A., Loukkalahti, M.: Improving reliability by focusing on the quality and condition of medium voltage cables and cable accessories. In: 24th International Conference on Electricity Distribution (CIRED), Glasgow, pp. 229–232. (2017)
5. Broussard, B.: Specifying Cable System Reliability. Pure Power, pp. 26–30 (2007)
6. Kharchenko, V.S., Lysenko, I.V., Sklyar, V.V., Herasimenko, O.D.: Safety and reliability assessment and choice of the redundant structures of control safety systems. In: Proceedings of IEEE East-West Design & Test Workshop (EWDTW 2005), pp. 212–218. Kharkov National University of Radio Electronics, Kharkov (2005)
7. Babeshko, E., Kharchenko, V., Leontiiev, K., Ruchkov, E., Sklyar, V.: Reliability assessment of safety critical system considering different communication architectures. In: Proceedings of 2018 IEEE 9th International Conference on Dependable Systems, Services and Technologies, Kyiv, Ukraine, pp. 18–21. (2018)
8. Babeshko, E., Illiashenko, O., Kharchenko, V., Ruchkov, E.: NPP Instrumentation and Control systems safety and reliability assessment considering different communication architectures. Nucl. Radiat. Saf. **2**(86), 38–43 (2020)
9. Threats to Undersea Cable Communications. https://www.dni.gov/files/PE/Documents/1---2017-AEP-Threats-to-Undersea-Cable-Communications.pdf. Accessed 12 Jan 2021
10. Ahangari, H., Atik, F., Ozkok, Y., Yildirim, A., Ata, S., Ozturk, O.: Analysis of design parameters in safety-critical computers. IEEE Trans. Emerg. Top. Comput. **8**(3), 712–723 (2020). https://doi.org/10.1109/TETC.2018.2801463
11. Farias, M., Nedjah, N., de Carvalho, P.V.R.: Resilient hardware design for critical systems. In: 2019 IEEE 10th Latin American Symposium on Circuits & Systems (LASCAS), pp. 237–240 (2019). https://doi.org/10.1109/LASCAS.2019.8667549

Evolution Process for SOA Systems as a Part of the MAD4SOA Methodology

Szymon Kijas(✉) and Klara Borowa

Institute of Control and Computation Engineering, Warsaw University of Technology, Warsaw, Poland

{sszymon.kijas,klara.borowa}@pw.edu.pl

Abstract. In this paper we propose a process for evolving service-oriented systems. The evolution comprises a series of changes made to the system structure. Starting from the moment when a request for change is defined, and ending with the change review, we define a set of phases that take place during the course of a single evolution step. For each phase, we explain in detail the activities that are performed in its course. We also define a set of artefacts needed by each activity and produced as a result of it. The proposed evolution process is an integral part of the evolution and development methodology for service-oriented systems (MAD4SOA) developed by our team.

Keywords: Evolution process · Architectural decisions · SOA

1 Introduction

Systems based on distributed architectures, such as service-oriented systems, are extremely popular due to their scalability, flexibility and ease of introducing changes. Because of their increasing complexity and the frequent requirement changes, the continuous and rapid evolution of systems is inevitable. Service-oriented architecture (SOA) is one of the possible ways to address the challenge of rapid system's evolution. The specific properties of a SOA systems, namely: being composed out of a set of loosely coupled, distributed, stateless services, support keeping up with the pace of changes to the requirements as well as with the emergence of new ones.

The typical evolution process of such a system consists usually of adding new services, removing, modifying the existing ones as well as changing the order in which they are invoked. However, the evolution of service-oriented system involves changes made to many layers of such systems (business processes, business services, etc.). This challenge has been addressed neither by the existing development nor maintenance methodologies. Therefore, a dedicated evolution methodology is needed in order to address the specific properties of service-oriented systems as well as the specific challenges connected with the evolution of such systems.

Most of the development methodologies for SOA systems described so far do not take into account their further evolution, or consider only the initial stage of maintenance. Our goal is to propose a methodology with the aim of documenting both the current

W. Zamojski et al. (Eds.): DepCoS-RELCOMEX 2021, AISC 1389, pp. 189–202, 2021.
https://doi.org/10.1007/978-3-030-76773-0_19

architecture of the service-oriented system, and the context of its development. All of this can be achieved by using architectural decisions [20, 21] as a means of documenting architecture. Our methodology, MAD4SOA (Maps of Architectural Decisions for SOA), consists of the following elements:

- The evolution process – which determines how a change should be made to the service-oriented system, and the artefacts produced during this process.
- The method of documenting the system's evolution – by means of architectural decisions, the problems, the solution variants and the chosen solutions are described. As we proposed in [1].
- The formalised model of evolution – including the model of the service-oriented system, the model change introduction, and the model of the architectural decisions. This is complemented by the correctness verification of the models. As proposed in [1].
- The early assessment method of distributed systems. As proposed in [2].

The goal of this paper is to introduce an evolution process model, based on the process defined by the ISO/IEC 20000-2 standard [3]. We have adapted this process to suit service-oriented systems, and have improved it by adding activities related to building service architecture and documenting architectural decisions.

The work on this methodology was initiated by research project No 5321/B/T02/2010/39.

2 Related Work

An evolution step of a service-oriented system consists of multiple activities: service modification, replacement, removal and even a change in the way that independent services communicate with each other. Nevertheless, so far the evolution of this class of systems was understood only as introducing changes to individual services, i.e. their interfaces, functionalities, etc. [4, 5, 13]. Detailed research on the maintenance and evolution of service-oriented systems was performed mainly to specific issues. This includes change tracking [6], change propagation [7, 8], versioning [9], impact analysis [10] and model-driven-approaches to service composition [14].

In general, the topic of service-oriented systems maintenance has barely been breached. Only the early post-deployment phase is discussed [8, 12], and properly managed using previously defined methodologies [14]. The evolution of services has been accounted for in the fractal process of SOMA methodology [4] with the concept of successive iterations. In [11], the authors propose the use of change-management mechanisms to control the evolution of service compositions. The evolution of systems based on microservices, which belong to a similar class of systems to service systems is also a topic of interest for researchers [15].

Therefore, the development of an evolutionary process and methodology adapted to service-oriented systems, understood as a set of service compositions, is still an open research problem.

3 Evolution Process for Service-Oriented Systems

As part of the MAD4SOA methodology, we define and formalise the process of introducing changes to service-oriented systems. This covers the key information necessary to manage the evolution process: the sequences of activities, decision points, artefacts necessary to carry out activities and artefacts created as a result of executing the activities. The primary purpose of the approach that we propose is to allow the evolution of the service-oriented system, along with its entire context, to be documented. Therefore, the final documentation not only contains the final systems' architecture, it additionally contains alternative variants that were considered during the design phase. When using the MAD4SOA methodology, the documentation of the system's evolution is in fact a documentation of architectural decisions made during the course of the project, including descriptions of the encountered architectural problems, considered variants for solving these problems and finally the solutions that were selected.

The process of introducing changes to a service-oriented system proposed in this paper is consistent with the abstract process defined in the ISO/IEC 20000-2 standard [3]. However, it is distinguished by two crucial differences:

1. An additional phase, the purpose of which is to prepare the project for a new release.
2. The decision to implement or abandon the change is made during the first phase of the process – the "change approval" phase has been removed.
3. The activities performed during particular phases have been adapted to the specifics of the service-oriented system evolution process.

The model of the proposed change implementation process is presented in Fig. 1. The individual phases of the change implementation process are described in detail in the next sections.

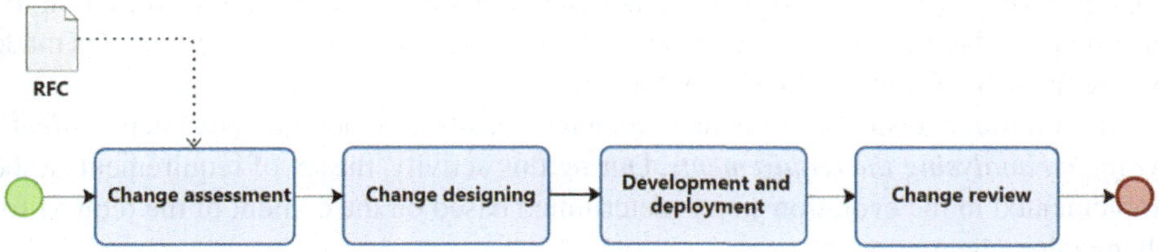

Fig. 1. Implementation change process.

3.1 Change Assessment Phase

The change assessment phase starts when a request for change document (RFC) is created. During this phase, the validity and scope of the change are assessed, the requirements are identified and analysed, and the time, cost and labour intensity of the change implementation are assessed. Then, on the basis of this assessment, the decision on whether the change will be implemented or not is taken. The change assessment phase comprises eight activities (Fig. 2): an assessment of the legitimacy of the change and its scope; the initial approval of the change; the identification and analysis of the requirements; the development of a prototype design for a new system release; the implementation of the prototype of the new system release; an assessment of the time, cost and labour intensity of the change; the acceptance of the change; and the schedule development for the implementation of the new release.

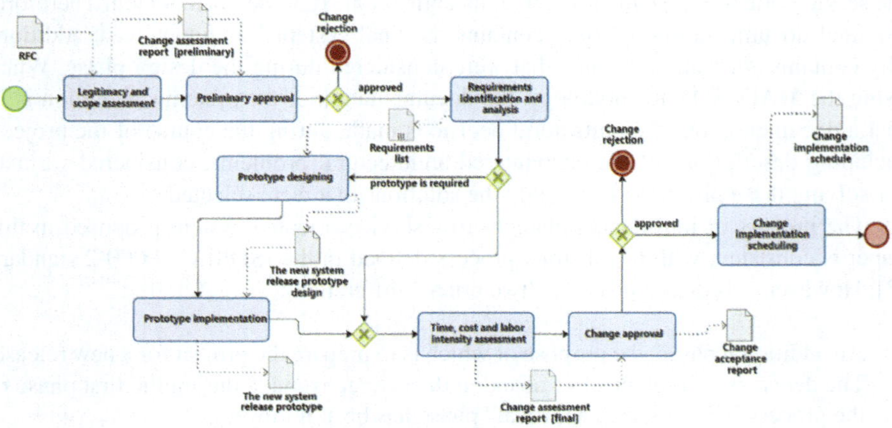

Fig. 2. Change assessment phase.

The first activity of this phase is the *legitimacy and scope assessment* of the change, during which the feasibility study of the change is carried out – since the first question that needs answering is "Do we need the change at all?" Following that, the scope of the impact of the change on the current system release is estimated (system functionalities that will have to be modified and new functionalities are identified). Based on the information obtained during this phase, in the next step, a *preliminary approval* is made before the further implementation of the change.

If an initial decision to implement a change has been made, the next step *is identifying and analysing the requirements.* During this activity, the set of requirements to be implemented in the evolution step is determined based on the content of the request for change document.

After identifying and analysing the requirements, it may be necessary to *develop a prototype* for the new system release. This enables the further evaluation of the change. A prototype consists of system elements defined in the business process layer and in the service layer (new versions of business processes, models of complex services, simple services and external services). The prototype is created in two stages: the creation of the *prototype design* and its *implementation.* Architectural problems are solved and architectural decisions are made during the *prototype design phase.* These decisions will shape the target architecture of the new system release.

Regardless of whether a prototype was created made or not, the next step is to *assess the time, cost and labour intensity of the change.* The impact of the change on the system's functionalities is covered by this detailed assessment. Then, the final decision is made on *whether to implement the change or not*. If the change was accepted, a *schedule for the implementation of a new system release* should be prepared.

Artefacts appearing in the change assessment phase are summarised in Table 1.

Table 1. Artefacts of the change assessment phase.

Artefact name	Artefact description
Request for Change (RFC)	A document describing the change from a business perspective. It describes the business goals and the effects of the change, but does not indicate how it should be implemented. The change request may contain various drawings and diagrams (e.g. BPMN or UML diagrams)
Requirements list	This document includes the set of identified requirements that are supposed be implemented in order to implement the change (create a new system release)
Change assessment report. Which has two variants, depending on the assessment's stage: *preliminary* and *detailed*	This document includes, in both the preliminary and detailed state versions: 1. The scope of change: a list of business processes and services to be added, removed or modified in the context of the change; 2. The impact analysis of the change – a description of the impact of the change on: quality attributes of the software (including SLA) e.g. reliability, performance, business continuity, etc.; list of affected business processes (e.g. requiring revision); parallel changes; estimated costs; estimated labour intensity; identified risks; and attachments (other documents used or produced during the change assessment); In the case of a detailed change assessment report, the impact analysis of the change will set out a prototype. On the other hand, the preliminary analysis does not have to contain all of the listed elements
The new system release prototype design	The design of the prototype of the new system release
The new system release prototype	The prototype of the new system release

(*continued*)

Table 1. (*continued*)

Artefact name	Artefact description
A collection of business processes and services	The collection includes: • modified business processes and complex services, • new business processes and complex services, • a list of deleted business processes and services: complex, simple and external, • a list of new simple and external services, • a list of simple and external services to be changed, • a list of service components to be modified and removed, and new components that will be implemented
Change acceptance report	The document includes: • justification of the decision to accept or reject the change, • labour intensity/estimated costs, • source of funding for the change implementation, • change implementation schedule, attachments: RFC, change assessment report and change prototype
Change implementation schedule	A schedule for the implementation of a new system release

3.2 Change Designing Phase

The *change designing* phase may start after the *approval phase,* as long as the change was approved and implementation plans were created. During the *change designing* phase, a design for the architecture of the new release is prepared (taking into account changes in the relevant layers of the service system). This phase consists of four activities (Fig. 3): *identification of architectural problems, fine-tuning business processes, identification of services,* and *design preparation* for a new system release.

In the first step, on the basis of the requirements, *architectural problems are identified.*

Then the existing business processes are *fine-tuned.* At this step, the architectural problems that shape the business have to be solved. If a prototype was prepared during *the change assessment phase*, it is expanded here.

During the *services identification* phase, the following services are identified: new services (including the selection of third party service providers), services to be modified and services to be removed. Therefore, the architectural problems defined in the previous step of the project, are solved.

Finally, based on the analysis documents performed during the previous steps of the *change designing* phase, the *new system release design* gets prepared. The project of the new system release contains a complete set of documents needed to initiate the implementation and deployment phases. This means that design decisions determining the implementation of individual service components, and the communication between them, are made during the design preparation phase. Finally, the test cases that will be needed in order to validate the new system release are developed.

The *change designing* phase is summarised in Table 2.

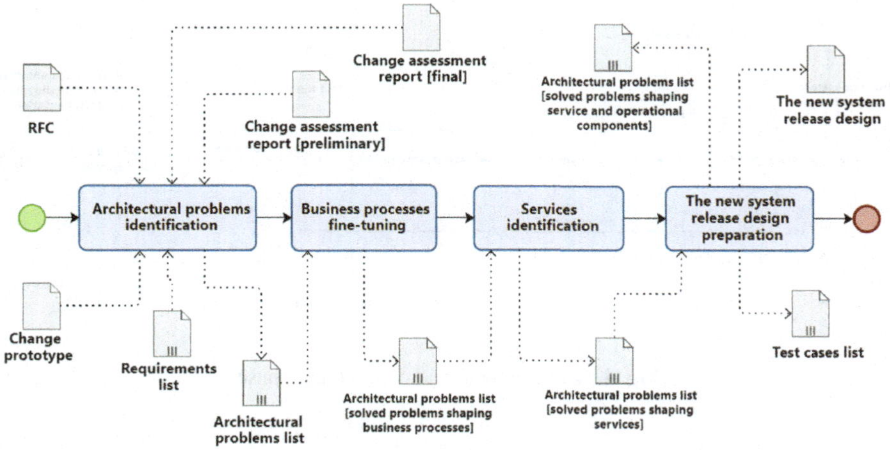

Fig. 3. Change designing phase.

Table 2. Artefacts of the change designing phase.

Artefact name	Artefact description
Architectural problems list	It describes the architectural problems identified and solved during the *change designing* phase. Even after the initial identification of architectural problems, this list remains open and may be expanded during the course of subsequent activities, since new problems may arise as a result of solving previously identified problems
The new system release design	This document includes all the information related to the change implementation: • New, changed and deleted business processes • New, changed and deleted services: complex (composition of other services), simple and foreign services • Service components to be changed, added and removed. Along with implementation descriptions and graphic models (e.g. UML) • Operational components that need to be modified or added • A set of architectural problems together, along with their chosen solutions and alternative solution variants
Test cases list	Test cases necessary for the approval of the new system release. Test cases should enable the verification of all the requirements defined in the *change assessment* phase

3.3 Development and Deployment Phase

The *development and deployment* phase is when the programming and deployment tasks are performed. This phase has been divided into four activities (Fig. 4): *implementation, test deployment, acceptance testing* and *production deployment*.

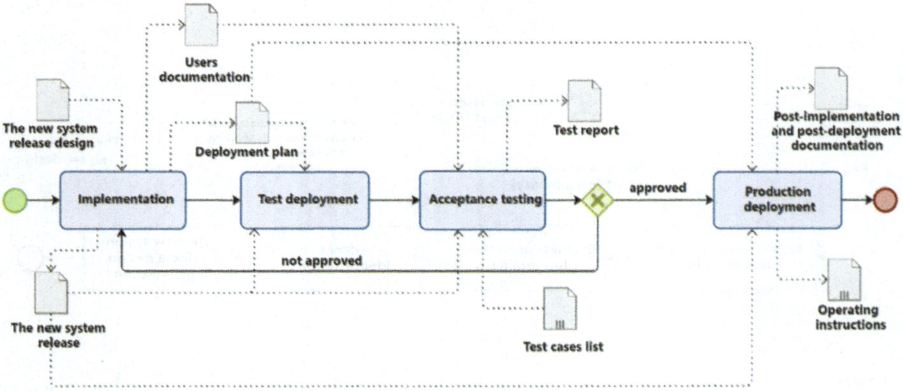

Fig. 4. Development and deployment phase.

At the *implementation* phase, the new service components and the changes to existing components are developed. At this point, service components are integrated with the operational components. During this phase, a new system release is created based on the design from the *change designing* phase. Additionally, this phase leads to the creation of the deployment plan and user documentation. We did not specify how programming tasks should be performed. During this step, almost any software development methodology can be used. Both agile [16] (e.g. Feature Driven Development, Scrum, Extreme Programming), and classical methods (e.g. the waterfall process [17] or RUP [18, 19]) can be applied. The choice of the approach is left to the development team.

After completing the implementation, the new release is goes to the acceptance testing environment (UAT). The *test deployment* is carried out according to the deployment plan developed during the implementation phase.

The next step in this process is *acceptance testing*, during which the acceptance tests are performed. These tests are supposed to ascertain whether all the features described by the requirements have been developed, and whether all the non-functional requirements have been met. If errors that prevent a production deployment are detected, the implementation phase is carried out again, in order to perform necessary corrections.

If a new release is approved at the test deployment phase, the *production deployment* takes place. As in the case of the test deployment, the production deployment is performed as described in the deployment plan. Finally, the new release is launched, the operating instructions are made available and the post-implementation documentation is drafted.

Artefacts appearing in the development and deployment phase are summarised in Table 3.

Table 3. Artefacts of the development and deployment phase.

Artefact name	Artefact description
The new system release	This is a new version of the system, updated with the implemented changes
User documentation	Documentation describing how to use the features of the new system release
Deployment plan	A plan that describes how the new release should be deployed. This includes the configuration and installation of all relevant components
Test report	The acceptance test report, including the test-performed scenarios and detected defects
Post-implementation and post-deployment documentation	Technical documents containing the deployment scenarios and describing the changes made to the source code
Post-implementation report	A list of errors, corrected and uncorrected, detected during the implementation phase
Operating instructions	Workstation instructions for the users and operators of the system, as well as administrative instructions meant for system administrators

3.4 Change Review Phase

After the development and deployment phase, the final evolution step takes place – the *change review* – during which two activities are performed (Fig. 5): the *change review* and *best design practices' documentation.* According to the ISO 20000, the standard change review phase is optional, though it is recommended. This phase concludes the whole evolution step of the service-oriented system.

During the *change review* phase, the changed artefacts are analysed and the architectural decisions that were made are validated. The purpose of this review is to spot best practices and mistakes, in order to then draw valuable conclusions that will be useful in the future. The change review report is written down in order to document the analysis of this phase. Finally, the best architectural practices document is updated and made available for future use.

We did not define a workflow for the *change review.* Any organisation that chooses to perform this phase should do so in accordance with their internal policies and practices.

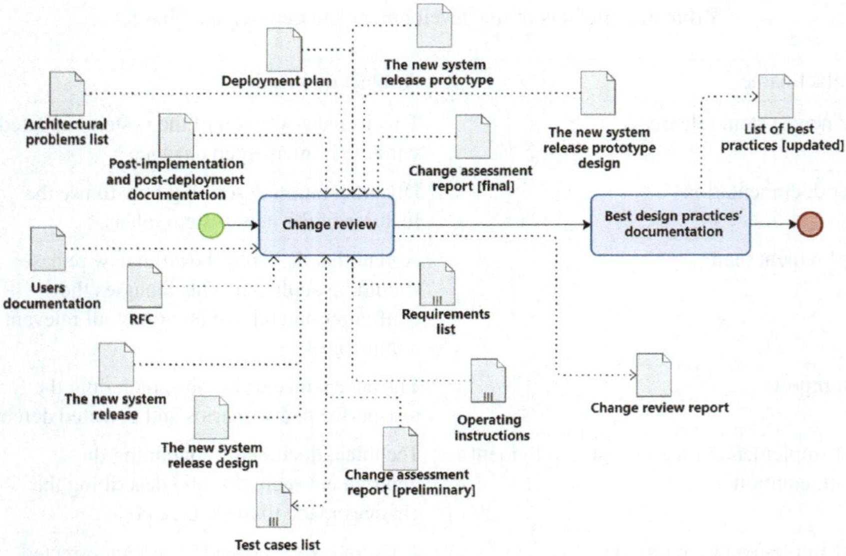

Fig. 5. Change review phase.

Artefacts appearing in the change review phase are summarised in Table 4.

Table 4. Artefacts of the changer review phase.

Artefact name	Artefact description
Change review report	A document developed individually by the organisation, depending on its needs
List of best practices	A list of best architectural practices used by the organisation. This should be updated after each evolution step. This document should be used globally, so it should be an all-encompassing document used in all of the organisation's projects

4 The Case Study

In this section the evolution process for service-oriented systems has been validated on an example of a real service-oriented system. This case study is based on the clearing system for instant payments operating in the Polish banking system. This system interacts with banks' systems participating in a transaction (the orderer and the receiver of the transfer) and with the system of the Central Bank. Basically, it works as follows:

- the clearing system verifies payment order,
- the clearing system confirms the consent of both banks,

- and finally the clearing system registers the change in the account balances of both banks and in the Central Bank system.

Our case study includes six releases of the clearing system (in fact initial release and five evolution steps):

- development of the functionality of instant payments between commercial banks as the first release,
- adding instant tax payments to the Central Bank in the second release,
- adding complaints handling functionality in the third release,
- adding a Back Office module in the fourth release,
- adding postponed payments in the fifth release,
- removing one type of the tax payments in the sixth release.

Due to the limited space we present only one evolution step of the evolution process of the clearing system that is adding of instant tax payments (the second release of the system):

Change Assessment Phase
The RFC described that we have to add possibility of making instant tax payments to Central Bank. In the first activity (legitimacy and scope assessment of the change) we have decided that this change is justified. However, it was possible to add the second role for Central Bank (receiver rule) and to achieve this business requirement without any changes but we have decided to additionally simplify the workflow. Next, our proposal gained preliminary approval. After identifying the key requirements we decided to develop a prototype. As the prototype served the business process model of the new workflow. Finally, labour and cost was estimated and our proposal was finally approved and scheduled.

Change Design Phase
At the beginning, we analysed and refined all of the requirements and then we identified the list of architectural problems that we need to solve (in fact we defined the model of architectural decisions that is not included in this article). Next, the business process prototype has been fine-tuned. Then, we identified new services that had to be implemented (services supporting making payments directly to Central Bank). Finally, the design of the second release of the system was prepared.

Development and Deployment Phase
At the beginning of this phase new functionality was implemented and deployed to the test environment (we have made internal tests). Next the acceptance tests was made. We had to fix some bugs that occurred after integration with new services in Central Bank. Finally the second release of the system was deployed to production environment.

Change Review Phase
The clearing house keeps the practice of documenting best architectural practices. Therefore, the optional change review phase was carried out. It was about reviewing the changes made and updating the list of best practices.

5 Discussion

Methodologies that focus on the development of service-oriented systems [4, 11], do not, in principle, take into account the aspects of their further evolution. Moreover, as explained in Sect. 2, research into the evolution of service-oriented systems is rather scarse. Therefore, developing an evolution and development methodology for service-oriented systems is an important challenge.

Such methodology should provide for the option of documenting all the valuable information about the evolution process, along with the context of the introduced changes. Since the decisions made during every evolution step potentially change the shape of the system architecture. Those decisions, along with alternative solutions considered should be preserved for future use. The creation development of such a methodology was the goal of our research.

The most important element of the proposed methodology is the process of introducing changes to the service-oriented system. A generic change process model was previously proposed in the ISO/IEC 20000-2 standard [3]. However, this model turned out to be insufficient and had to be adapted to the evolution of service-oriented systems.

The following changes in the organisation of the process phases have been introduced:

- The approval of the implementation of the change was moved to the first phase (change assessment) and split into two stages – an initial assessment of the change, and a detailed assessment (often preceded by prototype development).
- The change-designing phase has been changed to include specific activities carried out while designing service-oriented systems (service identification, business process fine-tuning).
- A new way of documenting architectural problems that must be identified, analysed, with solution alternatives and the chosen solution rationale has been introduced.
- Finally, the set of artefacts to be created during the course of each of the evolution step.

In contrast to the other service-oriented methodologies, such as those proposed in [4, 22], we are not trying to develop an entirely new approach. Our evolution process is designed so as to be compliant with the popular industry standard ISO 20000:2005 [3]. Firstly, organisations that would like to use MAD4SOA don't have to reorganise their maintenance procedures as there must be a really important reason to change established, proven development and maintenance practices. Secondly, our process is very flexible as it allows an organisation to use any of the most popular development process for developing their service-oriented systems (i.e. waterfall [17], RUP [18, 19] or agile [16]). This is another factor making industrial adoption easier, as an organisation can use its own development team without enforcing any unknown methodology. Finally our evolution process provides traceability features by linking architectural decisions modified in consecutive evolution steps. It allows to minimize the risks associated with repeating past mistakes and enhances capability of re-using proven system components in the future.

6 Conclusion

This article proposes an evolution process that determines how to introduce change into a service-oriented system. This process is an integral part of the evolution and development methodology dedicated to service-oriented systems (MAD4SOA).

We define a set of phases, the activities included in these phases and the artefacts produced during each evolution step of a service-oriented system. Finally, we have validated our process on the real-world service-oriented system being operated in Polish banking sector.

Further research will focus on adapting the process to the evolution of twin-based microservice-based systems.

References

1. Kijas, S., Zalewski, A.: Capturing the evolution of service-oriented systems with architectural decisions. In: Communication Papers of the 2020 Federated Conference on Computer Science and Information Systems (2020). https://doi.org/10.15439/2020f177
2. Zalewski, A., Kijas, S.: Beyond ATAM: early architecture evaluation method for large-scale distributed systems. J. Syst. Softw. (2013). https://doi.org/10.1016/j.jss.2012.10.923
3. ISO/IEC 20000-2:2005: International Standard - Information Technology - Service Management - Part 2: Code of practice (2005)
4. Arsanjani, A., Ghosh, S., Allam, A., Abdollah, T., Ganapathy, S., Holley, K.: SOMA: a method for developing service-oriented solutions. IBM Syst. J. (2008). https://doi.org/10.1147/sj.473.0377
5. Zuo, W.: Managing and modeling web service evolution in SOA architecture (2018)
6. Sindhgatta, R., Sengupta, B.: An extensible framework for tracing model evolution in SOA solution design. In: Proceedings of the Conference on Object-Oriented Programming Systems, Languages, and Applications, OOPSLA (2009). https://doi.org/10.1145/1639950.1639960
7. Dam, H.K., Ghose, A.: Supporting change propagation in the maintenance and evolution of service-oriented architectures. In: Proceedings - Asia-Pacific Software Engineering Conference, APSEC (2010). https://doi.org/10.1109/APSEC.2010.27
8. Ravichandar, R., Narendra, N.C., Ponnalagu, K., Gangopadhyay, D.: Morpheus: semantics-based incremental change propagation in SOA-based solutions. In: Proceedings - 2008 IEEE International Conference on Services Computing, SCC 2008 (2008). https://doi.org/10.1109/SCC.2008.16
9. Laskey, K.: Considerations for SOA versioning. In: Proceedings - IEEE International Enterprise Distributed Object Computing Workshop, EDOC (2008). https://doi.org/10.1109/EDOCW.2008.25
10. Hirzalla, M.A., Zisman, A., Cleland-Huang, J.: Using traceability to support SOA impact analysis. In: Proceedings - 2011 IEEE World Congress on Services, SERVICES 2011 (2011). https://doi.org/10.1109/SERVICES.2011.103
11. Orriëns, B., Yang, J., Papazoglou, M.P.: Model driven service composition. In: Orlowska, M.E., Weerawarana, S., Papazoglou, M.P., Yang, J. (eds.) ICSOC 2003. LNCS, vol. 2910, pp. 75–90. Springer, Heidelberg (2003). https://doi.org/10.1007/978-3-540-24593-3_6
12. High, R., Kinder, S., Graham, S.: IBM's SOA Foundation: An Architectural Introduction and Overview (2005)
13. Naji, H., Mikki, M.: A survey of service oriented architecture systems maintenance approaches. Int. J. Comput. Sci. Inf. Technol. **8** (2016). https://doi.org/10.5121/ijcsit.2016.8302

14. Mittal, K.: Build Your SOA Part 1: Maturity and Methodology. IBM (2005)
15. Sampaio, A.R., Kadiyala, H., Hu, B., Steinbacher, J., Erwin, T., Rosa, N., Beschastnikh, I., Rubin, J.: Supporting microservice evolution. In: Proceedings - 2017 IEEE International Conference on Software Maintenance and Evolution, ICSME 2017, pp. 539–543 (2017). https://doi.org/10.1109/ICSME.2017.63
16. Abrahamsson, P., Salo, O., Ronkainen, J., Warsta, J.: Agile software development methods: Review and analysis (2002)
17. Sommerville, I.: Software Engineering, 9th edn., p. 18 (2011). ISBN-10 137035152
18. Kruchten, P.: Rational Unified Process-An Introduction, 3rd edn. Addison-Wesley, Boston (2004)
19. Kruchten P.: The Rational Unified Process Made Easy: A Practitioner's Guide to the RUP: A Practitioner's Guide to the RUP Paperback (2003). ISBN-13 078-5342166095
20. Jansen, A., Bosch, J.: Software architecture as a set of architectural design decisions. In: Proceedings - 5th Working IEEE/IFIP Conference on Software Architecture, WICSA 2005, pp. 109–120 (2005). https://doi.org/10.1109/WICSA.2005.61
21. Bosch, J.: Software architecture: the next step. In: Oquendo, F., Warboys, B.C., Morrison, R. (eds.) EWSA 2004. LNCS, vol. 3047, pp. 194–199. Springer, Heidelberg (2004). https://doi.org/10.1007/978-3-540-24769-2_14
22. Bell, M.: Service-Oriented Modeling: Service Analysis and Design and Architecture. Wiley Publishing, Hoboken (2008)

Optimizations for Fast Wireless Image Transfer Using H.264 Codec to Android Mobile Devices for Virtual Reality Applications

Maciej Kopczynski(✉)

Faculty of Computer Science, Bialystok University of Technology, Wiejska 45A, 15-351 Białystok, Poland
m.kopczynski@pb.edu.pl

Abstract. This paper shows results and describes solution allowing for sending high quality image (video) rendered on high-end PC running Windows OS to Android-based mobile devices. Main purpose of this approach is to create versatile system devoted to **V**irtual **R**eality (**VR**) applications accessible by wireless communication, which needs no cables connecting PC and mobile device, therefore giving freedom of movements. Obtained results show possibility of achieving good image quality transmission with very low latency, including small delay in head movements tracking.

Keywords: Virtual reality · Image transmission · Small delay · Optimization · Mobile device · H.264

1 Introduction

One of the main challenges of the VR industry is its price. Most high-quality applications, like games are made for PC because of their computing power and ways of interaction in virtual space with available peripherals. However, the limited power of the phone, translates into fairly low quality of experience offered in this medium. The VR image transmission from the PC to the phone is crucial for virtual reality to achieve satisfactory levels of immersion. The solution proposed in this paper will allow to use the phone for virtual reality games without unnecessarily burdening computing power to both computer and mobile phone.

The next problem in the mobile VR helmet simulation is the excessive delay of the image in relation to the head movements, which results in motion sickness and prevents the comfortable gameplay. In order to overcome this effect, the strong focus have to be put on optimization of the VR image transmission algorithms from the computer to the phone to reduce image delay. Motion sickness can be seen for most people, when delay is bigger than 50 ms [3].

W. Zamojski et al. (Eds.): DepCoS-RELCOMEX 2021, AISC 1389, pp. 203–212, 2021.
https://doi.org/10.1007/978-3-030-76773-0_20

At the moment there are not many ready-to-use solutions for sending image from PC to mobile device by wireless network focusing on VR and considering low CPU usage, low latency head movement tracking and high image quality. In the literature one can find mainly descriptions of concepts or partial solutions for mentioned purpose. The linear model for the intra-refresh cycle-size selection adapting to the network packet loss rates and the motions in the video content was presented in [1]. High performance video encoding methods for Nvidia GPUs can be found in [5] and [4]. General concepts and descriptions of coders and decoders for fast video encoding is in [10], [6] and [11]. Latency analysis for H.264 coding standard is described in [9] and [8]. IP networks usage as a basis for sending encoded image can be found in [12] and [13].

It should be noted, that Trinus, as one of the competing companies, uses MJPEG encoding. MJPEG, which stands for Motion JPEG, is a streaming technique which encodes each individual frame as an independent JPEG-encoded image. Upside of MJPEG, is its widespread support, because JPEG has been in use as picture compression algorithm for several decades. Not all Android mobile devices support hardware accelerated H.264 codec, but such devices are usually very outdated and are not adequate for real time VR streaming. Another upside of MJPEG, is its resilience to transmission loss events. With every picture being decodable independently, there is no extended picture quality degradation caused by reference frame being corrupted, therefore leading to following frame using missing frame as a reference.

Another solution is a x264 developed by VideoLAN. It has an upside of being hardware-independent, which translates into wider compatibility. Unfortunately such algorithms are computationally very costly. Running x264 on the CPU usually required one full core being dedicated to the encoding process to sustain 60 FPS which is the minimum acceptable baseline to keep rendering responsive in VR.

The paper is organized as follows. In Sect. 2 some information about the basic definitions related to network and rendering quality are presented. The Sect. 3 focuses on description of implemented solution and optimizations made in test application, while Sect. 4 is devoted to presentation of the experimental results.

2 Basic Definitions

For the purpose of measuring obtained results related to latency of movement tracking and image change on mobile device and CPU load some definitions have to be introduced. Selected formulas are presented below. An important aspect of image transmission technology is the smallest possible delay between the head movements and the corresponding image change visible on mobile device. Too big delay can cause nausea and motion sickness. Despite of delay measured in milliseconds, two additional values are used here: **R**ender **F**rame **D**elivery **R**ate (**RFDR**) and **N**etwork **F**rame **D**elivery **R**ate (**NFDR**). RFDR presents loss on decoding compressed frames, while NFDR describes loss on network transport layer.

RFDR is calculated as:

$$RFDR = \frac{N_{MR}}{N_s} \cdot 100\% \tag{1}$$

where N_{MR} is number of frames in mobile device's display buffer, that where received and N_s is number of all frames rendered on the PC side.

NFDR is calculated as:

$$NFDR = \frac{N_{MS}}{N_s} \cdot 100\% \tag{2}$$

where N_{MS} is number of frames received on the mobile device network socket side and defragmented from partial datagrams, while N_s is number of all frames rendered on the PC side.

CPU usage on PC was measured using Microsoft Visual Studio profiler and is presented as an percentage of used CPU resources taking into account all cores involved in multitasking processes related to ran application.

3 Solution Description

Development of test application was divided into two tasks. First technological challenge was to identify bottleneck locations within the code for CPU processing, while second challenge was to reduce the time taken to transfer video frames from PC to mobile device over the standard wireless network.

Test application was created in .NET technology using C# language and Microsoft Visual Studio 2017 on the PC side. Mobile device running Android operating system part was created in Java using Android Studio 3 IDE. Technologies that were used for image processing based extensively on low-level APIs for Direct3D 11, H.264 and H.265 [2,5]. Figure 1 presents architecture of implemented test application.

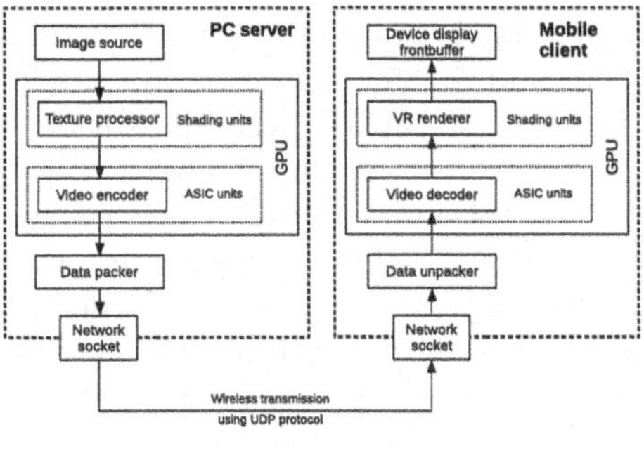

Fig. 1. Architecture of solution implemented in test application.

System consists of two parts - server running on PC and client running on mobile device. Main functional parts of PC server are:

- *Image source* – graphic rendering part of test application,
- *Texture processor* – responsible for transforming RGB pixels into NV12 format,
- *Video encoder* – transforms NV12 frames into H.264 bitstream,
- *Data packer* – responsible for splitting bitstream into MTU-sized datagrams.

Texture processor and *Video encoder* are implemented on GPU using shading units and **A**pplication-**S**pecific **I**ntegrated **C**ircuit (**ASIC**) encoding units (NVENC for Nvidia and VCE for AMD). *Data packer* is running on CPU.

Main functional parts of mobile client are:

- *Data unpacker* – responsible for defragmentation of individual datagrams into H.264 bitstream.
- *Video decoder* – transforms H.264 bitstream into RGB textures,
- *VR renderer* – transforms textures into stereoscopic view distored to match target device screen physical dimensions and lens parameters,
- *Device display frontbuffer* – responsible for storing data to be directly displayed.

Video decoder and *VR renderer* are implemented on GPU using shading units and ASIC decoding units, while *Data unpacker* is running on CPU.

Datagrams are sent from PC server to mobile client over wireless network using UDP protocol. Most of the optimization areas were related to encoding and decoding.

3.1 Architecture Optimizations

After deep analysis of first simple prototype of application, some important needs for optimization were identified for reducing CPU usage. Essential part was to implement algorithm that reduced CPU usage and speeded up the encoding (including frame encoding) process. Main idea was to use efficient multithreaded programming and explore graphic cards vendor specific optimizations. Additionally, native APIs available for H.264 modules were used. There are several key differences between H.264 and H.265 implementation provided by GPU hardware vendors such as Nvidia, Intel and AMD, who all provide software development kits to utilize their hardware as a significant improvement to encoding performance. Reducing the time taken to transfer video frames from PC to mobile device was related to optimization of the implementation. Author explored multithreaded approaches, but improvements were only marginal on the CPU side, where parallelism was limited by a relatively low amount of CPU cores that are available in consumer grade CPUs. ASICs available on gaming-grade GPUs were found to be considerably better than most CPU approaches in terms of overall frame throughput [5]. Device latency was variable across different hardware vendors but there was an overall improvement over CPU-based encoding, especially considering that CPU-based algorithms may experience suboptimal and uneven wait times caused by high load originating from the games, depending on scene complexity. ASIC-generated bitstream was generally compatible with mobile device decoding units but some non-critical behaviours that didn't conform to the MPEG-issued standard were observed. This led to creation of workarounds to limit number of graphical artefacts.

Author also changed the network architecture of image transmission. UDP protocol was used to maximally shorten communication control overhead between devices. The "Send and forget" technique allowed for a significant reduction of delays, but required compensation in terms of frame delivery [12]. Certain H.264 implementations (such as NVENC) implement reference frame invalidation. This technique is not a H.264 standard, but is very helpful in terms of low latency and lossy transmission. With this technique, *Mobile client* (acting as receiver and decoder) can notify *PC server* (acting as encoder and sender) that certain frames were lost during the transmission process, especially wireless one. Encoder can then attempt to use other stored frames as references, instead of trying to delta-encode in reference to frames, which do not exist on the client side. This would result in image corruption, which would be increasing process in next frames, until full refresh is received.

3.2 Methods and Data Structures Optimizations

Many of the algorithms used in test application allowed the use of better optimized alternatives without side effects. An example of such an optimization is matching dedicated H.264 compression parameters to the appropriate graphics cards instead of using one universal and generic solution. This enabled significant optimization of CPU usage and delays depending on the equipment used. One specific example of such optimization having big impact on processing speed, was changing intermediate texture processing to match expected pixel format and resolution alignment to match input format of given H.264 implementation. Some encoders can work directly on RGB texture, while other require conversion to NV12 (biplanar YUV pixel format). In cases, where encoder can't work directly on RGB (e.g. in software-implemented encoders), a partial hardware acceleration was created by using GPU as highly parallelized RGB to NV12 converter. Such transformation can be implemented in GPU as two-pass render, where planes are calculated. This general solution was used by the author to increase processing speed for the encoders with no direct RGB texture support. This is the main part of *Texture processor* which is responsible for pixel format conversion and feeded by *Image source* on the *PC server* side. Presented transformation uses GPU's vertex and fragment shader capabilities to achieve three tasks at the same time:

- color space is converted from RGBA to YUV in fragment shader,
- memory layout is changed from RGBA component-packed to separated into luminance (Y) and chrominance (UV) planes in 2-pass render, making the layout more beneficial to compression algorithms,
- chrominance channels are subsampled with linear filtering which reduces total frame byte size.

Pseudocode describing mentioned transformation is presented below.

INPUT: Frame $F_{RGBA} = [[R_0, G_0, B_0, A_0], [R_1, G_1, B_1, A_1], ..., [R_n, G_n, B_n, A_n]]$
OUTPUT: Frame $F_{YUV} = [[Y_0, Y_1, ..., Y_n], [U_0, V_0, U_1, V_1, ..., U_{\frac{n}{2}}, V_{\frac{n}{2}}]]$
1: *PlaneY* $\leftarrow \emptyset$ using *R8_UNorm* pixel format
2: *PlaneUV* $\leftarrow \emptyset$ using *R8G8_UNorm* pixel format
3: Configure *Triangle List* as primitive topology
4: Set GPU vertex shader to *VSFullScreen*
5: Set GPU fragment shader to *PSCalculateY*
6: Set *PlaneY* as output merger's render format
7: Configure rasterizer viewport dimensions to n
8: **for** each pixel in F_{RGBA} **do**
9:     $PlaneY[i] \leftarrow PSCalculateY(F_{RGBA}[i])$
10: **end for**
11: Set GPU fragment shader to *PSCalculateUV*
12: Set *PlaneUV* as output merger's render format
13: Configure rasterizer viewport dimensions to $\frac{n}{4}$
14: **for** each pixel in F_{RGBA} **do**
15:     $PlaneUV[i] \leftarrow PSCalculateUV(F_{RGBA}[i])$
16: **end for**
17: $F_{YUV} \leftarrow PlaneY \bigcup PlaneUV$

Input to the transformation is texture represented as F_{RGBA} frame built of n pixels in RGBA color space laid out sequentially in packed format. Output is the F_{YUV} frame representing n pixels in YUV color space laid out in biplanar format with chroma channels subsampled both vertically and horizontally (as in NV12). At the beginning, in lines 1–2, *PlaneY* and *PlaneUV* are initialized as 2D textures in *R8_UNorm* and *R8G8_UNorm* data format respectively. *R8_UNorm* describes one normalized 8-bit float value, while *R8G8_UNorm* describes two normalized 8-bit float values. Line 3 makes *Triangle List* as primitive topology using *ID3D11DeviceContext:: IASetPrimitiveTopology* call. Lines 4 and 5 bind GPU shaders, respectively vertex and fragment, to use defined function calls, which are presented below in source code part. Line 6 sets *PlaneY* as output merger's render format. Rasterizer viewport dimension is set to number of pixels of F_{RGBA} frame encoded in RGBA color space in line 7. Loop in lines 8–10 iterates over all pixels in input frame via execution of the shader program *PSCalculateY* responsible for fragment calculation, which is presented in details in source code part below. Results are stored in *PlaneY* data structure. Line 12 sets *PlaneUV* as output merger's render format. Rasterizer viewport dimension is set to quarter of number of pixels of F_{RGBA} frame in line 13. Loop in lines 14–16 iterates over all pixels in input frame via execution of the shader program *PSCalculateUV* responsible for fragment calculation, which is presented in details in source code part below. Interleaved *UV* values are stored in *PlaneUV* data structure. Finally, in line 17, calculated YUV values are stored in F_{YUV} data structure representing n pixels in YUV color space.

Delays were also reduced by changing the frame transport method from reliable and ordered to ordered but not reliable for the channel that carries video-encoded signal from PC to the mobile device [12]. It was decided to change the technique only to ensure the delivery of key frames (intra-coded pictures, I-frames) instead of all, which allowed to significantly reduce delays. Predicted frames are delivered in unreliable manner and losing them due to packet loss no longer blocks the stream, resulting in more smooth stream at the cost of minor quality drop when used in busy wireless networks with heavily used RF channel.

4 Experimental Results

Presented results were obtained using a PC equipped with an 8 GB RAM, 4-core Intel Core i5-6600k processor and Nvidia GeForce GTX 970 graphics card running Windows 10 operating system. Mobile device used in tests was Google Pixel XL with pure Android 8.0 operating system. Wireless network connection between PC and phone was established using TP-Link WDR3600 router configured to provide WiFi 802.11ac 5 GHz network. Image output settings were 1920×1080px with bitrate of 15 Mbit/s at 24bpp render color depth, limited to YUV 4:2:0 (chroma channels downsampled by 50%) during video encoding.

For the CPU usage, test application was rendering simple 3D scenery with stationary camera constantly rotating 360 degrees in horizontal axis. For the delay measurement, test application was showing rendered precise timers. Both output images were barrelled for purpose of proper display by mobile device placed in Google Cardboard.

4.1 CPU Resources Usage

CPU computational usage was verified using Microsoft Visual Studio profiler and is presented as percentage of used resources in comparison to all available resources. Tool was configured to track only processes related to test application and shows average usage of all CPU cores that were involved in calculations. All unnecessary applications on the PC were closed. Length of each test was about 120 s.

CPU usage results for early prototype of test application without any optimizations described in Sect. 3 are shown on Fig. 2.

It can be seen, that peak CPU usage is about 37%, while average is around 26%. Most of the processing time is consumed for preparing image stream to be sent to the mobile device. Therefore, with more sophisticated graphical applications, like computer games, decreased performance for end user is visible. This problem is related to background process for preprocessing and sending image to the phone.

CPU usage results for optimized test application are shown on Fig. 3.

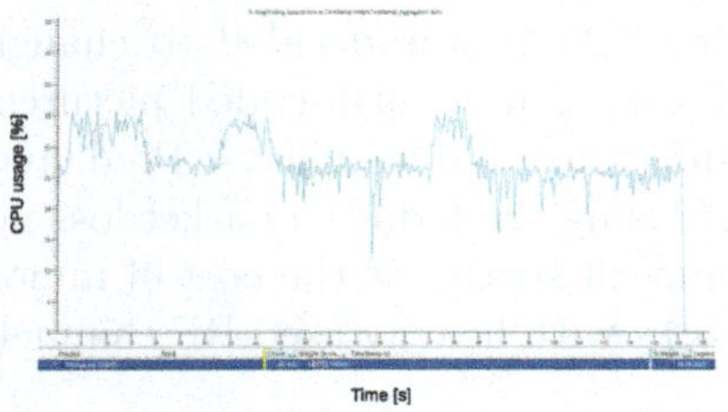

Fig. 2. CPU usage for test application before any optimizations

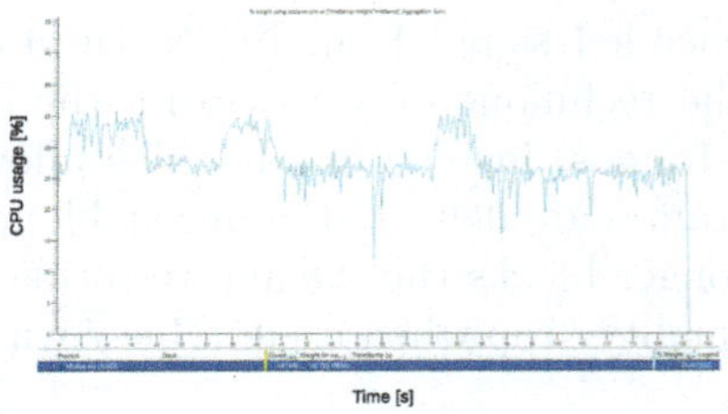

Fig. 3. CPU usage for test application after the optimizations

It can be seen, that peak CPU usage is about 8%, while average is around 5%. Introduced optimizations had huge impact on lowering CPU usage for background process related to sending image to phone. Using this approach, no performance loss is visible for end user, even when using high CPU demanding applications, especially newest titles of computer games.

4.2 Image and Tracking Delay

By using precise timing functions on both PC and mobile device, the average time for processes related to image transfer and head movements tracking with the greatest latencies were measured. On the PC side, time was measured starting from frame preprocessing moment to the moment of sending it. On the mobile device side time measurement was started at the moment of receiving the frame and finished at the moment of displaying it. The sensor measurement is based on the collection of time stamps at the moment of receiving sensory data from the device and comparing it with the marker after sending, rendering, encoding, return transport and decoding. Generally speaking, measured delay is from the moment the movement begins to the reception of the newly generated frame on the mobile device. The measured times were averaged to obtain a reliable result. This does not take into account the internal sensor delay, which is 1 ms for IMU in Google Pixel XL sampled at 1 kHz rate. This delay, however, does not significantly change the final result. Presented results were collected from 10 measurements and indicate average values for each measurement. Wireless network conditions were standard, what means, that no special actions were taken to isolate router from other RF sources, like different wireless networks.

Table 1 presents results acquired for test application running delay measurement mode before optimizations and after optimizations described in Sect. 3.

According to the results, a big improvement after the optimizations can be noticed. With no optimizations, average times were around 50 to 70 ms, what could cause motion sickness for many end users of such VR helmet simulation software. After the optimizations, average delay time is almost 2 times lower, what is 2 full frame cycles (each of them 16.6 ms) with a pessimistic case of improvement by only one full frame cycle.

Table 1. Results for delay measurement.

Test number	1	2	3	4	5	6	7	8	9	10
No optimizations										
Delay [ms]	52.12	58.54	65.33	69.34	59.34	56.22	64.47	59.98	61.19	58.36
All optimizations										
Delay [ms]	29.31	32.31	33.41	31.62	34.22	36.78	33.63	28.60	37.12	30.42

An important aspect was also maintaining acceptable level of image frame deliverability, which could have been problematic with the use of "Send and forget" approach, like in UDP protocol. By using the technique, where only intra-coded pictures are considered essential and therefore scheduled for retransmission in case of packet loss, RFDR and NFDR remained at good level while providing acceptable delays and high image quality. The measurements were performed based on a series of 3-minutes tests. Results show the efficiency of frames delivery sent by a PC to a mobile device (loss on network transport, NFDR) and efficiency of rendering (loss on decoding compressed frames, RFDR). Wireless network conditions were standard, what means, that no special actions were taken to isolate router from other RF sources, like different wireless networks. Achieved results for the optimized test application are presented in Table 2. There are no results for early version of test application, because previously used TCP protocol has internal algorithms for data retransmission, so NFDR parameter was 100%, but it had huge impact on delay, especially when some frames were lost during transmission.

Table 2. NFDR and RFDR results for test application after optimizations.

Test number	1	2	3
Sent frames [–]	10 978	11 549	10 747
Received frames [–]	10 919	11 504	10 682
NFDR [%]	99.46	99.61	99.40
Rendered frames [–]	10 733	11 308	10 489
RFDR [%]	97.78	97.91	97.60

5 Conclusions

Performed research shows, that creating efficient and fast method of sending VR-specific image from gaming PC to the mobile device is possible by using standard wireless networks. It opens the way to create fully functional solution, that gives possibility to properly simulate professional VR helmets, like HTC Vive or Oculus Rift, and maintaining at the same time low delay between head movement and corresponding image change on mobile device side increasing comfort of usage by end user, as well as low CPU usage on PC for supporting image encoding and transmission to mobile device.

Further research will focus on developing methods, that counteract shortcomings of network transmission, like forward error correction codes such as Reed-Solomon coding to improve NFDR further. It will act as a replacement to TCP-based stream quality control. Other optimization will include exploring possibility of removing bottlenecks by using customized assembly language inserts, what can help achieve significant and visible speed-up comparing to the code generated by the compiler.

Acknowledgments. The work was supported by the grant W/WI/2/2020, WZ/WI-IIT/2/2020 from Bialystok University of Technology and funded with resources for research by the Ministry of Science and Higher Education in Poland. Research results are based on the project "Development of new, innovative tools and interaction mechanisms in VRidge technology" financed by National Center of Research and Development.

References

1. Chen, H., Zhao, Ch., Sun, M.T., Drake, A.: Adaptive intra-refresh for low-delay error-resilient video coding. J. Vis. Commun. Image Represent. **31**, 293–304 (2019)
2. OpenH264 homepage. http://www.openh264.org. Accessed 02 Apr 2020
3. Hell, S., Argyriou, V.: Machine learning architectures to predict motion sickness using a virtual reality rollercoaster simulation tool. In: 2018 IEEE International Conference on Artificial Intelligence and Virtual Reality (AIVR), Taichung, Taiwan, pp. 153–156 (2018)
4. Keith, J.: Video Demystified: A Handbook for the Digital Engineer, 5th edn. (2007)
5. Patait, A., Young, E.: High performance video encoding with nVidia GPUs, GPU Technology Conference, nVidia, pp. 1–34. San Jose, CA (2016)
6. Ribas-Cobera, J., Chou, P.A., Regunathan, S.L.: A generalized hypothetical reference decoder for H.264/AVC. IEEE Trans. Circuits Syst. Video Technol. **13**(7), 674–687 (2003)
7. VRidge application homepage. https://riftcat.com/vridge. Accessed 02 Apr 2020
8. Schreier, R., Rothermel, A.: Motion adaptive intra refresh for the H.264 video coding standard, IEEE Trans. Consum. Electron. **52**(1), 249–253 (2006)
9. Schreier, R.M., Rothermel, A.: A latency analysis on H.264 video transmission systems. In: 2008 Digest of Technical Papers - International Conference on Consumer Electronics, pp. 1–2. IEEE, Las Vegas (2008)
10. Sullivan, G.J., Wiegand, T.: Video compression - from concepts to the H.264/AVC sandard. Proc. IEEE **93**(1), 18–31 (2005)
11. Sullivan, G.J., Ohm, J., Han, W., Wiegand, T.: Overview of the High Efficiency Video Coding (HEVC) standard. IEEE Trans. Circuits Syst. Video Technol. **22**(12), 1649–1668 (2012)
12. Wenger, S.: H.264/AVC over IP. IEEE Trans. Circuits Syst. Video Technol. **13**(7), 645–656 (2003)
13. Zheng, H., Boyce, J.: An improved UDP protocol for video transmission over Internet-to-wireless networks. IEEE Trans. Multimed **3**, 356–365 (2001)

Experimental Comparison of ML/DL Approaches for Cyberattacks Diagnostics

Aleksandr Krivchenkov, Boriss Misnevs, and Alexander Grakovski(✉)

Transport and Telecommunication Institute, Lomonosova Street 1, Riga, Latvia
{aak,bfm,avg}@tsi.lv

Abstract. The main goal of this article is experimental research on machine learning methodology for cyberattacks diagnosing. This study based on two publicly available datasets UNSW-NB15 and NSL-KDD. Its scope included the features reduction problem and calculations of the classification efficiency. We applied Machine Learning (ML) and Deep Learning (DL) methods to classify traffic. The methods of supervised *k*-nearest neighbours (*k*-NN) and artificial neural networks (ANN) were used. Accuracy, Precision, True Positive Rate (TPR), False Positive Rate (FPR) were calculated based on a series of numerical experiments for all types of attacks and for DoS (Deny of Service) attacks only. The features number reduction for observation is achieved, and the effect of this reduction on classification accuracy was investigated. We made some conclusions about the possibility of implementing ML/DL methods in intrusion detection systems (IDS).

Keywords: Network security · Denial of Service (DoS) · Intrusion Detection System (IDS) · Machine Learning (ML) · Deep Learning (DL)

1 Introduction

The European Union Network and Information Security Agency (ENISA) published a report in 2019, highlighting 15 different types of attacks (implemented threats) [1]. According to the report, most network attacks aimed at hitting their targets using packet streams. These attacks are known as Denial of Service (DoS) or Distributed DoD (DDoS) attacks.

In DDoS, as sources of such traffic can be many computers previously infected and controlled by an attacker. In this case, traffic comes from different IP addresses, including the Internet. The presence of many traffic sources makes it much more difficult for intrusion detection systems (IDS) to detect and for intrusion prevention systems (IPS) to filter the attacks. Existing IDS/IPS use many methods to detect packet streams for specific protocols (UDP, TCP, ICMP) [2].

The IDS can only detect an attack by analysing a stored significant dataset, and IPS allows to stop the attack using high-speed traffic analysis. Most IDS/IPS systems use several attack detection methodologies [3]. Machine learning (ML) and Deep Learning (DL) techniques are also used for attacks classification or clustering to provide broader and more accurate detection.

W. Zamojski et al. (Eds.): DepCoS-RELCOMEX 2021, AISC 1389, pp. 213–223, 2021.
https://doi.org/10.1007/978-3-030-76773-0_21

The effectiveness of IDS/IPS is based on efficient network event chain classifiers. This article examines classifiers' performance using supervised ML/DL learning methods and compares the classification performance for different types of attacks and different numbers of features in observations. We also offer recommendations for using ML/DL to detect attacks with a minimized number of features.

2 Testing of ML/DL Methods

The schema of ML/DL taxonomy methods for real-time systems securing is described in [4]. ML and DL can efficiently extract hidden information from big data with minimal human intervention, and many ML/DL techniques are used to detect attacks [5]. The platform for detecting attacks testing has the following structure (Fig. 1).

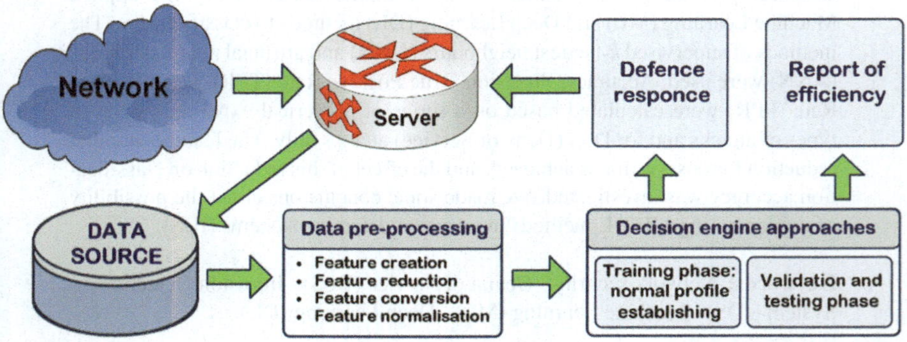

Fig. 1. The platform for ML/DL methods testing, signalling, and possible self-learning of IDS

In this research, we have used the *k*-Nearest Neighbours (*k*-NN) method for classifying the events for learning and testing. The *k*-NN is a simple, easy-to-implement supervised ML algorithm that can be applied to solve classification problems. We have also implemented supervised DL. The Artifice Neural Network (ANN) model with two hidden layers is described in Sect. 6.

3 The Efficiency of ML/DL Methods for Attack Detection

For ML/DL methods of a supervised type, the classic way of assessing the classification efficiency is based on the confusion matrix M_C:

$$M_C = \begin{bmatrix} TP & FN \\ FP & TN \end{bmatrix} \quad (1)$$

It consists of the number of correctly and incorrectly classified attacks. For the assessment of detection efficiency, the next parameters are used:

- **Accuracy** is a metric that estimates the overall percentages of detected attacks (*TP* - true positive, or attack presence detection) and normal events (*TN* – true negative, or attack absence detection). Accuracy reflects the proportion of success and is calculated as: $Accuracy = (TN + TP)/(TP + FP + TN + FN)$, where *FP* (false positive) is attack absence true detection, and *FN* (false negative) is attack absence false detection respectively;
- **Precision**: $Precision = TP/(TP + FP)$.
- **Recall,** also called the **True Positive Rate** *(TPR)* or sensitivity, is the proportion of correctly classified malicious instances of the total number ones and is computed as $TPR = TP/(FN + TP)$.
- **False Positive Rate** *(FPR)* is the proportion of normal instances of the total number of normals misclassified as attacks, and it is computed as $FPR = FP/(FP + TN)$.

ML/DL methods often are evaluated using the TPR-FPR plane. An ideal IDS system should have 100% TPR while 0% of FPR; that reflects that all the attacks detected without any misclassification. However, this is very difficult to demonstrate in a real environment.

The detection quality also can be characterised visually by the Receiver Operating Characteristic (ROC-curve), where the results of experiments are presented (Fig. 3).

Another essential characteristic of ML/DL methods is their applicability in "streaming" mode (this is very important for IPS). This fact is mentioned in many works, and there are some quantitative data on this matter. For example, the result [6, 7] describes some characteristics of complexity for ML/DL algorithms and makes it possible to choose a suitable method.

4 Testing Datasets

There are publicly available datasets that widely used to test IDS methods. Two of them are considered in this study, namely UNSW-NB15 [8, 9], which includes many new

Table 1. Datasets used for ML/DL methods testing

Name of data set	Attacks categories	Number of records training/testing	Number of features	Comment/Available
UNSW-NB15	9	82,332 /175,341	42	46 variants of DoS attacks types https://cloudstor.aarnet.edu.au/plus/index.php/s/2DhnLGDdEECo4ys?path=%2F
NSL-KDD	4	125,973 /22,544	42	11 DoS attack types https://www.unb.ca/cic/datasets/nsl.html

types of DoS attacks, and NSL-KDD, which is very popular for ML/DL testing. Some of the critical characteristics of datasets are shown in Table 1.

As part of the study, an in-depth analysis of the UNSW-NB15 and NSL-KDD datasets was carried out to study their structure and suitability for classifying all types of attacks, and DoS attacks only using the *k*-NN and ANN methods.

Before using datasets for classification, all the features must be converted to numbers or binary values. Therefore, the nominal values are no longer used for the features. After that, all the features values should be normalised. The NSL-KDD and UNSW-NB15 datasets after correction and normalisation were presented in [7].

5 Dataset Features Number Reduction

Big number of features (42 for UNSW-NB15 and NSL-KDD) complicates the IDS task and significantly increases processing time. Therefore, one of the frequently solved related problems is the problem of reducing the number of features. Various researchers have used different approaches of information gain (IG) and principal component analysis (PCA), which extracts the most salient features and removes irrelevant features from the original datasets [6]. After that, a support vector classifier (SVM) is usually used [8]. Hybrid evolutionary optimisation algorithms based on the black-smith method (GOA, GOSA) and simulated annealing (SA) are also manifested in some publications. There are several examples of using a two-phase hybrid method based on SVM for this purpose, together with a recursive neural network (RNN) [9] and many others. All of the above approaches can reduce the computation time with some insignificant loss of Accuracy.

The current work also presents an attempt to reduce the number of features in datasets based on the cross-correlation of features and to assess this actions's effect on the classification. We assume that the original features in UNSW-NB15 and NSL-KD contain repetitive and useless data for attacks diagnostics (in particular, for DoS attacks). Following the principles of "fuzzy" logic, we can divide the main alternatives of state of relation between the features ("independent" and "correlated") between every two features in four milder states: "independent", "weakly dependent", "weakly correlated", and "correlated". In the simplest case, uniform selection of thresholds ($\{t_{low}, t_{up}\}$) may be done by dividing the intervals from 0 to 1. For example, $t_{low} = 0, 25$, and $t_{up} = 0, 75$ respectively. Ideally, statistical analysis of big data is required to justify thresholds' determination, but in our case we will use this intuitive rule instead.

Thus, we use one of the possible variants of the statistical approach based on comparing the correlation of features with each other (cross-correlation matrix) with the following considerations::

- it is accepted that the mutual influence of two features is insignificant if the magnitude of their cross-correlation coefficient is less than t_{low};
- it is assumed that the features essentially depend on each other if the magnitude of their cross-correlation coefficient is greater than t_{up}.

Then, some features of the original dataset can be excluded, first based on their insignificant correlation with the attack event label ($C_{ij} < t_{low}$), and then exclude another

Table 2. Reduced correlation matrix for dataset NSL-KDD and DoS attacks (6 main features)

	logged_in	*count*	*serror_rate*	*same_srv_rate*	*dst_host_count*	*dst_host_ rerror_rate*	*DoS_vec*
logged_in	1,00						
count	−0,48	1,00					
serror_rate	−0,29	0,13	1,00				
same_srv_rate	0,53	−0,56	−0,29	1,00			
dst_host_count	−0,32	0,35	0,15	−0,35	1,00		
dst_host_rerror_rate	−0,39	0,41	−0,11	−0,67	0,30	1,00	
DoS_vec	−0,45	0,48	0,32	−0,69	0,34	0,48	1,00

part of remaining based on their strong dependence on each other ($C_{ij} > t_{up}$). This approach leaves a reduced set of 7 functions out of 42 original functions for the UNSW-NB15 [10].

For another dataset (NSL-KDD), some correlation coefficients of the functions with the DoS attack event label (*DoS_vec*) also fall below the 0.25 threshold. That is, some features will fall into the area $C_{ij} < t_{low}$ and will be excluded. After analyzing the data, only 6 functions remained, which are presented in Table 2.

Features remained after reduction:

- ***logged_in*** (F12: 1 if successfully logged in, 0 otherwise);
- ***count*** (F23: number of connections to the same host as the current connection in the past two seconds);
- ***serror_rate*** (F25: percent of connections that have SYN errors);
- ***same_srv_rate*** (F29: percent of connections to the same services);
- ***dst_host_count*** (F32: *count* for destination host);
- ***dst_host_rerror_rate*** (F38: *serror-rate* for destination host).

An indirect confirmation of the chosen rule is that all features remaining after reduction procedure indicate an increase in the number of connections (typical for attacks).

6 Artificial Neural Network (ANN) for Attack Classification

We took the normalised NSL-KDD dataset for ANN model construction. The dataset was in two versions: a full set of features (42) and a reduced set of features (6). The ANN model's output was a binary parameter assumed to be zero (0) if a DoS attack was not detected and one (1) if it took place.

The *Python* [12] language was used for the model implementation with the following libraries:

- ***pandas*** is an open-source, fast, powerful, flexible and easy-to-use data analysis and manipulation tool built on top of the Python programming language.
- ***numpy*** is a *Python* library used for working with arrays;
- ***sklearn*** is the most useful and reliable library for machine learning in *Python*;
- ***keras*** is a *Python* Deep Learning library;
- ***matplotlib*** is a comprehensive library for creating static, animated, and interactive visualisations in *Python*.

For the neural network, a dense neural network was chosen, where each neuron is connected to the other, and the output layer contains a single prediction neuron. The generalized structure of the ANN used in the study is shown in Fig. 2.

Two hidden layers were used, in each of which 16 neurons were defined, and a rectified linear activation function (*'relu'*) was used to obtain the best result. The usual sigmoid activation function was defined in the output layer to get a probability range [0, 1].

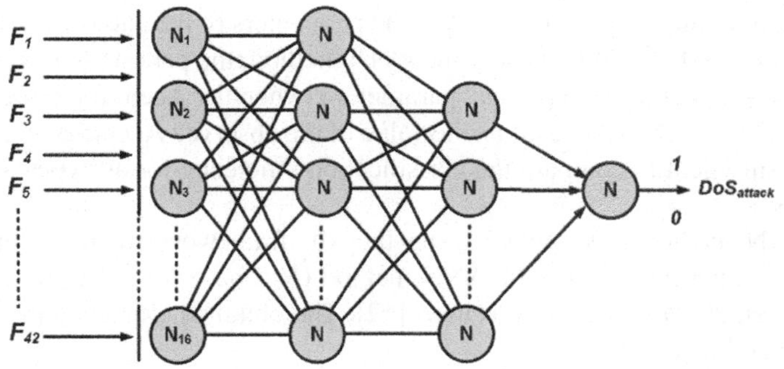

Fig. 2. Artificial Neural Network (ANN) structure for attack events classification

The following options used when compiling the ANN model:

- Method of stochastic optimization *RMSprop* [12] or root-mean-square propagation;
- The binary function of cross-entropy losses;
- Metric Accuracy for classification tasks.

7 Experimental Results for ML/DL Classification

The next series of experiments was carried out in the study: with UNSW-NB15 using the *k*-NN method (5 experiments), with NSL-KDD using *k*-NN (5) and ANN (2). Each experiment has its own unique 'Experiment code' (*U_all, U_DoS* etc.). The list and classification parameters of the experiments are presented in Table 3.

Table 3. List and parameters for classification experiments using the *k*-NN method

Experiment code	Data set	Number of features	Number of attack categories	Classification
U_all	UNSW-NB15	42	9	All attacks
U_DoS				DoS attacks
U_cat				Category attacks
U_7_all		7		All attacks
U_7_DoS				DoS attacks
N_all	NSL-KDD	42	4	All attacks
N_DoS				DoS attacks
N_DoS_ANN				DoS attacks
N_cat				Category attacks
N_6_all		6		All attacks
N_6_DoS				DoS attacks
N_6_DoS_ANN				DoS attacks

According to the "Experiment Code", the parameters of the classification efficiency are presented in Table 4. When using the *k*-NN method, the parameter *k* was set to 5.

We did not look for the "optimal" parameter k when the Accuracy value reaches its maximum. This is due to the fact that the value of the observed Accuracy in experiments at $k = 5$ is sufficient to compare the classification efficiency for all types and only for DoS attacks.

The *k*-NN method was applied using the ML framework for the *R* environment tool. For the experiments with the ANN method (*N_DoS_ANN, N_6_DoS_ANN),* the DL framework (*Python, Keras*) was used [12]. The obtained efficiency parameters are presented in Table 4.

Table 4. Classification efficiency for performed experiments

Experiment code	Classification characteristics			
	Accuracy	Precision	TPR	FPR
U_all	0,87	0,98	0,83	0,03
U_DoS	0,92	0,93	0,99	0,95
U_cat	0,72			
U_7_all	0,85	0,95	0,81	0,08
U_7_DoS	0,92	0,93	0,99	0,97
N_all	0,76	0,64	0,98	0,4
N_DoS	0,96	0,98	0,97	0,06
N_DoS_ANN	0,9598	0,9799	0,9658	0,0578
N_cat	0,81			
N_6_all	0,71	0,61	0,89	0,43
N_6_DoS	0,96	0,97	0,97	0,08
N_6_DoS_ANN	0,9293	0,9471	0,9587	0,1566

The experiments results on the *TPR* – *FPR* plane, or Receiver Operating Characteristics (ROC) curve, are presented in Fig. 3. The result of every experiment is shown as the plane's point in correspondence with its 'Experiment code' in Table 4.

DNN (Deep Neural Network) and DT (Decision Tree) curves for ML/DL methods, borrowed from [11], are added here for further discussion and comparison.

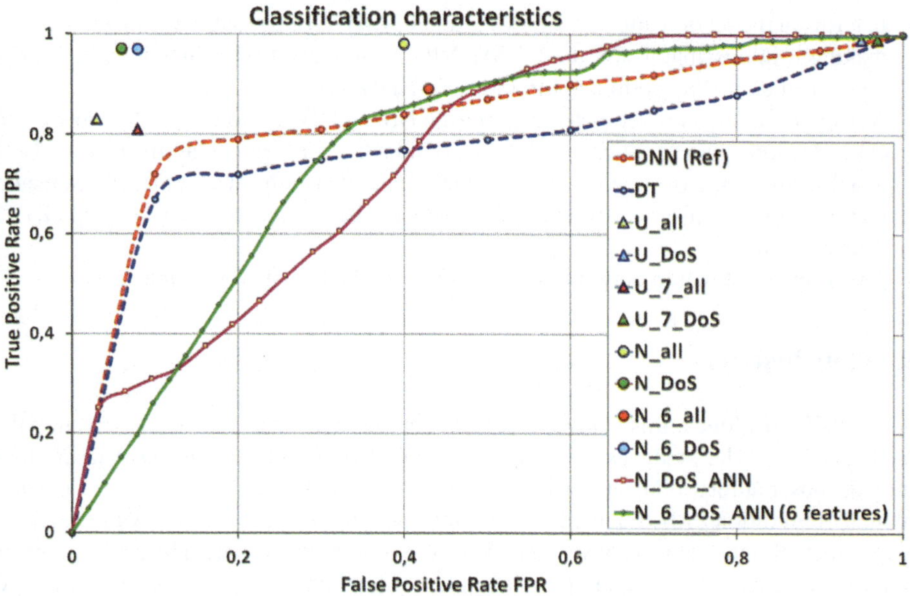

Fig. 3. Experimental Receiver Operating Characteristics (ROC) curve

8 Discussion and Results

The following statements were made after analyzing the results for previous experiments with the UNSW_NB15 dataset and extended experiments using ANN with NSL-KDD datasets:

- The use of these datasets for various ML/DL methods confirmed that the classification Accuracy was sufficient (in the 0.85–0.92 range). The Accuracy values demonstrate the ML/DL applicability in hybrid IDS. More research is required to recognise the relevance of ML/DL techniques in IPS.
- ML/DL techniques for attacks detecting demonstrate "sensitivity" for selecting training and testing data subsets, the set of features used in the observation, the type of features variables. In this regard, it becomes more problematic to compare different ML/DL methods. According to the authors of [10], the "simplest" methods seems more reasonable.
- The ANN experiments have also confirmed that the decrease in the number of features slightly reduces the classification quality. In some cases, this loss is small (1–2%), but the classification time is significantly reduced.
- Machine learning techniques for detecting attacks show an increased FPR value for all attack types in the dataset. However, for DoS-only attacks, there is a decrease in the number of false positives. In this case, the experimental results give TPR and FPR values of the order of 97% and 6%, respectively (almost ideal result). So, we again observe the sensitivity of ML/DL methods to the used dataset.

- In terms of the set of features, NSL-KDD and UNSW-NB15 practically do not overlap, making it impossible to use the UNSW-NB15 training set (presumably with a large number of attack types) to classify the NSL-KDD data.
- A decrease in the number of features for the NSL-KDD set from 42 to 6 worsens the classification characteristics for all types of attacks, but for the classification of DoS attacks, this decrease is only 1–2%. That means that only these correctly selected features contain information about DoS attacks. But this is true for the NSL-KDD dataset only.
- The problem of neural network overtraining for ANN exists for both datasets.

9 Conclusions

As already mentioned, this study is a direct continuation of the authors' research, published in [10]. The study uses the same UNSW-NB15 and NSL-KDD datasets that include observations bild on network traffic events analysis. Based on these experimental observations, ML/DL methods' effectiveness for detecting attacks was analysed. Experimental datasets were investigated for the possibility of reducing the number of features to reduce the computational complexity for IDS. Also, to classify traffic, we implemented supervised deep learning via ANN. The Accuracy and other characteristics of the classification were calculated after a series of numerical experiments.

Key conclusions about usage of ML/DL techniques with supervised learning in Intrusion Detection Systems (IDS) are as follows:

- Numerical experiments have shown that standard machine learning methods (in our case, the *k*-NN algorithm) demonstrate good quality for detecting attacks; this was confirmed, in particular, and for DoS attacks detecting. The achieved quality will be sufficient if ML methods are used in conjunction with some other detection methods.
- As in the previous study [10], the standard statistical approach was also implemented for the NLS-KDD dataset to reduce the number of processed features using correlation analysis. The authors convinced that for the practical use of machine learning methods, it is advisable to filter traffic and process only selected characteristic features for a particular type of attack. This approach also simplifies the creation of an Intrusion Prevention System (IPS).

However, based on the results obtained, the authors consider it appropriate to continue research using the available UNSW-NB15 and NSL-KDD datasets to identify common features set for several existing datasets. It can be based on pre-trained ANNs to improve the accuracy and reduce the model's training time for specific traffic of a specific network for use in IPS.

References

1. ENISA European Union Agency for Network and Information Security (ENISA): Threat Landscape Report 2018, 15 Top Cyberthreats and Trends (2019). https://doi.org/10.2824/622757. https://www.enisa.europa.eu. ISBN 978-929204-286-8, ISSN 2363-3050

2. Muniz, J., Lakhani, A.: Investigating the Cyber Breach: The Digital Forensics Guide for the Network Engineer. Pearson Education Inc, Indianapolis (2018)
3. Abdelhameed, M.: Designing an online and reliable statistical anomaly detection framework for dealing with large high-speed network traffic. A thesis for the degree of Doctor of Philosophy. University of New South Wales, Australia, June 2017. https://www.researchgate.net/publication/328784548_Designing_an_online_and_reliable_statistical_anomaly_detection_framework_for_dealing_with_large_high-speed_network_traffic#fullTextFileContent. Accessed 05 July 2020
4. Al-Garadi, M., Mohamed, A., Al-Ali, A., Du, X., Guizani, M.: A Survey of Machine and Deep Learning Methods for Internet of Things (IoT) Security (2018). https://www.researchgate.net/publication/326696402_A_Survey_of_Machine_and_Deep_Learning_Methods_for_Internet_of_Things_IoT_Security. Accessed 05 July 2020
5. Buczak, A.L., Guven, E.: A survey of data mining and machine learning methods for cyber security intrusion detection. IEEE Commun. Surv. Tutor. **18**(2) (2016). https://www.academia.edu/33112124/Data_Mining_and_Machine_Learning_Methods_for_Cyber_Security_Intrusion_Detection. Accessed 05 July 2020
6. Dwivedi, S., Vardhan, M., Tripathi, S.: Incorporating evolutionary computation for securing wireless network against cyberthreats. J. Supercomput. **76**(3) (2020). https://doi.org/10.1007/s11227-020-03161-w. https://www.researchgate.net/publication/338699460_Incorporating_evolutionary_computation_for_securing_wireless_network_against_cyberthreats. Accessed 06 Jan 2021
7. NSL-KDD and UNSW-NB15 datasets, csv files. https://drive.google.com/drive/folders/1y6vNHhFo9TegDES4UegqwBe_YkxMvfp9?usp=sharing. Accessed 25 Nov 2020
8. Salo, F., Nassif, A.B., Essex, A.: Dimensionality reduction with IG-PCA and ensemble classifier for network intrusion detection. Comput. Netw. **148**, 164–175 (2019)
9. Narendra Kumar, B., Bhadri Raju, M.S., Vishnu Vardhan, B.: A novel approach for selective feature mechanism for two-phase intrusion detection system. Indonesian J. Electr. Eng. Comput. Sci. **14**(1), 101 (2019)
10. Krivchenkov, A., Misnevs, B., Grakovski, A.: Using machine learning for DoS attacks diagnostics. In: Kabashkin, Igor, Yatskiv, Irina, Prentkovskis, Olegas (eds.) Reliability and Statistics in Transportation and Communication. LNNS, vol. 195, pp. 45–53. Springer, Cham (2021). https://doi.org/10.1007/978-3-030-68476-1_4
11. Tang, T.A., Mhamdi., L, McLernon, D., et al.: Deep learning approach for network intrusion detection in software defined networking. In: 2016 International Conference on Wireless Networks and Mobile Communications (WINCOM). IEEE (2016). https://doi.org/10.1109/WINCOM.2016.7777224.ISBN 978-1-090-3837-4
12. Chollet, F.: Deep Learning with Python, 1st edn. Manning Publications, New York (2017). 384 p.

Data Sparsity and Cold-Start Problems in $M - CCF$ Recommender System

Urszula Kużelewska(✉)

Faculty of Computer Science, Bialystok University of Technology, Wiejska 45a, 15-351 Bialystok, Poland
u.kuzelewska@pb.edu.pl

Abstract. Precise identification of users' neighbourhood has a direct impact on the performance of recommender systems. Clustering techniques are the solution that is often used for this purpose, however, they negatively affect the accuracy of recommendations.

In this article, a new version of the algorithm based on multi-clustering, $M - CCF$, is described. Instead of one clustering scheme, it works on a set of multi-clusters, therefore it selects the most appropriate one that models the neighbourhood of a target user most precisely. This article confirms the robustness of $M - CCF$ over the traditional methods against data sparsity and cold-start problems due to its ability to generate more accurate recommendations even if the number of ratings is insufficient.

Keywords: Multi-clustering · Collaborative filtering · Cold-start

1 Introduction

The rapid development of the Internet has been noticed in recent years. It is connected with a large expansion of data. To help users to cope with the information overload, Recommender Systems (RSs) were designed. They are computer applications with the purpose to provide relevant information to a user and as a consequence reduce his/her time spent on searching and increase the personal customer's satisfaction. The form of such relevant information is a list of items that are interesting and useful to the user [7, 15].

Collaborative filtering methods (CF) are the most popular types of RSs [3, 7]. They are based on users' past behaviour data: search history, visited web sites, and rated items, and use them for similarity searching, with an assumption that users with corresponding interests prefer the same items. CF approach has been very successful due to its precise prediction ability [18].

Data sparsity and cold-start problems are still open research challenges [19]. Data used in recommender systems is often very sparse due to the non-availability of users' ratings on every item present in the database. Users rate a very small percentage (less than 1%) of item collection [16]. A cold-start problem

W. Zamojski et al. (Eds.): DepCoS-RELCOMEX 2021, AISC 1389, pp. 224–233, 2021.
https://doi.org/10.1007/978-3-030-76773-0_22

appears when a target user (a user to whom the recommendations are generated) is new - that is, the number of ratings present in his vector is very few [1].

The article is organised as follows: the first section presents the background of sparsity and cold-start problems in the field of Recommender Systems. This section discusses common solutions with their advantages and disadvantages as well. Next section describes the proposed multi-clustering algorithm, $M-CCF$, whereas the following section contains the results of the performed experiments to compare multi-clustering, traditional collaborative filtering ($IBCF$) and single-clustering approaches ($SCCF$). The last section concludes the paper.

2 Background and Related Work

Input data in RSs is usually the ratings of users on a set of items. If a set of users is denoted as $X = \{x_1, \ldots, x_n\}$ and a set of items as $A = \{a_1, \ldots, a_k\}$, the matrix of the input data can be represented by a matrix $U = (X, A, V)$, where $V = \{v_1, \ldots, v_{nk}\}$ and is a set of ratings values [15]. The value v_{ij} stands for a rating given by the user x_i on the item a_j.

Although the number of users is usually greater than a number of items, the vectors of ratings remain very sparse, because often users rate a small subset of items. For instance example, in a library, an avid reader who reads 2–3 books per week, is able to rate about 150 volumes during the year, however, the average readership is considerably lower, that is 18 books per year [24].

A cold-start problem [3] occurs when it is not possible to generate accurate recommendations due to a lack of enough ratings. Typically, it occurs when a) a new item is entered in RS and does not contain initial ratings (*new item*) b) a new user starts to use a RS and does not rate many items (*new user*) [3]. The last type is the most important and occurs most frequently in systems that are already in operation. This article concerns a cold-start problem with a new user.

The common solution to a new user problem is to deliver, despite the ratings, additional external information, e.g., content information in hybrid RSs. In [10] the authors used items' description and references among them in association rules. Loh et al. [11] created user profiles based on information extracted from their scientific publications. Martinez et al. [12] used a combination of the traditional CF approach with a module acquiring domain knowledge of the recommendation background, e.g., user input queries. The model proposed in [5] was used for new user profiling. It directly interviewed newcomers eliciting their opinions about items.

2.1 A Cold-Start Problem and Neighbourhood Identification

Precise identification of a neighbourhood of a target user contributes towards data sparsity and cold-start problem solving. The traditional method for this purpose is k Nearest Neighbours (kNN) [18]. It calculates all user-user or item-item similarities and identifies k most similar objects (users or items) to the target object as its neighbourhood. The kNN algorithm is a reference method used for the CF recommendation process [3]. Simplicity and reasonably accurate results are the advantages of kNN approach; its disadvantages are low scalability and vulnerability to sparsity in data [19].

Clustering algorithms can be an efficient solution to the disadvantages of kNN approach due to the neighbourhood being shared by all cluster members [17]. The algorithm $k - means$ was used in [13] for collaborative filtering optimization, enhancing scalability and tackling a cold-start issue. In [4] a novel efficient association cluster filtering (ACF) algorithm was proposed. It used clustering on ratings to relieve a cold-start problem. The ratings within one cluster were used to evaluate the cluster's opinion, which was employed to predict the unknown ratings. ACF was particularly efficient in the case of very sparse data.

The following problems may arise when one applies clustering algorithms to neighbourhood identification: significant loss of prediction accuracy and different every recommendation outcome. The diversity of results is related to the fact that most of the clustering methods are non-deterministic and therefore several runs of the algorithms can effect obtaining various clustering schemes.

2.2 Multi-clustering Approach to Recommendations

The disadvantages described above can be solved by alternate clustering techniques, multi-view, multi-clustering, or co(bi)-clustering. They include a wide range of methods which are based on multiple runs of clustering algorithms or multiple applications of a clustering process on different input data [2].

Multi-clustering or co-clustering has been applied to improve scalability and tackling data sparsity problems in the domain of RSs. Co-clustering discovers samples that are similar to one another with respect to a subset of features. As a result, interesting patterns (co-clusters) are identified unable to be found by traditional one-way clustering [21]. The method $BiFu$ [22] was particularly efficient in a cold-start problem in terms of accuracy and scalability. The system uses bi-clustering to reduce dimensionality of the rating matrix, whereas the smoothing and fusion technique - to overcome the data sparsity.

The method described in [14] uses a multi-clustering technique, however, it is interpreted as clustering of a single scheme for both techniques. It groups the ratings to create an item group-rating matrix and a user group-rating matrix. As a clustering algorithm, it uses $k - means$ combined with a fuzzy set theory. The results confirmed the good performance of the algorithm when the data sparsity was high and the ability to make predictions for a new item.

Other applications are: recommendation of tourist attractions based on co-clustering and bipartite graph theory [20] and $OCuLaR$ (Overlapping co-CLuster Recommendation) [6] - an algorithm for processing very large databases, detecting co-clusters among users and items as well as providing interpretable recommendations. The system can handle the following problems: cold-start and data sparsity.

2.3 Contribution of Proposed Work

A novel recommender system with neighbourhood identification based on multi-clustering - $M-CCF$ - is evaluated in this paper in the context of cold-start and data sparsity problems. The following are the major contributions of $M-CCF$:

1. The neighbourhood of a target user is modelled precisely due to the fact that the system's overall neighbourhood is formed by a set of cluster schemes, thereby improving the recommendation accuracy.
2. Precise neighbourhood has a positive impact on sparse data managing, therefore a cold-start problem occurs rarely.

A former version of $M-CCF$ is described in [9], in which the input data come from multi-clustering approach, however a value of an input parameter in $k-means$ was the same when building one $M-CCF$ RS system. The article refers to time efficiency comparison with other methods, as well.

3 Description of M-CCF Algorithm

The novel solution consists of multiple types of clustering schemes that are provided for the method's input. It is implemented in the following way (for the original version, with one type of a clustering scheme, check in [8,9]).

Step I. Multiple clustering
The first step of $M-CCF$ is to perform clustering on the input data. The process is conducted several times and all results are stored in order to deliver them to the algorithm. In the experiments described in this paper, $k-means$ was selected as a clustering method, which was executed for $k = 10, 20, 50$ to generate input schemes (denoted by C set) for one $M-CCF$ RS system.

Step II. Building M-CCF RS system
It is a vital issue to have precise neighbourhood modelling for all input data. In $M-CCF$ it is performed by iterating every input object and selection of the best cluster from C set for it. The term *best* refers to the cluster which center is the most similar to the particular input object. Then, when all input data have their connected clusters, a traditional CF systems are built on these clusters. As a result, $M-CCF$ algorithm is created - a complex of recommender systems formed on their clusters as recommender data. A general formula of a neighbourhood calculated by $M-CCF$ method can be described by (1).

$$N_{mcl}(x_i) = C_j(t), \Rightarrow C_j(t) \in C, C = \{C_1(1), \ldots, C_j(1), C_1(2), \ldots, C_g(h)\} \quad (1)$$

where $C_j(t)$ is j-th cluster from t-th clustering scheme, and C are all clustering schemes generated by a clustering algorithm in several runs of different values of its input parameters. In this case, the metric of classification a particular object into a particular cluster is different than the similarity used in recommender systems, as well.

Step III. Recommendation generation
When generating recommendations for a target user, first of all, a relevant RS from $M-CCF$ is selected. It is also based on the similarity between the target user's and cluster centers' ratings. Then, the process of recommendation generation is performed as it is implemented in the traditional collaborative filtering approach, however, searching for similar objects is limited to the cluster connected to the particular recommender in $M-CCF$ algorithm.

When a neighbourhood is modelled by a single-clustering method, the border objects have fewer neighbours in their closest area than the objects located in the middle of a cluster. Multi-clustering prevents such situations, as it identifies clusters in which particular users are very close to its center. A major advantage of $M-CCF$ algorithm is a better quality of a target user's neighbourhood modelling, therefore resulting in high precision of recommendations, including highly sparse cases.

4 Experiments

The experiments concerned performance evaluation of the following algorithms: $M-CCF$, classical item-based collaborative filtering $IBCF$ and RS with the neighbourhood modelled by a single $k-means$ clustering. This is the most common evaluation framework used in articles. Direct comparison of recent original recommender systems might be very difficult due to the lack of access to their source code and problems with providing relevant experiment conditions. The main goal of the evaluation was to compare the performance on sparse data - user vectors containing 1, 2, or 3 ratings.

The evaluation was conducted on two MovieLens datasets: a small one containing 534 users, 11 109 items, and 100 415 ratings ($100k$), and a big dataset consisting of 4537 users, 16767 items and 1 000 794 ratings ($1M$) [23]. Table 1 describes the datasets' statistics. Note that the small set is more sparse (contains fewer ratings per user and per item) than the big one.

Table 1. Description of the datasets used in the experiments.

Dataset	Number of ratings	Number of users	Number of items	Ratings\users ratio	Ratings\items ratio
small dataset - $100k$	100 415	534	11109	188	9
big dataset - $1M$	1 000 794	4537	16767	220	60

During the experiments, attention was paid to the precision and completeness of the recommendation lists generated by the systems. The evaluation criteria were related to the following baselines: Root Mean Squared Error ($RMSE$) described by (2) and $Coverage$ described by (3). The symbols in the equations, as well as the method of calculation are characterised in detail below.

$$RMSE = \sqrt{\frac{1}{n \cdot k}\sum_{i=1}^{nk}\left(r_{real}(v_i) - r_{est}(v_i)\right)^2}, r_{real} \in [2,3,4,5], r_{est} \in \mathbb{R}_+ \quad (2)$$

$$Coverage = \left(\frac{1}{N}\sum_{i=1}^{N} r_{est}(v_i) > 0\right) \cdot 100\%, r_{est} \in \mathbb{R}_+ \quad (3)$$

where $\mathbb{R}_+$ stands for the set of positive real numbers and N is a number of required rating values for estimation.

The performance of the approaches was evaluated in the following way. Before the clustering step, the whole input dataset was split into two parts: training and testing. In the case of $100k$ set, the parameters of the testing part were as follows: 393 ratings, 48 users, 354 items, whereas the case of $1M$: 432 ratings, 44 users and 383 items. This step provided the same testing data during all experiments presented in this paper, therefore making the comparison more objective.

In the evaluation process, the values of ratings from the testing part were removed and estimated by the recommender system. The difference between the original and the calculated value (represented, respectively, as $r_{real}(v_i)$ and $r_{est}(v_i)$) was taken for $RMSE$ calculation. The lower value of $RMSE$ stands for a better prediction ability.

During the evaluation process, there were cases in which estimation of ratings was not possible. It occurs when the 2 items for which the calculations are performed, are not present in the same cluster. It is considered in *Coverage* index (3). In every experiment, it was assumed that $RMSE$ is significant if the value of *Coverage* is greater than 90%.

The cold-start problem was examined by verification whether the algorithm generates accurate recommendations when there are few ratings (1,2, and 3) in a user's vector. That is, if a user had excess values, they were removed randomly and finally, only cases in the correct number remained.

The experiments started from the evaluation of classical $IBCF$ RS. Table 2 reports the results on both datasets. The comparison takes into consideration the following similarity measures: $Cosine - based$, $LogLikelihood$, $Pearson$ $correlation$, both $Euclidean$ and $CityBlock$ distance-based and $Tanimoto$ coefficients. The values are averages of $RMSE$ values obtained from 10 runs of the system. This rule was applied in all experiments. Analysing the values in Table 2 it can be stated that the classical approach is not efficient in a cold-start problem (the values of $RMSE$ are around 1), particularly when users have only 1 or 2 ratings. A slight improvement can be observed in the case of $1M$ dataset, which is more dense than $100k$ data. There is a Tanimoto similarity measure that performs the best in nearly all cases in this recommender system.

The following experiment was performed on the recommender with the neighbourhood modelled by a single-clustering approach. $k - means$ was taken as a clustering method. A proper number of clusters and the most optimal distance measure among the points were additional issues in the case of $k - means$ algorithm. In the experiment, the number of groups was equal 10, 20, or 50, however in the article the results are presented in a compact way, as a range of

Table 2. Results of data sparsity testing (RMSE) - *IBCF* algorithm evaluated on both datasets. The best values are in bold.

Similarity Measure	100k dataset			1M dataset		
	1 rating	2 ratings	3 ratings	1 rating	2 ratings	3 ratings
Cosine	1.08	1.03	0.99	1.02	0.98	0.98
LogLikelihood	1.06	1.01	0.97	1.02	0.98	0.98
Pearson	1.60	2.27	1.97	**0.96**	**0.94**	1.01
Euclidean	1.10	1.03	0.99	1.01	0.96	0.97
CityBlock	1.13	1.05	1.01	0.97	**0.94**	**0.95**
Tanimoto	**1.02**	**0.98**	**0.94**	0.99	**0.94**	**0.95**

all obtained values without a particular case distinction (see Table 3). Finally, this experiment was performed 54 times (and repeated 10 times to decrease the influence of non-determinism of $k - means$) per input dataset. A Cosine-based distance was selected as a clustering measure due to the best overall performance in comparison to Euclidean and Chebyshev measures. Note that despite many repetitions of runs for a particular k value, there is no guarantee that the scheme selected for the recommendation process is optimal.

Table 3. Results of data sparsity testing (RMSE) - *SCCF* algorithm evaluated on both datasets. The best values are in bold.

Similarity Measure	100k dataset			1M dataset		
	1 rating	2 ratings	3 ratings	1 rating	2 ratings	3 ratings
Cosine	**0.90-0.97**	0.90-0.96	**0.91-0.97**	**0.89-0.94**	0.91-0.93	0.89-0.92
LogLikelihood	**0.90-0.98**	0.90-0.96	**0.91-0.96**	**0.89-0.94**	**0.90-0.93**	**0.89-0.91**
Pearson	0.99-2.12	1.01-2.44	0.99-8.83	1.07-1.81	2.11-2.86	2.29-2.56
Euclidean	**0.90-0.97**	**0.89-0.95**	**0.91-0.96**	**0.89-0.93**	0.91-0.93	**0.89-0.91**
CityBlock	0.93-0.97	0.92-0.97	0.93-0.96	0.94-0.98	0.95-0.96	0.93-0.96
Tanimoto	**0.90-0.97**	**0.89-0.95**	**0.91-0.96**	**0.89-0.93**	**0.90-0.93**	**0.89-0.91**

The results are better in comparison with the outcome from the previous experiment - all values are lower than 1. Furthermore, $RMSE$ is lower in the case of the big dataset. There are 3 similarity measures that generate the most accurate recommendations: LogLikelihood, Euclidean, and Tanimoto index. The lowest values are: for 100k dataset - 0.9, 0.89, 0.91 (for 1,2 and 3 ratings in user's data) and for 1M dataset - 0.89, 0.9, 0,89 (analogically, for 1,2 and 3 ratings in user's data). However, the range of values is wide: from 0.01 to 0.08 (excluding an outsider - Pearson correlation). It means that accurate recommendations are not achieved in each case. They depend on the number of clusters (k in $k - means$) as well as a particular run of the clustering algorithm.

The following experiment was performed using $M - CCF$ recommender. Table 4 contains the results: they are average values from 10 different runs of the method. In every case, all the clustering schemes were obtained using $k-means$ algorithm for $k = 10, 20, 50$ and given on $M - CCF$ input in this form.

Table 4. Results of data sparsity testing (RMSE) - $M-CCF$ algorithm evaluated on both datasets. The best values are in bold.

Similarity Measure	100k dataset			1M dataset		
	1 rating	2 ratings	3 ratings	1 rating	2 ratings	3 ratings
Cosine	**0.88**	-	-	0.83	0.92	0.97
LogLikelihood	0.94	0.95	1.02	0.84	0.94	0.97
Pearson	1.52	1.29	-	0.83	1.78	0.98
Euclidean	0.91	**0.92**	1.02	**0.78**	**0.88**	**0.94**
CityBlock	**0.88**	**0.92**	**0.92**	0.83	0.92	**0.94**
Tanimoto	**0.86**	**0.91**	0.94	**0.81**	**0.90**	**0.93**

There are 3 cases (marked as "-" in the table) when $Coverage$ values were lower than 90%. Otherwise, $RMSE$ values are present. The best similarity measure was Tanimoto index: $RMSE = 0.86$, 0.91 and 0.94 (for respectively 1,2,3 ratings in user's data) for 100k dataset and 0.81, 0.90 and 0.93 (for respectively 1,2,3 ratings in user's data) for 1M dataset. They are generally lower than the values from the previous experiments and the most distinct difference is for the small set, which is more sparse than the big one. It means that $M - CCF$ generates more accurate recommendations in the case of a cold-start problem with sparse data. Moreover, the final values are always unambiguous, hence the recommender system is able to generate the most optimal propositions.

In the case of 1M dataset, $M - CCF$ outperforms the previous solution only for 1 rating present in the user's vector. For 2 and 3 ratings, the values were comparable or slightly worse. However, note that there is not any range of values, but explicit numbers in every case. It points to elimination of the ambiguity of clustering scheme selection by the multi-clustering approach.

5 Conclusions

This paper describes a developed version of a collaborative filtering recommender system, $M - CCF$ and its ability to generate accurate recommendations in the case of sparse data and cold-start problems. The presented method models the neighbourhood using a multi-clustering algorithm. The algorithm eliminates a disadvantage appearing when the neighbourhood is modelled by a single-clustering method - dependence of the final performance of a recommender system on a clustering scheme selected for the recommendation process. Additionally, the preparation of input data was improved. Data come from many clustering schemes obtained from $k-means$ algorithm, however, its input parameters (k) were diversified.

We plan to extend the work in several directions. Further experiments will focus on the preparation of a mixture of clustering schemes that are selected from all partitionings obtained using different clustering algorithms. The selection will be performed on clusters' quality calculated by evaluation indices. It should improve the overall performance of recommendation: accuracy as well as scalability. It is also planned to collect source codes of other multi-clustering techniques and perform mutual comparison.

Acknowledgment. The work was supported by the grant from Bialystok University of Technology nr WZ/WI-IIT/2/2020 and funded with resources for research by the Ministry of Science and Higher Education in Poland.

References

1. Adomavicius, G., Tuzhilin, A.: Toward the next generation of recommender systems: a survey of the state-of-the-art and possible extensions. IEEE Trans. Knowl. Data Eng. **17**(6), 734–749 (2005)
2. Bailey, J.: Alternative Clustering Analysis: a Review. Intelligent Decision Technologies: Data Clustering: Algorithms and Applications, pp. 533–548. Chapman and Hall/CRC (2014)
3. Bobadilla, J., Ortega, F., Hernando, A., Gutiérrez, A.: Recommender systems survey. Knowl. Based Syst. **46**, 109–132 (2013)
4. Huang, Ch., Yin, J.: Effective association clusters filtering to cold-start recommendations. In: Proceedings of 7th International Conference on Fuzzy Systems and Knowledge Discovery, pp. 2461–2464 (2010)
5. Golbandi, N., Koren, Y., Lempel, R.: Adaptive bootstrapping of recommender systems using decision trees. In: Proceedings of the 4th International Conference on Web Search and Web Data Mining, pp. 595–604 (2011)
6. Heckel, R., Vlachos, M., Parnell, T., Duenner, C.: Scalable and interpretable product recommendations via overlapping co-clustering. In: IEEE 33rd International Conference on Data Engineering, pp. 1033–1044 (2017)
7. Jannach, D.: Recommender Systems: An Introduction. Cambridge University Press, Cambridge (2010)
8. Kużelewska, U.: Dynamic neighbourhood identification based on multi-clustering in collaborative filtering recommender systems. In: International Conference on Dependability and Complex Systems, pp. 410–419. Springer, Cham (2020)
9. Kużelewska, U.: Effect of dataset size on efficiency of collaborative filtering recommender systems with multi-clustering as a neighbourhood identification strategy. In: International Conference on Computational Science, pp. 342–354. Springer, Cham (2020)
10. Leung, C.W., Chan, S.C., Chung, F.L.: An empirical study of a cross-level association rule mining approach to cold-start recommendations. Knowl. Based Syst. **21**(7), 515–529 (2008)
11. Loh, S., Lorenzi, F., Granada, R., Lichtnow, D., Wives, L.K., Oliveira, J.P.: Identifying similar users by their scientific publications to reduce cold start in recommender systems. In: Proceedings of the 5th International Conference on Web Information Systems and Technologies, pp. 593–600 (2009)

12. Martinez, L., Perez, L.G., Barranco, M.J.: Incomplete preference relations to smooth out the cold-start in collaborative recommender systems. In: Proceedings of the 28th North American Fuzzy Information Processing Society Annual Conference, pp. 1–6 (2009)
13. Ntoutsi, E., Stefanidis, K., Nørvåg, K., Kriegel, H.P.: Fast group recommendations by applying user clustering. In: International Conference on Conceptual Modeling, pp. 126–140. Springer, Berlin (2012)
14. Puntheeranurak, S., Tsuji, H.: A multi-clustering hybrid recommender system. In: Proceedings of the 7th IEEE International Conference on Computer and Information Technology, pp. 223–238 (2007)
15. Ricci, F., Rokach, L., Shapira, B.: Recommender systems: introduction and challenges. In: Recommender Systems Handbook, pp. 1–34. Springer, Boston (2015)
16. Sarwar, B., Karypis, G., Konstan, J., Riedl, J.: Application of dimensionality reduction in recommender system—a case study. Minnesota University Minneapolis, Department of Computer Science (2000)
17. Sarwar, B.: recommender systems for large-scale e-commerce: scalable neighborhood formation using clustering. In: Proceedings of the 5th International Conference on Computer and Information Technology (2002)
18. Schafer, J.B., Frankowski, D., Herlocker, J., Sen, S.: collaborative filtering recommender systems. In: The Adaptive Web, pp. 291–324 (2007)
19. Singh, M.: Scalability and sparsity issues in recommender datasets: a survey. Knowl. Inf. Syst. **62**, 1–43 (2018). Springer
20. Xiong, H., Zhou, Y., Hu, C., Wei, X., Li, L.: A novel recommendation algorithm frame for tourist spots based on multi - clustering bipartite graphs. In: Proceedings of the 2nd IEEE International Conference on Cloud Computing and Big Data Analysis, pp. 276–282 (2017)
21. Yaoy, S., Yuy, G., Wangy, X., Wangy, J., Domeniconiz, C., Guox, M.: Discovering multiple co-clusterings in subspaces. In: Proceedings of the 2019 SIAM International Conference on Data Mining, pp. 423–431 (2019)
22. Zhang, D., Hsu, C., Chen, M., Chen, Q., Xiong, N., Lloret, J.: Cold-start recommendation using bi-clustering and fusion for large-scale social recommender systems. IEEE Trans. Emerg. Top. Comput. **2**(2), 239–250 (2014)
23. MovieLens Datasets. https://grouplens.org/datasets/movielens/25m/. Accessed 10 Oct 2020
24. State of readership in Poland in 2019. Report. https://www.bn.org.pl/download/document/1587585168.pdf. Accessed 19 Jan 2021

Explaining Predictions of the X-Vector Speaker Age and Gender Classifier

Damian Kwaśny, Paweł Jemioło(✉), and Daria Hemmerling

AGH University of Science and Technology, A. Mickiewicza 30, 30-059 Krakow, Poland
pawljmlo@agh.edu.pl

Abstract. In this paper, we assess the applicability of two Explainable Artificial Intelligence methods: model-agnostic Feature Ablation and Neural Networks-specific Integrated Gradients, to explain predictions of Deep Neural Network-based speech classification models. We use these techniques to explain predictions of two Deep Learning x-vector models trained for age and gender classification from speech. Our results show that both methods can be successfully used for speech classification related tasks, providing a deeper understanding of the model's behaviour. In particular, we confirm that the features highlighted by the explored methods are fundamental to the performance of the models and their removal results in a rapid performance degradation compared to the random baseline. We also show that the highlighted characteristics align with the theoretical fundamentals regarding age- and gender-based changes in the speech production process.

Keywords: Explainability · Artificial Intelligence · Integrated Gradients · Feature Ablation · Age classification · Deep Learning · Speech

1 Introduction

Speech processing is an area of research that has been undergoing dynamic development in recent years, mainly due to advancements in the field of Deep Learning. It covers a wide range of applications, from speaker recognition, identification through age and gender estimation to illness detection. Some of these application domains, such as public security and medicine, not only require models to perform well but also to be interpretable. It is a problem since most of the Deep Learning models are treated as *black-boxes*, systems of which internal workings we do not know.

To bridge this gap, researchers have developed various explainability methods such as Local Interpretable Model-Agnostic Explanations (LIME), Shapley Additive Explanations (SHAP) and Integrated Gradients (IG), to name just a few [1,17]. These methods have been successfully applied in tasks such as image or text classification [17]. However, their application to the problem of speech classification remains overlooked [4].

W. Zamojski et al. (Eds.): DepCoS-RELCOMEX 2021, AISC 1389, pp. 234–243, 2021.
https://doi.org/10.1007/978-3-030-76773-0_23

The contribution of this work is as follows. First, we implement a baseline x-vector model [8], which has gained popularity in various speech classification tasks [19] and train it for age and gender classification on a publicly available Common Voice dataset [3]. We then apply two XAI methods, Integrated Gradients [17] and Feature Ablation [1], to explain the predictions of the model on the evaluation dataset. Finally, we show that the features highlighted by both methods are indeed crucial for the performance of the model and confront them with theoretical knowledge on age- and gender-related speech differences.

The rest of the paper is organized as follows. In Sect. 2, we discuss background methods and technologies we utilize, and concepts we base on. Next, in Sect. 3, we present architecture details. What is more, we provide information about the explainability methods we use. Then in Sect. 4, we report on preliminary results. Finally, in Sect. 5, we summarize the paper and provide directions for future work.

2 Background

2.1 Speech Classification

Besides the linguistic content, the speech signal carries unique, individual characteristics of a single person's voice, such as the speaker's identity, emotional state, age or gender. The research on extracting this para-linguistic content has increased in recent years, with speaker recognition being the most explored area [8]. The currently dominant approach for speaker recognition is to represent the spoken utterance as a fixed-dimensional embedding vector. This vector is then used in an external classifier, such as Probabilistic Linear Discriminant Analysis (PLDA) [6,19].

Before the Deep Learning era came, the most used methods had based on i-vectors [6]. In this framework, a Universal Background Model (UBM) and a projection matrix are learned and used in a PLDA classifier [6].

However, recently a new Deep Neural Network (DNN) based embedding framework called x-vector was introduced. Instead of using the UBM, the authors propose to train a Time-Delay Neural Network (TDNN) for a speaker classification task and use the embedding produced by this pre-trained network in an external PLDA classifier. This method is shown to outperform the i-vector baseline, e.g. on the Speaker in the Wild (SITW) Core [15] dataset. In the article summarizing the results of the NIST Speaker Recognition Evaluation 2018 (SRE18) [19], the authors have explored various DNN-based speech embedding methods, including multiple extensions of the baseline x-vector framework. They have shown that these embeddings consistently outperform the i-vector baseline by a large margin.

There have also been attempts at applying the x-vector framework to the age estimation task. In [8], a method based on x-vectors was used, achieving a Mean Absolute Error (MAE) of 4.92 compared with 5.82 of the baseline i-vector system on the NIST SRE10 dataset.

2.2 Explainable Artificial Intelligence (XAI)

The field of the XAI has experienced dynamic progress in the last few years. The explanation method can be either intrinsic or post-hoc and can have either a local or a global scope. An intrinsic method is inherently model-specific and characterizes algorithms that are explainable by design, whereas post-hoc methods are usually model-agnostic. Global interpretability means understanding the logic of a model, following the entire reasoning leading to all the different outcomes. On the other hand, local interpretability focuses on explaining the reasoning behind a single prediction. In practice, global interpretability is practically not achievable when several parameters grow, and it is not well suited for complex models such as DNN [1].

Nowadays, one of the most important and popular local, model-agnostic method is SHapley Additive exPlanation (SHAP) and Local Interpretable Model-agnostic Explanations (LIME). While robust and well-grounded in the XAI world, their application to the time-series classification is not straightforward [14].

The favoured class of methods, especially in the context of ANNs, is sensitivity analysis. It explores how various perturbation of input features affects the results produced by the network. Visualizing the results of this analysis helps to understand how different input features influence the network and is especially popular in Convolutional Neural Networks (CNN) [10] for Image Recognition [1].

Another technique has been recently presented in [17]. The authors introduced a new gradient-based method called Integrated Gradients. IG is defined as the path integral of the gradients along the straight-line path from the baseline x' to the input x. For an input x and a baseline x', the integrated gradient along the ith dimension is calculated with the Eq. 1. $F(x)$ is a function that represents the Neural Network, while $\frac{\partial F(x)}{\partial x_i}$ states for the gradient of $F(x)$ along the ith dimension.

$$IG_i(x) :: = (x_i - x_i') \int_{\alpha=0}^{1} \frac{\partial F(x' + \alpha(x - x'))}{\partial x_i} d\alpha \tag{1}$$

The authors have shown its application to explain predictions in the tasks such as Image Recognition, which is a sequence classification problem and Neural Machine Translation, which is a sequence-to-sequence problem.

With all that said, applying the mentioned methods to explain predictions of speech classifiers is still rarely explored. Authors have attempted tackling this area in [4]. Authors showed that the features marked by their method as significant highly influence the prediction accuracy for spoken digit and gender classification problem.

3 Our Approach

3.1 Dataset

We have decided to use Kaggle's English subset of the Common Voice dataset [3] to conduct our experiments. The Common Voice dataset is the largest open-source, multi-lingual dataset. It contains more than 2500 h of transcribed speech in 40 languages. On top of the audio recordings and the corresponding transcriptions, it also contains metadata about the speaker, such as age group (teens, twenties, ..., eighties), gender (female, male, other) and accent.

3.2 Neural Network Architecture

Encouraged by the results presented in [8], we also decided to use the x-vector framework to perform our experiment. The summary of the architecture is shown in Table 1. N depicts the number of age classes, while T states for the length of an utterance. Since our training dataset contains only information about the age-bin of the speaker, not the exact age, we did not use any extra layers aimed for a regression task. Both the age and gender classifiers accept the 512-dimensional embedding as inputs. For details about the TDNN layer, please adhere to the relevant paper, for example [16].

Table 1. X-vector architecture summary.

Layer	Layer context	Total context	Size
TDNN1	$[-2, 0, 2]$	5	512
TDNN2	$[-2, 0, 2]$	9	512
TDNN3	$[-3, 0, 3]$	15	512
TDNN4	0	15	512
TDNN5	0	15	512
Stats pooling (Mean + STD)	$[0, T)$	T	1500 + 1500
Dense + ReLu	0	T	512
Dense + ReLu + BatchNorm	0	T	512
Dense + Softmax (age)	0	T	N = 8
Dense + Sigmoid (gender)	0	T	1

3.3 Explainability Methods

To explain the predictions obtained using a pre-trained x-vector network, we decided to use two local, model-agnostic XAI methods: Integrated Gradients and Feature Ablation. The choice of IG is motivated by the fact that it has been successfully applied to explaining the output of sequence-based models in the Natural Language Processing and Neural Machine Translation domains [17]. On the other hand, we decided to use FA because it is typically used in image recognition. Therefore, it could potentially be well suited for dealing with the kind of inputs the x-vector network accepts. Such inputs can be visualized as variable-width images depicting the values of the Mel-frequency cepstrum (MFCC) features along the time-axis.

Feature Ablation is an example of the sensitivity analysis method [1]. It involves replacing each input feature with an arbitrary baseline and computing the difference in results. The output is an attribution matrix of the same size as the input tensor, with high values for data points or features that highly influence the network prediction and low ones for those features that have little impact on the output.

Integrated Gradients have been already introduced in Sect. 2. Since the continuous integral shown in Eq. 1 is not numerically computable, it is in practice approximated by a summation presented in Eq. 2, where m is the number of steps in the Riemann approximation of the integral [17]. Other values have the same meaning as in Eq. 1.

$$IG_i^{approx}(x) :: = (x_i - x_i') \sum_{k=1}^{m} \frac{\partial F(x' + \frac{k}{m}(x - x'))}{\partial x_i m} \tag{2}$$

4 Preliminary Results

4.1 Experimental Setup

We run the training of our network using the *cv-valid-train* subset of the English Common Voice dataset and evaluate the results on *cv-valid-test* and *cv-valid-dev*. All the subsets were filtered to only contain utterances with both age and gender metadata present. The resulting dataset contains 73466 train examples that total around 84 h of data and 1535 test examples which correspond to 1.7 h of data.

All the recordings were converted to the *wav* format and re-sampled to 16 kHz. We used 30 MFCC [9] computed with a hamming window of 25 ms length and 10 ms stride, the higher cut-off frequency of 8 kHz and the lower cut-off frequency 40 Hz as the input features to our network. We trained the network for 100 epochs with a batchsize of 16 using *novograd* optimizer with $lr = 0.001$, $beta_1 = 0.95$, $beta_2 = 0.5$ and $weightDecay = 0.001$. We used a Cosine Annealing learning rate policy with 1000 warmup steps. The whole feature-extraction and training pipeline was conducted using the NeMo toolkit [12]. For the explainability experiments, we used the implementations of FA and IG provided in the Captum XAI framework [11]. For both methods, we have used zero-valued tensors as the baseline vector.

Table 2. Accuracy summary.

Task	cv-valid-dev	cv-valid-test
Age classification	80.74%	82.5%
Gender classification	98.14%	98.75%

4.2 Classification Accuracy

Even though the performance of the classifiers is not of primary concern in this work, we report the results of the classification with regards to the accuracy metric (see Table 2).

4.3 Explainability Experiments

Sample attribution matrices produced with both explored methods and tasks are shown in Figs. 1 and 2. For age classification, warmer colours indicate higher values corresponding to data points and features that are significant when generating correct prediction. For gender classification, where the output is a real value between 0 (male) and 1 (female), the warmer colours correspond to input data points leading to higher output value (resulting in prediction the of female).

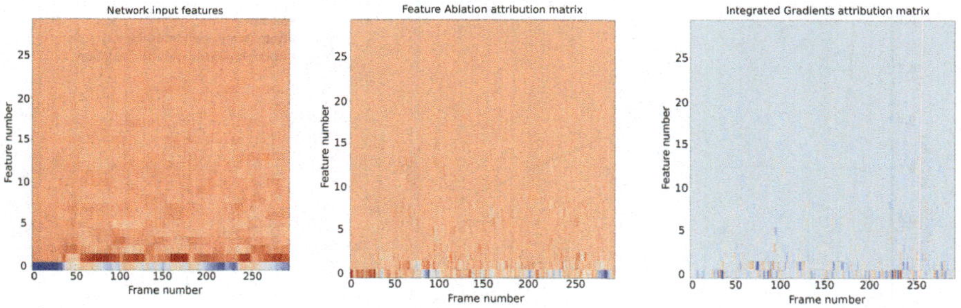

(a) Input features. (b) Feature Ablation. (c) Integrated Gradients.

Fig. 1. Network input features (left) and importance attribution matrices generated with Feature Ablation (center) and Integrated Gradients (right) for age cls.

Using stratified sampling, we extracted a representative 100-elements test subset from the test set used to evaluate the x-vector network performance. To assess that the features highlighted by both the algorithms are indeed relevant, we ran an experiment to see what happens to the network accuracy when a dropout mask is applied on the network input features, zeroing out 1, 2, 4, 8, 16 and 32% of the data points. The mask was generated either randomly (baseline) or according to the importance matrix generated using Feature Ablation or Integrated Gradients, in order of descending importance.

Figures 3 and 4 show the results of the experiment. While using a randomly generated mask, the network was able to retain the accuracy up until 8% of dropped points, and even with 16%, the results were still well above a chance level for age classification. On the other hand, removing the features according to the importance matrix caused a drastic degradation of performance of the model with as little as 1–2% of the data missing. Similarly, for gender classification, randomly generated masks still assured accuracy at the level above 90% with 16% of dropped points, while a significant degradation was observed when

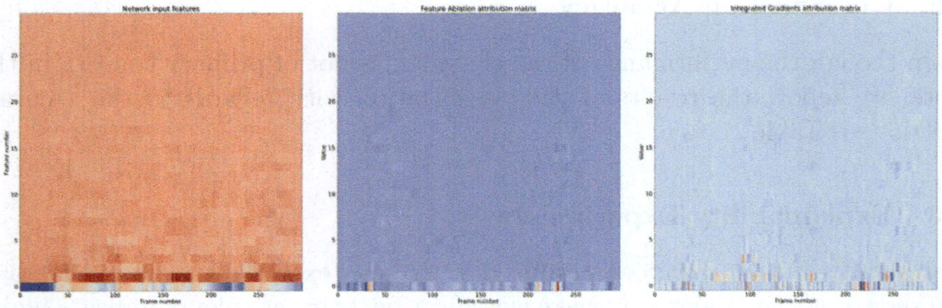

(a) Input features. (b) Feature Ablation. (c) Integrated Gradients.

Fig. 2. Network input features (left) and importance attribution matrices generated with Feature Ablation (center) and Integrated Gradients (right) for gender cls.

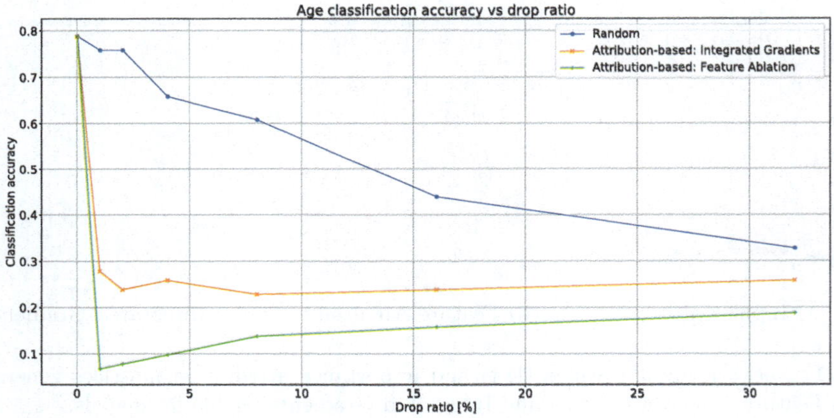

Fig. 3. Accuracy of the age classifier for different dropout mask generation methods.

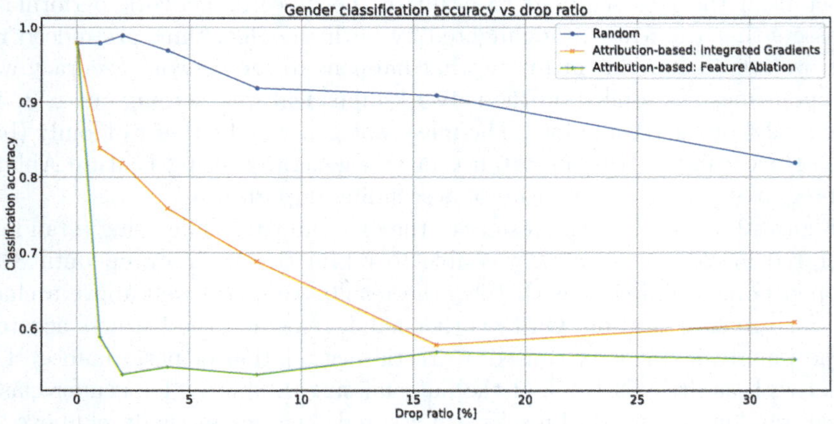

Fig. 4. Accuracy of the gender classifier for different dropout mask generation methods.

features were removed according to the importance matrix. Growing accuracy of IG (orange line) and FA (green line) based methods after a percentage of most essential inputs are set to 0 reaches a certain point comes from the fact that the classifier is more often outputting the majority class, indicating a potential bias of the model.

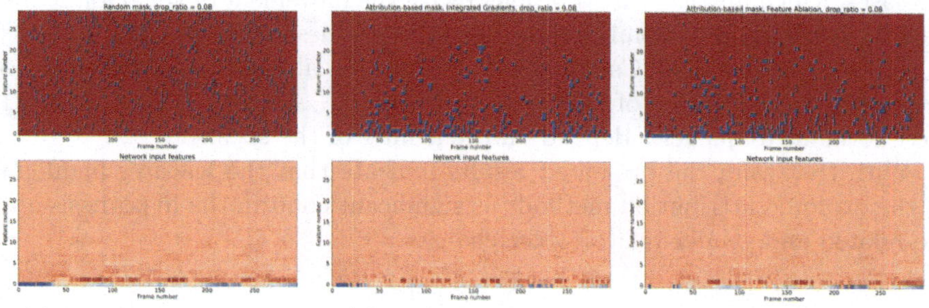

(a) Randomly selected. (b) Selected with IG. (c) Selected with FA.

Fig. 5. Age classification. Resulting dropout mask generated by different methods and the corresponding input features for dropout ratio = 8%.

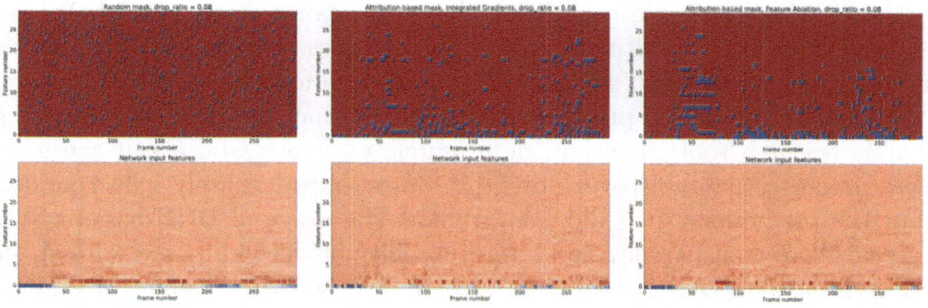

(a) Randomly selected. (b) Selected with IG. (c) Selected with FA.

Fig. 6. Gender classification. Resulting dropout mask generated by different methods and the corresponding input features for dropout ratio = 8%.

To better understand the abrupt performance degradation, we decided to visualize the drop masks obtained by the three explored methods. These are shown in Figs. 5 and 6. The blue colour indicates that a given data point was set to zero. According to Figs. 5 and 6, while the random mask discarded scattered data points across the whole spectrum of input features, both IG and FA removed full patterns corresponding to low MFCC coefficients and their multiples. Due to the nature of the MFCC features, the lower coefficients correspond to the energy and slow changes in the filterbank energies of the speech signal in the lower frequency spectrum. They also contain most of the information about the overall spectral shape of the vocal cords (source)-the vocal tract(filter) transfer function. With ageing, several changes occur in the human speech production mechanism, which results in acoustic parameters changes [5,7,18,20]. For older people, it becomes more challenging to move the tongue freely due to the loss of muscle strength. It causes changes in resonance frequency and affects the formants frequency [5]. Along with the ageing process, the stiffness of cartilages increases, which might create a problem with the movement of vocal folds, which leads to voice signal becoming thinner, stiffer and less pliable.

Finally, the low MFCCs carry the information about the formants [9], corresponding to the shape of the vocal tract (filter). According to [2,5,7,13,18], these formant frequencies depend on the gender of the speaker.

With that said, the presented results indicate that the features highlighted by the explored attribution methods as significant confirm the hypothesis about age-related and gender-related voice changes.

5 Conclusions and Future Work

In this paper, we have applied two popular XAI methods: Integrated Gradients and Feature Ablation to the currently rarely explored Speech Classification topic, in particular speech-based age and gender classification with a Deep Neural Network model. We have shown that the features highlighted by both of these methods as necessary are crucial for the performance of the model. We have also shown that the resulting contribution matrices align well with theoretical knowledge about age- and gender-related speech differences.

We believe the results presented here indicate that the explored methods are valuable research tools that provide a unique way to reason about the model, assess its correctness, discover potential biases and ultimately gain confidence about the predictions. It could also promote the usage of DNN-based systems in areas where transparency and interpretability of the results are critical, such as speech-based speaker verification in biometric security systems or diagnosis of illness from the voice signal in the healthcare industry.

It could be interesting to explore whether FA and IG are suited to compare the performance and robustness of different models trained for the same purpose as future work. To that end, we are currently implementing the FTDNN extension of the x-vector framework discussed in this paper. In the future, it would also be interesting to investigate the possibility of applying Integrated Gradients directly to different layers of the network, which could help us learn more about the inner workings of the model.

Acknowledgment. The paper is supported by the AGH UST research grant.

References

1. Adadi, A., Berrada, M.: Peeking inside the black-box: a survey on explainable artificial intelligence (XAI). IEEE Access **6**, 52138–52160 (2018)

2. Alhussein, M., et al.: Automatic gender detection based on characteristics of vocal folds for mobile healthcare system. Mob. Inf. Syst. **2016**, 1–12 (2016)
3. Ardila, R., Branson, M., Davis, K., Henretty, M., Kohler, M., Meyer, J., Morais, R., Saunders, L., Tyers, F.M., Weber, G.: Common voice: A massively-multilingual speech corpus. arXiv preprint arXiv:1912.06670 (2019)
4. Becker, S., Ackermann, M., Lapuschkin, S., Müller, K.R., Samek, W.: Interpreting and explaining deep neural networks for classification of audio signals. arXiv preprint arXiv:1807.03418 (2018)
5. Das, B., Mandal, S., Mitra, P., Basu, A.: Effect of aging on speech features and phoneme recognition: a study on Bengali voicing vowels. Int. J. Speech Technol. **16**(1), 19–31 (2013)
6. Dehak, N., Kenny, P.J., Dehak, R., Dumouchel, P., Ouellet, P.: Front-end factor analysis for speaker verification. IEEE Trans. Audio Speech Lang. **19**(4), 788–798 (2011)
7. Deliyski, Steve An Xue, D.: Effects of aging on selected acoustic voice parameters: preliminary normative data and educational implications. Educ. Gerontol. **27**(2), 159–168 (2001)
8. Ghahremani, P., Nidadavolu, P.S., Chen, N., , Villalba, J., Povey, D., Khudanpur, S., Dehak, N.: End-to-end deep neural network age estimation. In: Interspeech, pp. 277–281 (2018)
9. Gold, B., Morgan, N., Ellis, D.: Speech and Audio Signal Processing: Processing and Perception of Speech. John Wiley & Sons, Hoboken (2011)
10. Goodfellow, I., Bengio, Y., Courville, A.: Deep Learning. MIT Press, Cambridge (2016)
11. Kokhlikyan, N., Miglani, V., Martin, M., Wang, E., Reynolds, J., Melnikov, A., Lunova, N., Reblitz-Richardson, O.: Pytorch captum (2019). https://github.com/pytorch/captum
12. Kuchaiev, O., Li, J., Nguyen, H., Hrinchuk, O., et al.: Nemo: a toolkit for building AI applications using neural modules (2019)
13. Levitan, S.I., Mishra, T., Bangalore, S.: Automatic identification of gender from speech. In: Proceeding of Speech Prosody, pp. 84–88 (2016)
14. Lundberg, S.M., Lee, S.I.: A unified approach to interpreting model predictions. Adv. Neural Inf. Process. Syst. **30**, 4765–4774 (2017)
15. McLaren, M., Lawson, A., Ferrer, L., Castan, D., Graciarena, M.: The speakers in the wild speaker recognition challenge plan. Interspeech **2016** (2015)
16. Peddinti, V., Povey, D., Khudanpur, S.: A time delay neural network architecture for efficient modeling of long temporal contexts. In: Sixteenth Annual Conference of the International Speech Communication Association (2015)
17. Sundararajan, M., Taly, A., Yan, Q.: Axiomatic attribution for deep networks. arXiv abs/1703.01365 (2017)
18. Torre III, P., Barlow, J.A.: Age-related changes in acoustic characteristics of adult speech. J. Commun. Disord. **42**(5), 324–333 (2009)
19. Villalba, J., Chen, N., Snyder, D., Garcia-Romero, D., McCree, A., Sell, G., Borgstrom, J., Richardson, F., Shon, S., Grondin, F., et al.: State-of-the-art speaker recognition for telephone and video speech. In: Interspeech, pp. 1488–1492 (2019)
20. Vipperla, R., Renals, S., Frankel, J.: Ageing voices: The effect of changes in voice parameters on asr performance. EURASIP J. Audio Speech Music Process. **2010**(1), 525783 (2010)

Application of the Closed-Loop PI Controller as the Low-Pass Filter

Michal Lower(✉) and Pawel Dobrowolski

Faculty of Electronics, Wroclaw University of Science and Technology, Wroclaw, Poland
{Michal.Lower,Pawel.Dobrowolski}@pwr.edu.pl

Abstract. The quality of the measurement signal is often the basis for the reliable operation of the automatic control system. Real measurements are always affected by disturbances and an accuracy of measurements is within a certain range. The reliability of the automatic control system is based on its stable operation in the operating space of the system. Too high disturbance of the measurement signal may destabilize the control system. High frequency disturbances are one of the most common problem. Low-pass filters are used to eliminate high-frequency disturbances. However, these filters introduce a delay in the measurement signal. In control of dynamic systems, the delay of the measurement signal may cause the system destabilization. Such a problem was observed in flight stabilization system of the multi-rotor UAV. The paper proposes a low-pass filter based on PI controller in closed-loop feedback. In this filter, the values of the parameter of the PI controller are automatically changed based on the assessment of external factors. The goal of filtration is to eliminate disturbances while minimizing measurement signals delays. The proposed filter was compared with the traditional known solutions. The obtained results showed that the proposed solution introduces lower signal delays at the similar filtration range.

Keywords: Low-pass filter · PI controller · Reliability · UAV

1 Introduction

The problem of signal filtering is one of the elements influencing the reliability of the automatic control system. Ignorance of the true value of the process variable measuring signal (process value) due to high frequency disturbances results in oscillations in the control error, which in turn translates into the control value, the oscillations are transferred to the actuator. In extreme situations, the system may become destabilized and the reliability of the automatic control system depends on its stable operation in the system's operating space.

In order to minimize the impact of interferences, various types of filtration methods are used [8,9]. To eliminate high-frequency interference, low-pass filters are most often used. However, the filters introduce a delay in the measurement signal. The papers [3,4] present tests of the control system with a disturbed

W. Zamojski et al. (Eds.): DepCoS-RELCOMEX 2021, AISC 1389, pp. 244–253, 2021.
https://doi.org/10.1007/978-3-030-76773-0_24

measurement signal. The authors analyzed examples of various control methods, additionally showing the influence of the second order low-pass filter. Studies in which the authors embed classic filtration systems into the structure of the regulator can also be found [11,14].

Sometimes it is possible to minimize the influence of disturbances on the stability of the system by appropriate selection of the object controller settings. Such a method has been proposed by the authors [2,6,10]. However, such actions will reduce the sensitivity of the control system and delay the controller's response. Thus, the final effect of these actions is similar to the solutions with signal filtering.

The authors of the presented papers mainly focus on the quality of filtration. However, in a situation where the difference between the frequency of disturbances of the measuring signal and the operating range of the control object is not too large, the delay of the measuring signal resulting from signal filtering may lead to destabilization of the system. In such a situation, usually the measurement method is changed or filtering based on different measurement methods is applied and data fusion is performed [14]. However, such actions are not always possible or sufficient. Such a problem was observed in the unmanned aerial vehicle (UAV) multi-rotor flight stabilization system during Sky Tronic SkyNav autopilot tests. Examples presenting the problems of UAV multi-rotor flight stabilization can be found in [1,5,7,12,13]. The observed problem related to the filtration of the measuring signal forced the authors to extend the possibilities of the available methods of measuring signal filtration. In this paper, a low-pass filter working in a closed-loop with feedback is proposed. The filter is built based on the PI controller. On the grounds of the assessment of external factors, the parameter values of the PI controller are automatically changed. The purpose of filtration is to eliminate disturbances while minimizing measurement delays. The proposed filter was compared with the traditional known solutions. The obtained results showed that the proposed solution introduces lower signal delays at a similar filtration range.

2 Model of the Low-Pass Filter Based on the Closed-Loop PI Controller

High-frequency disturbances of the measurement signal are one of the factors that can significantly reduce the reliability of the control system by destabilizing this system. The high-frequency disturbance is understood as the indications summed up with the measurement, the rate of increase and frequency of changes of which is higher than the dynamics of the object's operation. Thanks to this property, in many situations, it is possible to remove such interference by means of classic low-pass filters. However, low-pass filters introduce delays in the measurement signal. If the difference between the disturbance frequency and the dynamics of the object's operation is small, then the delay resulting from filtration becomes important. This situation was observed in a real multi-rotor UAV facility, in an inertial tilt angle measurement system. Due to the specificity of the system, it was not possible to perform this measurement with a more effective method. The source of interference is the measuring system itself and the mechanical vibrations of the structure.

To perform simulation tests, the measurement signal was defined by the formula (1). According to the formula, the measuring signal is the sum of the correct signal value and the disturbance.

$$\widehat{y(t)} = y(t) + n(t) \tag{1}$$

Where: $y(t)$ – the true value of the signal, $n(t)$ – disturbance caused by multiple factors.

A low-pass filter with transmittance defined by formula (2) can be used for the interference filtering.

$$G_f(s) = \frac{k}{Ts + 1} \tag{2}$$

Where: k – filter enhancement, T – filter time constant.

The above transmittance is mathematically equivalent to the transmittance of the integrating controller I with the feedback and the structure shown in Fig. 1

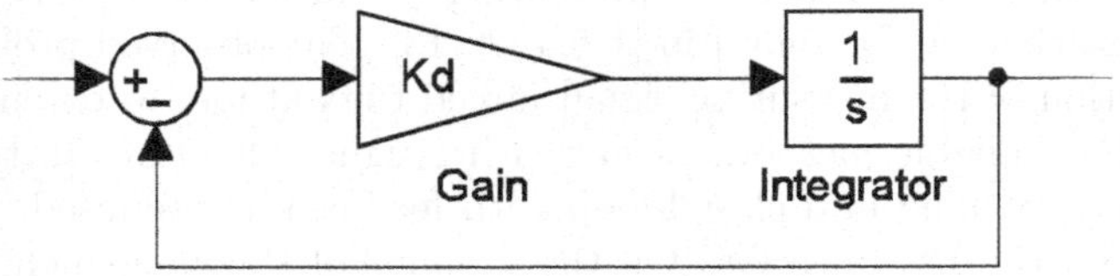

Fig. 1. Low-pass filter as a controller closed-loop with feedback

In the result it gives the transmittance according to the formula (3)

$$G(s) = \frac{1}{\frac{1}{K_d}s + 1} \tag{3}$$

This form of presenting the low-pass filter allows to introduce modifications based on the knowledge of control theory concerning classic PID control systems. Based on the observations of the phenomenon, it is also possible to introduce into the system mechanisms correcting the controller settings based on the identified state of the system, e.g. the magnitude of the process value (PV) change.

Thus, the controller I with feedback is a low-pass filter with the gain $K_d = 1$ and the time constant $T = 1/K_d$. The disadvantage of such a filter is a long output rise time proportional to the value of the time constant T, which makes it impossible to use in systems with high dynamics. In order to increase the speed of the controller output response to a sudden change of the setpoint value a proportional P term was introduced in the IND standard. The schematic diagram is shown in Fig. 2

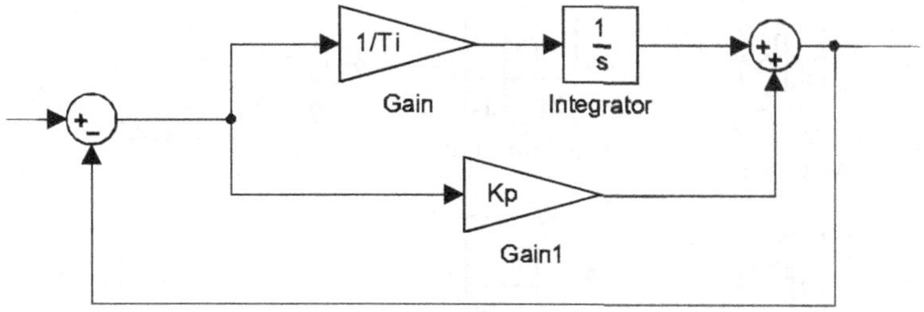

Fig. 2. Low-pass filter as a expanded controller closed-loop with feedback

Thanks to this procedure, transmittance (4) was obtained.

$$G_k\left(s\right)=\frac{K_pT_is+1}{T_i(K_p+1)s+1} \tag{4}$$

Due to the presence of the P term, the value on the controller output will reach the expected setpoint much faster, because the value of the Y output will be constantly increased by $K_p * e(t)$. When analyzing the transmittance above, the parameters K_p and $K_i = 1/T_i$ should be determined. The values of these parameters determine how the given term (P or I) will affect the value of the output signal depending on the value of the error, which can be described by the formula (5).

$$Y\left(t\right)=K_pe(t)+K_i\int e(t)dt \tag{5}$$

Two white noise generators with the time base fulfilling the condition (6) (one noise at twice the frequency than the other) were used for the simulation of disturbances. The disturbance obtained was summed and the sum was multiplied by the scale factor k (Gain2).

$$t_{G1}=2t_{G2} \tag{6}$$

The disturbances generated in this way were added to the set value signal (SP) in accordance with the Fig. 3. In addition, it was assumed that the disturbance amplitude must be smaller than the amplitude modulus of the set value changes according to the relationship (7).

$$\varDelta A_{noise}<|\varDelta SP| \tag{7}$$

The result of the operation of the disturbed measuring signal generator is shown in Fig. 4. The blue graph is the setpoint signal plus disturbance. For the first 100 s, the true value of the signal was set to 20, then it was lowered to 15.

The generated disturbance has a non-zero standard deviation and a non-zero mean value. For the correctness of the calculations, the simulation frequency was assumed to be 100 times greater than the highest frequency of disturbances. Numerical integration by the Euler method (with a constant integration step) was used for the tests.

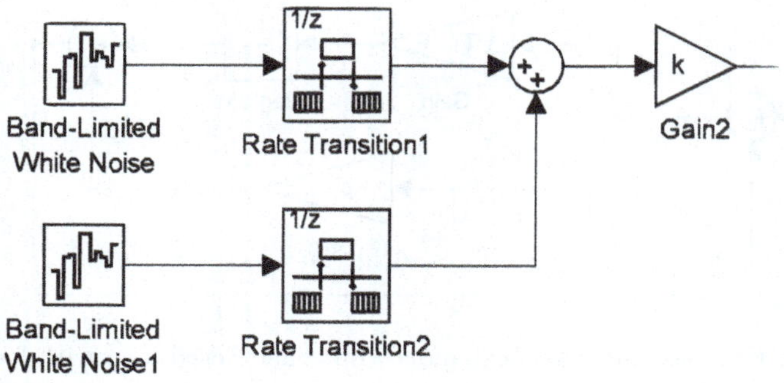

Fig. 3. Model of the signal generator with the disturbances in simulation tests

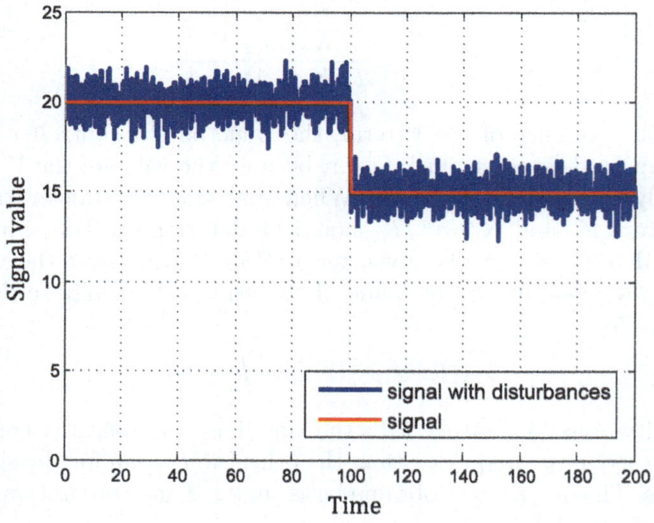

Fig. 4. Original signal and input signal with disturbances in simulation tests

3 Simulation Test Results

In order to verify the effectiveness of the developed model simulation tests were performed. In the first stage the system was tested for different values of K_p and K_i. Examples of simulation results are presented in the graphs in Figs. 5, 6. During each simulation, the values of K_p and K_i were determined. The simulation results were assessed on the basis of:

- rise time of the measuring signal,
- the magnitude of the disturbance amplitude of the output signal,
- Integral Absolute Error (IAE).

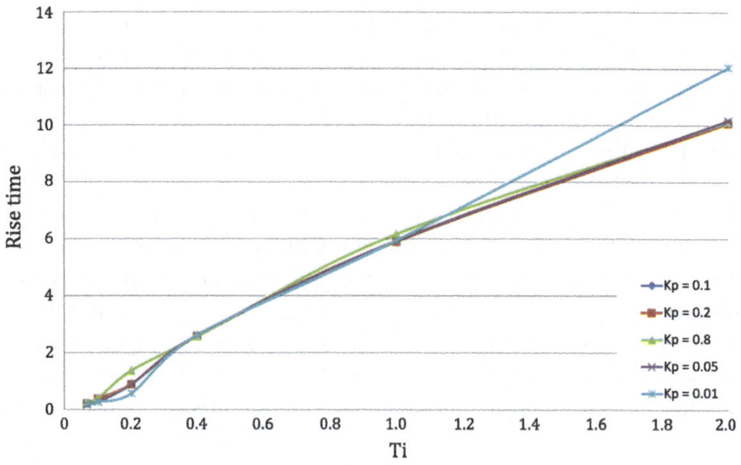

Fig. 5. Rise time of the output signal depends on T_i value for different K_p values

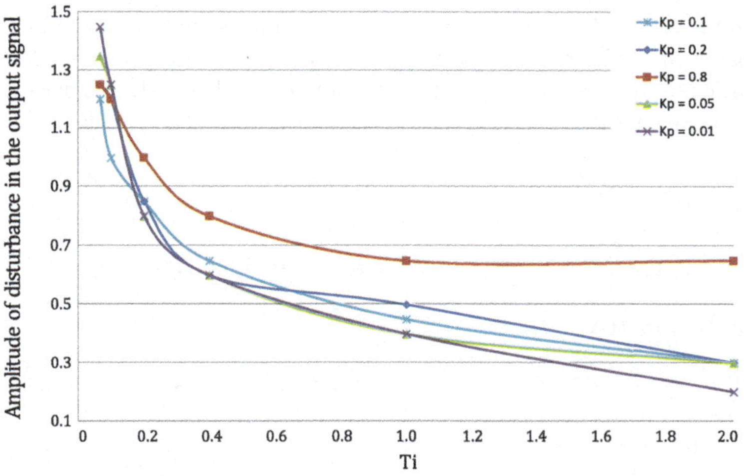

Fig. 6. Amplitude of disturbance in the output signal depends on T_i value for different K_p values

Based on the obtained results, an analysis was carried out to select the parameters of the target filtration system. The K_p parameter (proportional part gain) must be as small as possible. Based on the simulation, the range $K_p \in< 0.05, 0.8 >$ was determined. When it is in this range, it does not have a significant influence on the function accretion time (the graphs overlap). In the case of smaller values of K_p e.g. $K_p = 0.01$, the rise time begins to increase. The I term is dominant and the filter characteristic approaches the low-pass filter. The K_p parameter also affects the amplitude of disturbances in the output signal. For $K_p \in< 0.01, 0.2 >$ the diagrams of change of amplitudes (T_i) are very similar. For larger $K_p (K_p = 0.8)$, the amplitude of disturbances is kept at a much higher level (twice the amplitude). On this basis, it was concluded that the greater the K_p, the greater the signal amplitude will be. The interaction of the P term starts to be harmful, the disturbances are no longer suppressed. On the basis of the above, it was concluded that the appropriate range for K_p will be $K_p \in < 0.05, 0.2 >$.

T_i parameter affects the rise time of the signal. The greater the value of the integration constant T_i, the greater the rise time, i.e. the smaller the influence of the integrator on the value of the output signal. The larger the value of the integration constant T_i, the smaller the amplitude of the output signal disturbances, i.e. the larger the time constant effectively suppresses high-frequency signals.

Based on the above conclusions and test results, it was found that in order to obtain a filter capable of filtering fast-varying signals, the following should be used:

- fast rise time for small T_i values,
- low amplitude of the output signal for large values of T_i ($k_p = const$).

Therefore, the integration constant T_i was replaced by the function $T_i(|e(t)|)$, where the value of T_i depends on the value of the error modulus. It was checked what influence on the course of the output signal have the characteristics of changes of parameter $T_i(|e(t)|)$. The change of parameter $T_i(|e(t)|)$ was checked for:

- linear function (8)

$$T_i(|e(t)|) = a * |e(t)| + b \quad where \quad a = 3.625, \quad b = -3.125 \tag{8}$$

- quadratic function (9)

$$T_i(|e(t)|) = a * |e(t)|^2 + b * |e(t)| + c \quad where \quad a = 1.383, \quad b = -4.216, \quad c = 0.5 \tag{9}$$

- sigmoid function (10)

$$T_i(|e(t)|) = of_y + \frac{k}{1 + e^{-\alpha(|e(t)| + of_x)}} \tag{10}$$

where of_x, of_y – offset of function, k – gain is between T_{max} and T_{min}
- step function (11)

$$T_i(|e(t)|) = \begin{cases} 15 \; for \; |e(t)| \geq \sigma \; (\sigma \; - \; standard \; deviation) \\ 0.5 \; for \; |e(t)| < \sigma \end{cases} \tag{11}$$

In order to determine the efficiency of the filtration, the developed filter based on the PI controller was compared with the classic low-pass filter with transmittance (2) where $T = 2$. The adopted values had been considered optimal for this filter. Increasing the time constant T would increase the measurement delay, while reducing the time constant would increase the amplitude of disturbances. It should be added that this form of transfer function can be represented as an I controller with $K_d = 0.5$, therefore this test additionally shows the influence of the P term of the PI controller on the speed and quality of filtration. The filtration results for different values of the parameter T_i are compared in the Table 1 and the diagram of Fig. 8. The IAE index was used as a criterion, defining it as the absolute value of the difference between the signal after filtration and the original signal (before filtration).

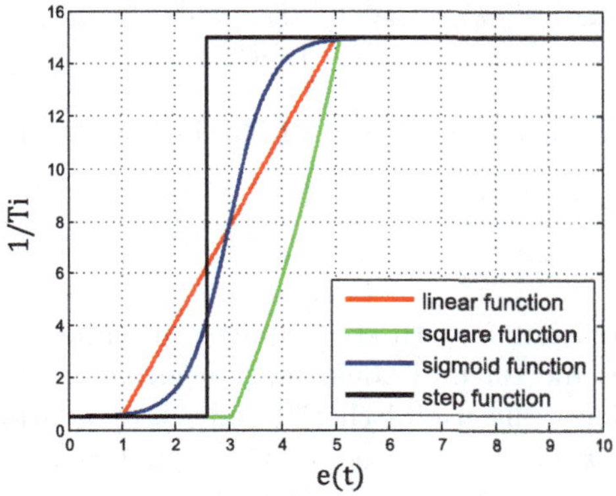

Fig. 7. T_i function used in simulation tests

Based on the data contained in the Table 1 it can be observed that the IAE index for the low-pass filter is even 2.6 times greater than the result for the modified filter. This is due to a delay in the basic version of the low-pass filter. In the steady state, the filtration with the low-pass filter and modified filters is similar.

Table 1. Comparison IAE index for different types Ti functions

Function Ti	IAE	Function Ti	IAE	Filter	IAE
Linear function	37680	Step function	25226	Low-pass filter	66775
Square function	27814	Sigmoid function	27819		

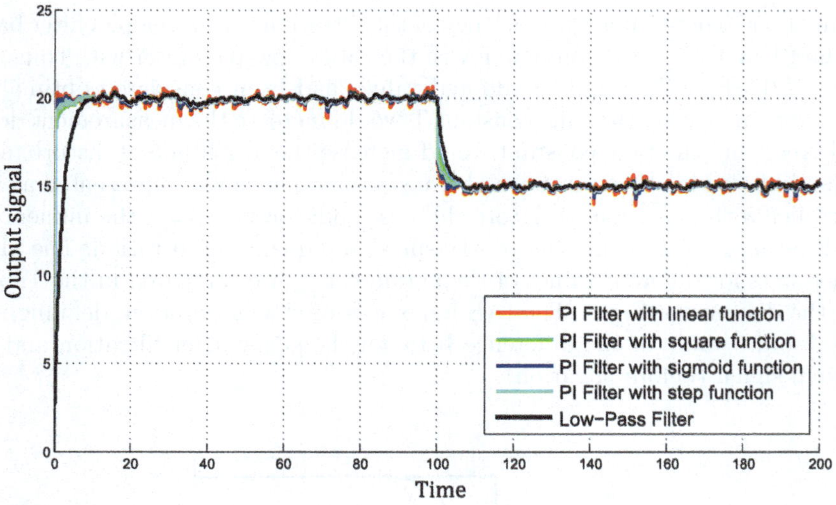

Fig. 8. Comparison between PI filtering and low-pass filter

4 Summary

Low-pass filtering for signals with step changes in values in systems with high dynamics introduces inertia and thus, may negatively affect the control. The proposed modification - the use of the PI controller as a low-pass filter with a variable value of $T_i(k_i)$ allows to reduce the signal delay (up to 50 times) so that the filter can be used for objects with high dynamics. The choice of the filter and its transition function depends on the effects we want to achieve. In the case of very short rise times, there will be greater amplitudes of steady-state oscillations. The T_i change characteristic must be more sensitive to step changes in the signal value which can be observed for linear and sigmoidal changes of T_i. In the event that we want a smaller amplitude of disturbances in the steady state, the signal delay will be longer (less sensitivity to small step changes in the signal value).

The presented low-pass filter solution has been simulated and compared to a classic solution with low computational complexity. At a later stage of the work, simulations will be made to assess the effectiveness of the proposed solution compared to more complex filtration systems, and additionally tests will be performed on a real UAV object.

References

1. Bouabdallah, S., Noth, A., Siegwart, R.: PID vs LQ control techniques applied to an indoor micro quadrotor. In: 2004 IEEE/RSJ International Conference on Intelligent Robots and Systems (IROS) (IEEE Cat. No.04CH37566), vol. 3, pp. 2451–2456. IEEE, Sendai, Japan (2004). https://doi.org/10.1109/IROS.2004.1389776

2. Kristiansson, B., Lennartson, B.: Evaluation and simple tuning of PID controllers with high-frequency robustness. J. Process Control **16**(2), 91–102 (2006)
3. Larsson, P., Hägglund, T.: Control signal constraints and filter order selection for PI and PID controllers. In: Proceedings of the 2011 American Control Conference, pp. 4994–4999 (2011)
4. Larsson, P., Hägglund, T.: Comparison between robust PID and predictive PI controllers with constrained control signal noise sensitivity. IFAC Proc. Vol. **45**(3), 175–180 (2012). 2nd IFAC Conference on Advances in PID Control
5. Lower, M., Szlachetko, B., Krol, D.: Fuzzy flight control system for helicopter intelligence in hover. In: 5th International Conference on Intelligent Systems Design and Applications (ISDA 2005), pp. 370–374. IEEE, Warsaw, Poland (2005). https://doi.org/10.1109/ISDA.2005.48
6. Micić, A.D., Mataušek, M.R.: Optimization of PID controller with higher-order noise filter. J. Process Control **24**(5), 694–700 (2014)
7. Nguyen Duc, M., Trong, T.N., Xuan, Y.S.: The quadrotor MAV system using PID control. In: 2015 IEEE International Conference on Mechatronics and Automation (ICMA), pp. 506–510. IEEE, Beijing, China (Aug 2015). https://doi.org/10.1109/ICMA.2015.7237537
8. Schreiber, T., Grassberger, P.: A simple noise-reduction method for real data. Phys. Lett. A **160**(5), 411–418 (1991)
9. Segovia, V.R., Hägglund, T., Åström, K.: Measurement noise filtering for PID controllers. J. Process Control **24**(4), 299–313 (2014)
10. Sekara, T.B., Matausek, M.R.: Optimization of PID controller based on maximization of the proportional gain under constraints on robustness and sensitivity to measurement noise. IEEE Trans. Autom. Control **54**(1), 184–189 (2009)
11. Shamsuzzoha, M., Lee, M.: Analytical design of enhanced PID filter controller for integrating and first order unstable processes with time delay. Chem. Eng. Sci. **63**(10), 2717–2731 (2008)
12. Szlachetko, B., Lower, M.: On Quadrotor Navigation Using Fuzzy Logic Regulators. In: Computational Collective Intelligence. Technologies and Applications, vol. 7653, pp. 210–219. Springer, Berlin, Heidelberg (2012). Series Title: Lecture Notes in Computer Science. https://doi.org/10.1007/978-3-642-34630-9_22
13. Szlachetko, B., Lower, M.: Stabilisation and steering of quadrocopters using fuzzy logic regulators. In: Artificial Intelligence and Soft Computing, vol. 7267, pp. 691–698. Springer, Berlin, Heidelberg (2012). Series Title: Lecture Notes in Computer Science. https://doi.org/10.1007/978-3-642-29347-4_80
14. Moon, Y.H., Ryu, H.S., Lee, J.G., Kim, S: Power system load frequency control using noise-tolerable PID feedback. In: ISIE 2001, 2001 IEEE International Symposium on Industrial Electronics Proceedings (Cat. No.01TH8570), vol. 3, pp. 1714–1718 (2001)

Reliability of Multi-rotor UAV's Flight Stabilization Algorithm in Case of Object's Working Point Changes

Michal Lower[1,2(✉)] and Boguslaw Szlachetko[1,2]

[1] Faculty of Electronics, Wroclaw University of Science and Technology, Wroclaw, Poland
{Michal.Lower,Boguslaw.Szlachetko}@pwr.edu.pl
[2] Sky Tronic sp z o.o., Wroclaw, Poland

Abstract. The subject of the paper is the dependability of a UAV stabilization system. The dependability depends on many factors among which the paper focuses on the accidental change of the working point of the system, e.g. by a change of the equilibrium point or change of the UAV mass. Both of these factors are neglected in publications because it is usually assumed that deviations from the designed operating point cannot occur during normal flight. This paper addresses the problem of designing a flight stabilizer that is robust to such random events. Adaptation of a traditional stabilizer based on a PID controller proved to be a non-trivial task. For this reason, in the paper, the use of the SkyNav controller developed at the Faculty of Electronics of the Wroclaw University of Science and Technology is proposed. The SkyNav controller uses the fuzzy logic rules to determine the correct values of flight stabilizing propellers control. Thanks to the application of fuzzy rules, a nonlinear controller was created, which much better reflects the nonlinear nature of the controlled object, such as a UAV. As a result, a much wider range of operating conditions was obtained, which makes the controlled UAV more reliable.

Keywords: Flight stabilization · Fuzzy logic controller · Reliability · UAV

1 Introduction

Flight stabilization of multi-rotor UAVs is a task sufficiently complex and demanding in terms of response time that this task is normally delegated to a computerized system called a flight controller. Virtually all off-the-shelf controllers use a fusion of measurement data from inertial, laser and other sensors, as well as GPS signals to determine the state of the UAV during flight. By this term we mean basic information such as UAV position (in local or global coordinates), UAV linear velocity, attitude (roll, pitch and yaw angles, and angular rotation speeds [1–3,6,7]. This information in turn allows the development of

W. Zamojski et al. (Eds.): DepCoS-RELCOMEX 2021, AISC 1389, pp. 254–263, 2021.
https://doi.org/10.1007/978-3-030-76773-0_25

UAV stabilization and state control algorithms, which are a key component of a flight controller. Unfortunately, controllers in use today use a classical technique based on PID controllers [2,4], which only work properly for objects with linear behavior. The UAV is a strongly nonlinear object. For example, in order to block the rotation of the UAV about the vertical axis (i.e. yaw angle), it is sufficient to ensure that the sum of the rotational speeds of the propellers spinning in opposite directions is equal to zero. In other words, there is a linear relationship here. However, the lift force produced by a single propeller is proportional to the square of the propeller's rotational speed. When the UAV tilts sideways or forward/backward, additional stabilizing lift force must be produced on the corresponding side of the UAV by changing the propeller speed. Unfortunately, it affects the speed balance of the other propellers and causes rotation about yaw axis. Moreover, the dynamic equations of UAV motion depend nonlinearly on the position of the center of gravity and the torque of the UAV treated as a rigid body. The standard solution is to linearize the dynamics of UAV motion at a particular operating point. Every change of mass, size of UAV, shift of the center of gravity causes significant changes of the operating point and the direction of these changes is difficult to predict in many cases. For this reason, a very important issue is to develop a flight controller that will be reliable and credible during normal UAV operations.

The UAV is a non-linear object, so it can be expected that the development of a non-linear flight controller will improve the reliability of flight stabilization. The use of fuzzy logic enables the implementation of a nonlinear controller. The authors of the paper developed such a controller, the process of reaching a solution is presented in [5,12,13]. It should be emphasized that the controller can be performed in many different ways with the aid of fuzzy logic. In the literature the other suggestions for solutions can be found, e.g. [8,10].

The paper presents the results of testing the SkyNav flight controller from Sky Tronic. The tests using SkyNav were carried out on a prototype Hoverbike air scooter manufactured by Skynamo. The innovative solution of the flight stabilizer was developed at the Faculty of Electronics of the Wroclaw University of Technology and as a result of technology transfer has been commercialized in the Sky Tronic spin-off company established at the Wroclaw University of Technology. The main emphasis has been put on the dependability aspect of the UAV associated with the use of a flight controller using Fuzzy Logic technology.

2 Standard and Non Standard Multi-rotor UAV

The most common UAV is the quadcopter. It is a multi-rotor object that is characterized by simplicity of design, lack of moving parts in the structure (except, of course, for the propellers, whose movement is required to generate lift force), and a very simple geometric configuration that allows for simple calculation of the lift force distribution. These features make the quadcopter the most widely used multi-rotor design. Other regular designs are the hexacopter and octocopter which have six or eight propellers respectively, distributed regularly in a circular geometry. A larger number of propellers is a more survivable solution for a single engine failure. The control and stabilization principle of the hexacopter and octocopter is the same as that of the quadcopter. For this reason, the quadcopter will be considered as the standard design of a multi-rotor UAV.

There exists irregular designs in which the propellers are arranged in a rectangular or cross geometry. Additionally, it is possible to separate the function of the propellers that produce the main lift force of the UAV from the stabilizing propellers. Such a construction is the flying scooter called Hoverbike. Like the quadcopter, it is a strongly nonlinear control object. The Hoverbike is an object, ultimately intended to carry people and transport cargo. Its intended mass is in the range of 350–450 kg, while its horizontal size is approximately 4 m long and 2 m wide. The Hoverbike's lift force system allowing climb and flight is composed of six horizontal propellers. The propeller arrangement is as shown in Fig. 1. The solution uses two large propellers, approximately 1.5 m in diameter, providing the main lift force of the object and four small stabilizing propellers, placed in a rectangular arrangement, on the edges of the scooter. The lift propellers rotate in opposite directions to each other and are driven directly by an internal combustion engine. The control propellers have electric motors and their rotation directions are the same as the classic quadcopter arrangement - two propellers rotate to the left and two to the right, with the diagonally arranged propellers rotating in the same direction. In terms of stabilizing the Hovebike in the horizontal plane, the algorithm performing this task does not differ from the solutions used in the quadcopter.

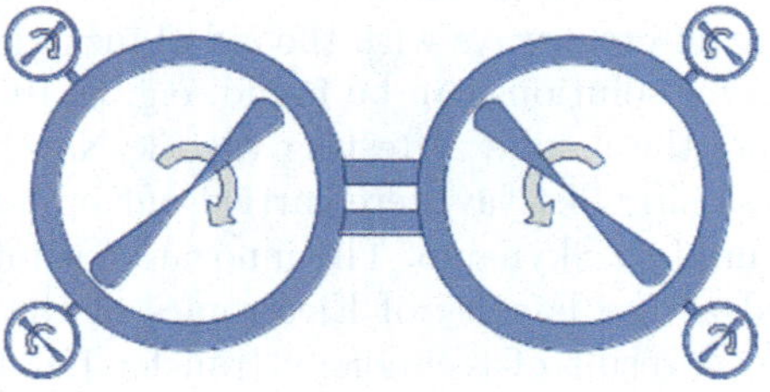

Fig. 1. Schema of the placement of propellers: main lift propeller (the big one), stabilizing propeller (a small one).

The major differences in the control system, comparing to quadcopter, are in vertical axis yaw and altitude stabilization. The four control propellers in the corners of the Hoverbike are too small in relation to the size of the object to achieve the expected effect related to yaw rotation or altitude stabilization. The altitude stabilization of the Hoverbike is accomplished by varying the angular velocity of the large lift propellers. Controlling the speed of these propellers is achieved by the internal combustion engine throttle and during the tests was set up remotely by the operator. Stabilization of the Hoverbike's yaw angle is accomplished by means of jets located directly under the lift propellers.

The curvature of the airstream underneath the propeller allows for a torque acting on the object.

3 PID Controller

Flight controller built on the basis of hierarchical - usually - structure of PID controllers is a linear system and therefore requires linearization of the UAV model around its operating point. The basic requirement for an object equipped with such a controller is a weakly disturbed space (e.g. by wind gusts), constant or predictable properties of the flying object, i.e. constant mass and invariable center of gravity. The quality of control obtained in this type of systems depends on the proper selection of the settings of individual members of the PID controller. After the controller is tuned in the operating point, a stable flight near this point is achieved. The more the conditions change the operating point, the worse the results of stabilization are, up to the point where it is impossible. A change of the operating point can occur, among other things, as a result of a change of the UAV weight, a strong gust forcing a violent reaction of the controller, a change of the power of the engines, a change of the battery voltage. A special case is the replacement of a typical 3S LiPo battery for a 4S which results in a change of nominal voltage from 11.1 V to 14.4 V. Powering the motors with a higher voltage results in a proportionally higher motor RPM, which significantly changes the operating point - in this case, the motors RPM would increase by about 30%.

There are no simple, clear-cut rules for determining the settings of a PID controller [9,11]. The tuning must be based on operator experience and a series of experiments performed directly at the operating point. Finding the right settings is a process that requires a lot of experience and does not always end with a satisfactory result. Therefore, such a controller fulfills its tasks to a relatively limited extent. Successful use of this type of algorithm is based on good knowledge of the UAV in a wide range of states. At the same time, the developed and tuned algorithm is closely related to the specific design of the UAV. In other words, its versatility is limited to a specific UAV model and most changes in design or in the electrical supply forces a re-tuning of the controller.

4 Fuzzy Logic Based Controller - the SkyNav

Dependability of the SkyNav flight stabilization algorithm results from the use of fuzzy logic rules. The benefits of such a solution can be observed when changes in UAV parameters or changes in the surroundings occur under normal flight conditions. The controller calculates the process variable based on the knowledge base written in the form of rules. The SkyNav controller was developed as a result of the authors' scientific work. The first steps of the creative process can be found in the literature [5]. Next, in further works, fuzzy control algorithm for a quadcopter was developed and presented in [12,13]. Based on the algorithm the commercial SkyNav autopilot was developed. Our tests conducted on the Hoverbike showed that the SkyNav controller has a much larger range of stable operation for the same controller settings than one would expect from a PID controller. During several months of testing, the designers of the Hoverbike changed the mass of the object, the load distribution, and the power of the stabilizing propellers several times. Changing the mass of the system by 40% was unnoticeable in stabilization quality, and similarly, changing the power of the drive motors (by as much as 300%).

The SkyNav controller requires tuning of two parameters linked to the direct physical properties of the UAV. One parameter is the gain of the controller proportional to the moments of inertia of the UAV. The other parameter is the acceleration factor, which depends on the available power and dynamics of stabilizing motors coupled with the propellers.

5 Experiments Results with the Use of the SkyNav

5.1 Experiments on a Test Bench, Under Suspension Conditions

Tests of flight control system in the first phase were performed on a specially prepared for this purpose test bench. The test bench consisted of a special arm, on which the Hoverbike scooter was suspended. The suspension was made with a steel rope and a hook capable of withstanding the force of over 500 kg. The suspension point was chosen at the center of gravity of the scooter. As a result, the steel rope connecting the hook and the scooter at rest marked a vertical line directed along the gravitational force, and the scooter hung horizontally. In fact, the frictional forces between the rope and the hook caused a small - on the order of single degrees - deflection of the scooter in the roll and pitch axes at rest. Yaw axis remained completely free.

As mentioned earlier, SkyNav requires tuning of two parameters. One parameter is the controller gain and the second one is the acceleration factor called E_{opti}. This coefficient can be set arbitrarily taking into account the power of the control motors. The maximum angular acceleration that can be braked in the system depends on the maximum torque of the braking force that is available. The parameter E_{opti} determines the maximum acceleration that is allowed in the system. Thus, this parameter, within the range of allowable values, determines the response rate of the controller. Setting E_{opti} too small will cause the system to be slower to reach the setpoint. In the experimental study, $E_{opti} = 5$ was set up.

The controller gains called I_{roll} and I_{pitch} should be proportional to the moment of inertia of the object in a given axis of rotation. Determining these parameters from theoretical calculations is a complicated task, so the authors set the parameter values a priori based on the visual appearance of the Hoverbike and approximate knowledge of its mass. These values were corrected according to results observed during several experiments.

The following figures Fig. 2 and Fig. 3 show plots of the measured angles in the roll and pitch axes. The changes in yaw angle were highly random since the yaw axis controller was not considered at all in these experiments. For this reason, yaw angle plots are not included in the figures. The experiment was conducted with the internal combustion engines turned on, which produced the main lifting force of the scooter due to the installed propellers. The rotational speed was chosen so that the generated lift force was close to that which would allow the scooter to fly freely. Such a conduct of the experiment was necessary because it was noticed that there is a large influence of the operation of internal combustion engines on the interference of measurements from inertial sensors, i.e. gyroscopes and accelerometers.

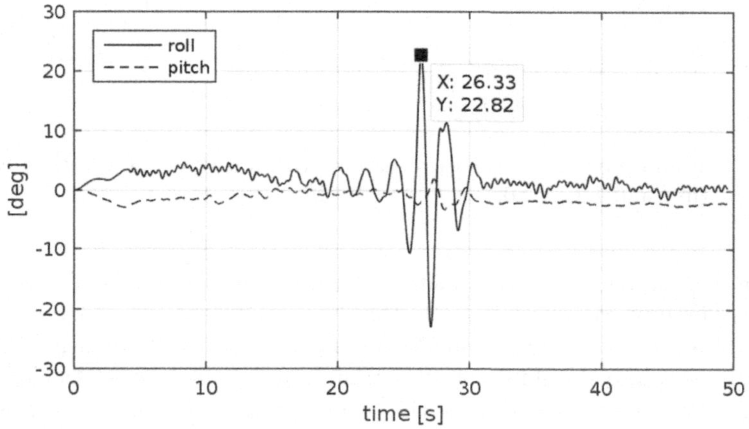

Fig. 2. A roll and pitch angle values registered with a set-point for both axes equals zero all the time and $I_{roll} = 30$.

In Fig. 2, the corresponding solid and dashed lines indicate the estimated values of the scooter's pitch and roll angles, respectively. The moment when the controller is turned on is visible at time $t = 4$ [s], which manifests itself by small oscillations in the roll axis and stabilization of the Pitch angle to the preset zero value. Before the stabilizer was turned on, the scooter was freely suspended from the rope and was positioned at roll $= 1°$ and pitch $= -1.5°$. At time $t = 26$ [s] a forced disturbance occurred which threw the controller out of equilibrium. As a result, after the disturbance the oscillations are dampened in about 4 s.

In Fig. 3 a plot of roll and pitch angles are presented, but with the correctly chosen value of $I_{roll} = 80$. Other conditions of the experiment are identical to the case presented in Fig. 2. Disturbances were forced two time: at about $t = 13$ [s] and $t = 23$ [s]. The disturbance was forced by the operator, but the operator was not able to force it only on the roll axis in a controlled manner. Hence, simultaneous disturbances on both roll and pitch axes are observed. The suppression of the disturbance occurred in less than 1 [s] and no overshoot was observed. Further increase of the $I + roll$ parameter led to a reduction of the controller response time, but this was at the expense of an increase in overshoot and for this reason it was decided to leave $I_{roll} = 80$ for further experimental work.

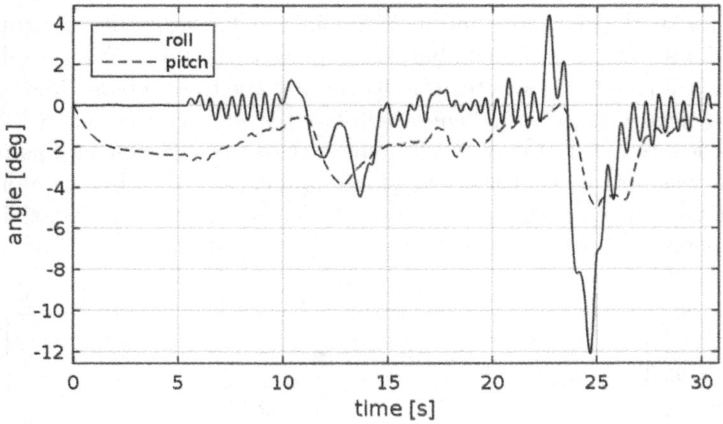

Fig. 3. A roll and pitch angle values registered with a set-point for both axes equals zero all the time and $I_{roll} = 80$.

A similar tuning procedure was performed for the pitch axis. The parameter I_{pitch} was mainly changed. I_{pitch} is the gain coefficient corresponding to the moment of inertia of the object in the pitch axis. The results presented in Fig. 2 and Fig. 3 were already obtained after tuning in the pitch axis. As can be seen in both figures, the stabilization of the scooter in the pitch axis performs better than in the roll axis. This is due to the much higher moment of force obtained from the same motors in the longitudinal direction of the scooter. During the tests, the disturbances were forced by the operator by lifting the front part of the scooter, which required a very high force with both the combustion and electric propulsion systems on, when all six propellers were generating lift forces.

5.2 Experimental Study of the SkyNav Controller During the Flight

After completing the research and experiments on the hook, the behavior of the SkyNav fuzzy flight controller during flight was studied. A very important conclusion of the experiments is that the operation of the stabilizer, whose settings were selected on the test bench in the hovering state of the scooter on the hook, was immediately correct and did not require any additional tuning of the parameters. The results of the first experiment after taking the scooter out of the hangar are shown in Fig. 4. Clearly visible are the effects of a light wind disturbing the airflow under the lifting engines, despite which, the scooter maintained a level within ± a few degrees, which is a very good result.

A few months later, SkyNamo undertook further research on a modified version of the Hoverbike scooter. The main changes from the original design were the internal combustion power unit and the stabilizer unit motors and propellers. Changing the internal combustion engine to a more powerful one had little effect

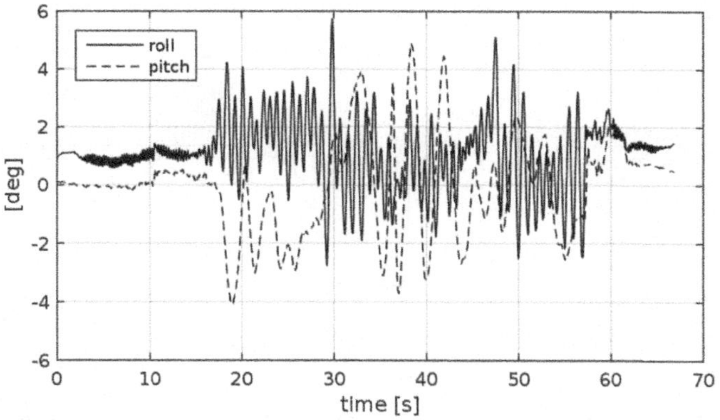

Fig. 4. A roll and pitch angles registered during hovering on a low altitude with a set-point equals zero for both axes all the time.

on the stabilizer, as both the center of gravity and mass did not change significantly. However, replacing the electric motors with larger and more powerful ones and changing the diameter of the propellers mounted on these motors, which involved increasing the wheelbase of the stabilizer motors, resulted in fundamental changes in the operating point of the system. The using of a PID controller in this case would require re-tuning the controller on a laboratory bench in a hover state and then conducting many tedious flight tests. Many of these would have ended in crashes and equipment destruction. The manufacturer of the Hoverbike scooter decided to undertake flight tests with our fuzzy controller right away, without prior tuning on the hook. In other words, it was assumed that the fuzzy inference used in the developed controller would allow the scooter's flight stabilizer to work properly even despite such significant changes.

After the design changes were made to the Hoverbike scooter, a series of experiments were performed. The results of stabilization and control during flight at low altitude above the ground are presented in Fig. 5. It is worth noting that the flight of the multirotor scooter just above the ground is characterized by high air turbulence caused by the air stream reflected from the ground. For this reason, the controller works here under very difficult conditions and has to stabilize the random disturbances all the time.

The first problem visible in Fig. 5 is the large oscillations in the roll axis caused by the controller overshooting, which is related to the incorrect I_{roll} parameter. The oscillations were so large that it was not possible to achieve set-point in roll equals 5° which was set by the operator at time $t \in (50, 60)$ [s]. On the other hand, the result of adjustment and stabilization for the pitch axis was surprising. The graph clearly shows almost smooth curve in the moments of stabilization of the pitch = 0°. Also, there are visible very fast reactions of the controller to the set-point value of pitch = 5° at time $t \in (17, 28)$ [s] and pitch = −10° at time $t \in (32, 44)$ [s] of the flight. It is worth noting that the results

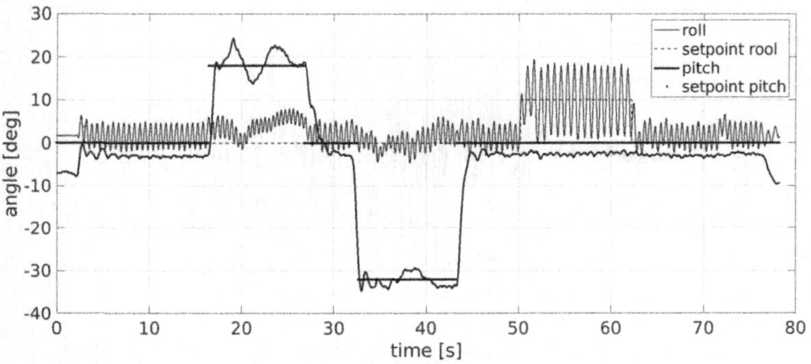

Fig. 5. A roll and pitch angles registered during flight on a low altitude with a setpoints imposed by the operator.

presented in Fig. 5. Were obtained after significant design changes, without any modification and tuning of the stabilization algorithm based on fuzzy inference.

6 Conclusions

The conducted experiments showed that changing the operating point (changing the mass of the object, changing the center of gravity, changing the power of the control motors), did not influence the reliability of the system. Moreover, the use of the proposed stabilization algorithm increases the dependability and the safety of multirotor UAV use. As a result, it makes it possible to use such an object to transport cargo with variable mass and with unknown distribution of this mass. A change in the operating point of a UAV can occur as a result of a number of different factors, e.g., emergency states, variable violent weather conditions, etc. Many of these factors are difficult to test under controlled conditions. As an example of an unexpected change, consider the addition of an elongated fuel tank about 1 m long to the Hoverbike scooter, which was placed in the horizontal plane. Such modification was introduced by Skynamo engineers without informing the authors. The tank caused a significant change in the center of gravity with each tilt, as it was not equipped with internal gores to slow the flow of fuel. Nevertheless, without any modifications or controller tuning, the scooter maintained stable flight. Another example of an unexpected modification involved in the scooter construction is the introduction of an additional aperture mounted asymmetrically in the space of the scooter's lift propellers. No effect of these modifications on stability was observed during the experiments, either.

The tests also showed the SkyNav controller to be highly resistant to some measurement system failures. An accidental programmer's error resulted in a patch that reduced the measured values of roll and pitch angles by a factor of ten. As a result of this error, the Hoverbike sometimes rapidly lost stability. However, the instability phenomenon was observed infrequently enough, with the other stabilization results being of little concern, that testing of the faulty system could be conducted fairly safely for several months.

In summary, it should be noted that the SkyNav system has been tested in an early prototype version and appropriate adjustments are being made to the algorithm based on these tests. One of the elements to be eliminated or reduced are oscillations near the equilibrium point. To achieve this, additional inference rules are planned to modify the algorithm responses near the equilibrium point.

References

1. Bouabdallah, S., Noth, A., Siegwart, R.: PID vs LQ control techniques applied to an indoor micro quadrotor. In: 2004 IEEE/RSJ International Conference on Intelligent Robots and Systems (IROS) (IEEE Cat. No.04CH37566), vol. 3, pp. 2451–2456. IEEE, Sendai, Japan (2004)
2. Bouabdallah, S.: Design and control of quadrotors with application to autonomous flying. PhD Thesis, Lausanne, EPFL (2006)
3. Cavallo, A., Cirillo, A., Cirillo, P., De Maria, G., Falco, P., Natale, C., Pirozzi, S.: Experimental comparison of sensor fusion algorithms for attitude estimation. IFAC Proc. Vol. **47**(3), 7585–7591 (2014)
4. Li, J., Li, Y.: Dynamic analysis and PID control for a quadrotor. In: 2011 IEEE International Conference on Mechatronics and Automation, pp. 573–578. IEEE, Beijing, China (Aug 2011)
5. Lower, M., Szlachetko, B., Krol, D.: Fuzzy flight control system for helicopter intelligence in hover. In: 5th International Conference on Intelligent Systems Design and Applications (ISDA 2005), pp. 370–374. IEEE, Warsaw, Poland (2005)
6. Madgwick, S.O.H., Harrison, A.J.L., Vaidyanathan, R.: Estimation of IMU and MARG orientation using a gradient descent algorithm. In: 2011 IEEE International Conference on Rehabilitation Robotics, pp. 1–7. IEEE, Zurich (Jun 2011)
7. Mahony, R., Hamel, T., Pflimlin, J.M.: Nonlinear complementary filters on the special orthogonal group. IEEE Trans. Autom. Control **53**(5), 1203–1218 (2008)
8. Marcu, E., Berbente, C.: UAV fuzzy logic control system stability analysis in the sense of Lyapunov. Sci. Bull. **76**(2), 37–48 (2014)
9. Noshahri, H., Kharrati, H.: PID controller design for unmanned aerial vehicle using genetic algorithm. In: 2014 IEEE 23rd International Symposium on Industrial Electronics (ISIE), pp. 213–217. IEEE, Istanbul, Turkey (Jun 2014)
10. Precup, R.E., Tomescu, M.L., Preitl, S.: Fuzzy logic control system stability analysis based on Lyapunov's direct method. Int. J. Comput. Commun. Control **IV**(4), 415–426 (2009)
11. Salih, A.L., Moghavvemi, M., Mohamed, H.A.F., Gaeid, K.S.: Modelling and PID controller design for a quadrotor unmanned air vehicle. In: 2010 IEEE International Conference on Automation, Quality and Testing, Robotics (AQTR), vol. 1, pp. 1–5 (2010)
12. Szlachetko, B., Lower, M.: On quadrotor navigation using fuzzy logic regulators. In: Computational Collective Intelligence, Technologies and Applications, vol. 7653, pp. 210–219. Springer, Berlin, Heidelberg. Series Title: Lecture Notes in Computer Science (2012)
13. Szlachetko, B., Lower, M.: Stabilisation and steering of quadrocopters using fuzzy logic regulators. In: Artificial Intelligence and Soft Computing, vol. 7267, pp. 691–698. Springer, Berlin, Heidelberg. Series Title: Lecture Notes in Computer Science (2012)

Semi-Markov Model of Processing Requests Reliability and Availability in Mobile Cloud Computing Systems

Jerzy Martyna(✉)

Institute of Computer Science, Faculty of Mathematics and Computer Science
Jagiellonian University, ul. Prof. S. Lojasiewicza 6, 30-348 Cracow, Poland
jerzy.martyna@uj.edu.pl

Abstract. The article presents the semi-Markov process as a model of processing requests in mobile cloud computing (MCC) systems. MCC is defined as a mobile device-based infrastructure for running standalone applications and/or accessing remote applications over wireless networks. This article defines a multi-state model based on semi-Markov processes for predicting the future availability and reliability processing requests in an MCC system. This allows for performance impact and makes the decision to increase the speed, disk space of the system, etc. Moreover, the numerical results for the model presented allow the efficiency of the request processing in the MCC system to be evaluated.

Keywords: Semi-markov process model · Mobile cloud computing system · Reliability and availability of processing requests

1 Introduction

Mobile cloud computing (MCC) refers to the cloud computing services in a mobile environment. Thus, it is a combination of mobile wireless networks and a computing cloud providing computing services, computing power, and a series of software resources for mobile users. It is a leading technology that has been developed for several years. Everyone, from cloud computing owners and mobile network operators, is interested in maintaining and expanding MCC systems. It is in everyone's interest to ensure the highest reliability and availability of processing requests in these systems. Mobile users are also interested in the quality and capabilities of MCC systems.

MCC systems have been the subject of many publications and research studies. Among others, a survey of MCC, which helps general readers have an overview of the MCC including the definition, architecture, and applications was presented by Dinh *et al.* [1].

In the paper by Sanari *et al.* [2] the impacts of heterogeneity in MCC are investigated, and related opportunities and challenges are identified. Moreover, predominant heterogeneity handling approaches like virtualisation, middleware,

W. Zamojski et al. (Eds.): DepCoS-RELCOMEX 2021, AISC 1389, pp. 264–272, 2021.
https://doi.org/10.1007/978-3-030-76773-0_26

and service oriented architecture (SOA) are also discussed. The issues of energy-efficient transmission in MCC were the subject of the work presented by Liu *et al.* [3]. In current papers, for mobile/cloud computing an application to management based on process state synchronisation was presented by Ejaz *et al.* [4]. The security and privacy aspects relevant to MCC are classified in the paper by Al-Omary *et al.* [5].

Providing high reliability and availability in cloud computing is a desirable goal for most IT investments. But despite this, reliability and high availability in cloud computing services are a great challenge. It has been the subject of many studies and articles. Among others, a systematic review and research challenges to improve the availability of a service, such as checkpointing, load balancing, and redundancy in cloud computing systems was given by Endo *et al.* [6]. An algorithm that makes the system highly fault tolerant by considering forward and backward recovery using diverse software tools is presented by Mohammed *et al.* [7]. A reference roadmap of reliability and availability in cloud computing environments was presented by Mesbahi *et al.* [8]. Despite this, according to the author's knowledge, none of these studies took into account the reliability and availability of MCC systems.

Semi-Markov processes are not as memoryless as Markov processes. Instead, the semi-Markov processes depend not only on the current states, but also on their duration in the current states. Therefore, several authors have used semi-Markov processes to model system reliability. Among others, Barbu *et al.* [9] has used a discrete time semi-Markov model for reliability and survival analysis. Some models based on semi-Markov processes were also developed by employing the aggregated stochastic process. For example, in the paper by Wang *et al.* [10], the performance of semi-Markov repairable systems with history-dependent up and down states was analysed. Cui *et al.* [11] proposed a single-unit repairable system consisting of an operating subsystem and a maintenance subsystem. Barbu *et al.* [9] used a discrete time semi-Markov model for reliability and survival analysis. In the paper by Bei *et al.* [12], a competing risk model with a restriction on transition times for semi-Markov multi-state repairable systems is proposed.

The main goal of the paper is to present the model of reliability and reachability of processing requests in the MCC system. With this model, it is possible to predict the future behaviour of the processing request in the MCC system for various parameters of the system. This allows for the calculation of MCC behaviour under different circumstances for known data for a given MCC system. It can also be helpful in the design of such systems.

The paper has been organised as follows: Sect. 2 presents the mobile cloud computing system. Section 3 presents the multi-state semi-Markov process based modelling of reliability and availability. Section 4 presents the model of the proposed approach using mathematical analysis and its implementation details with processing requests in the MCC system. Section 5 shows a numerical example of the application of the model presented. Section 6 concludes the paper.

2 Mobile Cloud Computing System

A mobile cloud computing (MCC) system can be described as the ability of cloud system services in mobile devices belonging to a number of mobile networks. The architecture of an MCC (see Fig. 1) is composed from the components: mobile users, mobile operators, Internet service provider (ISP), cloud providers [13].

The operation of the MCC of the system is to submit a request to the system by a mobile user. The response time is a random variable that depends on a number of factors, such as: finding free cache memory in the system controller, finding the appropriate cloud, gaining access to the appropriate storage device, and retrieving the requested data. If the data sought has been found in the controller cache memory, it is immediately sent to the mobile user. On the other hand, when data is retrieved from the computing cloud, it is again placed in the memory cache of the controller to be then sent to the mobile user. Additionally, during the processing of the request, the radio link with the base station may be lost or radio interference may occur due to interference. This either causes the request to stop processing completely or the connection is re-established and the interrupted request continues.

The response time depends on the reliability and availability of the system MCC. Establishing a value for reliability and availability is critical to determining the effectiveness of an MCC.

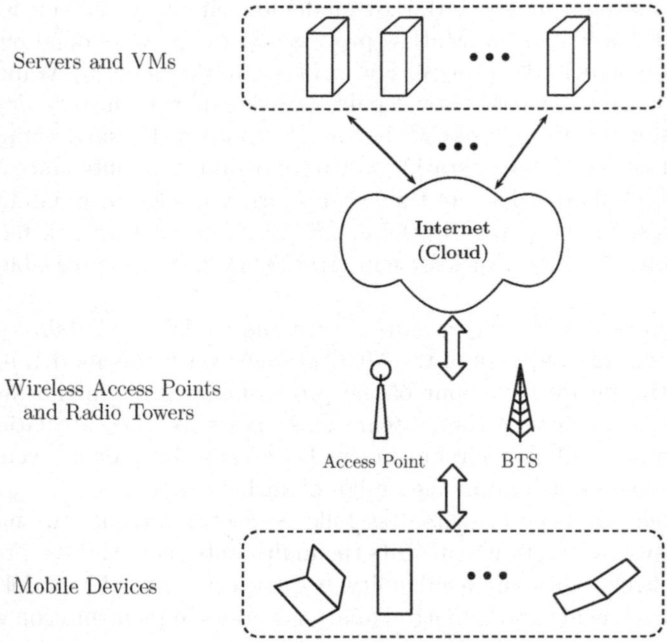

Fig. 1. Basic architecture of mobile cloud computing system.

3 Multi-state Semi-Markov Process Based Modelling of Reliability and Availability

The Multi-state semi-Markov (MSSM) process was chosen to predict the reliability and availability of the MCC system. Using MSSM for modelling reliability and reachability does not require application of machine learning, which allows for obtaining results with higher accuracy compared to a linear time series. The general formulation of the continuous-time discrete-state semi-Markov process models are presented by Howard [14] and Rausand *et al.* [15].

The semi-Markov process (SMP) is described by states and transitions between them managed by a matrix of transition probability. SMP can therefore be defined by the transition probability matrix and the initial distribution. Time spent in any state after entering it is a random variable with any distribution.

Thus, a kernel matrix $Q(t)$ is given by:

$$Q(t) = [Q_{ij} : i, j \in S] \tag{1}$$

where Q_{ij} is the probability that the process stays in a state i for time less than or equal to t. before it moves to state j and S is a set of states.

Additionally, the initial state vector is defined as follow:

$$\mathbb{P}_0 = [p_i^{(0)} : i \in S] \tag{2}$$

where $p_i^{(0)}$ defines the initial distribution of SMP process and $\sum_{i \in S} p_i^{(0)} = 1$.

The Eqs. (1) and (2) given above define the stochastic behaviour of SMP. Time can be referred as the sojourn time and it has an arbitrary distribution. Future states of processes are not dependent on previous states and sojourn times.

The *probability density function* (pdf) given by $f_{ij}(t)$ is defined as the sojourn time corresponding to the transition from state i to state j at time t. Analogously, the *cumulative distribution function* (cdf) is indicated by $F_{ij}(t)$. Thus, the probability that the next transition state is j and not any other state k reachable from state i is given by:

$$C_{ij}(t) = f_{ij} \prod_{k \neq j} (1 - F_{ik}(t)) \tag{3}$$

The kernel matrix is defined by:

$$Q(t) = [C_{ij}(t)] \tag{4}$$

The waiting time density function for state i is defined by

$$w_i(t) = \sum_{j \in S} C_{ij}(t) \tag{5}$$

The probability that the system does not in leave state i by time t is given by

$$w_i(t) = 1 - \int_0^t w_i(t)dt \tag{6}$$

This probability is also referred to as the *complementary cumulative waiting time probability*. Given the Eq. (3) and calculated above $w_i(t)$ the probability can be found of being in each state j assuming that the system is in state i, namely:

$$\Phi_{ij} = \delta_{ij} w_i(t) + \sum_k \int_0^t C_{ik}(\tau)\Phi_{kj}(t-\tau)d\tau \tag{7}$$

where $\delta_{ij} = 1$ if $i = j$, 0 otherwise and $i, j, k = 1, 2, \ldots, \mid S \mid -1$.

The system of equations given by the Eq. (7) can also be presented in a simpler form, namely

$$\phi(t) = diag(W(t)) + \int_0^t C(\tau)\phi(t-\tau)d\tau \tag{8}$$

System reliability can be expressed in terms of its time-to-failure distribution, which can be represented by respective *pdf*, *cdf* [16]. If the system is non-repairable, then the lifetime of a nonrepairable element last until the first entrance in the subset of unacceptable states called oft absorbing states and described by A. Assuming that the process started in state i at time zero, the first passage time to the subset of unaccepted state j is given by $\phi_{ij}(t)$ and its reliability is given by Lisnianski [17], namely:

$$R(t) = 1 - \phi_{ij}(t) \tag{9}$$

The instantaneous availability of the system can be computed by summing ϕ_{ij} over the set of acceptable states B that meet a preset demand, namely

$$A(t) = \sum_{j \in B} \phi_{ij}(t), A \cup B = S \tag{10}$$

where S is the set of all states in the system. Thus, the instantaneous unavailability is given by $1 - A(t)$.

4 SMP Model of Processing Requests in an MCC System

Consider the MCC model for a processing request in a mobile cellular network environment (see Fig. 2). This model takes into account four states: *Fully active*, *Partially active*, *Repair*, *Failure*. A processing request may become degraded ($3 \rightarrow 1$), that is, become *Partially active*, when signal loss occurs due to moving too far away from the nearest base transceiver station (BTS), access point or due to radio interference. A processing request may go into the *Repair* state if radio interference is reduced. However, if the processing request is not corrected, it goes into *Failure* state. A processing request may also be in the *Failure* state when the system is completely damaged, caused by a system or hardware error. In the *Partially active* state, the processing request either next state ($1 \rightarrow 2$) or fails ($1 \rightarrow 0$) completely. If the request processing in *Repair* state regains access to the radio resource, it will revert to the *Fully active* state.

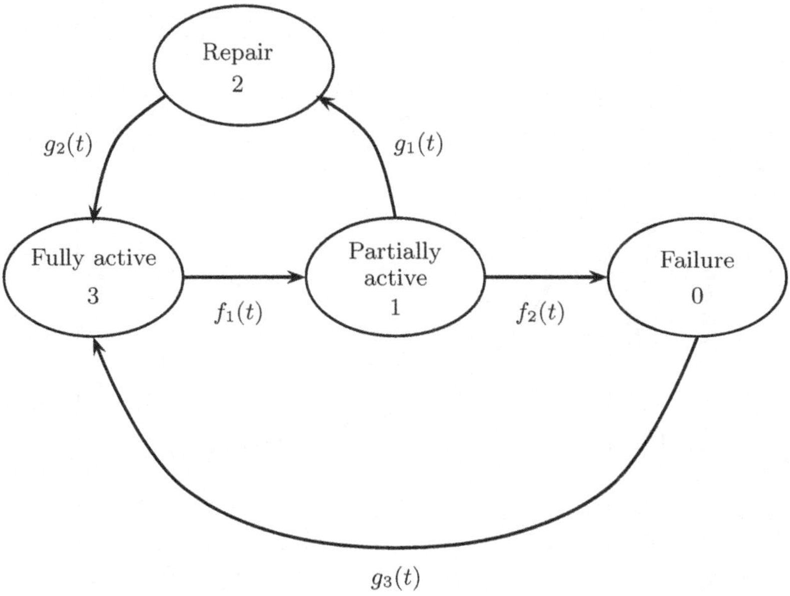

Fig. 2. Four-state transition model with repair for MCC system.

Let time to check the connection quality in *Partially active* state have an exponential distribution with mean μ and its *pdf* be given by $g_1(t)$. Let the *cdf* of time to repair be given by $G_1(t)$. The Weibull parameters for the time to *Failure* $(1 \to 0)$ are λ and γ with a *pdf* indicated by $f_2(t)$ and the *cdf* is given by $F_2(t)$. Let the *pdf* of the distribution time to *Repair* state $(2 \to 3)$ be denoted by $g_2(t)$ and the *cdf* be $G_2(t)$. After the failure occurs, the system can be repaired $(0 \to 3)$, and the repair time is exponential distribution with *pdf* given by $g_3(t)$. Let the *cdf* is denoted by $G_3(t)$.

The kernel matrix $C(t)$ for the processing request consists is given by:

$$C(t) = \begin{bmatrix} 0 & 0 & 0 & g_3(t) \\ f_2(t)(1 - G_1(t)) & 0 & g_1(t)(1 - F_2(t)) & g_2(t) \\ 0 & 0 & 0 & 0 \\ 0 & f_1(t) & 0 & 0 \end{bmatrix} \tag{11}$$

The matrix $W(t)$ consists of the closed form expressions and is as follows:

$$W(t) = \begin{bmatrix} 1 - G_3(t) \\ e^{-\mu t - (\lambda t)^{\gamma}} \\ 1 - G_2(t) \\ 1 - F_1(t) \end{bmatrix} \tag{12}$$

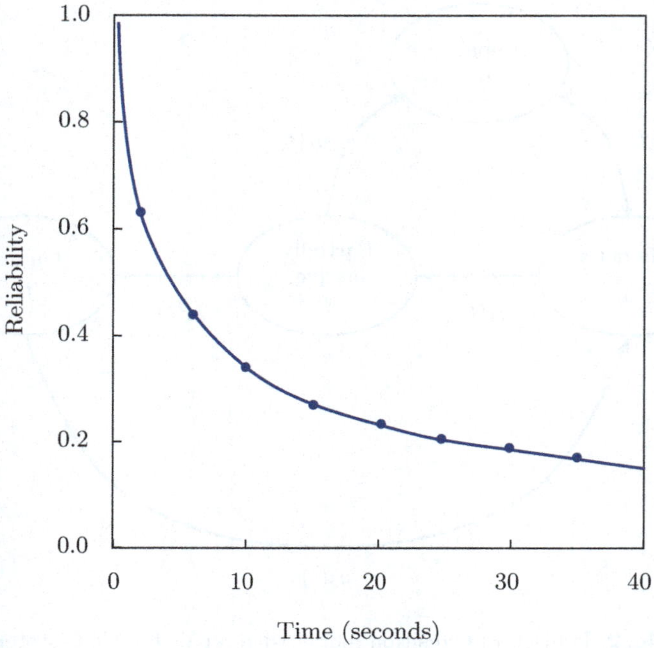

Fig. 3. Reliability of processing requests in MCC system.

Then, by substituting the above matrices in Eq. (8) can be obtained the state probabilities, namely

$$\phi(t) = \begin{bmatrix} 1 - G_3(t) \\ e^{-\mu t - (\lambda t)^{\gamma}} \\ 1 - G_2(t) \\ 1 - F_1(t) \end{bmatrix} + \int_0^t C(t)\phi(t-\tau)d\tau \qquad (13)$$

5 Numerical Example

In this section, the presented discrete-time semi-Markov model of the MCC system was applied to obtain some numerical results.

The following data was adopted: the average time to failure of a processing request occurs once every month and the value of *cov* is equal to 0.1. The average repair time is 2 s every month. In turn, the average time for partial degradation is 0.6 s per 1 day. Let the mean time to failure state from the partially active state be 2 years with the value of *cov* equal to 0.25. The system fails when it is in state 0, which will be treated as an absorption state. The system is assumed to start at time $t = 0$.

Figure 3 presents the reliability for processing requests in the MCC system calculated for the given parameters. It can be observed that the reliability of the processing request decreases significantly with the passage of time.

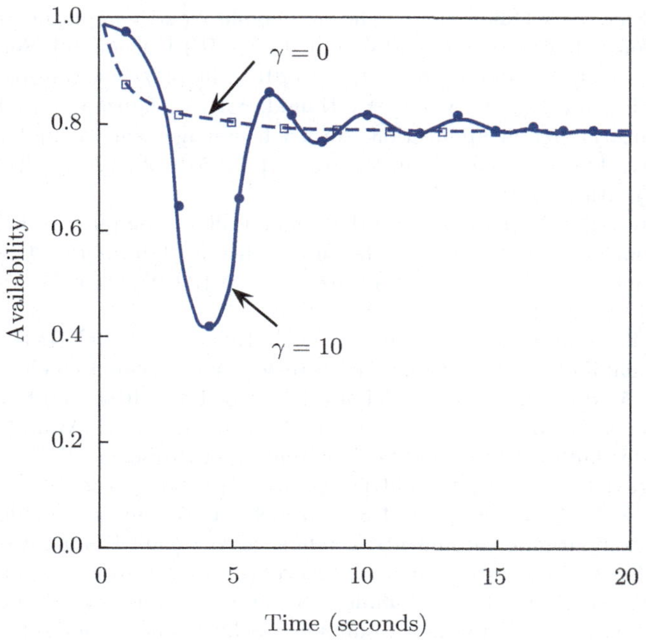

Fig. 4. Availability of processing requests in MCC system.

Figure 4 shows the availability of the processing requests in MCC system for the abovementioned values and the changing values of the γ parameter. It can be seen from the graphs that an increase in the value of this parameter causes an increasing number to have reachability values.

6 Conclusion

The article presents a method of finding the reliability and reachability values for a processing request in a mobile cloud computing system. This method uses the discrete-time semi-Markov model that can be used in many practical situations. The example presented illustrates the technique used, which enables its use in mobile cloud computing system applications.

References

1. Dinh, H.T., Lee, Ch., Niyato, D., Wang, P.: A survey of mobile cloud computing: architecture, applications, and approaches. Wireless Commun. Mob. Comput. **13**(18), 1587–1611 (2013). https://doi.org/10.1002/wcm.1203
2. Sanaei, Z., Abolfazli, S., Gani, A., Buyya, R.: Heterogeneity in mobile cloud computing: taxonomy and open challenges. IEEE Commun. Surv. Tutorials **16**(1), 369–392 (2014) https://doi.org/10.1109/SURV.2013.050113.00090

3. Liu F., Shu, P.: eTime: energy-efficient mobile cloud computing for rich-media applications. IEEE COMSOC MMTC E-Letter IEEE Commun. Soc. Multimedia Commun. Tech. Committee **8**(1), 12–14 (2013). http://www.comsoc.org/~mmc/
4. Ahmed, E., Naveed, A., Gani, A., Hamid, S.H.A., Guizani, M.: Process state synchronization-based application execution management for mobile edge/cloud computing. Future Gener. Comput. Syst. **91**(2), 579–589 (2019). https://doi.org/10.1016/j.future.2018.09.018
5. Al-Omary, A.Y.: A secure framework for mobile cloud computing. In: 2019 International Conference on Innovation and Intelligence for Informatics, Computing, and Technologies (3ICT), Sakhier, Bahrain, pp. 1–6 (2019). https://doi.org/10.1109/3ICT.2019.8910294
6. Endo, P.T., Rodrigues, M., Gonçalves, G.E., Kelner, J., Sadok, D.H., Curescu, C.: High availability in clouds: systematic review and research challenges. J. Cloud Comput. **5**(16), (2016) https://doi.org/10.1186/s13677-016-0066-8
7. Mohammed, B., Kiran, M., Maiyama, K.M., Kamala, M.M., Awan I.-U.: Fail over strategy for fault tolerance in cloud computing environment. Softw. Pract. Experience **47**(9) 1243–1274 (2017) https://doi.org/10.1002/spe.2491
8. Mesbahi, M.R., Rahmani, A.M., Hosseinzadeh, M.: Reliability and high availability in cloud computing environments: a reference roadmap. Human-centric Comput. Inf. Sci. **8**(20), 1–31 (2018). https://doi.org/10.1186/s13673-018-0143-8
9. Barbu, V., Boussemart, M., Limnios, N.: Discrete time semi-Markov model for reliability and survival analysis. Commun. Stochast. Theor. Methods **33**(11), 2833–2868 (2004). https://doi.org/10.1007/978-0-8176-8206-4_30
10. Wang, L.Y., Cui, L.R.: Aggregated semi-Markov repairable systems with history-dependent up and down states. Math. Comput. Model. **53**(5–6), 883–895 (2011). https://doi.org/10.1016/j.mcm.2010.10.025
11. Cui, L.R., Du, S.J., Hawkes, A.G.: A study on a single-unit repairable system with state aggregations. IIE Trans. **44**(11), 1022–1032 (2012) https://doi.org/10.1080/0740817X.2012.662309
12. Lirong, W.B.C., Chen, F.: Reliability analysis of semi-Markov systems with restriction on transition times. Reliab. Eng. Syst. Saf. **190** 1–10 (2019) https://doi.org/10.1016/j.ress.2019.106516
13. Shamim, S.M., Sarker, A., Bahar, A.N., Rahman, M.A.: A review on mobile cloud computing. Int. J. Comput. Appl. **113**(16), 4–9 (2015). https://doi.org/10.5120/19908-1883
14. Howard, R.A.: Dynamic Probabilistic Systems, Markov Models. vol. 1, Wiley, New York (2007)
15. Rausand, M., Høyland, A.: System Reliability Theory. Models, Statistical Methods, and Applications. Second Edition, Wiley, New York (2004)
16. Modarres, M., Kaminskiy, M., Krivtsov, V.: Reliability Engineering and Risk Analysis: A Practical Guide. CRC Press, Boca Raton (1999)
17. Lisnianski, A., Levitin, G.: Multi-state System Reliability: Assessment Optimization and Applications. World Scientic, Singapore (2003)

Softcomputing Approach to Sarcasm Analysis

Jacek Mazurkiewicz[1(✉)] and Jakub Woszczyna[2]

[1] Faculty of Electronics, Wrocław University of Science and Technology, ul. Wybrzeże Wyspiańskiego 27, 50-370 Wrocław, Poland
jacek.mazurkiewicz@pwr.edu.pl

[2] Unit4 Poland Sp. z o.o., ul. Powstańców Śląskich 7A, 53-332 Wrocław, Poland

Abstract. The goal of the paper is to create a system for sarcasm recognition in texts. The system effectiveness is significantly better than a random guess and it is functional in chosen types of sarcasm. The system is tested and created using datasets based on input from Twitter users but also on headlines of online news magazines. Used datasets contain different type of sarcasm in form appropriate for neural network to train on. The limitation of the system is recognition of specialized sarcasm which requires a unique knowledge to understand it and consider it to be an instance of sarcasm. The system determines sarcasm only through the website and seek the proper context, so the system determines sarcasm only in the given sentence. Two basic solutions had been developed: a neural network with different configurations of layers and a convolutional neural network. The implemented solutions give very satisfactory results.

Keywords: Sarcasm recogniton · ANN · CNN · NLP

1 Introduction

The goal of this paper is to create a system for sarcasm recognition in texts. The system effectiveness should be significantly better than a random guess and should be functional in chosen types of sarcasm. The system is tested and created using datasets based on input from Twitter users but also on headlines of online news magazines. Existing datasets are not devoid of a defect. Some of the data are not correctly parsed, missing words or the sarcasm which is very uncommon even for human to detect. Also, each of these datasets contains different type of sarcasm in form appropriate for neural network to train on. Even though the data are comprehensive and merging datasets can give very satisfying results. The limitation of the system is recognition of specialized sarcasm which requires a unique knowledge to understand it and consider it to be an instance of sarcasm. Such type of ironic texts is rare and very hard to comprehend by a human so that system would have to be "trained" in almost every field of science, sociology, slang etc. Another constraint is the context of sarcasm. Sometimes one sentence is considered sarcastic only with the knowledge of the previous message of another user. The system determines sarcasm only through the website and seek the proper context, so the system determines sarcasm only in the given sentence.

W. Zamojski et al. (Eds.): DepCoS-RELCOMEX 2021, AISC 1389, pp. 273–283, 2021.
https://doi.org/10.1007/978-3-030-76773-0_27

Sarcasm is a sharp semantic expression of the sentiment in which the author expresses the point of view opposite to the literal meaning of the expression. It is used often to show disadvantages of a solution or an opinion. Sarcasm has many definitions; as shown in Cambridge Dictionary [4] it is "the use of remarks in way that says the opposite of what you mean in order to insult someone or show them that you are annoyed". Another definition gives Macmillan English Dictionary [7] – "the activity of saying or writing the opposite of what you mean, or of speaking in a way intended to make someone else feel stupid or show them that you are angry". According to Chaudhari and Chandankhede [2] types and subtypes of sarcasm occurring in texts can be distinguished depending on many factors and situation of usage. The following specification is based on Zafarani [13], Mondher Bouazizi and Tomoaki Ohtsuki [1] works: *Type 1* - Sarcasm as a disparity of sentiments, *Type 2* - Sarcasm as a means of conveying emotion, *Type 3* - Sarcasm as a form of written expression, *Type 4* - Sarcasm as a function of expertise, *Type 5* - Behavior-based sarcasm.

Next section briefly presents the state of art. Section 3 introduces the features of used datasets. Two proposed sarcasm detectors MLP-type ANN and CNN-type are described in Sect. 4 and Sect. 5 respectively. Section 6 discusses the results just before the conclusions available in the final Sect. 7.

2 State of Art

2.1 Semi-supervised Recognition of Sarcastic Sentences Twitter, Amazon

As first robust system used for sarcasm detection is considered algorithm implemented by Davidov and other researchers [3]. Authors use two different data sets: a collection of 5.9 million tweets collected from Twitter, and 66 000 product reviews collected from Amazon. It is semi-supervised sarcasm identification algorithm (SASI) for sarcasm detection in Amazon product reviews or in Twitter, which implements two modules. First, semi-supervised pattern acquisition for identifying sarcastic patterns which is used to match features for a classifier, and second, a classification part that classifies each sentence to a proper class. The input is relatively small, labeled sentences with discrete range annotation of intensity of sarcasm. Data is preprocessed, so objects like product, author, company etc. are replaced by appropriate tag, like'[PRODUCT]','[USER]' etc. Presented solution consists of five feature types: punctuation, pattern extraction, pattern and punctuation, data enrichment with punctuation, data enrichment with pattern extraction. The main feature type is pattern extraction. Its defined by sequence of high-frequency words (HFWs) and content words (CWs). Pattern is an ordered sequence of high frequency words and slots for content words. Each pattern allow 2–6 HFWs and 1–6 slots for CWs. Next, for each sentence a feature value is calculated for each pattern and according to the determined parameters scores are being computed. To build a classification model for new examples a weighted k-nearest neighbor punctuation-based features are used. A weighted average of the "k-closest" training set vectors is the score. For cases with no matching vectors default value is being assigned. Punctuation-based features like length of the sentence, number of exclamation, question or quotation marks, number of capital letters were normalized by dividing maximal weight of each group. Results presented by the researchers were very promising, evaluation for SASI algorithm that contains all features combined accuracy totals 81.3%, 76.6% of precision for Amazon dataset and 86.3% of accuracy, 79.4% for Twitter dataset.

2.2 Sarcasm Detection on Twitter: A Behavioral Modeling Approach

Researchers in the paper called "Sarcasm Detection on Twitter: A Behavioral Modeling Approach" basic on works of Davidov [3], Rockwell [14], psychological and behavioral studies tries to combine proper features to train sarcasm detecting, self-learning algorithm. They present different forms of sarcasm, introduce so-called SCUBA framework – Sarcasm Classification Using a Behavioral modelling Approach and make demonstration of the importance of past information accessible from historical tweets for sarcasm detection. The document pays attention to the sarcasm division. Extracted features are utilized in the learning framework. Each feature is computed using different algorithms to obtain affect score. It means the way of calculating contrasting connotations is based on SentiStrength lexicon-based tool optimized for detecting tweet sentiment. Similar system is used to find contrasting present tweet of the user with the previous ones using SMOG [11] to determine readability. Complexity of the expression is analyzed by constructing probability distribution over length of words in the current but also past tweets. Difference between them is computed using Jensen-Shannon divergence. SentiStrength is used also in computing another feature which are means of conveying emotion. Sentiment of the user past tweets are divided into overlapping buckets (depending on the number of prior answers to the current tweet) – each bucket consists of the past n tweets posted by the user. Next feature is based on prosodic variations and analyzing structure of the sentence. Often repeated letters, exclamation marks, ellipsis or changing letters to capitalized can prove emotion charge added to the statement. Authors obtain part-of-speech (POS) tags using TweetNLP and compute the probability distribution. Obtained by the researchers results are surely satisfactory because system with implementation of all features has accuracy of 83.46%.

3 Datasets

Most of the past studies about sarcasm detection use training data for neural network models collected from Twitter. Twitter databases are usually collected using hashtags "#sarcasm", "#sarcastic", "#irony" etc. Such datasets are not examined once more by human to check if all data is correctly parsed. These are often contaminated with random words or characters and it can be noisy in terms of defining sarcasm. Somebody can use hashtag "#sarcasm" in other context and that tweet is not being sarcastic, thus, such sentence will be placed in the dataset. Alternative to that is dataset made of collection of news headlines, written by professionals. Both types of datasets will be used in experiments.

3.1 iSarcasm

Described dataset had been collected using Twitter data by S. Oprea and W. Magdy [12]. So-called tweets, usually short messages posted by users of Twitter can be tagged as sarcastic or ironic by themselves. Dataset contains 3150 samples in total. Such collection presents sarcasm as form of conversational language used commonly by people, represented by huge amount of users from different environments and communities. Disadvantage of such solution is contamination of unwanted words or graphic signs, emojis and paradoxically colloquial language which uses word shortcuts and shorthands (brachylogies).

3.2 News Headlines Dataset

Rishabh Misra, creator of the "News Headlines Dataset For Sarcasm Detection", decided to collect sarcastic sentences and prepare data for sarcasm detection from other source than Twitter [6]. The dataset had been established using two news website. TheOnion – produces sarcastic versions of events happening in the world, collected was all the headlines from sarcastic categories. Second news website – HuffPost – presents real, non-sarcastic headlines. Total number of samples remains 55 328. Advantages over the Twitter datasets: news headlines does not contain any spelling mistakes, as the sole purpose of TheOnion website is to publish sarcastic content, noise related to mistaken labeling of the sarcasm in the sentence is minimized, News headlines are "self-contained", unlike tweets which can be a reply to other tweets.

4 MLP-Type ANN Sarcasm Detector

This neural network consists of different hidden layers depending on particular configuration. Solution is implemented using Tensorflow 2.0 and Keras basing on the neural network made by Aaditya Bhat [5]. Proportion of training data to testing data is 3:1.

4.1 Data Preparation

Firstly, data are parsed to the two arrays – sentences and labels. Each item of one array responds to the second one with the same number. After dividing to the training and testing sets data are being tokenized and sequenced. Tokenization allows to vectorize data by turning each text corpus into a chain of integers, simply turning words into numbers. Such action significantly improves ease and time of processing texts. Two parameters were set for the Keras Tokenizer – *num_words* and *oov_token*. First one determines maximum number of words to keep which is set to 10000, *oov_token* takes passed token and replace with it words which were not encountered before – in this case $< oov >$. Creating sequences is simply building arrays made of tokens defined by tokenization. In Keras *texts_to_sequences* method is used to do that. Sequenced data need to be transformed (padded) into sequences of the same length. Three padding properties are set: maximum length, padding type and truncating type. Maximum length totals 100 characters as this is the number which should not be exceeded. Padding and truncating types are set to "post". This pads or truncates values at the end of the sequence. After this, data are correctly prepared for the neural network input.

4.2 Network Topology

Model in the neural network is sequential so it enables to build a network made of many hidden layers in custom configuration. Neural network in the first configuration consists of: Embedding layer, 1-dimensional Global Average Pooling layer, Dense layer with ReLU activation function, Dense layer with Sigmoid activation function.

Embedding layer is used as the first layer in the model as it turns positive integers into dense vectors of fixed size. It takes 2D tensor as input with shape *(batch_size, input_length)* and 3D output tensor *(batch_size, input_length, output_dim)*. In this model it totals (None, 100) as input, and (None, 100, 16) as output, while "None" batch size means unbounded batch size for the network. Batch size with not fixed size provides better flexibility of the network while testing. It has vocabulary size 10 000, embedding dimension 16 and input length same as the length of the padded sequences – 100. Vocabulary size says that input word index should be no greater than 9 999 while embedding dimension determines size of the output array. Global Average Pooling operation is used to sum up vectors generated in the embedding layer and reduce the spatial dimensions of a tensor. Global Average Pooling 1D in Keras reduces 3D tensor of $h \times w \times d$ to $1 \times 1 \times d$. Dense layer is a basic layer used in the ML environment. It simply connects each input neuron to the output neuron. As first argument size for the output shape is passed, in this example – 32, input size is not determined. Activation function of the dense layer positions the weight passed to the hidden layer on the level depending on type of the activation function. Last hidden layer in the network with output of 1 dimension is dense layer with sigmoid activation function.

4.3 Alternative Network Topology

The second configuration - the same sequential model is used. This neural network differs from the first configuration on one additional layer after the embedding layer and activation function in the first Dense layer. After embedding layer the convolutional layer is placed and softmax activation function is implemented into first dense layer. Convolutional layer creates a convolution kernel that is convolved with the input layer. Scheme of the neural network in the second configuration: Embedding layer, Convolutional layer with 1D, 1-dimensional Global Average Pooling layer, Dense layer with softmax activation function, Dense layer with sigmoid activation function.

4.4 Network Compilation and Training

After building the model it is compiled with two parameters required in Keras: loss function and optimizer. Mean squared error loss function is used and Adam optimizer. While training, model strives to the minimum and by mean squared error, loss function penalize bigger mistakes more than the smaller ones. Several algorithms are used for optimizing weights in the neural network and minimizing loss function. In this neural network, Adam optimizer is used. As Kingma et al. [10], claims, the optimizing method is "computationally efficient, has little memory requirement, invariant to diagonal rescaling of gradients, and is well suited for problems that are large in terms of data parameters". At the end, prepared training and testing data is served to the model and number of epochs is set. The more epochs are performed on the data the more accurate it becomes but with too many epochs, model can be overfitted. Overfitting phenomenon means matching the model to the very narrow amount of data, whereas model will not work properly on data which was not seen before.

5 CNN-Type Sarcasm Detector

Implemented model is based on the Convolutional Neural Network (CNN) described by Denny Britz, on WILDML website [8]. That model was originally designed to recognize review sentences sentiment – positive or negative [9]. Whereas, that kind of model is originally designed by Yoon Kim – Harvard student. Yoon Kim proposed usage of Convolutional Neural Network for sentence classification [9]. CNN operates on data formed in vectors, or in general tensors, widely used in image recognition as it automatically extract distinctive features for each recognized class. Because of that, usage of CNN in Natural Language Processing is not a standard – written data has to be adjusted to take form of matrix suitable to CNN processing. The original model uses pre-trained word vectors for sentence classification. Word vectors uses *1-of-V* encoding, where V is the vocabulary size. Features are encoded onto a lower dimensional vector space using a hidden layer, extracting essential features of words in their dimensions. In result semantically close words are close also in cosine/Euclidean distance. Analogically sarcastic features, by extracting proper words, found by the neural network in training sentences are placed in their dimensions. Using neural network of this architecture to sarcasm recognition give surprisingly satisfactory results.

5.1 Data Preprocessing

Analogously to the previous solution, data is padded into the length of the longest sentence. Vocabulary index maps each word to an integer between and the vocabulary size. Sentence becomes a vector of integers, and all data is turned into $x \times y$ tensor. Indices of the input sentences are randomly shuffled. After data splitting into training and testing set, data preparation is over.

5.2 Architecture

Embedded representation of the sentences in form of a tensor are passed to the convolutional layer with filters and feature maps where convolutions on the sentences are performed. After this, result is max-pooling into the feature vector, dropout regularization is added and classification is realized using softmax layer. To create architecture of the CNN model the most important arguments are being set: Length of the input sequence – to have proper input size and maxpooling matrix, Number of output classes – in this model, two classes, true or false, Size of the embeddings and vocabulary – set to determine the size of the embedding layer. Shape of the embedding layer: (*vocabulary_size, embedding_size*), Filters – number of filters and their sizes, to convolutional filters covering the given number of words. Embedding layer is the first layer in the model and it simply maps vocabulary into low dimensional vectors. Embedding matrix is initialized using random uniform distribution and it is learned during the training. Convolutional layer is combined with the max- pooling one because after the convolution output, applying filters, the output is max-pooled and result tensor is obtained. At the end, it has to be reshaped and transformed into one long feature vector. The last one, dropout layer performs regularization. The dropout method is very popular and efficient. It forces neurons to learn individually by stochastic disabling a fraction of the neurons. By doing that neurons cannot adapt to the predominant conditions and work out own value. Prediction given from the neural network is obtained by doing matrix multiplication and choosing the class with the biggest number. To properly train the network, loss function is essential. The loss measures how wrong the network's results were. It is calculated after each epoch of training and the goal is to minimize it by setting the weights of the neurons. Cross-entropy loss function is used to measure the loss in the presented model, it is standardly used for models whose output is a number between 0 and 1. This loss function is based on logarithm and penalizes both types of errors, false positives and false negatives, depending on how big the error was. Optimizing the loss value is performed using popular and effective Adam optimizer [10]. The name Adam is derived from adaptive moment estimation. This optimizer maintains a single learning rate for all weight updates which does not change during the training.

6 Results

To train and test created systems and neural network models, three datasets had been used. First two are established from the same source – News Headlines Dataset [6], third one contains data obtained from iSarcasm [12]: Dataset no. 1 – 26709 samples – "Sarcastic headlines", dataset used for network training, Dataset no. 2 – 28619 samples – "Sarcastic headlines" dataset not seen by the network before, Dataset no. 3 – 3150 samples – Twitter dataset – dataset with different characteristic/attributes comparing to the "Sarcastic headlines".

6.1 Sarcastic Headlines Dataset Experiments

Base topology neural network is made of four hidden layers: embedding, global average pooling and two dense layer with ReLU and sigmoid activation functions, 30 epochs of training had been performed, proportion of training to testing data totals 3:1. Neural network testing for all datasets had been executed. On the grounds that score from the neural network is presented in range from 0 to 1, results are divided into certainty over 60 and 80 percent: sentence with score 0.64, while sentence should be considered as sarcastic, would not be counted to the "certainty over 80%" collection. Analogically, not sarcastic sentence with score about 0.32 would not be determined as "80% certain" (Tables 1 and 2).

Alternative topology neural network consists of the following layers: embedding, convolutional, global average pooling and two dense layers with softmax and sigmoid activation functions. Comparing to the previous neural network, convolutional layer is added and one activation function is switched to softmax (Tables 3 and 4).

Table 1. Base topology neural network parameters

Loss	Accuracy	Validation loss	Validation accuracy
1.25%	98.72%	17.27%	81.05%

Table 2. Base topology neural network results for all datasets

	Dataset no.1	Dataset no.2	Dataset no.3
Certainty over 60%	94.4%	85.2%	87.8%
Certainty over 80%	93.8%	82.3%	86.0%

Table 3. Alternative topology neural network parameters

Loss	Accuracy	Validation loss	Validation accuracy
0.89%	99.10%	15.61%	82.45%

Table 4. Alternative topology neural network results for all datasets

	Dataset no.1	Dataset no.2	Dataset no.3
Certainty over 60%	95.3%	88.2%	91.2%
Certainty over 80%	94.1%	86.3%	82.7%

6.2 Sarcastic Headlines and Twitter Dataset Experiments

The same topologies of the neural networks are in use for this stage of experiments. The parameters and results are present in tables below. Base topology network parameters are seemingly worse than the first configuration, but the difference is nonsignificant (Tables 5 and 7).

Table 5. Base topology neural network parameters

Loss	Accuracy	Validation loss	Validation accuracy
1.77%	98.20%	16.87%	81.14%

Results (Tables 6 and 8) are very similar to the previous stage of experiments. Here, neural network seems to be more certain but disparities are vestigial.

Table 6. Base topology neural network results for all datasets

	Dataset no.1	Dataset no.2	Dataset no.3
Certainty over 60%	88.8%	86.7%	81.2%
Certainty over 80%	85.3%	83.1%	88.2%

Table 7. Alternative topology neural network parameters

Loss	Accuracy	Validation loss	Validation accuracy
2.09%	97.84%	17.58%	80.48%

Table 8. Alternative topology neural network results for all datasets

	Dataset no.1	Dataset no.2	Dataset no.3
Certainty over 60%	99.0%	86.3%	89.6%
Certainty over 80%	97.3%	84.5%	88.3%

6.3 Experiments with CNN-Type Sarcasm Detector

This model's result is binary, so output gives information only about if the sentence is sarcastic or not (Table 9).

Table 9. CNN-type sarcasm detector – results, training – dataset no.1

Number of epochs	Dataset no.1 accuracy	Dataset no.2 accuracy	Dataset no.3 accuracy
10	98.3%	96.7%	87.0%
15	98.4%	97.0%	82.0%
20	98.3%	96.5%	84.3%
25	98.3%	97.0%	81.8%
30	98.3%	97.0%	80.3%
40	98.4%	97.0%	78.8%

Confusion matrix (Table 10) helps to see properly what are the results obtained from the network. It shows what data are considered misguidedly and what are considered correctly. Number of true positives (TP), true negatives (TN), false positive (FP) and false negatives (FN) is presented below for network trained with 25 epochs (Table 11).

Table 10. CNN-type sarcasm detector – confusion matrix, training – dataset no.1

	TP	TN	FP	FN
Dataset no.1	43.0%	55.3%	0.8%	0.8%
Dataset no.2	45.4%	51.6%	2.3%	0.7%
Dataset no.3	43.6%	48.3%	4.0%	4.1%

Table 11. CNN-type sarcasm detector – results, training – dataset no.3

Number of epochs	Dataset no.1 accuracy	Dataset no.2 accuracy	Dataset no.3 accuracy
15	96.1%	92.4%	96.6%
20	95.8%	92.4%	97.1%
25	95.8%	92.1%	97.2%
30	95.8%	92.3%	97.0%

With arising number of epochs, models trained on the first dataset seems to be more fitted to the characteristic of the dataset. Accuracy for Sarcasm Headlines sentences arises with number of epochs. Inversely proportional dependence can be seen on Twitter dataset accuracy – the best score is for the lowest number of epochs model training. Confusion matrix present very good results for all datasets. It is much easier to guess the non-sarcastic sentence than the sarcastic one. There is also a slight difference in proportions of training data in positive and negative examples what can indicates better match of the network to recognizing non-sarcastic sentences.

7 Conclusions

The paper creates softcomputing foundations for sarcasm recognition in texts, shows topicality of the Natural Language Processing and conduct experiments on sarcasm recognition. The created software fulfilled assumptions and allowed to recognize sarcasm in texts on a satisfactory level. The work classified multiple types of sarcasm and presented a view on the matter based on existing works on sarcasm recognition. The applied technologies, and datasets used for the neural network training have been described in detail. Two basic solutions had been developed: a neural network with different configurations of layers and a convolutional neural network. The implemented solutions give very satisfactory results. The main advantage of neural networks trained on Sarcastic Headlines dataset over Twitter dataset is amount of data. It can be seen that network trained on the greater amount of data has better accuracy and is more comprehensive. More data would also resolve very popular problem of overfitting. Conducted researches allow to conclude that the solutions recognize something similar to the sarcasm because it is not effective on each type of dataset. Further developing created system can carry many interesting conclusions by adjusting network parameters or adding another hidden layers, more responsive for some features of sarcasm. A great improvement, making the system more useful in the business world would be context analyzer. Tool for sarcasm recognition is necessary in marketing companies but analyzing circumstances of the written sentence is crucial in the issue of sarcasm.

References

1. Bouazizi, M., Ohtsuki, T.: A pattern-based approach for sarcasm detection on Twitter. IEEE Access **4**(1–1), 5477–5488 (2016)
2. Chaudhari, P., Chandankhede, C.: Literature Survey of Sarcasm Detection. In: IEEE WiSPNET 2017 Conference, pp. 2041–2046 (2017)
3. Davidov, D., Tsur, O., Rappoport, A.: Semi-supervised recognition of sarcastic sentences on Twitter and Amazon. In: Proceedings of the Fourteenth Conference on Computational Natural Language Learning, Uppsala, Sweden, 15–16 July 2010, Association for Computational Linguistics, pp. 107–116 (2010)
4. https://www.dictionary.cambridge.org. Accessed 15 Mar 2021
5. https://www.github.com/aadityaubhat/medium_articles/tree/master/sarcasm_detection. Accessed 18 Mar 2021
6. https://www.kaggle.com/rmisra/news-headlines-dataset-for-sarcasm-detection. Accessed 21 Mar 2021
7. https://www.macmillandictionary.com. Accessed 10 Mar 2021
8. https://www.wildml.com/2015/12/implementing-a-cnn-for-text-classification-in-tensorflow. Accessed 10 Mar 2021
9. Kim Y.: Convolutional neural networks for sentence classification. In: Proceedings of the 2014 Conference on Empirical Methods in Natural Language Processing (EMNLP), October 2014, Doha, Qatar. Association for Computational Linguistics, pp. 1746–1751 (2014)
10. Kingma, D.P., Ba, J.L.: Adam: a method for stochastic optimization. In: 3rd International Conference for Learning Representations ICLR, San Diego, pp. 1–15 (2015)
11. McLaughlin, H.: Smog grading: a new readability formula. J. Reading **12**(8), 639–646 (1969)
12. Oprea S., Magdy W.: iSarcasm: A Dataset of Intended Sarcasm (2019)
13. Rajadesingan, A., Zafarani, R., Huan Liu, H.: Sarcasm detection on Twitter: a behavioral modeling approach. In: Proceedings of the Eighth ACM International Conference on Web Search and Data Mining February 2015, WSDM 2015, pp. 97–106 (2015)
14. Rockwell, P.: The Effects of Cognitive Complexity and Communication Apprehension on the Expression and Recognition of Sarcasm, Nova Science Publishers, Hauppauge (2007)

Open–source–based Environment for Network Traffic Anomaly Detection

Marcin Michalak[1(✉)], Łukasz Wawrowski[1], Marek Sikora[2], Rafał Kurianowicz[1], Artur Kozłowski[1], and Andrzej Białas[1]

[1] Research Network Łukasiewicz — Institute of Innovative Technologies EMAG, ul. Leopolda 31, 40–189 Katowice, Poland
Marcin.Michalak@emag.lukasiewicz.gov.pl

[2] Department of Computer Networks and Systems, Silesian University of Technology, ul. Akademicka 16, 44–100 Gliwice, Poland

Abstract. The paper presents an open–source–based environment for network traffic anomaly detection. The system complements the well known network security platforms as it tries to detect unexplained descriptions of the traffic. For this purpose several anomaly detection algorithms were applied. To assure better system performance, the moving history approach is also applied.

Keywords: Network traffic analysis · Anomaly detection · R environment · Elasticsearch · Security Operation Center

1 Introduction

In recent years more and more aspects of our work and living moved to the Internet. Such a trend is especially visible since the Spring 2020, when almost all branches of our personal or professional activities were enforced to become remote due to the CoViD–19 pandemic lockdowns. On the one hand, such a migration brings a lot of benefits: floating working hours or time and fuel saving due to the home office working mode. But on the other hand, we have to make a short reflection about the general aspects of network traffic safety and reliability.

The paper concerns the research project RegSOC (Regional Security Operation Center), whose objective is to develop a prototype of the specialized Security Operations Center (SOC), mainly for public institutions and smaller business organizations. Its general requirements and structure were already presented in [3]. However, it is also worth to consider its smaller version which will not communicate with external services.

SOC is based on three pillars: people, processes and technology, ensuring the confidentiality, integrity and availability of today's information technology enterprise. Highly qualified cybersecurity specialists of different competences embraced by the proper management structure are responsible for defence

W. Zamojski et al. (Eds.): DepCoS-RELCOMEX 2021, AISC 1389, pp. 284–295, 2021.
https://doi.org/10.1007/978-3-030-76773-0_28

against the unauthorized activity within computer networks, performing different kinds of monitoring–, detection–, analysis–, response– and restoration processes—with the use of specialized technological equipment.

The paper concerns technological issues, especially the threats detection technology. The signature-based threat analysis uses knowledge about early detected threats and is not able to detect emerging threats. For this reason this analysis is supported by the detection of network traffic anomalies.

The research results related to the anomaly detection methods and tools are presented in the paper. The research is performed in the open-source-based environment, which is very suitable for the future RegSOC users.

The aim of the research is to evaluate the preselected anomaly detection algorithms implemented in the open source environment. The knowledge acquired during research can be applied for sensors developed in the RegSOC project.

The paper is organized as follows: Sect. 2 includes the research context, especially the project RegSOC requirements; Sect. 3 presents the research environment, embracing methods, tools and implementations; Sect. 4 presents the content of the reports generated by the system; finally, the last section includes the summary and some perspectives of further works.

2 Research Context

The detection of anomalies and threats in network traffic is now a standard element of infrastructure in practically every network. Commonly for this purpose we use systems that detect threats and anomalies on the basis of defined rule sets (e.g. Snort, Suricata). The efficiency of such systems depends mostly on currently used rules which describe attacks. It means, in practice, that such systems will be useless once unknown attacks (i.e. not yet described by rules) or dedicated attacks (with the use of tools dedicated to one purpose) occur. The use of anomaly-detection algorithms can allow to detect new, still undescribed attacks. An example of such an attack is the one that occurred in the Polish banking system [16]. Its detection was delayed due to the use of dedicated tools by the attacker. The attack was finally detected after an analysis of network traffic in the attacked locations.

Still, it is important to note that the detection of threats based only on finding anomalies is related to another undesirable phenomenon. If unsuitable methods or improperly tuned algorithms are employed, such an approach may lead to generating too many false positive alerts. This, in turn, will make it difficult to use such tools in practice. Therefore, if a network traffic monitoring system is implemented, the primary stage of tuning and configuration of the whole system cannot be overestimated.

The environment presented in the paper is the result of the RegSOC project and may be dedicated to a public institution as a free complementary tool for network security assurance.

The RegSOC project is aimed at the development of certain components needed to create the RegSOC system, including:

- the hardware and software equipment, such as the sensors (NIDS—Network-based Intrusion Detection Systems), communication infrastructure, specialized applications and the monitoring platform (SIEM—Security Information and Event Management),
- the set of operational processes and procedures responsible for the SOC services and their internal support,
- the organizational model of operation of the regional centres in cooperation with the national cybersecurity structure.

RegSOC, as many other SOCs, is based on well-defined processes focused on the security monitoring, security incident management, threat identification, digital forensics, risk management, vulnerability management, security analysis, etc. [15,21]. The processes are foundation of the RegSOC services offered to future customers.

The threat identification and analysis is the key issue of the RegSOC processes. Network traffic data are permanently sampled, ordered and analysed to detect possible attacks. Attacks previously known are detected using the signatures of attack (a unique arrangement of information related to the attacker's attempt to exploit an IT asset or its vulnerability). This is called a rule-based correlation.

The anomaly detection correlation based on machine learning is able to detect both kinds of attacks: known or unknown, especially if it is supported by the rule based system. The research concerns mixed-mode approach combining the anomaly detection and attack signature analysis.

The effectiveness of such mixed-mode approach depends on the preciseness and quality of the knowledge about network traffic acquired by the machine learning process. The normal behaviour and the suspected behaviour should be distinguished. It is based on the context of the network traffic.

3 Environment Description

For the purpose of the work a research environment was established within the Łukasiewicz–EMAG LAN network. The environment was used in the RegSOC project. The logical scheme of the network components is presented in Fig. 1.

To acquire the network traffic, port mirroring was used on the first switch placed behind the edge router. Such a solution makes it possible to track the whole incoming and outcoming traffic of Łukasiewicz–EMAG. This seems to be the most important aspect when it is necessary to detect anomalies and potential threats in the traffic. On the other hand, however, the adopted solution makes it impossible to track the whole of the internal traffic in the LAN network—what we can see is only a part of the traffic coming through the monitored switch. In addition, the solution does not allow to observe attacks launched from the outside (we monitor the situation behind the edge router which takes on the external traffic and is the first line of defence). This location of the network traffic measurement was practically the only possible solution to be implemented in the Łukasiewicz–EMAG infrastructure. Moreover, it seems to be an optimal

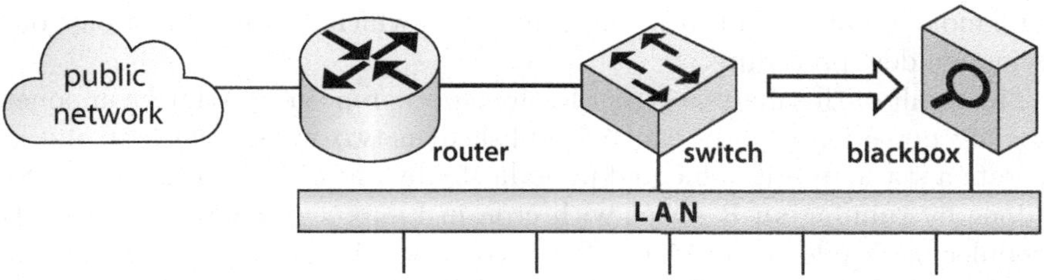

Fig. 1. Diagram with a section of the Łukasiewicz–EMAG LAN infrastructure, including the elements of the research environment. Blackbox is equipped with two interfaces: LAN, to access the blackbox content and mirroring (arrow) to sniff the incoming/outcoming transfer.

choice as it allows to observe the most important anomaly-detection point and, at the same time, significantly limits the volume of data to be analyzed (we do not register the router–filtered traffic and the majority of the LAN traffic).

The research was conducted on the black box server functioning as a sensor and a data analyzer at the same time.

3.1 System Architecture

The general scheme of the system is presented in Fig. 2.

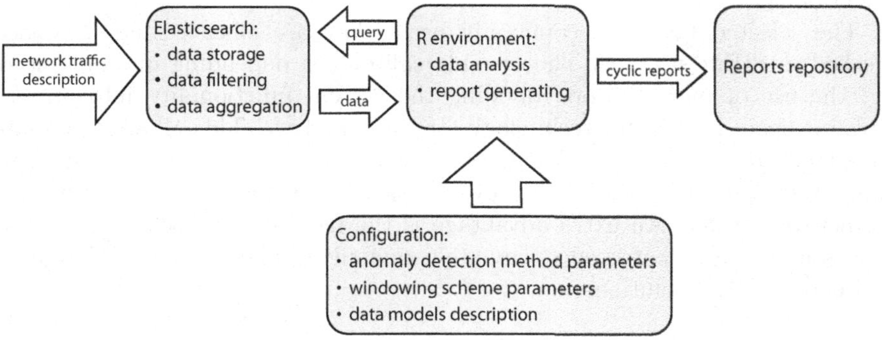

Fig. 2. A general scheme of the system.

The two main components are two environments: Elasticsearch and R. The network traffic description is loaded into Elasticsearch all the time and stored in the inner repositories. The main role of the second component (R) is to send proper queries to Elasticsearch cyclically and analyze the response to these queries. Apart from Elasticsearchr package [11], which is essential to establish the Elasticsearch $\Leftrightarrow$ R environment data transfer, also additional packages are the most crucial: tidyverse [20] for data manipulation, lubridate [9] to allow to easy handle of dates and times, and runner [12] which contains functions for the moving window procedure.

The result of the analysis is sent to the reports repository (third component). Anomaly detection reports can be published in two ways. The first one is to generate a static website rendered periodically by an R Markdown file [2]. It can be done by running an R script with cron in Linux systems or Windows Task Scheduler in Windows machines. The second way is the use a dedicated shiny [5] application where, apart from the anomalies list, an additional data explorer can be implemented. In this solution a report can be generated on demand and a more advanced configuration is possible, however it is dedicated to Linux systems.

The last component is the configuration block. This module allows the operator to tune the anomaly detection parameters values as well as to switch between different data models (the set of network traffic description variables) and change parameters of the moving window procedure.

3.2 Components

The black box sensor configuration presents as follows: Ubuntu 18.04 (operating system), Xeon E–2124@3.30 GHz, 32 GB RAM, 4 NIC, HDD 1 TB (hardware). Moreover, the following tools were used for data registration and collection [1]: Packetbeat, Elasticsearch and Kibana (for network traffic monitoring and registration, collecting registered data on the network traffic and the collected data reporting and visuzalizing respectively [1]), R and RStudio Server (anomaly detection [17]).

The selected tools are commonly used, they are efficient and have been checked in SIEM systems. They enable calibration depending on one's needs, e.g. the use of more sensors, dividing the system functionality into separate tools, or using the cluster technology for their analysis. This allows to monitor infrastructures of any size. Additionally, when using the network traffic sampling technique, it is possible to use the solution in infrastructures with very big network traffic. An extra advantage is the fact that the selected tools are open-source ones. That is why they are commonly applied in free-of-charge and commercial SIEM solutions.

3.3 Methods

The platform implements several methods used for anomaly detection with the general scheme suggested in [14]. The first two are the typical ranks of observations in respect of the possibility of being an anomaly: LOF—Local Outlier Factor [4] and RKOF—Robust Kernel–Based Local Outlier Factor [8]. Another method applied for anomaly detection is the density–based clustering algorithm called DBSCAN [7]. Moreover, a statistical test–based approach was applied: Generalized Extreme Studentized Deviate (GESD) test [18]. The method is one–dimensional, however, in the paper [14] we provided two approaches to the multi-dimensional data analysis: the weak one consists in considering the observation as an anomaly if at least one of the dimensions manifests such a behaviour (logical OR of all dimensions abnormality) while the strong one requires all dimensions to behave as anomalies to consider the observation in this way (logical AND of all dimensions abnormality).

The presented platform implements the above mentioned methods with the following packets: [13] (LOF), [19] (RKOF), [10] (DBSCAN), and [6] (GESD).

3.4 Configurability

The presented system is flexible and may be configured in many aspects to provide the demanded final functionality. These aspects will be described below.

Elasticsearch. Network packets analytics using Elasticsearch and Packetbeat allow to specify the network devices and protocols to sniff. Parsed packets data are sent to Elasticsearch and can be visualised by Kibana which may strengthen the system operator's knowledge about the network traffic characteristics. It is possible to limit or extend the size of embedded Elasticsearch repositories taking into account the dedicated server resources.

R Environment. The main role of this module is to perform analysies cyclically. Apart from the moving window procedure parameters, it is possible to modify anomaly detection algorithms parameters and the analysis frequency to fit the server resources. Moreover, the Elasticsearchr package [11]—used to download data—provides a variety of specific query language abilities to move some filtering and preprocessing techniques from R to Elasticsearch, which may increase the R environment efficiency.

Moving Window Procedure. The moving window procedure allows to analyze only the selected—the newest—part of historic data and update them cyclically. The update consists in deleting the oldest ones and adding the latest. In the case of anomaly detection the data are represented as two separate sets: the history of the length h and the present of the length p. The sum of these two sets is the input for the analysis, however, only the interpretation of the present is the result of the analysis. In other words, we do not take into consideration whether some outliers were detected in history or not, only outliers found in the present become significant.

Let the sequence t of a consecutive observation be given (Fig. 3), the length of the history $h = 10$ and the length of the present $p = 4$.

Let us also assume that the first analysis is performed after 15th observation. That means that the outlier analysis bases on the set of observations $\{t_4, t_5, \ldots, t_{15}\}$ (1st history) and only the interpretation of observations $t_{12}, \ldots, t_{15}$ (1st present) is further reported. Afterwards, the update of the data is applied. It consists in deleting observations from t_4 up to t_7, joining the remaining elements from the 1st history and the 1st present as the 2nd history, and

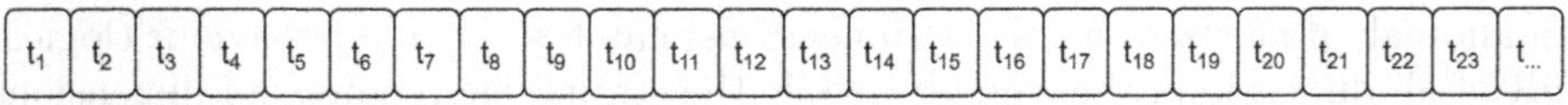

Fig. 3. Interval aggregated data referring to network traffic description (the increasing index means the newly incoming data).

setting the observations from t_{16} to t_{19} as new—the 2nd—present. A simple visualization of this procedure is presented in Fig. 4.

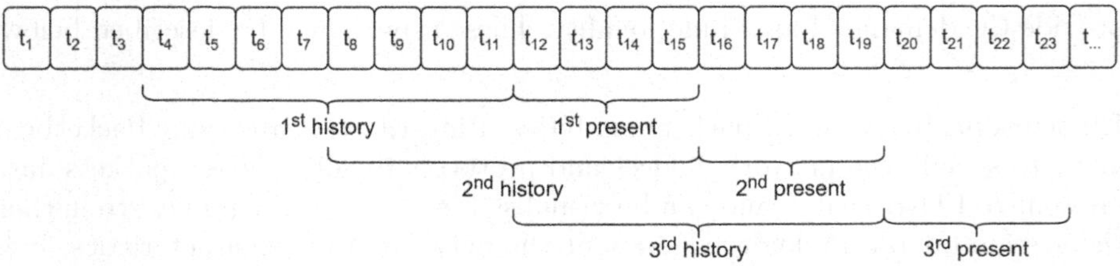

Fig. 4. Three iterations of window–based anomaly detection in network traffic data.

Intuitively, the procedure may be applied for raw data as well as for aggregated data.

4 Application

In this section the results of practical applications of the system are presented and described. The first application referred to normal network traffic monitoring. The main aim of this analysis was to check the operational capabilities of the system, including the reporting module. The second one referred to monitoring normal traffic with several attacks introduced artificially. Here, it was planned to check the outlier detection methods on the data with simulated attacks and non–typical situations.

4.1 The General Data Description

The data were gathered by the Packetbeat software, which aggregates sent packages into sessions. The further analysis was performed on such aggregated data. We assumed to apply the common step of data aggregation and the aggregation time was equal to 1 min. As it was possible to have Packetbeat aggregates referring to the sessions much longer than one minute, we decided to filter the sessions only to such ones whose duration did not exceed the mentioned interval. It is worth to mention that such a step may be performed on the Elasticsearch site with the proper query definition.

From the original Packetbeat session description we took the following variables as the raw ones: number of packets (sent and received), number of bytes (also sent and received) and number of documents related to the session (the information about the inner—Elasticsearch—representation of the session). Based on these variables the R–site time–based aggregation provided the following set of derived variables: minimum, quartiles (Q1, median and Q3), arithmetical mean, maximum, sum of all raw variables which results in a final set of 21 derived variables.

4.2 Normal Traffic Analysis

Below the report summarizing a single network traffic anomaly detection performance is presented (Fig. 5). This report is an example of an hourly generated report, based on minutely aggregated data.

The report consists of four sections. The first section (Parameters) provides information about the time of the report generation, the used data model (the set of derived variables taken into account) as well as the values of moving window procedure parameters. Moreover, the full range of the analyzed data is presented (start and end date).

The second section (Time complexity)—especially the last column: elapsed—provides technical information about the time complexity of the analysis. The time is divided into two aspects: gathering the data (sending the query to Elasticsearch and getting the results) and analyzing the data. Such information may become very useful for the system operator, especially in the system start–up phase when all parameters should be tuned.

The third section—Anomaly frequency—provides information about the total number of detected anomalies. The number of methods that reported a given anomaly is also provided in the table below.

The last section provides a detailed list of all detected anomalies. A detected anomaly description consists of the date/time of the aggregate, the derived variables values (for this aggregate), the flag columns indicating whether the method reported the aggregate as an anomaly, and—finally—the total number of anomaly reporting methods. It is worth to notice that the data presented in this section can be downloaded for some further analytical tasks.

The system provides also the ability to perform off–line analyses of past data, whose results are presented in a similar way.

4.3 Enforced Attacks Monitoring

During the 72 h experiment several following anomalies were manually introduced to the monitored network traffic: bruteforce attacks on ftp service and ssh protocol, over 5 GB file download, bruteforce attack on rdp service, DOS attack on http service, another ftp bruteforce attack and finally another rdp service bruteforce attack. Each attack was planned for 40 min.

The system was working on one minute aggregates and 24 h history. The input data were two–dimensional, including only minute aggregated number of flows and number of bytes. The same four methods were used as previously. Each

Anomaly detection report

Parameters

Report generated on: 2021-02-08 12:14:44

Features in analysis: median number of packets, median number of bytes

History duration: 1440

Present duration: 60

Start date: 2021-02-07T11:14:00.000Z

End date: 2021-02-08T12:14:00.000Z

Time complexity

Downloading data from elasticsearch time:

```
user  system elapsed
0.16    0.03   24.26
```

Anomaly detection time:

```
user  system elapsed
2.47    0.23    2.81
```

Anomaly frequence

Total detected anomalies: 5

number_of_reporting_methods	number_of_anomalies
1	4
2	1

Reported anomalies

dateTime	doc_count	packets_boxplot.q2	bytes_boxplot.q2	lof	rkof	gesd1	gesd2	n
2021-02-08 12:07:00	21817	3	867.5284	Anomaly	Anomaly	Typical	Typical	2
2021-02-08 12:00:00	77108	3	1689.7658	Typical	Typical	Anomaly	Typical	1
2021-02-08 12:03:00	62904	3	1724.1282	Typical	Typical	Anomaly	Typical	1
2021-02-08 12:05:00	54675	4	1746.0000	Typical	Typical	Anomaly	Typical	1
2021-02-08 12:06:00	72962	4	1765.4641	Typical	Typical	Anomaly	Typical	1

Fig. 5. The example of an hourly generated report.

hour a report, similar as presented in Fig. 5, was generated. In Fig. 6 one can see reported anomalies on the background of the introduced attacks.

The X axis represents time while on the Y axis the applied outlier detection methods are placed. Black dots point which minute aggregate was reported by the method as an anomaly.

It occurs that LOF and RKOF parameters set in the first experiment were not proper for use in the present one. Anomalies were reported almost with the same frequency through all time of the analysis, independently of the introduced disturbances occurrence. GESD1 was able to detect only one type of the anomaly, referring to the DOS attack.

Having the results as presented in Fig. 6 and on the basis of the knowledge of the incident meaning we decided to extended the data model of minute aggregates with additional variables for LOF and RKOF. We took into consideration also the number of flows to specified ports, strongly connected with specified

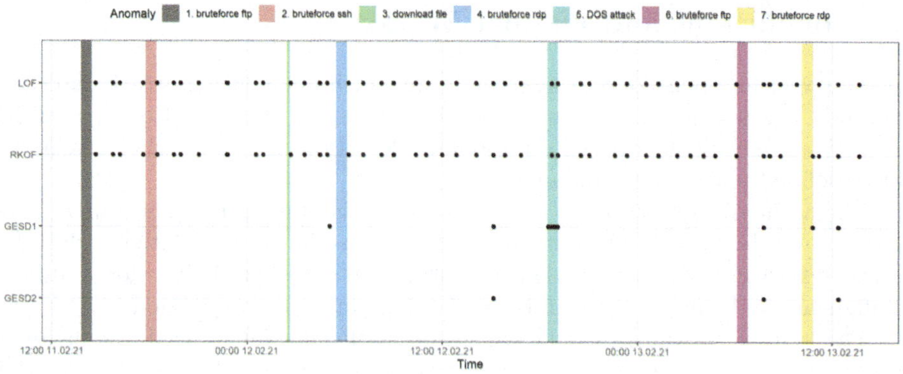

Fig. 6. Detected anomalies in the context of artificially introduced non–typical situations and attacks.

services. On the other hand, we developed several onedimensional GESD models, designed for specified network traffic interventions. The results of improved experiments are presented in Fig. 7.

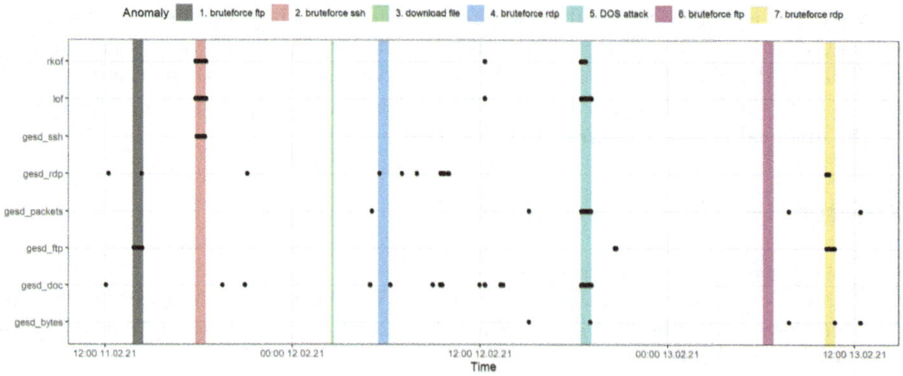

Fig. 7. Detected anomalies in the context of artificially introduced non–typical situations and attacks after data model change.

There are two important effects: firstly, LOF and RKOF reported fewer anomalies during the "normal" traffic and, secondly, more attacks were quite well reported. That points that the suggested direction of models improvement was proper.

However, some incidents were not identified by any method. This refers to the file download and the second brute–force attack on the ftp server. These situations require further data analysis or new data model building.

5 Conclusions and Further Works

In the paper an open–source–based environment implementing attack detection methods was presented. Its functionality extends the abilities of other network security engines as it is based on anomaly detection methods, where no previously defined patterns are required.

The main advantage of the system is that it is built from open source modules. It may be very significant for organizations that have limited budget for network security issues. The other advantage of the system is its wide ability of configuring and tuning, depending on the domain of its application.

The system capabilities were proved on artificially introduced anomalies in the network traffic and most of them were properly detected by one or more applied methods.

The present version of the system does not provide the ability to report detected anomalies to external services. However, this is one of the goals of the system further development. Our future works will focus also on providing implementation guides (a well known issue called "cold start problem" in data analysis) as well as on developing more advanced collaboration between Elasticsearch and R environment in data preprocessing.

Acknowledgements. RegSOC—Regional Center for Cybersecurity. The project is financed by the Polish National Centre for Research and Development as part of the second CyberSecIdent—Cybersecurity and e-Identity competition (agreement number: CYBERSECIDENT/381690 /II/NCBR/2018). We also would like to thank our colleagues from the Wroclaw Univeristy of Technology—our project partner—for preparing the artificial attack infrastructure and script.

References

1. Elasticsearch. https://www.elastic.co/
2. Allaire, J., Horner, J., Xie, Y., Marti, V., Porte, N.: Markdown: Render Markdown with the C Library Sundown (2019). https://CRAN.R-project.org/package=markdown. r package version 1.1
3. Bialas, A., Michalak, M., Flisiuk, B.: Anomaly detection in network traffic security assurance. In: Zamojski, W., et al. (eds.) Engineering in Dependability of Computer Systems and Networks. pp. 46–56. Springer International Publishing, Cham (2020). https://doi.org/10.1007/978-3-030-19501-4_5
4. Breunig, M.M., et al.: LOF: identifying density-based local outliers. In: Proceedings of the 2000 ACM SIGMOD International Conference on Management of Data, pp. 93–104 (2000). https://doi.org/10.1145/342009.335388
5. Chang, W., Cheng, J., Allaire, J., et al.: shiny: Web Application Framework for R (2020). https://CRAN.R-project.org/package=shiny. r package version 1.5.0
6. Dancho, M., Vaughan, D.: anomalize: Tidy Anomaly Detection (2019). https://CRAN.R-project.org/package=anomalize. R package version 0.2.0
7. Ester, M., Kriegel, H.P., Sander, J., Xu, X.: A density-based algorithm for discovering clusters in large spatial databases with noise. In: Proceedings of the 2nd International Conference on Knowledge Discovery and Data Mining, pp. 226–231 (1996)

8. Gao, J., Hu, W., Zhang, Z.M., Zhang, X., Wu, O.: RKOF: Robust kernel–based local outlier detection. In: Advances in Knowledge Discovery and Data Mining, pp. 270–283 (2011). https://doi.org/10.1007/978-3-642-20847-8_23
9. Grolemund, G., Wickham, H.: Dates and times made easy with lubridate. J. Stat. Softw. **40**(3), 1–25 (2011). https://www.jstatsoft.org/v40/i03/
10. Hahsler, M., Piekenbrock, M., Doran, D.: dbscan: Fast density-based clustering with R. J. Stat. Soft. **91**(1), 1–30 (2019). https://doi.org/10.18637/jss.v091.i01
11. Ioannides, A.: elasticsearchr: A Lightweight Interface for Interacting with Elasticsearch from R (2019). https://cran.r-project.org/package=elasticsearchr. v. 0.3.1
12. Kałędkowski, D.: runner: Running Operations for Vectors (2020). https://CRAN.R-project.org/package=runner. R package version 0.3.7
13. Madsen, J.H.: DDoutlier: Distance & Density-Based Outlier Detection (2018). https://CRAN.R-project.org/package=DDoutlier. R package version 0.1.0
14. Michalak, M., et al.: Outlier detection in network traffic monitoring. In: 10th International Conference on Pattern Recognition Applications and Methods, vol. 1, pp. 523–530 (2021)
15. Muniz, J., McIntyre, G., AlFardan, N.: Security Operations Center: Building, Operating, and Maintaining Your SOC. Cisco Press, 1st edn. (2015)
16. Niebezpiecznik: How the Poland's Financial Supervision Authority attack wasperformed? (in Polish) (2020).https://niebezpiecznik.pl/post/jak-przeprowadzono-atak-na-knf-i-polskie-banki-oraz-kto-jeszcze-byl-na-celowniku-przestepcow/. Accessed 01 Jun 2020
17. R Core Team: R: A Language and Environment for Statistical Computing. R Foundation for Statistical Computing, Austria (2013). http://www.R-project.org/
18. Rosner, B.: Percentage points for a generalized esd many-outlier procedure. Technometrics **25**(2), 165–172 (1983)
19. Tiwari, V., Kashikar, A.: OutlierDetection: Outlier Detection (2019). https://CRAN.R-project.org/package=OutlierDetection. R package version 0.1.1
20. Wickham, H., et al.: Welcome to the tidyverse. J. Open Source Softw. **4**(43), 1686 (2019). https://doi.org/10.21105/joss.01686
21. Zimmerman, Z.: Ten Strategies of a World–Class Cybersecurity Operations Center. The MITRE Corp. (2104)

Building AFDX Networks of Minimal Complexity for Real-Time Systems

Andrey Morkvin, Valery Kostenko, and Vasily Balashov(✉)

Lomonosov Moscow State University, Leninskie Gory, MSU, 1, Bldg. 52, Room 764, 119991 Moscow, Russia
{s02170018,kost,hbd}@cs.msu.su

Abstract. The article formulates the problem of constructing an on-board switched network of minimal complexity (total length of cables) necessary for transmitting a set of periodic messages in real time, and proposes an algorithm for its solution. The algorithm combines greedy criteria with limited enumeration, and builds the network structure and the set of virtual links for messages transmission. Experimental evaluation results for the proposed algorithm are also presented.

Keywords: Real-time switched network · Virtual link · Cable length minimization

1 Introduction

Modern real-time information and control systems have federated architecture or integrated modular architecture. The second approach is known as integrated modular avionics (IMA). Several standards have been developed that regulate the construction of IMA systems: ARINC 651 [1] specifies basic principles of building IMA systems; ARINC 653 [2] is the specification of operating systems; FC-AE-ASM-RT is the specification of the information exchange network based on the Fibre Channel switched network [3]; ARINC 664 (AFDX) [4] is the Ethernet-based data exchange network specification. Switch-based network topology is typical for IMA-based on-board systems.

In [5], using the example of a radar system with phased antenna arrays, it is shown that the transition from federated architecture to IMA architecture leads to an increase in data flow by 10^3–10^5 times, depending on the characteristics of the radar. To provide sufficient bandwidth, complex switched on-board networks must be constructed. Minimization of network complexity, while keeping the network able to transfer the specified traffic in real time, is a relevant problem of IMA system design. From dependability point of view, the less is the total cable length in the network, the less is the electromagnetic interference, and thus the probability of data transfer errors [6].

In this paper, we formulate the problem of building an on-board AFDX network with minimal complexity (calculated as the total length of cables, or links) necessary for real-time transmission of a specified set of periodic messages; both the physical network structure and the system of virtual links must be constructed. Algorithm for solving this problem is presented, as well as results of its experimental evaluation.

W. Zamojski et al. (Eds.): DepCoS-RELCOMEX 2021, AISC 1389, pp. 296–305, 2021.
https://doi.org/10.1007/978-3-030-76773-0_29

2 The Problem of Building the AFDX Network with Minimal Complexity

A real-world target system like an aircraft or an automated production line usually provides limited possibilities for installation of network equipment, such as switches and network cables (links). Let us suppose that possible locations where the switches can be installed are known, as well as lengths of routes (link lengths) between such locations, and between these locations and locations of network end systems.

A hypothetical *maximum network* is the network in which in every possible location a switch is present, and every network link (switch-to-switch, switch-to-end system) is also present. We propose to solve the network complexity minimization problem by finding which subset of the maximum network is sufficient for real-time transfer of specified traffic; the minimized criterion is the total length of links.

The problem is formally stated as follows. Let $G = (N \bigcup K, E, V, R)$ represent the maximum network: N is the set of end systems, K is the set of switches, $E = E_{sw} \bigcup E_{end}$ is the set of links (E_{sw} represents inter-switch links, E_{end} represents links between switches and end systems), V specifies lengths of links from E, R specifies bandwidths for links from E. For every end system from N, a set of subscribers is specified.

The network workload is a set MSG of periodically transmitted messages, in which each message is characterized by: transmission period T_{msg}; size $size_{msg}$; maximum release jitter J_{msg} (determined by variability of data readiness time); sender subscriber src_{msg}; set of recipient subscribers dst_{msg}.

Following real-time constraints are imposed on message transmission: the message must be transmitted at least once during its period; end-to-end transmission time must not exceed t_{msg}; end-to-end jitter must not exceed J^*_{msg}.

A network $G^* = (N^* \bigcup K^*, E^*, V^*, R^*)$ is a sub-network of the maximum network G: $N^* = N$ is the full set of end systems (no end system can be discarded without breaking data transmission), $K^* \subseteq K$ is a subset of switches, $E^* \subseteq E_{sw} \bigcup E_{end}$ is a subset of links, $V^* \subseteq V$ and $R^* \subseteq R$ are lengths and bandwidths of links from E^*. The notation $G^* \subseteq G$ is used below for a sub-network of the maximum network G.

We introduce *network complexity* as the total length of all links:

$$S(G^*) = \sum_{e \in E^*} V^*(e)$$

Virtual links in AFDX network are means for sharing the physical bandwidth between multiple data flows, in our case corresponding to periodic messages. To build a virtual link vl means to define its properties: LM_{vl} is the maximum frame size (including frame header); BAG_{vl} is the minimum start-to-start time gap between frame transmissions for zero frame generation jitter; JM_{vl} is the maximum frame generation jitter on the sender end system; S_{vl} is the sender end system; D_{vl} is the set of recipient end systems; $Tree_{vl}$ is the set of frame transmission routes in the network (making up a tree with root in the sender end system and leaves in the receiver end systems); MSG_{vl} is the set of messages transmitted via the virtual link.

To design and route virtual links, the following constraints must be satisfied:

- The total bandwidth reserved for the virtual links passing through the physical data link e must not exceed its bandwidth:

$$\forall e \in E^* : \sum_{vl \in e} \frac{LM_{vl}}{BAG_{vl}} \le R_e \tag{1}$$

- For each virtual link, the frame sending frequency required to send its periodic messages (split into frames) must not exceed the maximum frame sending frequency:

$$\forall vl \in VL : \sum\nolimits_{msg \in MSG_{vl}} \left\lceil \frac{size_{msg}}{(LM_{vl} - c)} \right\rceil * \frac{1}{T_{msg}} \le \frac{1}{BAG_{vl}} \tag{2}$$

- The maximum frame jitter on the sender end systems must not exceed 0.5 milliseconds:

$$\forall vl \in VL : JM_{vl} \le 0.5\ \text{ms} \tag{3}$$

- The maximum message end-to-end transmission time and the maximum end-to-end jitter must not violate the following constraints:

$$\forall msg \in MSG : \begin{cases} Dur(msg) \le t_{msg} \\ Jit(msg) \le J^*_{msg} \end{cases} \tag{4}$$

  where $Dur(msg)$ и $Jit(msg)$ are end-to-end message transmission time (or duration) and jitter, estimated by corresponding procedures.

To solve the problem is to build a minimal complexity network G^*, and a set of virtual links VL, for the maximum subset of messages $MSG^{max} \subseteq MSG$ which can be transferred through the maximum network under constraints (1)–(4).

Thus, there are two criteria for optimization: $\max(|MSG^*|)$ and $\min(S(G^*))$. To make our problem unicriterial, we leave only one optimized criterion: $\min(S(G^*))$, and convert the other criterion to a constraint: $|MSG^*| = |MSG^{max}|$, where MSG^{max} is the maximum (by the number of messages) message set for the maximum network.

Therefore, the optimization problem is stated as follows:

$$\min(S(G^*))$$

$$G^* \subseteq G; \quad |MSG^*| = |MSG^{max}|; \qquad \text{constraints (1)} - \text{(4)}$$

3 Algorithm for Building the Minimal Network

3.1 General Scheme of the Algorithm

Step 1. Create the maximum network G.

Step 2. Create a virtual link (initially without a route) for each message; calculate the properties LM_{vl}, JM_{vl} and BAG_{vl} (Subsect. 3.2), considering the constraint (2).

Step 3. Check the constraint (3) for each virtual link. If the constraint is not satisfied for all virtual links, perform the links aggregation procedure (Subsect. 3.3) for sender subscriber of every virtual link that violates the constraint (3). This procedure aggregates messages from the same subscriber so that constraints (2) and (3) are satisfied.

Step 4. Build a route for each virtual link, considering the constraint (1) and the total length of links already used for data transmission (virtual links are processed in decreasing order by the required bandwidth):

Step 4.1. Perform the routing procedure (Subsect. 3.5) for the current virtual link to find route(s) with sufficient free bandwidth, connecting the sender subscriber to the receiver subscriber(s) of this virtual link.

Step 4.2. If routing is successful, proceed with the next virtual link.

Step 4.3. Else, perform the limited enumeration procedure (Subsect. 3.6) for routes of already assigned virtual links, to find alternate routes for them that allow routing the current virtual link.

Step 4.4. If the procedure on Step 4.3 finishes successfully (i.e. there are routes for both the current virtual link and the previously allocated ones), proceed with the next virtual link.

Step 4.5. Else, in case only one message is transmitted via the current virtual link, assignment of this virtual link is considered impossible, and the virtual link and its message are removed. Otherwise, remove the message with the highest LM/BAG ratio from the current virtual link, perform the aggregation procedure (Subsect. 3.3) on this virtual link, and return to Step 4.1.

Step 5. For each message (except those removed on Step 4.5):

Step 5.1. Estimate the message transmission time and jitter (Subsect. 3.7), check the constraint (4).

Step 5.2. If the constraint is satisfied, proceed with the next message.

Step 5.3. Else, perform the procedure for reconfiguring the virtual link for the current message (Subsect. 3.8).

Step 5.4. If, after performing the procedure, constraint (4) is satisfied, proceed with the next message.

Step 5.5. Else, perform the virtual links aggregation procedure (Subsect. 3.3) which processes the virtual links with the same sender end system as the virtual link for the current message.

Step 5.6. If after aggregation procedure execution the constraint (4) is still violated, remove the current message, and its virtual link (if it is the only message in this virtual link).

Step 6. Remove the unused links (those that are parts of no virtual link route) from the maximum network G. Remove unused switches (those with no connected links) from the network. The obtained network is the result of the algorithm.

The algorithm described above is based on the algorithm from [7], that maximizes the number of messages for which virtual links are constructed, without consideration of physical link lengths. In our algorithm, the efforts to minimize the total length of links by preferring shorter routes and non-empty links are made on steps 4.1, 4.3, 5.5

by performing the procedures for initial routing, re-routing by limited enumeration, and aggregation of virtual links. These procedures use the length-aware routing criterion, and thus they are described in detail below. For other procedures, only general description is provided, with more detail available in [7].

3.2 Calculating the Properties of Virtual Links

This procedure calculates the *LM* and *BAG* properties of a single-message virtual link by solving the problem of minimizing the virtual link bandwidth $bw_{vl}(n)$, where n is the number of frames into which the message is split. Solution of this problem is based on the analysis presented in [8]. After calculating *LM* and *BAG* for all virtual links, the procedure enumerates all end systems, and for each end system calculates *JM* for every virtual link originating from it, based on worst-case interference from frames of other virtual links with the same sender end system. Details on this procedure are provided in [7].

3.3 Virtual Links Aggregation Procedure

The algorithm for this procedure is based on [7]. Suppose that aggregation is performed for the virtual link vl originating from the end system es. Purpose of this procedure is to decrease the end-to-end frame transfer time for vl by decreasing the number of frames transferred through the network, and thus eliminate violation of constraint (4) for vl. The procedure essentially builds a virtual link, which transfers several messages, inherited from two or more virtual links that are aggregated. It is also possible to eliminate violation of constraint (3) using this procedure, as the frames of a single (aggregated) virtual link create less worst-case interference on the frames of vl than frames of two separate virtual links, so the value of JM_{vl} is decreased.

Step 1. For each subscriber src connected to es, compose the set Vl_{src} of (existing) virtual links with messages sent by src.

Step 2. If every Vl_{src} contains no more than one element, or aggregation has already been attempted for all pairs of virtual links within every Vl_{src}, then return a failure.

Step 3. Select a new (unexamined) pair of virtual links vl_1 and vl_2 from the same set Vl_{src} such that the value $r(vl_1)*r(vl_2)$ is maximal among all such pairs. Here $r(vl)$ is defined as follows:

$$r(vl) = \frac{LM_{vl}}{BAG_{vl*}|MSG_{vl}|}, \text{ or}$$

$$r(vl) = \frac{\alpha * \frac{LM_{vl}}{BAG_{vl}} + \beta * \min(t_{msg})}{|MSG_{vl}|}, \ \alpha + \beta = 1.$$

Step 4. Aggregate messages of vl_1 and vl_2 (Subsect. 3.4) and create a new (aggregated) virtual link with the resulting set of properties.

Step 5. If the bandwidth required for the aggregated virtual link exceeds the total bandwidth required for vl_1 and vl_2, or constraint (3) is not satisfied, keep the original virtual links, discard the aggregated one, and go to Step 2.

Step 7. If a route was previously found for at least one of the virtual links vl_1, vl_2, perform the routing procedure (Subsect. 3.5) for the aggregated virtual link. If the procedure fails, keep the original virtual links, discard the aggregated one, and go to Step 2.

Step 8. Remove vl_1 and vl_2 from Vl_{src} and insert the aggregated virtual link into Vl_{src}. If the violated constraints (3) and/or (4) are not yet satisfied for vl, then go to Step 2; otherwise, return a success.

The length-aware routing criterion is used on step 7 of this procedure.

3.4 Messages Aggregation Procedure

This procedure calculates the *LM*, *BAG* and *JM* properties for a virtual link that transfers a set of messages originating from the same subscriber. The procedure solves a bandwidth minimization problem similar to the one described in Subsect. 3.2, taking in account that several messages, in common case with different sizes and periods, must be fitted into one virtual link with a single set of properties. Details on this procedure are provided in [7].

3.5 Virtual Link Routing Procedure

This procedure constructs a route for the given virtual link. The route connects the sender end system to the receiver end system, and each link of the route has sufficient free bandwidth to transfer the virtual link's messages. If the virtual link has several receiver end systems, a set (or tree) of routes is constructed.

Algorithm for this procedure is based on [7], but we introduce the criteria that prefer routes, which are shorter and contain physical links with higher number of previously assigned virtual links.

Step 1. Transform G by eliminating all the links for which the free bandwidth is lower than the bandwidth required for the given virtual link.

Step 2. Run Yen's algorithm on the resulting subgraph. Stop the algorithm after finding the shortest path to one of the receiver end systems, to which the path has not yet been found.

Step 3. If the paths to all receiver end systems were found, return these paths as the routes for the virtual link. Otherwise, go to Step 2.

Criteria for Yen's algorithm (lower is better):

- $I(e) = \frac{V_e}{k+1}$, where V_e is the length of the link e, and k is the number of virtual links already assigned to e.
- $I(path) = C_1 * \sum_{path} I(e) + C_2 * cost(path,\ vl)$ is the criterion for comparing paths, where $cost(path,\ vl) = \sum_{e \in path(S_{vl},n)} \frac{LM_{vl}}{R_e} + \sum_{v \in path(S_{vl},n),\, v \neq vl} \frac{LM_{vl}}{\max\limits_{e \in path}(R_e)}$ is a heuristic that estimates the cost of a path, $C_1, C_2 \in [0,\ 1]$, $C_1 + C_2 = 1$.

3.6 Limited Enumeration Procedure

The algorithm for this procedure is based on [7]. Purpose of the procedure is to enable routing of a virtual link, which previously failed to be routed. The idea is that some already routed virtual links block the routing of the given virtual link, but if the given virtual link was routed "before" them, and these links "after" it, the problem would not emerge. So the procedure attempts to find such (sub)set of already routed virtual links by enumeration of sets with power (number of elements) not greater than the given limit, which equals 2 in this paper.

Step 1. For each set of routed virtual links, whose power does not exceed the given limit, perform Steps 1.1–1.7:
Step 1.1. Deallocate all routes of the virtual links from the current set.
Step 1.2. Perform the routing procedure (Subsect. 3.5) for the given virtual link vl.
Step 1.3. If procedure fails, go to Step 1.7.
Step 1.4. Else, perform the routing procedure for all virtual links deallocated on Step 1.1.
Step 1.5. If at least one of the virtual links could not be routed, deallocate the route for the virtual link vl and go to Step 1.7.
Step 1.6. Else, return a success.
Step 1.7. Restore the initial routes of the virtual links.
Step 2. Return a failure.

The length-aware routing criterion is used on steps 1.2 and 1.4 of this procedure.

3.7 Estimating the Maximum Message Transmission Time and Jitter

Maximum end-to-end message transmission time and end-to-end jitter are estimated according to the following scheme, based on [7]. Estimation is performed when parameters and routes of virtual links are already calculated.

$$Jit(msg) = Dur(msg) - Dur_{min}(msg)$$

$$Dur(msg) = \mu + \delta + \Delta \qquad Dur_{min}(msg) = \mu + \delta_{min} + \Delta_{min}$$

- μ is a constant characterizing the time required for splitting the message into frames and assembling the message from frames.
- $\delta = BAG_{vl} \sum_{msg \in MSG} \left\lceil \frac{size_{msg}}{LM_{vl} - c} \right\rceil$ is the maximum time from the instant the frames were put into the virtual link queue (on the sender end system) to the instant the last frame was taken from this queue, c is the frame header size.
- Δ is the maximum transmission time of the last frame. It can be found using techniques from [9–11].
- $\delta_{min} = BAG_{vl} \left| \frac{size_{msg}}{LM_{vl} - c} \right|$ is the minimum time from the instant the frames were put into the virtual link queue to the instant the last frame was taken from this queue; it is estimated as the message waiting time in case the queue was empty in the former instant.

- $\Delta_{min} = max_{path \in Tree_{vl}} \left(\sum_{e \in path} \frac{LM_{vl}}{R_e} + \sum_{k \in path} t_k \right)$ is the minimum transmission time of the last frame; it is estimated as the frame transmission time in case the frame, while passing through the switches along its route, always arrives at empty output port queue (of the switch) and is immediately transmitted farther; t_k is the frame processing time in the switch k for such case.

3.8 Virtual Link Reconfiguration Procedure

Purpose of the procedure is to calculate new parameters (and possibly choose a new route) for a virtual link that violates constraint (4) on message transmission time and/or jitter. The procedure is executed when parameters and routes of virtual links are already calculated, thus there are more data available in comparison to initial calculation of virtual link properties (Subsect. 3.2) or their recalculation during virtual links aggregation (Subsects. 3.3 and 3.4). With these new data, virtual link properties can be more precisely adjusted in accordance to properties of other virtual links. Details on this procedure are provided in [7].

As the procedure changes the virtual link properties, and possibly its route, its execution may lead to violation of constraint (4) for virtual links that previously did not violate this constraint. Thus, a re-check of constraint (4) is performed for virtual links already checked for meeting the constraint (4), and if a new violation is detected, the procedure returns a failure.

The length-aware routing criterion is used in the attempt to re-route the virtual link, if such attempt is made during virtual link reconfiguration.

4 Experimental Evaluation of the Algorithm

Goal of the experiments was to estimate efficiency of the algorithm in minimization of total cable length in the network while guaranteeing real-time transfer of all periodic messages. To reach this goal, it is reasonable to construct such input data sets that it is a priori known, which sub-network is sufficient to transfer all the messages.

Data sets for experiments were constructed as follows: (1) choose the basic network topology; (2) generate a message set that can be transferred through the basic network, creating a sufficiently high load on this network; (3) extend the basic network to the maximum network with extra switches and links, so that multiple extra routes are available for messages transfer. Algorithm operation on such data set is considered successful if it manages to cut the maximum network down to the basic network or another network with total cable length equal to that of the basic network.

Three pairs of (basic, maximum) network topologies were used in the experiments. Two of the topologies are shown in Fig. 1. Solid squares, circles and lines depict the switches, end systems and links of the basic network; together with dashed ones, they form the maximum network. Multiple end systems attached to the same switch are shown as one circle, for brevity. The third maximum topology was taken from [12] and cut down to the basic one by removing some redundant inter-switch links. The topologies are typical for onboard networks, in particular the third one is used in Airbus A380.

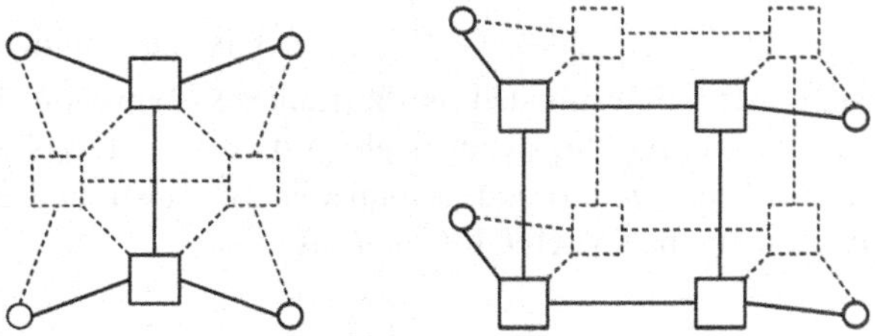

Fig. 1. Network topologies for the experiments

For each of the three network topologies, message sets of following types were generated: (1) size: 16 bytes – 1 Kbyte, period: 10 ms – 1 s, maximum end-to-end transmission time: 10 ms – 100 ms; (2) size: 1 Kbyte – 100 Kbytes, period: 100 ms – 10 s, maximum end-to-end transmission time: 10 ms – 1 s; (3) mix of these two types. Type 1 corresponds to control and navigation traffic in aircraft onboard networks, type 2 – to "media" traffic like images from a weather radar, type 3 represents heterogeneous traffic. For every type of message set, the message parameters were randomly selected in the specified ranges.

For every pair (network topology, message set type), 10 message sets were generated and used in the experiments. Depending on message set type and network topology, a message set included from 100 to 300 periodic messages, which is consistent with available data on traffic characteristics in onboard networks (from [12] as well as from authors' own experience in the industry). Experiment results are shown in Fig. 2.

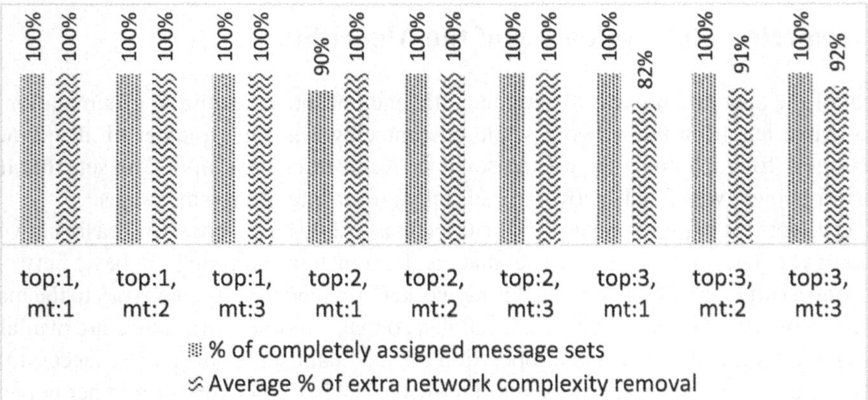

Fig. 2. Results of the experiments ("top" – topology, "mt" – message set type)

The experiments show that in almost all cases 100% of messages were successfully assigned, i.e. virtual links for them were constructed. The network was completely stripped of extra complexity also in most cases, except for topology 3, which is very rich in alternative routes; even for it, more than 80% of extra complexity was removed.

5 Conclusion

The algorithm proposed in this paper, for a given set of periodic messages, builds a switched exchange network of minimal complexity necessary for transmitting messages in real time, and a set of virtual links for messages transmission. Complexity is measured as the total length of cables in the network. Minimization of network complexity reduces electromagnetic interference on the cables and increases reliability of the network. Experimental evaluation results for the proposed algorithm are also presented, indicating that the algorithm successfully removes extra complexity from networks with realistic topology and workload.

Future research can be aimed at support for: (1) minimization of the number of switches, together with total cable length minimization; (2) choosing locations for the switches without changing the topology of the network (except for cable lengths).

Acknowledgments. This work was supported by the Russian Foundation for Basic Research, project №19–07-00614.

References

1. ARINC Specification 651. Design Guidance for Integrated Modular Avionics. Airlines Electronic Engineering Committee (1997)
2. ARINC Specification 653. Avionics Application Software Standard Interface. Airlines Electronic Engineering Committee (2007)
3. INCITS 373. Information Technology – Fibre Channel Framing and Signaling Interface (FC-FS). International Committee for Information Technology Standards (2003)
4. ARINC Specification 664. Aircraft Data Network, Part 7. Avionics Full Duplex Switched Ethernet (AFDX) Network. Airlines Electronic Engineering Committee (2005)
5. Kostenko, V.A.: Architecture of software and hardware complexes of onboard equipment. J. Inst. Eng. **60**(3), 229–233 (2017). (in Russian)
6. Al-Kuwaiti, M., Kyriakopoulos, N., Hussein, S.: A comparative analysis of network dependability, fault-tolerance, reliability, security, and survivability. IEEE Commun. Surv. Tutorials **11**(2), 106–124 (2009)
7. Vdovin, P.M., Kostenko, V.A.: Organizing message transmission in AFDX networks. Program. Comput. Softw. **43**(1), 1–12 (2017)
8. Al Sheikh, A., et al.: Optimal design of virtual links in AFDX networks. Real-Time Syst. **49**(3), 308–336 (2013)
9. Scharbarg, J.L., Fraboul, C.: Methods and tools for the temporal analysis of avionic networks. In: New Trends in Technologies: Control, Management, Computational Intelligence and Network Systems, pp. 413–438 (2010)
10. Frances, F., Fraboul, C., Grieu, J.: Using network calculus to optimize the AFDX network. In: Proceedings of European Congress on Embedded real-time software (ERTS), pp. 1–8 (2006)
11. Bauer, H., Scharbarg, J.L., Fraboul, C.: Applying Trajectory approach with static priority queuing for improving the use of available AFDX resources. Real-time systems. **48**(1), 101–133 (2012)
12. Amari, A., Mifdaoui, A.: Specification and performance indicators of AeroRing – a multiple-ring ethernet network for avionics embedded systems. Sensors **18**, 1–28 (2018)

Efficient Computation of the Best Controls in Complex Systems Under Global Constraints

Grzegorz Mzyk(✉)

Wrocław University of Science and Technology, 50-370 Wrocław, Poland
grzegorz.mzyk@pwr.edu.pl
http://staff.iiar.pwr.edu.pl/grzegorz.mzyk

Abstract. The problem of efficient calculation of optimal controls in large-scale systems is considered. The structure of interconnections is assumed to be known, whereas parameters of individual components remain unknown. Particular elements cannot be optimized independently of the remaining part of the system owing to some global constraints connected with limited resources. Using input-output observations all blocks are firstly identified by recursive instrumental variables method. Next, high-dimensional optimization problem is decomposed on series of levels. On each level one-dimensional problem is solved with the use of convex programming, taking into account the amount of available resources. Lower levels play supporting role for upper ones.

Keywords: Distributed computing · System identification · Optimal control · Decomposition and coordination · Multi-level optimization · Parallel computing · Dimension reduction

1 Introduction

Complex systems, including a lot of interconnected and mutually dependent elements, are considered in the paper. Presented idea of control is based on the theory of hierarchical approach, which has been intensively elaborated over the last four decades ([5,7,11,14]). Owing to the fact that interconnected elements are dependent, some specific problems arises for large scale nets. In particular:

- dimensionality: performance index of the whole system depends on the number of controls/decisions, which lead to high-dimensional algebraic problems ([9]),
- global constraints: full decomposition of the global optimization problem on smaller subproblems is usually not possible, since all elements are implicitly dependent through the resource limitations,
- identifiability: rich excitation for identification of some elements can be impossible, unless the system is disconnected,

W. Zamojski et al. (Eds.): DepCoS-RELCOMEX 2021, AISC 1389, pp. 306–315, 2021.
https://doi.org/10.1007/978-3-030-76773-0_30

- noise transfer: owing to existence of feedback connections, the random output noise is transferred to the inputs, which leads to asymptotic bias errors of mathematical models of the blocks ([19]),
- heterogeneity: some delays in measurement collecting procedure for particular blocks exclude application of global approach,
- nonlinearity: it is difficult to assure convexity of performance index when the system is not linear ([6,12]).

Our goal is to make optimal decisions (in real time) on local controls for particular blocks, provided that global resources can be limited. Proposed methodology is based on the following stages:

- on-line identification of block parameters by recursive *instrumental variables* method,
- parallel calculation of local optimal controls by *interaction balance method* with the use of identified parameters,
- *multi-level optimization* of the global cost function for active constraints concerning resource limitations.

For clarity of presentation we limited our problem to the class of static linear blocks. However, proposed methodology can be simply generalized for Hammerstein type nonlinear dynamic blocks ([8,13]). In a lot of commonly met hierarchical control problems, the accurate mathematical models of individual components are needed. Under the term 'complex' we understand the fact that the system is built of a number of interconnected components (subsystems), e.g., in the typical production system each element is excited by the outputs of other blocks (see [5]). In consequence of mutual interconnections, these components are dependent and their separation may be impossible or too expensive.

The concept of multi-level control means that some decisions are global (coordination variables) and influence all blocks, and some of them are local and can be set independently. Such philosophy allows for dimension reduction and application of parallel and distributed computation techniques ([4]).

In Sect. 2 the considered problem is formulated precisely. Next, in Sect. 3, recursive procedure of identification of components on the basis of external system measurements is presented. We emphasize that each model can be computed in parallel. Then, in Sect. 4, parallel procedure of local unconstrained control optimization by interaction balance method is described. Finally, in Sect. 5, multi-level convex programming is applied, when the global resource limitations requires coordination between particular blocks. Simple numerical example is also presented to illustrate practical effectiveness of the procedure.

2 The System

Let us consider the system, shown in Fig. 1, consisting of n static linear elements described as follows

$$y_i = a_i x_i + b_i u_i + \xi_i \qquad (i = 1, 2, ..., n), \tag{1}$$

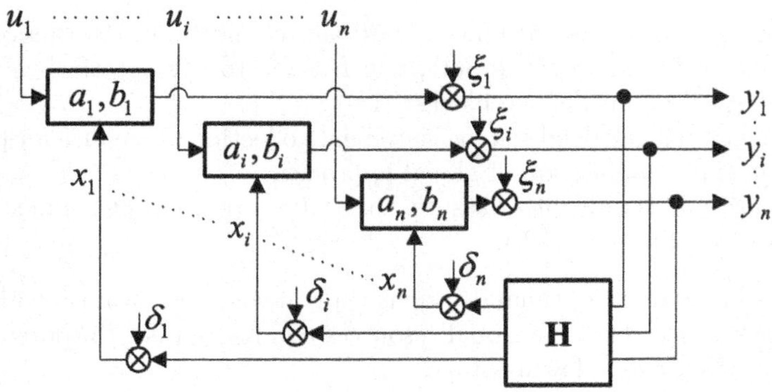

Fig. 1. The complex n-element linear static system

where

$$u = (u_1, , ...u_i, ..., u_n)^T, \qquad x = (x_1, ..., x_i, ..., x_n)^T, \qquad y = (y_1, ..., y_i, ..., y_n)^T \tag{2}$$

are the external controls, hidden interaction inputs, and measured outputs of the system, respectively. The signals

$$\delta = (\delta_1, ..., \delta_i, ..., \delta_n)^T \text{ and } \xi = (\xi_1, ..., \xi_i, ..., \xi_n)^T \tag{3}$$

are zero-mean, mutually independent random noises. Block H, which defines the system interconnection structure is *a priori* known binary matrix H (i.e. $H_{i,j} = 0$ – 'no connection', $H_{i,j} = 1$ – 'connection'), i.e.,

$$x_i = H_i y + \delta_i, \tag{4}$$

where H_i denotes ith row of H. Our aim is to find the optimal controls u^*, from the given set $u \in \mathcal{U}$ of admissible inputs, which minimizes some given cost function $Q(u)$

$$u^* = \arg\min_{u \in \mathcal{U}} Q(u). \tag{5}$$

Owing to the fact that system outputs $y(u)$ depend on u, and are noise corrupted (random), criterion $Q()$ can be treated as a risk function from the decision theory point of view. For desired output y_d and quadratic loss function $L(y, y_d) = \|y(u) - y_d\|^2$ we minimize $Q(u) = E\|y(u) - y_d\|$. To model relation between y and u, parameters $\{(a_i, b_i)\}_{i=1}^n$ of individual elements must be first estimated, using the set of data $\{(u^{(k)}, y^{(k)})\}_{k=1}^N$ collected in the experiment. We emphasize that internal excitations x cannot be measured. Introducing matrices

$$A = \text{diag}(a_1, a_2, ..., a_n), \qquad B = \text{diag}(b_1, b_2, ..., b_n), \qquad H = (H_1^T, H_2^T, ..., H_n^T)^T \tag{6}$$

the system, as a whole, can be described as follows

$$\begin{cases} y = Ax + Bu + \xi \\ x = Hy + \delta \end{cases}. \tag{7}$$

By inserting x to the first equation in (7) we get

$$y = A\,(Hy + \delta) + Bu + \xi,$$
$$(I - AH)\,y = Bu + A\delta + \xi,$$

which leads to

$$y = Ku + G\theta \tag{8}$$

where

$$G = (I - AH)^{-1}, \qquad K = (I - AH)^{-1}B = GB, \qquad \theta = A\delta + \xi. \tag{9}$$

Equation (8) resembles description of the simple object with the input u, output y, transfer matrix K, and additive zero-mean output noise $G\theta$.

3 On-line System Identification

In this section we focus on identification of one particular ith element. Obviously, the procedure should be run for each $i = 1, 2, ..., n$ independently (in parallel).

3.1 Least Squares Method

Introducing matrices of input and output data of ith element

$$Y_{iN} = \left(y_i^{(1)}, ..., y_i^{(k)}, ..., y_i^{(N)}\right)^T, \tag{10}$$
$$\Phi_{iN} = \left(\phi_i^{(1)}, ..., \phi_i^{(k)}, ..., \phi_i^{(N)}\right)^T,$$
$$\text{where } \phi_i^{(k)} = (x_i^{(k)}, u_i^{(k)})^T,$$

we get the following measurement equation

$$Y_{iN} = \Phi_{iN}(a_i, b_i)^T + \xi_{iN}. \tag{11}$$

Since the input x_i included in ϕ_i is not accessible (cannot be measured), traditional least squares methodology cannot be applied directly. Owing to (4), the most natural evaluation is

$$\widetilde{\phi}_i = (\widetilde{x}_i, u_i)^T, \text{ with } \widetilde{x}_i = H_i y = x_i - \delta_i.$$

It leads to the following estimates

$$(\widehat{a}_i^{l.s.}, \widehat{b}_i^{l.s.})^T = \left(\widetilde{\Phi}_{iN}^T \widetilde{\Phi}_{iN}\right)^{-1} \widetilde{\Phi}_{iN}^T Y_{iN}, \tag{12}$$

where

$$\widetilde{\Phi}_{iN} = \left(\widetilde{\phi}_i^{(1)}, ..., \widetilde{\phi}_i^{(k)}, ..., \widetilde{\phi}_i^{(N)}\right)^T.$$

Estimate (12) is based on modified version of measurement Eq. (11), in which unknown Φ_{iN} was substituted with $\widetilde{\Phi}_{iN}$, i.e.,

$$Y_{iN} = \widetilde{\Phi}_{iN}(a_i, b_i)^T + \Theta_{iN}. \tag{13}$$

Consequently, in (13) the disturbance

$$\Theta_{iN} = [\theta_i^{(1)}, ..., \theta_i^{(k)}, ..., \theta_i^{(N)}]$$

appears instead of ξ_{iN} in (11). Owing to the fact, that $\theta_i^{(k)}$ is generally correlated with interaction inputs included in vectors $\widetilde{\phi}_i^{(k)}$, least squares estimate is biased, even asymptotically, i.e., the error

$$(\widehat{a}_i^{l.s.}, \widehat{b}_i^{l.s.}) - (a_i, b_i) = \left(\widetilde{\Phi}_{iN}^T \widetilde{\Phi}_{iN}\right)^{-1} \widetilde{\Phi}_{iN}^T \Theta_{iN}$$

does not tend to zero, as $N \to \infty$, although $E\theta_i^{(k)} = 0$. Formally, the situation is analogous to SISO linear dynamic system identification corrupted by correlated noise. To cope with this problem instrumental variables technique can be successfully applied.

3.2 On-line Instrumental Variables Approach

Standard application of instrumental variables (see [18] and [15]) leads to the following generalization of (12)

$$(\widehat{a}_i^{i.v.}, \widehat{b}_i^{i.v.})^T = \left(\Psi_{iN}^T \widetilde{\Phi}_{iN}\right)^{-1} \Psi_{iN}^T Y_{iN}, \tag{14}$$

where

$$\Psi_{iN} = \left(\psi_i^{(1)}, ..., \psi_i^{(k)}, ..., \psi_i^{(N)}\right)^T$$

represents additional matrix of instrumental variables with the same dimensions as $\widetilde{\Phi}_{iN}$, i.e.

$$\psi_i^{(k)} = \left(\psi_{i,1}^{(k)}, \psi_{i,2}^{(k)}\right)^T,$$

which fulfills the following two conditions
(C1) $\psi_{i,1}$, and $\psi_{i,2}$ are correlated with the inputs u_i, such that

$$\frac{1}{N}\Psi_{iN}^T \widetilde{\Phi}_{iN} = \frac{1}{N}\sum_{k=1}^{N} \widetilde{\phi}_i^{(k)} \psi_i^{(k)T} \to E\widetilde{\phi}_i \psi_i^T,$$

with probability 1 as $N \to \infty$, and the limit matrix $E\widetilde{\phi}_i \psi_i^T$ is of full rank.
(C2) $\psi_{i,1}$, and $\psi_{i,2}$ are *not* correlated with the resulting output noise θ_i, i.e.,

$$\frac{1}{N}\Psi_{iN}^T \Theta_{iN} = \frac{1}{N}\sum_{k=1}^{N} \psi_i^{(k)} \theta_i^{(k)} \to E\psi_i \theta_i,$$

with probability 1 as $N \to \infty$, and the limit matrix $E\psi_i \theta_i = 0$.

On the basis of Theorems 6.1 and 6.2 in [13] we conclude that if the instrumental variables matrix Ψ_{iN} fulfills *(C1)* and *(C2)* then

$$(\widehat{a}_i^{i.v.}, \widehat{b}_i^{i.v.}) \to (a_i, b_i) \tag{15}$$

with probability 1, as $N \to \infty$, and the optimal choice of instruments with respect to covariance minimization of the estimate is

$$\psi_i^* = (\overline{x}_i, u_i)^T, \text{ with } \overline{x}_i = E(x_i|u) = H_i K u. \tag{16}$$

Since the matrix K is unknown, the least squares can be used as a pilot estimate supporting generation of proper instruments. Both (12) and (14) can be computed online (see [1,4]). Old measurements need not to be stored in memory, and matrix inversion operation is avoided, i.e.,

$$(\widehat{a}_i^{i.v.}, \widehat{b}_i^{i.v.})_{(k)}^T = (\widehat{a}_i^{i.v.}, \widehat{b}_i^{i.v.})_{(k-1)}^T + P_{i,k}\psi_i^{(k)} \left[y_i^{(k)} - \widetilde{\phi}_i^{(k)T} (\widehat{a}_i^{i.v.}, \widehat{b}_i^{i.v.})_{(k-1)}^T \right], \tag{17}$$

where

$$P_{i,k} = \frac{1}{\lambda} \left[P_{i,k-1} - \frac{P_{i,k-1}\widetilde{\phi}_i^{(k)} \psi_i^{(k)^T} P_{i,k-1}}{\lambda + \psi_i^{(k)^T} P_{i,k-1} \widetilde{\phi}_i^{(k)}} \right]. \tag{18}$$

and $\lambda = 1$ for stationary blocks, and $0 < \lambda < 1$ when tracking time-varying parameters.

4 Decomposition of Unconstrained Problem

Let

$$y_d = \left(y_{d,1}, ..., y_{d,i}, ..., y_{d,n} \right)^T \tag{19}$$

be desired vector of outputs, recommended with respect to performance of technological process. Owing to (8) one can write that

$$Q(u) = E \left\| Ku - y_d + G\theta \right\|_2^2 = \left\| Ku - y_d \right\|_2^2 + E \left\| G\theta \right\|_2^2.$$

Since the component $E \left\| G\theta \right\|_2^2 = \text{const}$, it suffices to minimize

$$\overline{Q}(u) = \left\| Ku - y_d \right\|_2^2,$$

which, for unconstrained problem leads to the solution of linear equation $Ku = y_d$ with respect to u. Vector y_d is usually provided by experts in the field of application, and matrix K(see (9)) is estimated as

$$\widehat{K} = (I - \widehat{A}H)^{-1}\widehat{B},$$

with $\widehat{A}$ and $\widehat{B}$ generated on-line, according to (17), i.e.,

$$\widehat{A} = \text{diag}(\widehat{a}_1^{i.v.}, ..., \widehat{a}_i^{i.v.}, ..., \widehat{a}_n^{i.v.}), \qquad \widehat{B} = \text{diag}(\widehat{b}_1^{i.v.}, ..., \widehat{b}_i^{i.v.}, ..., \widehat{b}_n^{i.v.}).$$

Direct computation of $u^* = \widehat{K}^{-1} y_d$ may be however numerically unstable and computationally complex, as the estimate $\widehat{K}$ is random and $\dim \widehat{K} = n \times n$ can be relatively large. For unlimited resources ($\mathcal{U}$=$\mathcal{R}^n$) optimization can be splitted into n (independent) local balancing tasks

$$u_i^* = \frac{y_{d,i} - \widehat{a}_i H_i y_d}{\widehat{b}_i}, \tag{20}$$

which can be computed in parallel. In (20) the component $H_i y_d$ represents approximation of the interaction x_i in the desired state and the local control/decision $u_i = u_i^*$ is computed to assure appropriate balance.

5 Multi-level Optimization with Limited Resources

In practice, application of fully decomposed strategy (20) may be excluded, owing to potential limitations imposed on global resources, possible to engage in the whole system. Usually, they are connected with founds or energy and can be formulated as follows

$$S = \{u : \|u\|_2^2 \leq U\}, \tag{21}$$

where S denotes acceptable set of solutions. Distribution of resources must be thus synchronized on upper level. Some of arguments $u_1, ..., u_i, ..., u_n$ are selected to play the role of coordination variables, whereas remaining ones are changed on the range determined by higher level. This method can be obviously duplicated on each resulting level. Finally, it is possible to decompose one complex n-dimensional problem on n one-dimensional layers

$$\min_{u_1, u_2, ..., u_n} Q(u_1, u_2, ..., u_n) = \min_{u_1} \left(\min_{u_2} ... \left(\min_{u_n} Q(u_1, u_2, ..., u_n) \right) \right). \tag{22}$$

For i-th layer, arguments $u_1, u_2, ..., u_{i-1}$ are treated as constants (determined by upper levels), and optimization is done with respect to $u_i, u_{i+1}, ..., u_n$. The lower level problem can be solved analogously, remembering that $u_1, u_2, ..., u_{i-1}$ occupy a part of resource amount U.

6 Example

Let us consider the two-element system including cross-feedbacks

$$A = \begin{bmatrix} 0.5 & 0 \\ 0 & 0.25 \end{bmatrix}, \quad B = \begin{bmatrix} 1 & 0 \\ 0 & 1 \end{bmatrix}, \quad H = \begin{bmatrix} 0 & 1 \\ 1 & 0 \end{bmatrix},$$

with the cost function $Q(u) = E \|y(u) - y_d\|_2^2$, and desired output $y_d = (4, 4)^T$. For unconstrained problem, with unlimited resources ($U = \infty$), in Sect. 6.1 we present the global approach, based on computation of matrix K and its inversion. Since the $n \times n$ random matrix inversion procedure can be numerically unstable and time consuming, parallel version of the optimization procedure is shown in Sect. 6.2. Finally, for limited resources, two-level strategy is proposed in Sect. 6.3.

6.1 Global Approach

For $U = \infty$ we simply get that

$$u^* = \arg\min_u \overline{Q}(u) = K^{-1} y_d = \begin{bmatrix} 1 & -0.5 \\ -0.25 & 1 \end{bmatrix} \begin{bmatrix} 4 \\ 4 \end{bmatrix} = \begin{bmatrix} 2 \\ 3 \end{bmatrix},$$

i.e., $u_1^* = 2$, $u_2^* = 3$, and $\overline{Q}\left(u^*\right) = 0$.

6.2 Parallel Computation for Unconstrained Problem

The optimal controls u_1^* and u_2^* can be also computed in parallel (independently) by interaction balance method

$$u_1^* = \frac{y_{d,1} - a_1 H_1 y_d}{b_1} = \frac{4 - 0.5 \begin{bmatrix} 0 \ 1 \end{bmatrix} \begin{bmatrix} 4 \\ 4 \end{bmatrix}}{1} = 2,$$

$$u_2^* = \frac{y_{d,2} - a_2 H_2 y_d}{b_2} = \frac{4 - 0.25 \begin{bmatrix} 1 \ 0 \end{bmatrix} \begin{bmatrix} 4 \\ 4 \end{bmatrix}}{1} = 3,$$

avoiding necessity of potentially high dimension matrix inversion. Obviously, in the direct approach, and $U < \infty$, it is not guaranteed that $u_1^{*2} + u_2^{*2} \leq U$.

6.3 Multi-level Convex Optimization Procedure

For limited global resources, i.e.

$$u_1^2 + u_2^2 \leq U,$$

the optimization problem becomes much more complicated. Resources consumed by control u_1 influences range of accepted variability of u_2. Our problem can be decomposed on two layers – upper

$$\min_{u_1, u_2 : u_1^2 + u_2^2 \leq U} Q\left(u_1, u_2\right) = \min_{u_1 : u_1^2 \leq U} \left(\min_{u_2 : u_2^2 \leq U - u_1^2} Q\left(u_1, u_2\right) \right) = \min_{u_1 : u_1^2 \leq U} P\left(u_1\right) \tag{23}$$

with the decision variable u_1, and lower

$$P\left(u_1\right) = \min_{u_2 : u_2^2 \leq U - u_1^2} Q\left(u_1, u_2\right) = \min_{u_2 : u_2^2 \leq U - u_1^2} R\left(u_2\right) \tag{24}$$

with the variable u_2. The one-argument function $R(u_2)$ is understand on lower level as $Q\left(u_1, u_2\right)$ with the fixed value $u_1 = \text{const}$, imposed from above (established on upper level). Minimization of both $P(u_1)$ and $R(u_2)$ can be done with the use of iterative convex programming technique (see [2]) described in the Appendix. Minimization of $R(u_2)$ plays the supporting role, and is equivalent, from the upper level point of view, to computation of $P(u_1)$ for a given u_1.

Thus, both procedures are nested. Optimal controls obtained in the computer experiment are

$$u_1^* = 0.625, \qquad u_2^* = 0.781,$$

and respectively

$$y_1^* = 1.160, \qquad y_2^* = 1.071, \qquad Q(u_1^*, u_2^*) = 16.64.$$

7 Summary

Hierarchical strategy presented above allows for universal approach to decision making in complex, interconnected systems. Proposed identification algorithms are passive in the sense that estimation of parameters can take place under random excitation of arbitrary probability density function. The system is thus observed during its functioning, without any interrupts. Moreover, recursive version of algorithms are provided, which can be applied for fast signal and parameter changes ([1]).

Modeling of individual blocks plays supporting role in control optimization. Parameter estimates obtained in the identification stage are plugged in to the cost function to be minimized.

The procedure is shown in convenient form, which can be potentially applied in many domains of science and technology, where hierarchical approach to complex system control is recommended, e.g. in chemistry [10], transportation systems [16], power engineering [20], or safety management [17].

Acknowledgements. The work was supported by the National Science Centre, Poland, grant No. 2016/21/B/ST7/02284.

Appendix

Iterative minimization algorithm of one-dimensional convex function f(x)

Initialization. Set the starting interval $[\kappa_0, \lambda_0]$, and some ε_0, such that $\kappa_0 < x^* < \lambda_0$, and $0 < \varepsilon_0 < \frac{\kappa_0+\lambda_0}{2}$.

Step n-th. If $f\left(\frac{\kappa_n+\lambda_n}{2} - \varepsilon_n\right) \geq f\left(\frac{\kappa_n+\lambda_n}{2} + \varepsilon_n\right)$ then set $\kappa_{n+1} := \frac{\kappa_n+\lambda_n}{2} - \varepsilon_n$, and $\lambda_{n+1} := \lambda_n$.

If $f\left(\frac{\kappa_n+\lambda_n}{2} - \varepsilon_n\right) < f\left(\frac{\kappa_n+\lambda_n}{2} + \varepsilon_n\right)$ then set $\kappa_{n+1} := \kappa_n$, and $\lambda_{n+1} := \frac{\kappa_n+\lambda_n}{2} + \varepsilon_n$.

Set $\varepsilon_{n+1} := \varepsilon_n/2$.

Stop Condition. Stop, if $|\lambda_n - \kappa_n|$ is appropriately small.

References

1. Astrom, K., Wittenmark, B.: Adaptive Control, 2nd Edition, Courier Corp., (2013)
2. Boyd, S., Boyd, S.P., Vandenberghe, L.: Convex Optimization. Cambridge University Press, Cambridge (2004)

3. Brown, R.G., Hwank, P.Y.: Introduction to Random Signals and Applied Kalman Filtering: with MATLAB Exercises, Wiley, New York (2012)
4. Deng, Y.: Applied Parallel Computing. World Scientific, Danvers (2013)
5. Findeisen, W., Bailey, F.N., Brdyś, M., Malinowski, K., Tatjewski, P., Woźniak, A.: Control and Coordination in Hierarchical Systems. Wiley, New York (1980)
6. Giri, F., Bai, E.W. (eds.): Block-Oriented Nonlinear System Identification. Springer, Berlin (2010)
7. Hasiewicz, Z.: Applicability of least-squares to the parameter estimation of large-scale no memory linear composite systems. Int. J. Syst. Sci. **12**, 2427–2449 (1989)
8. Hasiewicz, Z., Mzyk, G.: Hammerstein system identification by non-parametric instrumental variables. Int. J. Control **82**(3), 440–455 (2009)
9. Kincaid, D.R., Cheney, E.W.: Numerical Analysis: Mathematics of Scientific Computing, American Mathematical Society, Rhode Island (2009)
10. Lefkowitz, I.: Systems control of chemical and related process systems. IFAC Proc. volumes, **8**(1), 521–528 (1975) Part 2
11. Mesarowic, M.D., Macko, D., Takahara, Y.: Theory of Hierarchical. Academic Press, New York, Multilevel Systems (1970)
12. Mzyk, G.: Nonlinearity recovering in Hammerstein system from short measurement sequence. IEEE Signal Process. Lett. **16**(9), 762–765 (2009)
13. Mzyk, G.: Combined Parametric-Nonparametric Identification of Block-Oriented Systems. Springer, Berlin (2014)
14. Singh, M.G., Titli, A.: Systems: Decomposition, Optimisation and Control. Pergamon Press, Oxford (1978)
15. Stoica, P., Söderström, T.: Instrumental variable methods for system identification. Circuits Syst. Sig. Process. **21**(1), 1–9 (2002)
16. Vrancken, J., Soares, M.: Multi-level control of networks: the case of road traffic control. In: IEEE International Conference on Systems, Man and Cybernetics, pp. 1741–1745 (2007)
17. Wahlström, B., Rollenhagen, C.: Safety management-a multi-level control problem. Saf. Sci. **69**, 3–17 (2014)
18. Ward, R.: Notes on the instrumental variable method. IEEE Trans. Autom. Control **22**, 482–484 (1977)
19. Wong, K.Y., Polak, E.: Identification of linear discrete time systems using the instrumental variable method. IEEE Trans. Autom. Control, **AC-12**(6), 707–717 (1967)
20. Xiao, J., et al.: Multi-level energy management system for real-time scheduling of DC microgrids with multiple slack terminals. IEEE Trans. Energy Convers. **31**(1), 392–400 (2015)

Building of a Variable Context Key Enhancing the Security of Steganographic Algorithms

Łukasz Nozdrzykowski and Magdalena Nozdrzykowska(✉)

Faculty of Computer Science and Telecommunications, Maritime University of Szczecin, Szczecin, Poland
{l.nozdrzykowski,m.nozdrzykowska}@am.szczecin.pl

Abstract. This article presents a concept of a context key for steganographic algorithms. The proposed scheme of building the context key is designed to enhance the level of security of the applied algorithm. Reading a hidden message should be possible when certain requirements defined as the context are met. For example, such requirements include a specific level of lighting, location and a password protected access, while typing the password itself can be arranged to require appropriate context. This may be characterized by the speed of typing the characters etc. Therefore, the context key will be used in mobile devices such as a smartphone, tablet, etc. The context key building scheme has been proposed, allowing the key to be changed at any stage of the hiding algorithm. The proposed solution protects subsequent bits of a hidden message against retrieving or forward inference, which means that a key broken at a certain stage does not allow concluding what key will be used in the next algorithm step and what key was used previously. In addition, the presented key scheme allows reading a message while the context is not fully met.

The authors aimed to propose the possibility of using the context key as a type of strengthening the security of steganographic algorithms by adding new elements blocking the possibility of reading messages by unauthorized persons. The possibilities offered by sensors in mobile devices were used for this purpose. The novelty in the proposed key is the use of threshold methods of sharing secrets as a mechanism enabling the user to read a message even if not all environmental conditions are satisfied during the reading of the secured message.

Keywords: Steganography · Data protection · Context key · Security in mobile systems · Variable steganographic key

1 Introduction

The purpose of information protection systems is to provide a number of services, including: authentication, identification and non-repudiation of information origin, and the confidentiality and integrity of transferred data. By definition, an information protection system provides the service of hiding the information from an unauthorized party. In general, information protection systems are associated with the combination of different cryptographic algorithms in one system that provides different levels of protection. It

W. Zamojski et al. (Eds.): DepCoS-RELCOMEX 2021, AISC 1389, pp. 316–325, 2021.
https://doi.org/10.1007/978-3-030-76773-0_31

is assumed that cryptographic algorithms use various kinds of mathematical transformations (confusion and diffusion) dependent predominantly on the key, that convert an open message into a secret message that cannot be reconstructed without a key. The security of these methods is associated with the length of the key used, which is not entirely correct, because it is strong diffusion and confusion that are to ensure safety, not the key itself. Nevertheless, the key used in information protection systems is vital. Encryption by using a key ensures confidentiality, while a digital signature additionally provides authentication and non-repudiation of information origin. One-way hash functions are an exception, but their role is not encryption, but the provision of the mechanisms for verifying data integrity [1]. Current cryptographic algorithms, described by standards, guarantee high security, which is proven (symmetric cryptography detection) or based on difficult to compute problems, where their security is assumed (asymmetric cryptography).

A different concept is used in steganography. While in cryptography the fact of using an appropriate algorithm can be disclosed, because security is provided by an algorithm, in steganography the fact that a steganographic algorithm is used is hidden. Various media can be used as information carriers in steganography. These include images, video files, audio files, hidden files and partitions, micro prints, hidden fields in network protocols and noise in communication systems. The steganographic algorithm is intended to ensure security by appropriate encryption of information in the data carrier so that a third party does not suspect presence of a hidden message. Cryptography and steganography algorithms can be combined with each other. If a message is encrypted first, then steganographically hidden, we get a stegano-cryptographic method [2].

Many commonly known steganographic methods, mainly modifications of the method of least significant bit, do not use the steganographic key and offer low safety against the methods of stega-analysis. Such methods are part of the groups of algorithms of fragile steganography [3].

Robust steganography increases the security of a hidden message. Steganography is said to be robust if the algorithm provides protection against various classes of attack against a steganographic message, the hidden message is more difficult to get destroyed and where its decryption is not easy. Protection from removal is provided by the algorithm construction alone, or the manner of hiding the message, for example by choosing the frequency domain instead of space domain. For steganographic algorithms in digital images it is easier to secure a message against the effect of lossy compression. The protection against decryption is ensured by cryptography itself or the use of encoding dependent on the steganographic key. This article will present the mechanism of enhancing the confidentiality of information protection algorithms, which will get particular importance in steganographic algorithms with a key by embedding new mechanisms of the context key. By extending the standard key with a part of the context key, the message can be read only when additional requirements called the context are met, e.g. receiver's location. The context key is to strengthen security. Security itself must be assured by the steganographic method alone. The security of a particular steganographic method is not discussed in this article.

The context key solution can also be used for other safety purposes. It can authorize a user in any IT system, or be an access key to a database or selected application. The authors propose a system of context key construction for steganographic algorithms, especially for fragile steganography algorithms used in mobile devices such as smartphones and tablets. While the concept of the context key is known, the authors propose a new scheme for it. The scheme is based not on specific numbers read from device sensors, but on a range of values, which after encrypting will be part of the context key. The authors then propose the use of threshold methods of secret sharing in order to enable the message to be read when not all environmental conditions for reading the context key are met. In addition, the authors propose the use of the seed and one-way hash function in order to prevent an attacker from retrieving or forward inference on the key if they were able to read a key fragment at any time of the steganographic algorithm operation. An indirect novelty is the key change during the hiding of subsequent message blocks.

The article is divided into 6 main sections: Sect. 1 contains the introduction, Sect. 2 describes the use of keys in steganographic systems available in the current literature. Section 3 reviews the literature on the use of sensors in mobile devices and wearables in security systems. In Sect. 4, the authors propose their original idea of using sensors in steganographic systems as a context key. In Sect. 5 the authors suggest places where the proposed steganographic key could be located. The final section summarises the proposed steganographic key algorithm.

2 Keys in Steganographic Systems

Steganographic keys are mainly related to robust steganography algorithms. In recent years, keys have been frequently included in algorithms using digital images as carriers. However, these are mostly modifications of the least significant bit method. Keys used in a simple way indicate the location of the message [2], where the function with the key defines subsequent hiding sites. In the work [2] the key also contained a value specifying how many bits can be concealed in one pixel. In [4] a key is used in a pseudorandom generator indicating the order of the bytes in the carrier message. When a transform to the frequency domain is used, the authors build in a key to define coefficients in which a message is hidden. Such modification is applied to the I.G.D. method known for watermarks placed in the DCT transform [5], where the authors chose the wavelet transform DWT. The combination of hiding data, watermarks in that case, with a key for the pseudorandom generator is the method proposed in [6] where the cosine transform DCT and spectrum diffusion algorithm is used for hiding. The work [7] proposed to use decomposition by the Haar wavelet transform, where a message is hidden in the HH and HL decomposition bands by the least significant bit method, while the message itself is previously encrypted by the ACHTERBAHN-128 algorithm. The downside of these algorithms is non-smart approach that may reveal the fact that a message has been hidden. Most recent methods use algorithms of message hiding that indicate such places for hiding where a potential message will not cause visible changes in the carrier. For instance, messages are hidden in blocks that have the most points belonging to contour lines [8]. These authors' scheme of enhancing the confidentiality by the context key is independent of the steganographic algorithm used, and each algorithm with key can be extended with the demonstrated context key fragment. The context key will raise the level of the security of the algorithm, but it will not ensure security of the algorithm itself, unless it is secure already.

The context key is linked to the term context cryptography, where in the cryptographic algorithm additional security is applied, which for deciphering requires an additional criterion to be met, such as geographical position or light intensity in the room. This is directly related to modern mobile devices such as smartphones and tablets, which are equipped with a number of sensors. This approach is not about altering the cryptographic algorithm itself, it consists in creating additional safeguarding methods that require the system to meet extra conditions for readout from the sensors. The key used for encrypting/decrypting or its fragment is dependent on the context, thus it depends on satisfying those conditions. In this article, the authors propose the construction of a context key for security algorithms, mainly for steganographic algorithms, because they are often considered as insufficiently secure, which mainly refers to fragile steganography.

3 Use of Context in Security Systems

Many publications report on sensors fitted in a smartphone used to confirm identity. Researchers are currently testing the user's behaviour in a specific test environment. The research data will be used to train a neural network to indicate whether the device is controlled by the owner or another person. The sensors used are a touch screen, accelerometer, gyroscope, light sensor. The most commonly used sensors are accelerometer and gyroscope as they are fitted in almost every smartphone and in wearables.

Two accelerometer sensors and a gyroscope were used in the work [9]. Users in the first phase were buying goods for 10–13 min on the prepared website. In the meantime, the application was reading values from both sensors with the frequency of 100 Hz. After the synchronisation process, the data collected were used as a training dataset. DeepAuth identifies the user based on the accelerometer and gyroscope, i.e. by using unique behaviour of the phone user. The method is capable of determining, based on sensor readings, whether at a given instant the smartphone is used by the owner or a third party. The authors specified the accuracy of their method at 96.7%.

In the work of [10] the authors used an accelerometer and other sensors in a device for the verification of the user's identity based on his gait, or the manner of walking. The data from the sensor are sent to a server that, based on collected data from the sensor, decides whether to confirm or reject the verification process. The presented algorithm of gait-based verification had 80% accuracy and approximately 10% rate of false positives.

The authors in [11] proposed a solution called SmarterYou for the authentication using three items: smartphone, IoT wearable and a cloud server. Similarly to [9], the solution requires gathering a training dataset for correct user identification. For user authentication the authors use a gyroscope and accelerometer that collect information on the behaviour of the user. The system is designed to detect an unauthorised user, for example, when someone leaves their smartphone unattended, and a third party will start using it. In this case, the system will automatically log the user out from the applications in which he is logged. The authors compared their solutions to the existing ones. The application reached 98.1% accuracy and low rate of false negatives and false positives.

4 Composition of a Steganographic Context Key

This chapter presents a scheme of enhancing confidentiality by extending the standard steganographic key with a fragment of the context key. We propose a context key scheme, where the key alone will be reconverted before each use. The presented scheme is universal and designed so that it can be used in security systems in one device and during communication between two parties. On the whole, it is a new approach to enhancing the security of IT systems through a variable context key.

The proposed context key will be presented using sensors available in the Android operating system.

The sensors can be divided into hardware, software or hardware/software sensors. Sensors can be divided into three main groups: motion sensors, environment sensors and position sensors [12–14]. The latter group includes sensors the values of which can be calculated from information provided by other sensors. The accuracy of software sensors may be lower than hardware sensors. The accuracy of the measurements of each sensor depends on the electronic component used by the manufacturer of mobile device. Manufacturers of electronic components do not have imposed standards of sensor accuracy. In consumer electronics, low measuring error is not required. Some scientists attempt to determine measuring errors of sensors. [16]. Nevertheless, the software sensor will provide the same data as the hardware sensor. Given the principles of building security systems, repetitions should be avoided. Table 1 contains a list of sensors, their type and measurement units. The authors therefore propose to use only hardware sensors available in the device. Note that the parties involved in communication must inform each other about what sensors may be jointly used.

Another stage of data preparation for the context key is the filtration of data from sensors. Filtration is required because instantaneous readings are characterized by high data fluctuations. Several readings and averaging results is a better solution. It should be noted, however, that such data, especially when a device is held in hand, may have temporarily very high fluctuations. Extreme data should therefore be removed.

The data filtered in this way will be used for the construction of a context key for information protection systems. The general diagram of the proposed context key is given in Fig. 1.

Fig. 1. Preliminary context key

The key consists of three parts. These are encrypted context data, user password and a special seed that is to ensure adequate randomness. The first and each subsequent use of the key is preceded by the conversion of the key by one-way function to hinder the inference what key was used in the previous step and which key will be used in the next step, even if in the current stage the attack was successful. Therefore, the key has a special seed at the end, which by no means should be made available by the system to third parties. The specific key building stages will be presented further.

4.1 Encoding Data from Sensors

The first stage is the encoding of data from sensors. These data are characterised by continuous fluctuations of value. However, to read the message the key requires accurate values. Therefore, it is necessary to encode data from sensors. Depending on the type of sensor and its accuracy, the data derived from sensors can be divided into N ranges. Example ranges for a tested Samsung Galaxy Note 10 Lite are given in the table below:

Table 1. Example range for the tested device

No	Sensor name	Maximum value	Range	Code value
1	Light	6000lx	0–600	0
2	Light	6000lx	601–1200	1
3	Light	6000lx	1201–1800	2
…	…	…	…	…
10	Light	6000lx	5401–6000	9

The decision which sensors will be used for key construction should be made by the system user. For maximum diversity, it is recommended that different sensors be used, and their order changed in the initial key.

The data from sensors are intended to make the user ensure the same environmental conditions in which the message was encrypted. However, it is known that exactly the same environmental conditions are difficult to ensure. Lighting, for instance, may vary. At this stage encryption ranges may be increased. However, we suggest another solution: threshold secret sharing scheme. Threshold secret sharing allows splitting the main secret, e.g. a password, into a group of so-called shares. The threshold of secret reconstruction is also established. The secret is reconstructed when the user collects at least the number of shares making up the threshold. The authors propose adopting a scheme of the threshold secret sharing for the construction of the context key. The key will make it possible to read hidden messages despite the failure to meet all environmental conditions. The Shamir protocol with Lagrange polynomial for interpolation (Article] is proposed, or any of the protocol of secret sharing. The secret will be the properly encrypted values for particular sensors given in the order indicated by the user. User's ability to define the order in which each sensor performs sensing additionally increases the random character of this part of the key.

4.2 User Password

The user key is the second part of the initial key. This key can be stated by the user itself along with password principles, that is a minimum combination of 8 letters of different size with digits and special characters for increasing the key complexity. The proposed key algorithm assumes that the user has no maximum restrictions on the length of the stated password. The password can also be generated by any robust pseudo-random generator or determined by the key-setting protocol. Where protected data are to be hidden and read on the same device, the key can be generated based on the constant information obtained from the operating system itself, such as: model, user name, identification numbers of components and their manufacturers, serial numbers, hardware parameters of the device. The higher number of data obtained, the higher the key security is. At this stage, a password can be generated by a generator with linear feedback shift register with non-linearity elements for enhanced security. The generator input itself may also be based on the password given by the user, including the user's behaviour, e.g. measured time between typing single characters, the pressure force measurement and other random data. This approach will hinder the reading of hidden data by a non-authorised person.

4.3 The Seed as a Key Security Element

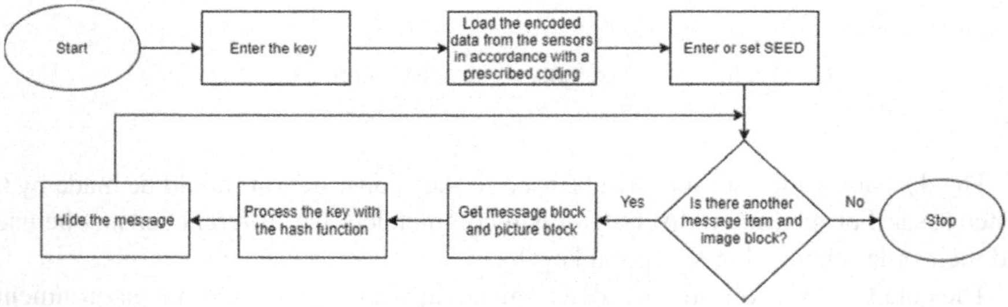

Fig. 2. A diagram of the proposed context key for steganographic algorithms, carried by a digital image.

Figure 2 depicts the initial context key diagram with a special box marked as seed. It presents the key conversions through the hash function preceding each subsequent data hiding. The proposed context key scheme can be adapted easily to any security systems using passwords. Unfortunately, not all of these systems may assure sufficient security for a hidden message. This applies to fragile steganography, in which it becomes possible to guess a password for a given place of message hiding. The authors therefore propose such key complexity that when the key is identified at a certain stage of the message hiding algorithm, it will not be possible to retrieve it and perform forward inference on the context key used.

Encoding is needed for this purpose. Therefore, we propose adding a seed at the end of the key and using a one-way hash function. The hash function must generate its desired length. Where the hash length is insufficient, the key can be processed again by the same function to obtain a length suitable for the message hiding algorithm. To assure a proper level of security, modern hash functions like SHA-3 [17] should be used, while older hash functions SHA-1 will require double hash conversion. We propose to use one-way conversion before first use of the key, so that if the key is broken after the first data hiding, reading the initial key will not be possible. Due to its operation, the one-way function will not permit retrieving the key. An additional secret seed will not permit forward inference on what newly generated key will be used there. This scheme only requires that the seed be generated by the system and kept in the place inaccessible for the attacker. Besides, the seed cannot be disclosed and should be a cryptographically strong pseudorandom string.

4.4 Data Exchange for the Key

The proposed scheme of the context key construction can be used to protect data kept inside the mobile device, but it can also be used in the process of communication between two parties by using the steganographic algorithm. However, in such case the communicating parties need to agree on the required data. While some of the data, such as the thresholds for particular sensors may be sent via a public channel or using cryptography, the data containing the seed and password should be protected. It should be noted, however, that cryptography includes protocols of secrets exchange, such as Diffie-Helmann protocol. It is suggested that in communication between two parties, sensitive data should be exchanged by using the secret/key exchange protocol.

5 The Place of Key Embedding in the Steganographic Algorithm

Many of the steganographic algorithms do not use a key or use simple keys. In this chapter, the authors indicate the place of the proposed context key. It is assumed that the key will be changed before each use, even when the key is repeatedly used during a single case of data hiding.

Places of the steganographic algorithm extension with a key:

1. The selection of the information carrier – each steganographic algorithm uses a carrier for storing data, such as digital images. The communicating parties may have a database of images useful to conceal information in. The purpose of the algorithm keep the fact of the message concealing undetected. Therefore, it is necessary to choose images that have high texture differentiation. The key used may indicate which image will be suitable for hiding information in it. For example, the key will indicate parameters such as entropy, kurtosis and histogram width, homogeneity and the contrast of the matrix, co-occurrence matrix, i.e. parameters giving the information about the texture differentiation.
2. Choice of the location of hidden message – the key will be the indicator of the locations of message hiding in a digital image. The key can indicate direct pixels and bits for hiding data in them, and may be an input for an algorithm of randomizing subsequent places of data hiding. The key can be used to indicate single pixels, pixel blocks, bit planes or transform coefficients in the case of algorithms based on the transform domain.
3. The selection of the decomposition algorithm in the wavelet transform; currently the wavelet transform is increasingly used in steganographic hiding of data. This is

due to the property of this transform which decrypts an image to its approximation and subsequent levels of detail that represent noise. Hiding takes place at the level of noise, invisible to the human eye. The key can indicate a selected type of wavelet transform, and allow the selection of transform coefficients.
4. The cipher key - the context key becomes an input to a selected cryptographic algorithm, which will encrypt a message before it is later hidden by the steganographic algorithm.

6 Summary

The authors present a scheme of building a context key for up-to-date information protection systems and access control in mobile device and IoT systems. The focus is on the use of the context key in steganography. The context key does not assure security by itself. Its purpose is to raise the level security. The proposed key is based on the data obtained from the user, constant data derived from the mobile device and readings from sensors that render the context. The context has to be provided for the message to be read. In practice, the context will be particular lighting intensity, location or a password entered in a specific manner. The key scheme itself allows for incomplete fulfilment of the context and enables communication between different parties. In addition, the context key is a variable key, thus data bits are protected when the key is identified at one stage of the hiding algorithm operation.

The novelty in the proposed approach is strengthening the security of the context keys. These authors suggest an improvement of this approach by using ranges of values instead of specific values read from a sensor, where these ranges would be encrypted as a fragment of the context key. They added the seed value to the key, which, combined with one-way hash function, is intended to secure the reading of the key backwards or forwards when the key fragment happens to be guessed at any stage of the data hiding algorithm operation. The authors also propose to adopt threshold methods for secret sharing in order to enable the message to be read when the recipient cannot satisfy all environmental conditions for reading the message using the context key. The authors propose a universal approach allowing the user to increase the security of any steganographic algorithm and other related security algorithms, especially those characterized by low security or those that do not use keys at all.

References

1. Tiwari, H., Asawa, K.: Cryptographic hash function: an elevated view. Eur. J. Sci. Res. **43**(4), 452–465 (2010)
2. Majunatha Reddy, H.S., Raja, K.B.: High capacity and security steganography using discrete wavelet transform. Int. J. Comput. Sci. Secur. (IJCSS) **3**(6), 462–472 (2010)
3. Umamaheswari, M., Sivasubramanian, S., Pandiarajan, S.: Analysis of different steganographic algorithms for secured data hiding. IJCSNS Int. J. Comput. Sci. Netw. Secur. **10**(8), 154–160 (2010)
4. Ker, A.D.: Resampling and the detection of LSB matching in colour bitmaps. N: Security, Steganography, and Watermarking of Multimedia Contents VII, San Jose, California, USA, 17–20 January 2005, pp. 1–15 (2005)

5. Hovančák, R., Foriš, P., Levický, D.: Steganography based on DWT transform. In: 16th International Conference Radioelektronika (2006). https://www.scribd.com/document/106326386/dwt38
6. Pitas, I.: Digital Image Processing Algorithms and Applications, Wiley, New York (2000)
7. Mohamed, M.I., Dessouky, M.I., Deyab, S., Elfouly, F.H.: Wavelet Data Hiding using Achterbahn-128 on FPGA Technology. In: UbiCC Journal - Special Issue of IKE'07 Conference, IKE 2007 (2008)
8. Cachin, C.: Digital steganography. In: van Tilborg, H.C.A. (eds.) Encyclopedia of Cryptography and Security. Springer, Boston (2005) https://doi.org/10.1007/0-387-23483-7_115
9. Amini, S., Noroozi, V., Pande, A., Gupte, S., Yu, P.S., Kanich, C.: DeepAuth: a framework for continuous user re-authentication in mobile apps. In: Proceedings of the 27th ACM International Conference on Information and Knowledge Management (CIKM 2018), Association for Computing Machinery, New York, pp. 2027–2035 (2018)
10. Ren, Y., Chen, Y., Chuah, M.C., Yang, J.: Smartphone based user verification leveraging gait recognition for mobile healthcare systems. In: 2013 IEEE International Conference on Sensing, Communications and Networking (SECON), New Orleans, LA, pp. 149–157 (2013)
11. Lee, W., Lee, R.B.: Implicit smartphone user authentication with sensors and contextual machine learning. In: 2017 47th Annual IEEE/IFIP International Conference on Dependable Systems and Networks (DSN), Denver, CO, pp. 297–308 (2017)
12. Android Sensors. https://developer.android.com/guide/topics/sensors/sensors_overview. Accessed 30 Jan 2021
13. Android sensors motion. https://developer.android.com/guide/topics/sensors/sensors_motion. Accessed 30 Jan 2021
14. Android sensors position. https://developer.android.com/guide/topics/sensors/sensors_position. Accessed 30 Jan 2021
15. Android sensors environment. https://developer.android.com/guide/topics/sensors/sensors_environment. Accessed 30 Jan 2021
16. Accuracy of an accelerometer. https://www.analog.com/en/analog-dialogue/articles/how-to-improve-the-accuracy-of-inclination-measurement-using-an-accelerometer.html. Accessed 15 Jan 2021
17. SHA-3 Standard: Permutation-Based Hash and Extendable-Output Functions. https://doi.org/10.6028/NIST.FIPS.202

Method of Visual Detection of the Horizon Line and Detection Assessment for Control Systems of Autonomous and Semi-autonomous Ships

Łukasz Nozdrzykowski and Magdalena Nozdrzykowska(✉)

Faculty of Computer Science and Telecommunications, Maritime University of Szczecin, Szczecin, Poland
{l.nozdrzykowski,m.nozdrzykowska}@am.szczecin.pl

Abstract. Autonomous and semiautonomous ships need visual systems for the identification of the horizon line and recognition of invisible obstacles on electronic charts or the radar - positioning and identification of other objects. Varying observation conditions are a major disadvantage in such systems, as they are related to the accuracy of horizon detection. The proposed method of horizon detection for varied observation conditions is based on a combination of selected filtrations and statistical calculations. The method enables the identification of a horizon line in difficult conditions of observation, e.g. clouded sky or sunset. The development of the method for horizon line identification was preceded by an analysis of strengths and weaknesses of relevant digital image processing methods. The novelties of the method include: 1) development of a method for assessing how useful an acquired image is in horizon line detection; 2) automating the acquisition of parameters used in proposed algorithms of image processing; 3) selection of algorithms for line detection for optimized speed of computing and reliability of the process. The proposed method combines filtration and statistical methods. As a result, the system detects the horizon in any observation conditions reliably and fast. This property is important for an autonomous system that must respond to events in real time, including sudden appearance of an obstruction.

Keywords: Horizon line detection · Autonomous ships · Filtering and statistical methods · Machine vision · Image processing

1 Introduction

The development of computer systems and algorithms of image processing enables seeking effective methods of visual image recognition to be used in smart, autonomous transport vehicles [1, 2]. While these systems are quite common in road vehicles, they are at a developing stage in shipping. These solutions are developed by modern e-navigation systems [3]. What stimulates progress most are the concepts of ship control for autonomous and semi-autonomous ships [4, 5]. On autonomous ships, expected to proceed without human participation, traditional IT systems and sensors mounted in ships are not sufficient and the need arises for computers to take over the function of

W. Zamojski et al. (Eds.): DepCoS-RELCOMEX 2021, AISC 1389, pp. 326–338, 2021.
https://doi.org/10.1007/978-3-030-76773-0_32

human eye [6, 7]. Even more interesting capability of computer systems to be improved is the recognition of obstacles that the radar cannot see or the AIS system does not report [8].

Machine vision comprises the capability of detecting objects that cannot be neglected in ship conduct, as well as recognizing them [9]. The classical process of machine vision is divided into a few steps: acquisition, assessment of image usefulness, improvement of quality, segmentation, recognition and decision how to respond to the recognized object. The crucial issue here is that an object to be visually recognized is a sea-going ship moving in the dynamic marine environment. Continuous waves, wind and currents act on the ship with forces causing to roll.

In practice it is not possible to develop a universal system of recognition, so it becomes necessary to extend recognition algorithms with image rotation phase, opposite to ship inclination to one side. Therefore, it is necessary to determine the horizon line and rotate the image to that line, only then algorithms of segmentation and recognition can be applied.

Autonomous ships currently under development are designed as mostly small, low-cost boats, not equipped with advanced systems found on traditional ships [10]. For such boats an ‘extra eye’ is particularly needed. For a vision system ability to recognize objects, the major requirement is to detect the horizon line to build effective recognition systems. The authors present a method of horizon line detection for digital colour images accounting for various visibility conditions.

2 Methods of Horizon Line Detection

Many publications deal with the problem of horizon line detection. At present, a set of methods is used for higher accuracy of horizon detection. Methods developed by various authors focus on edge detection [11, 12], other are based on colours [11, 13, 22], while those proposed in recent years make use of neural networks [14, 17–24]. Research on waterline detection [23, 24] and methods using infrared images [15, 16] also utilize the above mentioned methods.

To enhance accuracy, methods based on edge recognition use statistical methods [11] or analysis of image colours [13]. In the study [11] the authors divide main algorithms based on edge detection into five groups. The algorithms used were based on: local covariance, edge detection, Hough transform, median filter and regression. All the examined algorithms showed high accuracy and efficiency. In [13] the authors subjected an obtained image to edge detecting filtration, then the resultant image underwent a Hough transform. In the final image the horizon line height and inclination were determined. The authors obtained good results of horizon detection. Nevertheless, the analyzed images were made in good visibility conditions, the image resolution was below HD quality, and the authors did not test the efficiency of their algorithms.

In [12], the author presents all methods of edge detection based on colour, image statistics and filtration. All the methods were divided into four main groups, and subgroups. The selected groups include: intensity methods using local functions, intensity methods using global functions, methods using colour and local functions, methods using colour and global functions. For each type of horizon detection algorithm, the author used 21 colour images, taken mostly at sea. According to the author's research, the best results were achieved by algorithms CINRHT and CMVDHT, belonging to methods using colour and local functions.

Methods based on neural networks use two of the mentioned methods, additionally placing the results at the input of a neural network that takes the data as a training set. In a training set an accurate line of horizon is determined by the human. Neural networks in use include SVM, naïve Bayes or J48. The results obtained by this method were characterized by high accuracy. Like in the other above methods, the authors used images with low resolution. The method based on machine training proposed by [22] analyses images in 21 attributes – 7 different characteristics for 3 image channels, i.e., RGB. The authors used the following attributes: intensity, mean, standard deviation, smoothness, third central moment, uniformity and entropy. Each pixel in the image was considered as a separate dataset classified by three classifiers: SVM, J48 and naïve Bayes. In the study, the authors obtained a black and white image where black represented water or land and white represented the sky. The accuracy of the classifiers was 95.21% for good visibility conditions, and 76% for bad visibility conditions. With a low resolution, the horizon appeared as a straight line, failing to reflect the waves. In the study [10], the authors segmented the image into areas representing the sea, sky and other objects. The segmentation yielded a horizon line that was subjected to smoothing in case objects like ships or buoys appeared. The algorithm results are promising – the method accuracy was 94%. The authors in [21] used machine learning to create Maximally Stable Extremal Edges (MSEEs). The Canny algorithm for edge detection was modified to choose the largest and smallest thresholds. This operation generated N binary images that were placed at the SVM classifier input. The horizon line in the training sample was indicated manually. The results were highly divergent, from no error to 72%. One disadvantage of the method was long recognition time of the algorithm: 43 s.

Methods based on infrared images are used when insufficient light exists for colour photos - at dusk or night. Infrared images are monochromatic. Obtained images can be handled by all algorithms. In [15] three methods were combined: edge detection, Hough transforms and covariance or histogram. The results of horizon line detection were very good – maximum mean error was 4 pixels, although for XGA resolution images the processing of one picture took up to 61.1 s. Another method of horizon detection from infrared images was used in [16], where an image was divided into segments of the sea, land and sky. The division into regions was made by creating a map of segments and a map of curves. The accuracy of the method was estimated at 95.7%. A disadvantage of infrared cameras is their cost, increasing in proportion to camera visibility range.

All the presented methods had commonly accessible or own training bases, with pictures taken from the shore or ship. A majority of images were taken in good visibility conditions, i.e., on a sunny or partly cloudy day. Few images featured poor visibility conditions, i.e., overcast sky or poor light images. In good light, the biggest problems that resulted in wrong identification of horizon line was the sunshine reflected from the water surface [22] or the line of clouds adjacent to the horizon line [5]. In images with overcast sky, the line of clouds was mistaken for the horizon line [14, 22]. Studies [11–22] lack the testing of how proposed algorithms perform in low lighting conditions - twilight or bad visibility conditions, when the colour of the clouds blends with that of the sea. Besides, many authors use low resolution images, which are not useful in creating algorithms for autonomous ship control.

3 Assessment of Image Usefulness for Horizon Line Detection

These authors propose to use a method for assessing image quality assessment in view of the ability to recognize a horizon line in the image. The addition to the horizon line detection algorithm of the element related to the assessment of image suitability for detection is a new component not used by other authors in publications, referred to in Sect. 2 of this article. We add an element of image quality analysis in a manner consistent with the recognition by human vision. Notably, some of the acquired images may not be suitable for horizon line recognition. This limitation refers to images acquired in the evening and night time and in bad atmospheric conditions. Before creating algorithms for detection of a horizon line, it becomes necessary to assess image usefulness for a horizon detecting algorithm. This is required because poor visibility in severe hydrometeorological conditions makes the horizon line detection difficult. In normal conditions of observation, detection of an image is not feasible at maximum contrast, understood as the degree to which an observed element can be distinguished against its background: water against the sky and vice versa. Unfortunately, the global contrast of an image does not allow determining whether the horizon line can be detected based on a simple statement that the sky area is always the brightest region, while the sea area is the darkest. An image obtained in the evening will reveal a ship or another object emitting light on the horizon.

On the global scale, a colour histogram seems to be useful as a diagram presenting the tonal content of an image, or graphic representation of image luminance distribution. Examination of a colour histogram brings information on most frequent intensity levels of colours or colour groups; moreover, it is possible in further processing of an image to automate acquisition of data needed, for instance, for image segmentation.

An analysis of the histogram allows examining the existing areas and assessing their diversity. If this method may be performed automatically, it will be possible to automatically assess the feasibility of horizon line detection. The problem is that while the human can analyze a histogram quickly, the computer must operate on coefficients of measures obtained from the examined histogram. Based on a standardised colour histogram, the following basic parameters are calculated: mean, variance skewness, kurtosis, excess [25, 26].

It is worthwhile for histogram analysis to measure its range and width.

Table 1 presents three images with reduced brightness, with the impact of changing brightness on the histogram and measures calculated on the basis of the histogram. The most important measures are the width, range and kurtosis, the latter parameter indicating the peakedness (or alternatively, flatness) of the histogram. Large values mean that the histogram is closely concentrated about the mean, so it very peaked (leptokurtic). Based on these parameters and the peak top in the histogram we can easily assess the possibilities of horizon line detection. For these parameters, the authors will analyze the algorithms for horizon line detection. It should be noted that as visibility conditions deteriorate, the histogram gets narrower, its range increases and so does the kurtosis. Thus it becomes

Table 1. Sea images of decreasing brightness and corresponding colour histograms and measures of: variance, skewness and kurtosis.

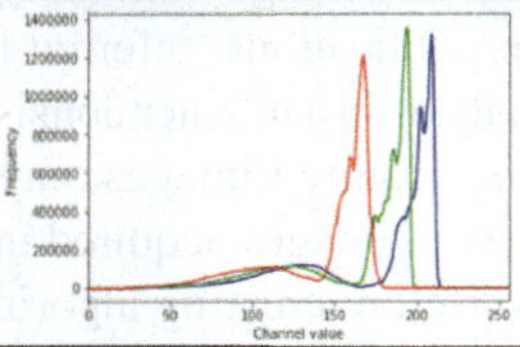	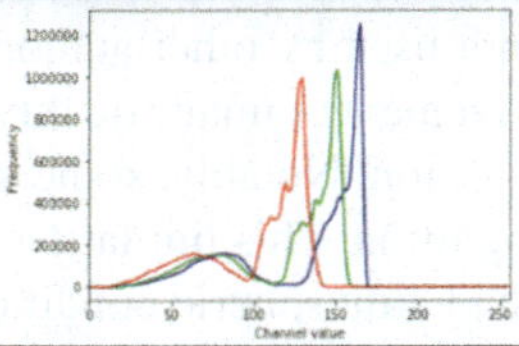	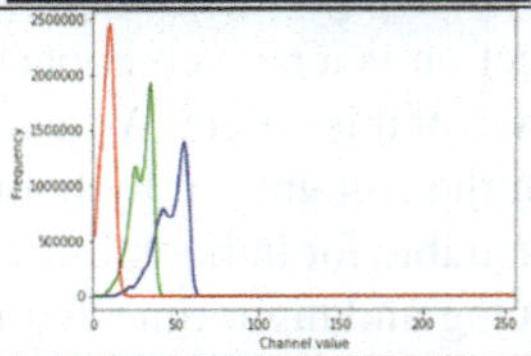
Variation: 2.294 Skewness: 3.730 Kurtosis: 14.832	Variation: 1.910 Skewness: 3.160 Kurtosis: 11.354	Variation: 3.523 Skewness: 3.923 Kurtosis: 15.168

possible to determine the limit to which the horizon line can be recognized by a selected algorithm.

4 The Proposed Methods of Horizon Line Detection

In this section the authors will discuss tests of three methods of horizon line detection, while in Sect. 5 the methods will be assessed. Besides, applicable horizontal line detection algorithms will be presented, analyzed and the effects of their operation will be referred to the possibility of horizon line detection in bad observation conditions.

One of the methods for the detection of horizon line is the linear convolution filter. Most of the researchers make a major error selecting a convolution filter. High-pass filters, suitable for use in this context, are of two types. There are first and second derivative filters. The former are used for the detection of the edge border of colour change region, the latter are applied for detecting the edge itself. The border is the surrounding of the edge. While border detection refers indirectly to the detection of horizon line, it should be noted that these filters detect the border lines of an edge ideally parallel, perpendicular or tilted at 45 degrees. That is how the Sobel filter is defined. Principally, such filters are not good at detecting edges oriented at other angles, and yet an autonomous ship affected by wind and waves will rarely be so conveniently positioned relative to the horizon line. For this reason the use of Sobel or Prewitt filters will not be effective in most cases. Another possibility is to use these filters combined, i.e., by combining the outcome of several filters in one image. In practice, however, there is no need to use more than two filters, for vertical and horizontal directions. In place of first derivative filters, a second derivative (Laplacian) filter can be used. It will provide information on edge points in the image, neglecting the surroundings.

Both filters yield good results, delivering good material for further processing. One drawback of the Sobel filter is that it strongly responds to any additional shapes and smooth transitions to the background, such as clouds, while a Laplacian filter cuts them out perfectly. Unfortunately, a Laplacian filter poorly performs where the horizon line merges between the sea and the sky. This is depicted in Fig. 1, where to enhance the effect the kernel was increased to 9 × 9 and the map of colours was reversed.

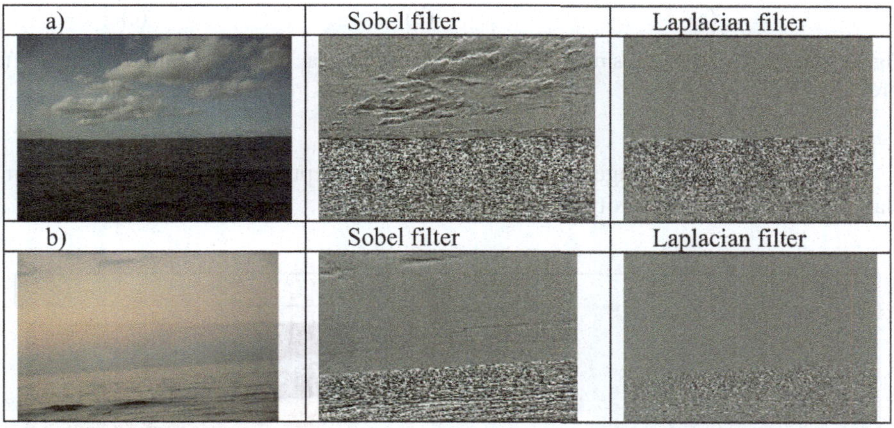

Fig. 1. Difficulties resulting from the application of Sobel and Laplacian filters: example image.

It is apparent that the Sobel and Laplacian filters hamper correct recognition of the horizon line. In Fig. 1a, the Sobel filter detected large numbers of cloud points, which in the case of large size clouds will cause errors in the detection of the horizon line. The Laplacian filter in Fig. 1b insufficiently detected edge points of the horizon line on the image boundaries. The disadvantage of both filters is that they poorly detect edges for lower kernel size. An increased kernel size, that cannot be too large, improves the situation. Kernels larger than 9 × 9 introduce a lot of undesirable noise. The choice of a kernel and filter should be matched to image character. Building a recognition algorithm, we should check whether the horizon line was detected for the smallest kernel. If not, the kernel has to be increased. Further herein the Canny filter will be additionally verified.

In algorithms of image segmentation, one basic step is to use simple thresholding, proceeding until contrast changes occur. Simple thresholding converts all pixels with colour above a specific threshold to white, and those below the threshold are changed to black. An advantage of the method is simplicity, and by using a histogram, the process can be easily automated. According to Sect. 2, the image histogram for a photo presenting the sea has at least two peaks. One peak stands for the colour of the water, the other for the sky. There are often more peaks, but two are always dominant. As the colour of the sky is brighter, we can easily interpret the colours of the histogram referring to the brighter part. It is demonstrated in Fig. 2.

In that example image the method used was simple thresholding with a threshold above 140 for the colour for which the second peak in the histogram appears in grey shade. This thresholding method enables quick separation of the sky and sea regions. The threshold value can be estimated based on peaks found in the histogram.

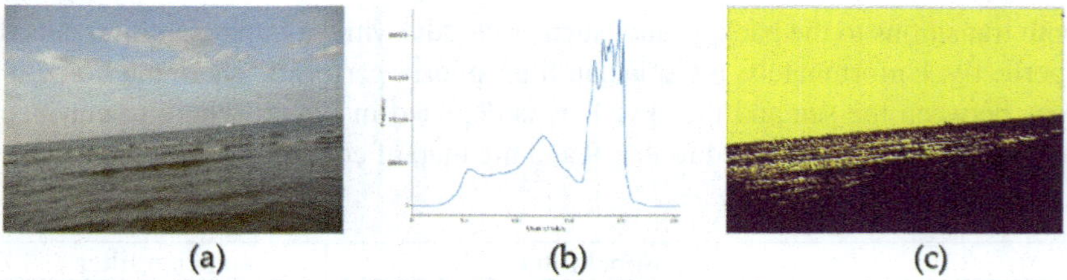

Fig. 2. Image subjected to simple thresholding: a) original, b) histogram, c) image after thresholding

The third option of image segmentation is the concept of image entropy, indicating places of the greatest diversity (the most information).

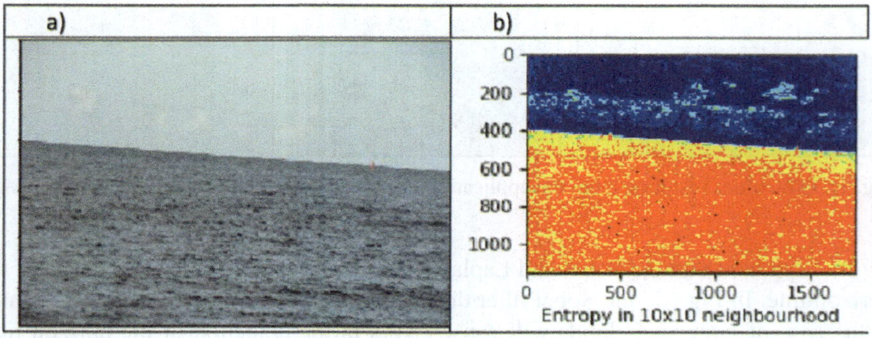

Fig. 3. Entropy calculated for the image a) original image, b) image entropy.

Figure 3 presents calculated entropy of an example image. As the sea area comprises the largest image diversity, while the sky mostly features areas of high smoothness, the image entropy diagram shows a visible boundary between its levels. The change in the entropy level determines the horizon line. The greatest disadvantage of such method is a long time of determining the entropy level for the whole image.

The methods will be assessed on the basis of a library of eight images of the sea, made in various hydrometeorological conditions and at various times of day, i.e., images of varied contrasts. They are presented in a tabular form in Fig. 4.

Table 2 presents the results of analyses for eight images of different classes. The analysis only aimed at verifying that a given algorithm enables finding out visually that the horizon line is made visible. The results showed that the entropy method attains the best results. Its main disadvantage is long calculation time, which clearly disqualifies it.

The fastest method, thresholding, is characterized by errors of recognition. Only 50% of all images were correctly recognized. In the case of simple thresholding the correct detection rate falls mainly when the value of histogram entropy decreases, histogram width decreases, and the kurtosis increases. At the same time, when the histogram width decreases, the range increases, while this range does not result from single files that

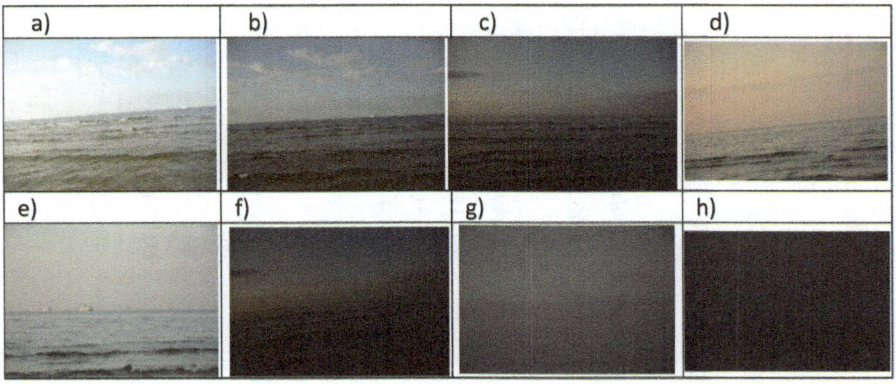

Fig. 4. Test images.

artificially raise its value. Therefore, the horizon line will be more difficult to detect in cases where the histogram width diminishes, and the range increases.

Much better results are obtained by using 7 × 7 Sobel and Laplacian filters. The correct determination of the horizon line was at 90% level. While the Sobel filter enables better detection of the horizon line, it notably also responds strongly to the presence of clouds. It should be borne in mind that the Sobel filter shows edge borders, not edges themselves, thus the visualized elements are rather thick lines that make further analysis difficult. Most problematic are long clouds with horizontal borders. The Canny filter has capabilities similar to the Sobel filter, as the former practically uses implements the latter. Its advantage is the elimination of cloud effects, similarly to the Laplace filter. For this reason, and the fact it smoothly co-operates with the Hough transform, the Canny filter will be used in further considerations. Section 5 proposes a method for raising the detection rate of the Canny filter. Image entropy has been neglected due to long computing time.

5 Methods of Improving the Quality of Horizon Line Detection and the Proposed Algorithm of Horizon Line Detection

In this section, the authors will propose a final horizon line detection algorithm based on the analyses provided before and suggest a method of improving line detectability. At the end, the correctness of detection by the proposed algorithm will be examined.

To improve the quality of operation of a linear high-pass filter, such as a Canny filter, the edges have to be peaked. Two types of algorithms can be applied for the purpose. One is a low-pass Gaussian filter, which permits to reduce noise while maintaining the edges. The other proposed method is a non-linear median filter.

An interesting characteristic of this filter is that for a large filtration window an image erosion takes place, where all edges of that image are left. As a result of erosion, the edges remain, while all small objects are removed. The larger the filtration window, the better the effect. However, this increases time of calculation, so the size of the window should be dependent on available computing power.

Table 2. Results of horizon visual indication correctness for Full HD images.

Image	7a	7b	7c	7d	7e	7f	7g	7h
Measured histogram statistics for a given image								
Contrast	0.81494	0.92828	0.97101	0.84905	0.80800	0.97826	0.82905	0.92307
Histogram avg.	11882.7326	13078.5326	18085.6541	13638.2159	12068.3165	0643.2413	24778.2680	49888.3125
Histogram width	201	184	133	176	90	116	97	48
Variance hist.	1.0808	1.13623	1.26403	1.52060	1.6702	1.72495	1.90494	2.8111
Entropy hist.	4.98167	4.87818	4.75736	4.65103	4.3048	4.28549	4.12140	3.47918
Kurtosis hist.	3.7259	1.6798	16.8083	5.9098	1.5942	1.5942	2.1181	9.1515
HISTmax-HISmin	56025	53429	107936	75853	682231	69494	69009	145448
Tested algorithm - correctness								
Sobel filter	Yes	Yes	Yes	Yes	Almost	Yes	Yes	Yes
Laplacian filter	Yes	Yes	Yes	Yes	Yes	No	Yes	Yes
Thresholding	Yes	Yes	Yes	Yes	No	No	No	No
Entropy	Yes	Yes	Yes	Yes	Yes	Yes	Yes	Yes
Canny 3 × 3	Yes	Yes	Yes	Yes	Almost	Half line	Half line	No

Fig. 5. Canny image filtration

The use of the Canny filter without additional filtration with the Sobel 3 × 3 window is shown in Fig. 5. The Canny filter makes the horizon line visible, although not perfectly, which leads to inaccuracies in poor contrast or incomplete detection.

Considering further use of the Canny filter, we will make an attempt to enhance the differences at the horizon line boundary to improve detection. A median filter will be used. With a large mask window, it causes image erosion, retaining the edges, and by a Gaussian filter smooths noise, leaving the edge points untouched. In case of low contrast images where the above actions will not be possible, an additional convolution filter for sharpening will be initially used, with the mask: array ([−1, −1, −1], [−1, 9, −1], [−1, −1, −1]).

Authors' tests have led to an observation that the Canny filter itself combined with the Hough transform did not offer in practice satisfactory results (Fig. 6). The adaptive form of Hough transform turned out better, although the correct horizon line detection remained problematic. Only after the use of preliminary filtration the horizon line detection was correct. The best results were yielded with the use of the median width 5 or larger. Owing to the median used, the number of false lines was reduced. If there were too many false lines, and the main line of horizon began to vanish, their number was reduced by Gaussian blur. If the median filter did not visualize the horizon line, it was necessary to apply a sharpening filter, followed by the procedure employing the median and Gaussian blur. Change of the threshold value for the Hough transform was helpful in cases of very low contrast between the water and the sky.

Based on the conclusions from the analysis, the following stages of the method of horizon line detection on a digital image were defined (Fig. 7):

1. change from the RGB space to the grayscale space;
2. determine a histogram in shades of gray;
3. determine the mean, place of local maxima, skewness and kurtosis for assessing the suitability of an image for horizon line detection;
4. perform a median filtration to obtain image erosion in order to defuzzify edges and remove details. A larger median window is required for images of high entropy of the histogram and its large width at larger kurtosis values. The procedure enhances major edges, including the horizon line;
5. perform Gaussian blur to remove excessive number of false lines due to noise and histogram entropy and many local maxima in the histogram plot;
6. perform convolution filtration sharpening edges thus retained in the cases of dark images, where the Canny filter brings no positive results;

a) Sharpen: No Median:7 Gauss size: 3x3 Sobel size: 3 threshold max for Hough: 255 Time: 0.045992 s	b) Sharpen: No Median:7 Gauss size: 5x5 Sobel size: 3 threshold max for Hough: 255 Time: 0.044612 s	c) Sharpen: No Median:13 Gauss size: 5x5 Sobel size: 3 threshold max for Hough: 255 Time: 0.061521 s	d) Sharpen: No Median: 9 lub 11 Gauss size: 3x3 Sobel size: 3 threshold max for Hough: 112 Time: 0.052502 s

e) Sharpen: No Median:11 Gauss size: 3x3 Sobel size: 3 threshold max for Hough: 100 Time: 0.044943 s	f) Sharpen: No Median:5 Gauss size: 3x3 Sobel size: 3 threshold max for Hough: 70 Time: 0.080824 s	g) Sharpen: Yes Median:3 Gauss size: 0x0 Sobel size: 3 threshold max for Hough: 255 Time: 0.011324 s	h) Sharpen: No Median:9 Gauss size: 3x3 Sobel size: 5 threshold max for Hough: 255 Time: 0.051593 s

Fig. 6. Use of Canny filter with adaptive Hough transform and preliminary processing

7. use the Canny filter;
8. use the adaptive Hough transform;
9. identify the longest line out of the determined ones;
10. (optionally) determine the inclination angle of the defined straight line relative to x-axis of the image

In the case of an image from a video camera the time of processing one frame is critical. With 25 frames per second, the time of calculation cannot exceed 0.04 s. Therefore, to accelerate calculations, it is worth using typical solutions, such as reduction of image resolution, histogram calculations performed on an incomplete number of frames, transfer of part of the calculations to another computer core or utilize the capacity of advanced graphic cards or stream processors. It follows from the presented computing time data that for Full HD images the proposed algorithm reaches 0.4 s (Fig. 6 and Table 2).

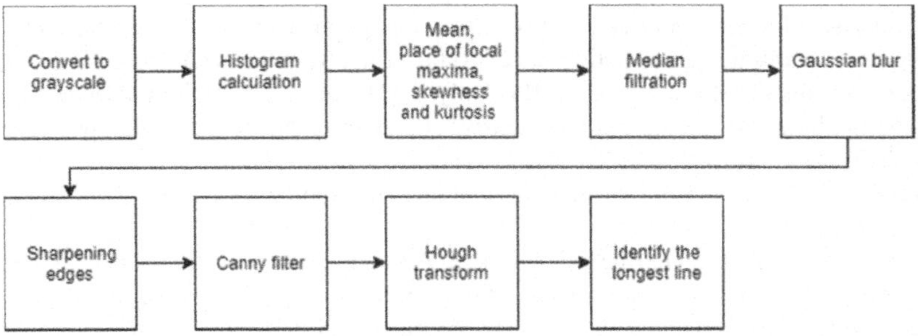

Fig. 7. Diagram of the proposed algorithm.

6 Summary

The paper proposes a method of horizon line detection that uses an image from an RGB camera or a black and white camera. To achieve this, algorithms of image processing were selected for horizon detection and a method of enhancing the effectiveness of the horizon line detection was developed. This was based on an analysis of the advantages and disadvantages of existing methods, including the correctness of detection and computing time. The novelties of the solution are the method for assessing the suitability of an image for horizon line detection and algorithms for quality improvement of horizon line detection. Besides, the authors' original contribution is the automation of the method for acquiring parameters used in the proposed method. The method may be used on autonomous and semi-autonomous ships for positioning and subsequent identification of objects around the ship. This enhances situational awareness at sea. The authors intend to focus on the automation of the acquisition of setting parameters for each phase of the proposed algorithm. It is important that during the detection of horizon line taking the defined steps, the authors encountered no problems in accomplishing that task, which allows them to expect that the said process of automation is on the right track.

References

1. Mapangaa, K., Veera, R.S.: Machine vision for intelligent semi-autonomous transport (MV-iSAT). Procedia Eng. **41**, 395–404 (2012)
2. Bergström, M., Hirdaris, S., Valdez Banda, O., Kujala, P., Sormunen, O.-V., Lappalainen, A.: Towards the unmanned ship code. In: Kujala, P., Lu, L. (eds.) 2018 Proceedings of the 13th International Marine Design Conference, IMDC 2018, vol. 2, pp. 881–886 (2018)
3. Burmeistera, H.-C., Bruhnb, W., Rødsethb, Ø., Porathec, T.: Autonomous unmanned merchant vessel and its contribution towards the e-navigation implementation: the MUNIN perspective. Int. J. e-Navig. Marit. Econ. **1**, 1–13 (2013)
4. Koikas, G., Papoutsidakis, M., Nikitakos, N.: New technology trends in the design of autonomous ships. Int. J. Comput. Appl. **178**(25), 4–7 (2019)
5. Bertozzia, M., Broggib, A., Fasciolia, A.: Vision-based intelligent vehicles: state of the art and perspectives. Robot. Autonom. Syst. **32**(1), 1–16 (2000)
6. Burmeister, H.-C., Bruhn, W.C., Rødseth, Ø.J., Porathe, T.: Can unmanned ships improve navigational safety? In: Proceedings Transport Research Arena, Paris (2014)

7. Wróbel, K., Montewka, J., Kujala, P.: Towards the assessment of potential impact of unmanned vessels on maritime transportation safety. Reliab. Eng. Syst. Saf. **165**, 155–169 (2017)
8. Tu, E., Zhang, G., Rachmawati, L., Rajabally, E., Huang, G.: Exploiting AIS data for intelligent maritime navigation: a comprehensive survey from data to methodology. IEEE Trans. Intell. Transp. Syst. **19**(5), 1559–1582 (2018)
9. Masaki, I.: Machine-vision systems for intelligent transportation systems. IEEE Intell. Syst. Appl. **13**(6), 24–31 (1998)
10. Steccanella, L., Bloisi, D., Blum, J., Farinelli, A.: Deep learning waterline detection for low-cost autonomous boats. In: Strand, M., Dillmann, R., Menegatti, E., Ghidoni, S. (eds.) Intelligent Autonomous Systems 15, IAS 2018. Advances in Intelligent Systems and Computing, vol. 867. Springer, Cham (2019)
11. Gershikov, E., Libe, T., Kosolapov, S.: Horizon line detection in marine images: which method to choose? Int. J. Adv. Intell. Syst. **6**(1), 79–88 (2013)
12. Zafarifar, B., Weda, H., de With, P.H.N.: Horizon detection based on sky-color and edge features. In: Society of Photo-Optical Instrumentation Engineers 2008 SPIE Conference Series, vol. 6822 (2008)
13. Gershikov, E.: Is color important for horizon line detection? In: 2014 International Conference on Advanced Technologies for Communications, ATC 2014, pp. 262–267 (2014)
14. Ahmad, T., Bebis, G., Nicolescu, M., Nefian, A., Fong, T.: An edge-less approach to horizon line detection. In: 2015 IEEE 14th International Conference on Machine Learning and Applications (ICMLA), Miami, FL, pp. 1095–1102 (2015)
15. Lipschutz, I., Gershikov, E., Milgrom, B.: New methods for horizon line detection in infrared and visible sea images. Int. J. Comput. Eng. Res. **3**(3), 226–233 (2013). ijceronline.com
16. Kim, S.: Sea-based infrared scene interpretation by background type classification and coastal region detection for small target detection. Sensors **15**(9), 24487–24513 (2015)
17. Mou, X., Shin, B., Wang, H.: Hierarchical RANSAC for accurate horizon detection. In: 2016 24th Mediterranean Conference on Control and Automation (MED), Athens, pp. 1158–1163 (2016)
18. Yan, Y., Shin, B.-S., Xiaozhengmou, Mou, W., Wang, H.: Efficient horizon detection on complex sea for sea surveillance. Int. J. Electr. Electron. Data Commun. **3**(12), 49–52 (2015)
19. Jeong, C.Y., Yang, H.S., Moon, K.D.: Horizon detection in maritime images using scene parsing network. Electron. Lett. **54**(12), 760–762 (2018)
20. Liang, D., Liang, Y.: Horizon detection from electro-optical sensors under maritime environment. IEEE Trans. Instrum. Meas. **69**(1), 45–53 (2020)
21. Ahmad, T., Bebis, G., Regentova, E.E., Nefian, A.: A machine learning approach to horizon line detection using local features. In: Bebis, G., et al. (eds.) Advances in Visual Computing, ISVC 2013. Lecture Notes in Computer Science, vol. 8033 (2013)
22. Fefilatyev, S., Smarodzinava, V., Hall, L.O., Goldgof, D.B.: Horizon detection using machine learning techniques. In: 2006 5th International Conference on Machine Learning and Applications, ICMLA 2006, Orlando, FL, pp. 17–21 (2006)
23. Ling, F., Xiao, F., Du, Y., Xue, H.P., Ren, X.Y.: Waterline mapping at the subpixel scale from remote sensing imagery with high-resolution digital elevation models. Int. J. Remote Sens. **29**(6), 1809–1815 (2008)
24. Wei, Y., Zhang, Y.: Effective waterline detection of unmanned surface vehicles based on optical images. Sensors **16**, 1590 (2016)
25. Rudnicki, Z.: Analiza sekwencji obrazów niejednorodnych. Informatyka w Technologii Materiałów **2**(3), 86–96 (2003)
26. Kim, H.-Y.: Statistical notes for clinical researchers: assessing normal distribution (1). Restor. Dent. endod. **37**, 245–248 (2012)

Influence of Various DLT Architectures on the CPU Resources

Patryk Pankiewicz[1,2](✉)

[1] Politechnika Śląska, 44-100 Gliwice, Poland
patryk.pankiewicz@globallogic.com
[2] GlobalLogic Sp. z o.o., Strzegomska 46B, 53-611 Wrocław, Poland

Abstract. Together with an increment of complexity level and amount of software in currently produced cars, the industry has to invest more effort in security and safety aspects. One way to increase quality, safety, and security is AUTOSAR - a partnership that focuses on the standardization of automotive software and its architecture.

The purpose of this paper is to present the results of research conducted on the influence of different setups of the AUTOSAR DLT module (Diagnostic Logging and Tracing) on various resources like CPU (Central Processing Unit) load, memory consumption, and communication bus load.

With the help of research results from this paper, the DLT can be configured and used in a way that saves resources while taking into consideration the system's limitations.

Keywords: AUTOSAR · DLT · Automotive

1 Introduction

AUTOSAR was founded in 2003 by major companies in the automotive industry. Intense development of AUTOSAR results in a standard present in almost every produced car. AUTOSAR influences and defines software in multiple layers, starting from the architectural view, going through modules' standardization and optimization, ending at software integration guidelines. AUTOSAR DLT module was created to standardize logging and tracing activities and integrate them with the AUTOSAR stack. It can use any available communication bus–e.g., Ethernet, UART (Universal asynchronous receiver-transmitter), CAN (Controller Area Network). The DLT allows the whole car to have one common approach to logging and tracing. It offers multiple configuration options concerning buffer sizes, log levels, messages structure, stack integration, and more. The main scope of research conducted for this paper concerns the data formatting and verbosity options of the DLT. There are multiple options and architecture decisions that have their advantages and disadvantages. The research can help to decide which option is more appropriate for concrete usage.

General description of the DLT can be found in AUTOSAR official documents – [1, 2], and [3]. Additional articles covering the definition of AUTOSAR DLT related topics are [4] and [5]. They show the possibility of employing description files (FIBEX)

W. Zamojski et al. (Eds.): DepCoS-RELCOMEX 2021, AISC 1389, pp. 339–348, 2021.
https://doi.org/10.1007/978-3-030-76773-0_33

together with the AUTOSAR stack, which is one of the methods used by the author and described in detail later.

The complexity of embedded systems creates a need for a significant amount of information about the software's execution. Mostly because of costs, developers must perform software optimization in terms of computational power. Such relation creates a problem of requiring lots of information while having limited hardware and software resources. There are existing papers [6, 7] that are proving that hardware and software resources are always desirable to be optimized in automotive projects. The existing literature on topics related to AUTOSAR OS (Operating System) optimization is very rich – e.g. [10, 11]. These methods are providing a way of minimizing the "communication cost" which, as a result, relates to the CPU load reduction. That being said, no papers concerning optimization and configuration of the AUTOSAR DLT module were found.

2 DLT Optimizations

The standard usage of DLT that is present in most projects is string-based verbose logging. That means the system sends information in the human-readable form, e.g., "Ignition signal has changed". Such configuration of DLT is easy to set up and maintain but far from optimal in terms of ECU resources. The author performed an analysis of DLT parameters and configuration options to identify ways to optimize the impact on ECU resources. The analysis provided certain areas that have a significant influence. The ones that have shown the most noticeable effect were:

- DLT description modes,
- DLT header modifications,
- binary logging,
- log level processing method.

All four areas described below are the subject of conducted research.

2.1 DLT Description Modes

AUTOSAR DLT offers two modes of transferring data: verbose and non-verbose.

In verbose mode - the data type information is embedded in the frame itself. This mode's main advantage is that no external description files are needed, making it much easier to set up and maintain. The disadvantage is that a higher amount of data is transmitted onto the communication bus. Verbose Mode frame is presented below (Fig. 1).

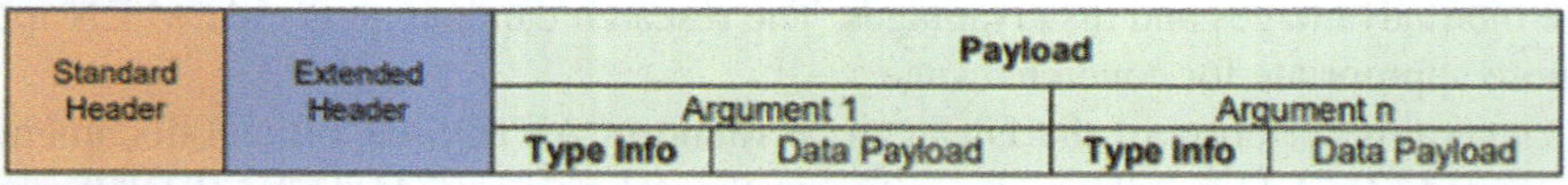

Fig. 1. Verbose mode message [2]. The type info precedes every payload present in the frame.

What is important is that the "Type Info" is a 32-bit field – so even the data payload is a binary value (1 bit), the each argument will have 33 bits.

In non-verbose mode - the description of the transmitted data is stored in an external XML file (a FIBEX file defined by [3]). The payload itself contains only the "Message ID" followed by the data. It saves transmission resources like CPU, memory, and communication bus usage; however, it is more difficult to maintain and requires integration with external tools (Fig. 2).

Fig. 2. Non-verbose mode message [2]. Single message ID is used for the following data payload.

There is no perfect setting. It depends on the project's characteristics and limitations – like the available memory, the throughput of the used communication bus, available CPU. Depending on available resources, the architect can define the desired mode. The more static data[1] is used, the more gain is produced by the non-verbose mode. Non-verbose mode has another advantage in terms of security – it provides basic data obfuscation. Without the description, FIBEX files the captured payloads are slightly harder to decode (they do not contain type information). However, if the priority is sending significant amounts of dynamic data[2], then the verbose mode provides better flexibility and maintainability.

Switching from verbose to non-verbose mode allows sending only the ID of each message instead of transmitting the whole payload. The result is a reduction of the required flash memory, CPU, and communication bus load.

2.2 Binary Logging

Binary logging is the next optimization suggested by the author. Binary logging requires the usage of the non-verbose mode of DLT. It means replacing static frames with empty non-verbose messages (Table 1).

Instead of using a standard verbose frame with a string message:

The ECU sends an empty frame containing just the Message-ID. The PC application uses the FIBEX file and translates it to the dedicated description. The FIBEX can contain multiple auxiliary information like the source file name where the messages originate and the line number in code. A detailed description of this mode can be found in [2] (Table 2).

Such modification reduces flash memory usage in systems where a lot of static data is transmitted.

[1] Static data – fixed, non-changeable information that is sent at given point of software's execution. For example "Error while trying to setup connection".

[2] Dynamic data – states, variables, payloads that are changing depending on the execution. For example "12:05:01, 12 V, 0.250 A, 22 °C".

Table 1. Exemplary string based verbose message.

Header	Type Information	Payload
	string	"Watchdog has been triggered in ignition management component."
	4 bytes	60 bytes

Table 2. Exemplary binary non-verbose message.

Header	Message ID	Payload
	00 00 00 01	-
	4 bytes	0 bytes

2.3 Log Levels Processing

DLT messages have a parameter (log level) that defines the category of a message. They vary from less important information, "info" or "debug", up to "error" and "fatal". The primary usage of log levels is filtering – at any given time, the user can change the currently desired log level. Any messages below the set level will not be transmitted. The filtering can be done in two ways – centrally by the DLT driver or in a distributed manner by each software component on its own.

The author analyzed both ways. In cases with a large number of logs, central filtering creates a strain on the system resources because of two reasons:

1. Every time a message is transmitted – it employs the RTE (Run-Time Environment), which queries the DLT driver. Even if the outcome is that the message should not be sent, multiple jumps to functions with several context changes are executed.
2. Every message is validated by checking if the given context was registered. In bigger systems with hundreds of applications and context IDs, DLT lookup tables can have several thousand elements, creating significant computational strain.

Considering these two reasons, it is better to process the log levels separately in each software component. In such a case, if the set log level is not sufficient, no action is taken, the RTE and DLT are not informed, the whole check is limited to a single if statement. The only important aspect is ensuring that each software component has information about the current set log level.

2.4 DLT Header

The full DLT Header can encode up to 26 bytes, including various information about message counters, message modes, timestamps, and other parameters. Not all header elements are required. The standard [3] defines the optional parameters. The PC application often creates constraints as per what header elements are needed for parsing the incoming frames.

Manually writing a simple DLT parsing application creates the best environment for customization. In such a scenario, the architect can choose the required elements. In such a case, a full 26 bytes header can be minimized to even 8 bytes (4 bytes of header and 4 bytes for message ID), which matters significantly in high-frequency logging and tracing.

3 Research Setup and Scope

The purposes of measuring DLT's influence on the ECU (Electronic Control Unit) resources were as follows:

- Measuring and presenting the magnitude of the ECU burden created by using a default DLT configuration.
- Measuring and presenting the potential of optimization on every AUTOSAR system that employs DLT logging.

3.1 Research Setup

The author performed the research on a real use case in an industrial project. The ECU architecture is 32-bit PowerPC, and the used communication bus is UART configured at 115200 bps. The project implements many software components and complex device drivers (CDDs) that are using the DLT.

Initially, the project used verbose mode with strings based logging. For the sake of this research, the author switched to the non-verbose binary mode, together with distributed logs' level processing and DLT header modification to 8 bytes (all described in Sect. 2).

Code analysis has shown the amount of static and dynamic usage of DLT. In this context, static usage sends fixed messages at a certain point of execution of the software – checkpoints, logs, and program flow information. The dynamic usage is tracing, sending the actual values of variables, states, registries.

Secondly, a runtime analysis with a Lauterbach Debugger was employed to measure the exact amount of data sent on the UART bus and the connected processing time. Such analysis allowed to calculate the communication bus load and the CPU load.

3.2 Research Results

Flash Usage

All strings transmitted in a DLT message occupy the flash memory of the microcontroller. It concerns all the constant messages sent during runtime. Considering the size of automotive projects and the number of modules, this creates a massive database of signals that are occupying flash memory.

The total amount of string constants in the analyzed software was over 22 kB embedded in over 1150 messages. All this was static constant data that could be fully transferred to the FIBEX description files. Out of the 22 kB, only 630 bytes were dynamic payload information dependent on the application's execution. That means that the reduction of flash usage by employing non-verbose mode in this particular use case was as high as 86.66%. As described before, in Sect. 2.3, the author used binary logging to achieve this. Each complex static payload was replaced with a 4-byte message (only the message ID). Additionally, originally the textual description of the logging information was minimized because of limited memory. With FIBEX file usage, the textual description can be full and can carry all the required information (Table 3).

Table 3. Comparing the original and optimized amount of static data (after introducing FIBEX and modifying headers)

	Original	After optimization
Header size	26 bytes	4 bytes + 4 bytes message ID
Amount of messages	1157	1157
Payload sent	36337 bytes	4845 bytes (including message ID)
Percent	100%	13.33%

CPU Usage

The CPU measurements used a hardware timer with a frequency of 80 MHz. The timer measured the execution time of the most consuming DLT routines – sending, payload processing, and buffering functions. Figure 3 presents the CPU load before and after introducing listed changes in DLT structure. The modifications most contributing to the reduction of CPU load were: non-verbose mode and decentralized log levels processing.

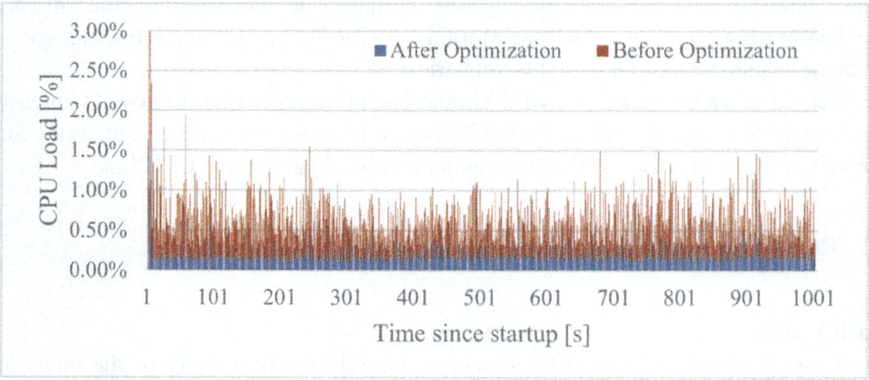

Fig. 3. CPU Load [%] in time [s].

As presented in the graph, the minimized data load has greatly reduced the CPU load. The average reduction of the relative CPU load over the measurement time was equal to **50.18%** (before 0.41%, after 0.20%). The result was obtained by calculating the arithmetic mean of CPU load usage every second within the measured timeframe (1000 s).

Communication Bus Usage

DLT can use any communication bus. Preferred buses are the faster ones like Ethernet or CAN, however because of costs, mostly used are the slower ones (e.g., UART). Ethernet or CAN-based data loggers are much more expensive, and the engineering teams need logging from multiple locations: from the factory, during road tests, in laboratories, and validation stations.

While Ethernet enables the user to create multiple virtual networks to split the throughput, not all buses offer such options. UART, LIN (Local Interconnect Network), or CAN have to send the logging and tracing information together with other non-DLT messages. Such solutions imply the obligation to limit the logging throughput, not to stress the bus too much. Suppose the receiving PC application does not require all fields of the DLT header. In that case, it is very beneficial to change the DLT configuration to minimize the DLT header size for slower communication buses. Combining multiple messages into a grouped one can also save a lot, provided that the messages are sent with high frequency.

The author has written a script that allows capturing all transmitted DLT data every second, which he used to measure the communication bus usage. The results are visible below. The plot clearly shows the peak usage of DLT at startup - at this time, all components are trying to send the required info to the bus. It poses a problem to the DLT buffers that can be too small to store all data, leading to data loss (Fig. 4).

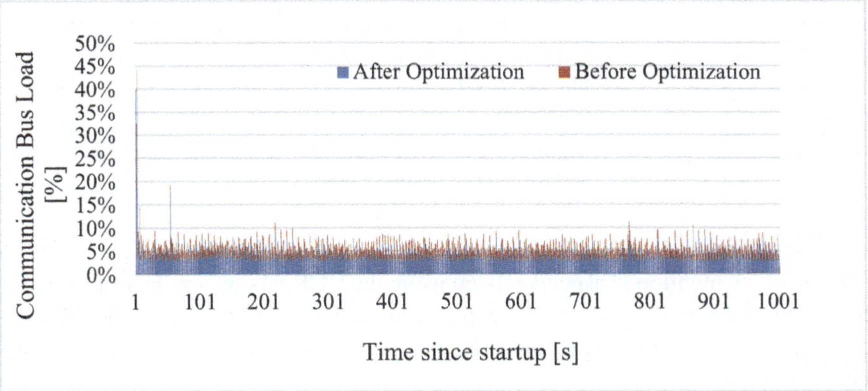

Fig. 4. Communication bus load [%] in time [s].

The actual reduction of the sent data during the measurement time is shown in the table below. The optimization results from the non-verbose mode, binary logging, and changing the DLT header (Table 4).

Table 4. Summary of the amount of sent data

Data sent original	Data sent optimized	Difference	Total reduction
664 896 bytes	197 384 bytes	467 512 bytes	**70,3%**

3.3 Discussion and Conclusions

Depending on the requirements of the development environment, the architect can prioritize different parameters. The presented results were obtained using the following means (Table 5):

Table 5. Summary of changes introduced to obtain final results.

Category	Description	Mostly affected resource
Architecture & configuration	Switching to the non-verbose mode of the DLT. Employing FIBEX files with a dedicated application to parse various data types within single frames (binary logging)	Flash usage & Communication bus load & CPU load
Architecture	Combining various frames in one cumulated payload send with lower frequency	Communication bus load
Architecture	Processing the Log Level in each component instead of in the DLT layer	CPU load
Configuration	Fully employing the log levels – splitting all communication into debug/error/log/etc. categories	Communication bus & CPU load
Configuration	Minimizing the DLT header and using only desired fields	Communication bus load

Some of the mentioned aspects rely on using the DLT parsing application that allows free modification of the header and embedding various data in one frame. Without a custom application, the configuration is strongly dependent on the constraints of the used PC software.

All listed optimizations allowed to reduce the DLT related flash usage by 86.66%, the average CPU load spent on processing the logging and tracing information by 50.18%, and the average communication bus load by 70.3%.

Taking into consideration automotive requirements and everlasting optimization needs due to cost reduction, obtained values are significant. The proposed approach saves hours of analyzing issues thanks to extended logs and traces while saving money on hardware by optimizing the required resources. Furthermore, the mentioned steps will allow obtaining the possibility to include logging and tracing in a much higher degree – with more information to be gained. Lastly, the proposed changes are also beneficial to the security aspect since they create an additional obfuscation layer, making it harder to reverse engineer the transmitted data. The research scope was general - the author made a comparison between a standard usage and including all described methods of optimization. A more detailed set of measurements separately inspecting each method's influence can be executed as a next step.

As mentioned, the first step of a good DLT architecture should be an initial plan that defines logging software requirements. Frequency of sending, amount of data, used communication bus, used PC application, amount of static and dynamic logs can change the desired logging strategy – as shown on the research results. Appropriate profiling and measuring scripts can be employed to continually improve the outcome of the configuration.

References

1. AUTOSAR Homepage. https://www.autosar.org/fileadmin/user_upload/standards/classic/4-2/AUTOSAR_SWS_DiagnosticLogAndTrace.pdf. Accessed 10 Apr 2020
2. AUTOSAR Homepage. https://www.autosar.org/fileadmin/user_upload/standards/foundation/1-0/AUTOSAR_PRS_DiagnosticLogAndTraceProtocol.pdf. Accessed 10 Apr 2020
3. ASAM e.V.: Data Model for ECU Network Systems (Field Bus Data Exchange Format) Version 4.0.0 (Sept 2011)
4. Widder, S.: ASAM und Autosar. ATZ Elektron **1**, 40–43 (2006). https://doi.org/10.1007/BF03223811
5. Langer, F., Eilers, D.: New standard simplifies autosar software testing and validation. ATZ Elektron Worldw **5**, 32–35 (2010). https://doi.org/10.1007/BF03242262
6. Long, R., Li, H., Peng, W., Zhang, Y., Zhao, M.: An approach to optimize intra-ECU communication based on mapping of AUTOSAR runnable entities. In: 2009 International Conference on Embedded Software and Systems, Zhejiang, pp. 138–143 (2009). https://doi.org/10.1109/ICESS.2009.63
7. Allden,T.: Developing a prediction model for estimating hardware requirements based on standardized embedded system characteristics in the automotive domain. https://hdl.handle.net/20.500.12380/300038
8. Khenfri, F., Chaaban, K., Chetto, M.: Efficient mapping of runnables to tasks for embedded AUTOSAR applications. Journal of Systems Architecture. https://doi.org/10.1016/j.sysarc.2020.101800. ASAM und Autosar, Standards verbessern, Systeme und Prozesse (2020). https://doi.org/10.1007/BF03223811
9. Gupta, P., Singh, N.P., Srinivasan, G.: An efficient approach for mapping AUTOSAR runnables in multi-core automotive systems to minimize communication cost. In: 2019 Innovations in Power and Advanced Computing Technologies (i-PACT) Vellore, India, pp. 1–4 (2019). https://doi.org/10.1109/i-PACT44901.2019.8960215

10. Wozniak, E., Mehiaoui, A., Mraidha, C., Tucci-Piergiovanni, S., Gerard, S.: An optimization approach for the synthesis of AUTOSAR architectures. In: 2013 IEEE 18th Conference on Emerging Technologies & Factory Automation (ETFA), Cagliari, pp. 1–10 (2013). https://doi.org/10.1109/ETFA.2013.6647952
11. Zhao, Q., Gu, Z., Zeng, H.: Design optimization for AUTOSAR models with preemption thresholds and mixed-criticality scheduling. J. Syst. Archit., **72**, 61–68 (2017). ISSN 1383-7621. https://doi.org/10.1016/j.sysarc.2016.08.003. (http://www.sciencedirect.com/science/article/pii/S1383762116300984)

Monitoring the Granulometric Composition on the Basis of Deep Neural Networks

Andrey Puchkov, Maksim Dli, Ekaterina Lobaneva, and Yaroslav Fedulov(✉)

National Research University "Moscow Power Engineering Institute" (Branch) in Smolensk, Energetichesky Proyezd 1, g., Smolensk 2014013, Russia

Abstract. A method for monitoring the granulometric composition of raw pellets obtained in the process of phosphorus production from apatite-nepheline ores wastes is proposed. The method is based on the use of convolutional neural networks to evaluate the current granulometric composition and recurrent neural networks to predict it. In the proposed prediction method, the convolutional neural network, in addition to estimating the current composition, is used as an encoder that reduces the input data dimension for their further processing by the recurrent neural network. The Python program implementing the monitoring method is written. The results of the model experiment performed using the developed program are presented. The architecture of neural networks applied in the model experiment is described.

Keywords: Granulometric analysis · Deep neural networks · Texture recognition

1 Introduction

The development of modern industry is moving towards the mass creation and implementation of cyber-physical systems (CPS), which are understood as the information-computing shell and production processes symbiosis. The CPS "brain", based on artificial intelligence and other technologies, accepts data from sensors in the real world and uses it to control physical elements. This makes it possible to continuously analyze polymodal information about the production process course and the environment, and to implement optimal control on this basis, achieving high levels of energy and resource efficiency of the process and product quality [1].

Despite the impressive success of artificial intelligence methods, at present, a universal, self-learning cybernetic technological processes support tool has not been created. Therefore, the task of adapting advanced data analysis methods for the needs of a particular production remains urgent.

The purpose of the present research was to develop a method for automated monitoring of the raw pellets granulometric composition in the CPS production of phosphorus from wastes of apatite-nepheline ores. These wastes occupy vast areas near the mining and processing plants of the Murmansk region in Russian Federation and have a significant negative impact on the ecology of the adjacent territories [2].

W. Zamojski et al. (Eds.): DepCoS-RELCOMEX 2021, AISC 1389, pp. 349–358, 2021.
https://doi.org/10.1007/978-3-030-76773-0_34

The object of the research was one of the CPS units – a granulator that prepares ore waste for further processing in CPS; the research subject was the procedure for analyzing the granulometric composition of the granulator output product – raw pellets.

The research task was to develop algorithms and software for automation of granulometric analysis of raw pellets in CPS for phosphorus production from apatite-nepheline ore wastes. The proposed method is based on the apparatus of deep neural networks. The results of the model experiments, which showed the efficiency of the created algorithmic and software tools for automated raw pellets granulometric analysis at the outlet of the granulator, are presented.

2 Materials and Methods

In many industrial productions, one of the factors affecting the technological process performance and the finished product quality is the granulometric composition of raw materials, which characterizes its structure and affects its physical properties [3, 4]. There are no unified classifications for the raw materials regimentation based on the granulometric analysis. This is explained by the difference in the activity areas where the identification of granulometric composition is required – building materials technology, lithology, mining, mineral processing, ground science, soil science and others. Each of them uses its own fractions size classes scales and methods of conducting granulometric analysis: sieve analysis; microscopic examination; particle sizes scaling according to their deposition rate; counting method; centrifugal separation method and others.

Modern methods of granulometric analysis include methods based on laser diffraction [5], X-ray diffraction [6], as well as the fast visualization method (Flashsizer, FS), showing monomodal and narrower distributions than sieve analysis [7]. For the granulometric analysis of nanomaterials and nanoparticles in suspensions, the particle trajectory analysis method described in the ISO 19430: 2016 Particle size analysis – Particle tracking analysis (PTA) method, published in December 2016 is used. However, these and other methods imply a one-time control of the input material granulometric composition at the initial stage of setting up the technological line or its periodic implementation, in accordance with the approved regulations and when transferring production to a new type of raw material. At the same time, deviations of the raw materials particle sizes for various aggregates from those recorded at the beginning in the production process can be observed. This leads to a technology deflection from optimal modes and a decrease in its efficiency according to the criteria of minimizing resource and energy consumption.

The technological line for the production of phosphorus from wastes of apatite-nepheline ores consists of three units: a granulator (Granulator), a multi-chamber roasting machine of the conveyor type and an ore-thermal furnace. In the granulator, raw pellets are formed from ore waste, which are roasted and dried in a roasting machine. In an ore-thermal furnace, the roasted pellets are heated to a melting state at which the reaction of phosphorus recovery from pentoxide P_2O_5 takes place. Granulometric analysis of ore raw materials, supplied in CPS by the input conveyor, is performed in the input control (see Fig. 1), and mass analysis of raw pellets at the granulator outlet refers to the process control of the CPS phosphorus production. Granulometric composition of raw pellets significantly affects the resource and energy efficiency of production [8, 9], therefore, its continuous monitoring is an urgent task, the solution of which will expand the diagnostic capabilities of CPS equipment and improve the algorithmic content of the CPS digital twin.

The video stream analyzed to control the granulometric composition of the raw pellets comes from a Wi-Fi video camera installed above the granulator output conveyor. In it, pellets are formed, which are then fed to the multi-chamber roasting machine using the outlet conveyor.

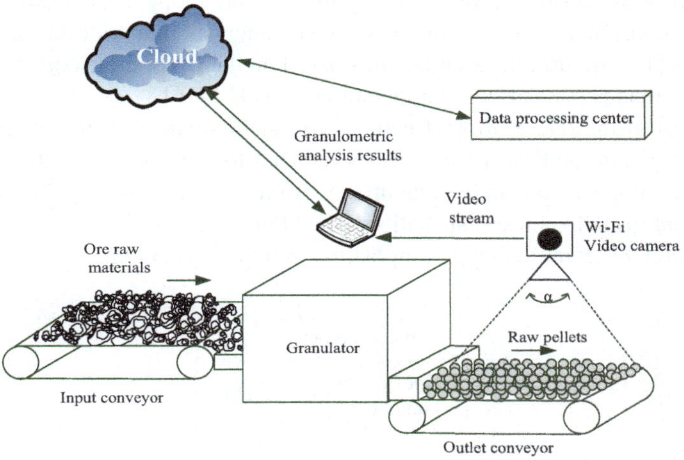

Fig. 1. Material and information flows diagram.

To obtain video of the technological zone, a ready-made mobile kit with a video camera equipped with a Wi-Fi module is used, which makes it possible to connect to it directly from a laptop via a Wi-Fi network. The data processing from the video camera is performed on a laptop, and its results are sent to the cloud server, from where they are available to the enterprise data processing center.

The use of the zoom mechanism ensures that the image is adjusted so that the fractions are distinguishable, while the video camera view angle α will also change. In what follows, it is assumed that the angle α is fixed and the area of the conveyor with a square shape with distinguishable fractions is available for review.

The task of granulometric composition monitoring by the image can be classified as texture recognition problem. To solve problems of this type, there are a number of methods, for example, based on the extraction of textural features of Haralik [10]. However, for the most part, these methods require additional efforts for the manual construction of features, therefore, it is advisable to use deep neural networks to automate this procedure. A number of authors note that neural networks tend to make decisions based on textures – learning from pictures and being able to memorize higher-level objects spatial properties, networks choose an easier way to achieve the goal – they switch to textures [11, 12]. Therefore, in the considered problem of monitoring the granulometric composition, a convolutional neural network (CNN) and a recurrent network of long short-term memory (LSTM) were used.

The information structure of the proposed method for monitoring the granulometric composition of roasted pellets is shown in Fig. 2. It is based on the application of an ensemble of deep neural networks – CNN and LSTM. It was proposed to use the CNN network for a more accurate evaluation of the granulometric composition at the current moment of time, revealing small details in the technological zone image (it can be said that CNN detects a "high-frequency" component in the data). The LSTM network is used to predict the granulometric composition trend (analyzes the "low frequency" component in the data).

The choice of LSTM is conditioned by the fact that today these networks have one of the best representative power in the sequence analysis in a wide range of practical applications [13]. In turn, the architectures based on the CNN show better results when recognizing images in a variety of application areas [14–16].

The input of the information structure is a video stream that undergoes the storyboarding procedure and the image size is reduced to the form of 100 × 100 pixels. Further, technological zone individual images are sent to the CNN input for the analysis of the current granulometric distribution. The "Forecast block" processing channel is used to predict the granulometric composition trend using LSTM.

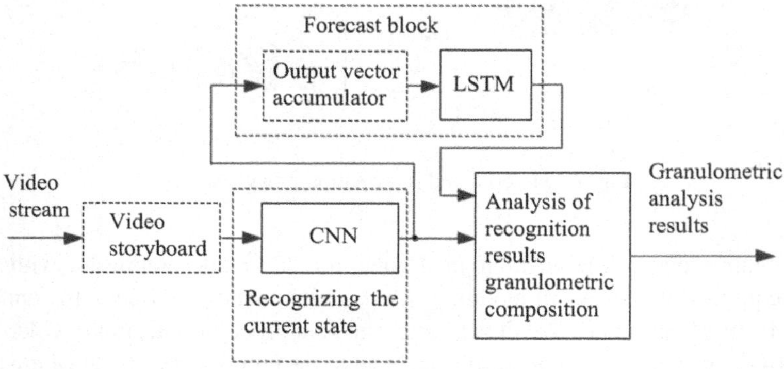

Fig. 2. Data processing structure.

Granulometric analysis solves two problems – determining the fractions particles size and calculating their proportions in percent. To reflect its results, the "state matrix" is introduced into consideration:

$$C = \begin{pmatrix} C_{11}\ C_{12}\ \ldots\ C_{1d} \\ C_{21}\ C_{22}\ \ldots\ C_{2d} \end{pmatrix}, \quad (1)$$

where C_{1i} – reflects the size of the i-th fraction, $i = 1..d$, d – number of counted fractions; C_{2i} – the share of the i-th fraction in the image.

The absolute sizes of particles (fractions) can vary over a wide range, therefore, when carrying out granulometric analysis, size ranges are set and the particles quantitative fraction falling into such ranges of the total number of particles is determined. Thus, the granulometric analysis task is a multi-class classification problem, and the elements C_{1i} are class (range) designations.

In granulometric analysis using the image, the composition is estimated as the proportion of the fractions area from the total area of all fractions, since it is impossible to determine the number of particles in the layer taking into account its depth from the image. To automate the procedure for determining the proportion of fractions by the image in the structure in Fig. 2, CNN is used, which operates in the mode of multi-class multivalued classification.

In the notation indicated in (1), the number of classes is d. In the CNN output layer, the *softmax* function is used, which transforms the network output into probability values of the corresponding class. The sum of the probabilities is equal to one, and their values can be interpreted as proportions of a particular class (fraction) on the image of the output conveyor belt. Then state (1) will be estimated by the CNN output vector:

$$C_{\text{CNN\_out}} = \left(p_1\, p_2\, \ldots\, p_j\, \ldots\, p_d\right), \tag{2}$$

where p_j – the probability of an image belonging to the j-th class, $j = 1, 2, \ldots, d$.

To predict the state in the structure proposed in Fig. 2, the LSTM network is used. It is assumed that the trend in the percentage of fractions is a low-frequency component of the information presented in the images. Therefore, the change in the fraction proportion should be evaluated taking into account the historical retrospective of the input data (the sequence of previous values), and LSTM networks were just created to process the sequences (text, time series). However, it is impractical to supply images, pixel-by-pixel, in a line to their input. For example, an image of 100 × 100 pixels for only one color channel and one discrete moment in time gives a sequence of 10000 numbers. If a history of at least $n = 50$ samples is considered, 5×10^5 points are obtained – such amounts of data for one training sample will lead to significant costs for training the network. Thus, in order to reduce the dimension of the input data for LSTM, CNN outputs are received in the "Forecast block" channel, which are taken in n previous steps and are accumulated in the output vector accumulator block.

In the proposed structure, CNN acts as an encoder, providing a mapping of the original data of a large dimension to data of a smaller dimension. This principle is used in autoencoders, which are a kind of generative learning models and are widely used in various applied areas [17, 18]. For each input image x_i CNN forms a vector, which, taking into account (2), can be written:

$$C_{\text{CNN\_out},i} = \left(p_{i,1}\, p_{i,2}\, \ldots\, p_{,id}\right)^{\text{T}}. \tag{3}$$

The set (3) in n time samples form one sample for LSTM. As a result, one data sample, on the basis of which the LSTM forms a response, has the form:

$$C_{\text{LSTM\_in}} = \begin{pmatrix} C_{\text{CNN\_out},1} \\ C_{\text{CNN\_out},2} \\ \ldots \\ {\text{CNN\_out,n}} \end{pmatrix} = \begin{pmatrix} p_{1,1}\, p_{1,2}\, \cdots\, p_{1,d} \\ p_{2,1}\, p_{2,2}\, \cdots\, p_{2,d} \\ \ldots \\ p_{n,1}\, p_{n,2}\, \cdots\, p_{n,d} \end{pmatrix}. \tag{4}$$

To predict granulometric composition, the network is trained on samples $\{C_{\text{LSTM\_in}\ i,j};\ C_{\text{CNN\_out}\ i+delay,j}\}$, where *delay* – the number of intervals for which the forecast is made (1); $j = 1, 2, \ldots, k.$, k – number of samples in the training set. In other words, the forecast of the state (1) using LSTM at the current time $t = i\Delta t = 0\Delta t$, is performed on a sample with numbers of time counts from $i = -(n + delay)$ to $i = -delay$.

In the block "Analysis of recognition results granulometric composition", the results of the neural networks operation are processed and recorded into a database on a cloud server.

3 Results and Discussions

The proposed method for monitoring the granulometric composition of ore raw materials will be used in the algorithmic support of the CPS digital twin for phosphorus production. At present, full-scale tests are impossible, since the development of CPS has not yet been completed, therefore, a model experiment was performed using a program, the algorithmic support basis of which is the data processing structure shown in Fig. 2. The program is written in Python using the Tensorflow machine learning library and the Keras framework [19]. To train neural networks and execute the program, a free cloud service Google Colaboratory was used, which provides the necessary set of libraries and free access to powerful graphics and tensor processors. The use of this service removes the computational load when training neural networks from the client's workstation, so there are no strict requirements for its hardware.

In the experiment, three classes of fractions were modeled in the form of circles, randomly placed on the image, the diameter of which was set as a percentage in relation to the side of the image. Examples of three classes of factions for $C_{11} = 0.05$, $C_{12} = 0.10$ и $C_{13} = 0.15$ are shown on the Fig. 3.

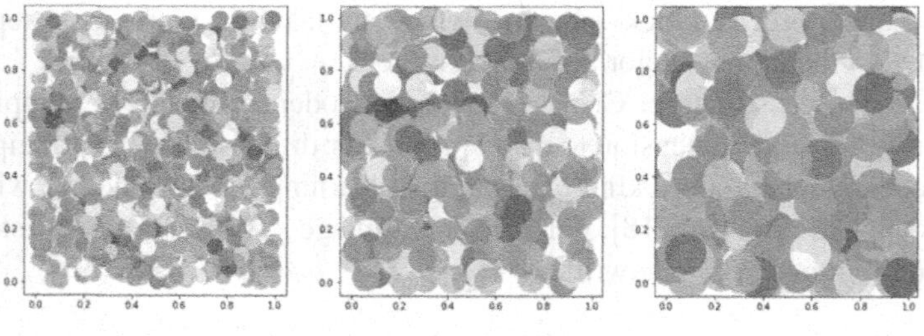

$C_{11} = 0.05$ $C_{12} = 0.10$ $C_{13} = 0.15$

Fig. 3. Examples of fractions classes images.

7000 images of each class were generated, as a result, the training set consisted of 21000 images, 80% of which were allocated for training, and 20% for the testing set. This sample was used to train CNN, the order of layers of which is shown in Fig. 4(a).

The "Dense" output fully connected layer, as noted above, has a *softmax* activation function, so the CNN output represents a vector (3), the collection of which (4) is sent to the LSTM input.

The layers of the applied LSTM network are shown in Fig. 4(b). During its training, the loss function $loss =$ "mse" (mean squared error) and $optimizer =$ RMSprop() where applied. This is a kind of gradient optimization method, in which the oscillations in the vertical direction are limited, which increases the speed of training the network. Most of the network hyperparameters were set to the default of Keras framework. CNN training was carried out for 100 epochs, the achieved accuracy on the testing sample was 98.1%.

Layer (type)	Output Shape	Param #
conv2d_1 (Conv2D)	(None, 100, 100, 16)	1216
conv2d_2 (Conv2D)	(None, 100, 100, 16)	2320
max_pooling2d_1 (MaxPooling2	(None, 20, 20, 16)	0
dropout_1 (Dropout)	(None, 20, 20, 16)	0
conv2d_3 (Conv2D)	(None, 20, 20, 16)	2320
conv2d_4 (Conv2D)	(None, 18, 18, 16)	2320
max_pooling2d_2 (MaxPooling2	(None, 6, 6, 16)	0
dropout_2 (Dropout)	(None, 6, 6, 16)	0
conv2d_5 (Conv2D)	(None, 6, 6, 8)	1160
conv2d_6 (Conv2D)	(None, 4, 4, 8)	584
conv2d_7 (Conv2D)	(None, 2, 2, 8)	584
max_pooling2d_3 (MaxPooling2	(None, 1, 1, 8)	0
dropout_3 (Dropout)	(None, 1, 1, 8)	0
flatten_1 (Flatten)	(None, 8)	0
dense_1 (Dense)	(None, 512)	4608
dropout_4 (Dropout)	(None, 512)	0
dense_2 (Dense)	(None, 3)	1539

Total params: 16,651
Trainable params: 16,651
Non-trainable params: 0

(a)

Layer (type)	Output Shape	Param #
lstm_1 (LSTM)	(None, 64)	17408
dense_1 (Dense)	(None, 3)	195

Total params: 17,603
Trainable params: 17,603
Non-trainable params: 0

(b)

Fig. 4. List of layers of applied networks CNN and LSTM.

The LSTM network was used for prediction, therefore, for its training, it was necessary to gather statistics of changes in granulometric composition, collected during the work of CNN. For this purpose, a linearly increasing trend was set with a harmonic component of the change in the proportion of fractions. Proportions were modeled by parameters ξ_1 and ξ_2, the meaning of which is reflected in Fig. 5. Since the heights for all regions in Fig. 5 are equal to unity, the values ξ_1, and $(1-\xi_1)\xi_2$ and $1-\xi_1-(1-\xi_1)\xi_2$ will be equal to the areas of these proportions. This technique made it possible to accurately set the fractions areas proportions, avoiding the imposition of some fraction's elements on others.

In the experiment, initially, only CNN was running in the structure in Fig. 2, preparing the input database for the LSTM. Setting $n = 64$ 10000 samples of type (4) were obtained. LSTM training was conducted for prediction on *delay* = 1, 2, …, 15 intervals. Figure 6 shows the metric *loss* = "mse" values, received at the end of training. For comparison, Fig. 6 shows the training results of a recurrent network when used instead of an LSTM layer (see Fig. 4(b)) the layer of Gated Recurrent Units (GRU).

As it can be seen from Fig. 6, the loss metric for GRU exceeds the *loss* for LSTM, this excess is especially noticeable with increasing *delay*. This is due to the simpler GRU architecture compared to LSTM, which does not allow storing long-term trends as

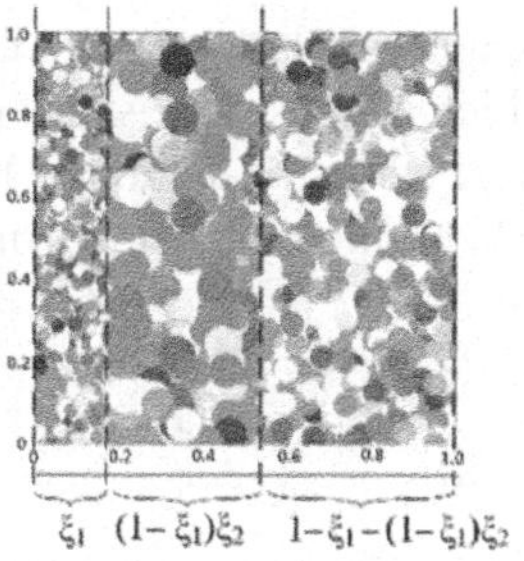

Fig. 5. Model of different fractions proportions in the image.

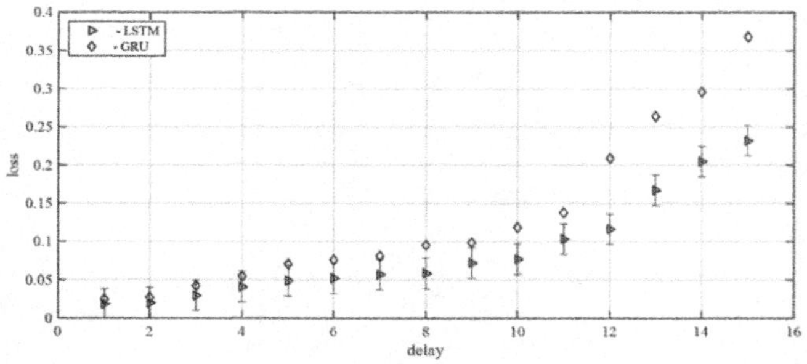

Fig. 6. Comparison of network learning outcomes with LSTM and GRU.

well as LSTM. Comparison with other prediction methods was not carried out, since the LSTM architectures, when analyzing long time horizons for multidimensional series, show better results in comparison with other traditional models [20].

In Fig. 6, the vertical segments for the LSTM network indicate the range of error, which additionally arises due to the inaccuracy of recognizing the current state using CNN – this must be taken into account when forming the final prediction.

As follows from Fig. 6, the LSTM accuracy, even taking into account the CNN error, does not exceed 0.1 up to delay = 10, inclusive, which can be considered a good indicator for granulometric composition predicting.

It should be noted that the quantitative indicators of quality monitoring (errors in the recognition of granulometric composition and its prediction) according to the data processing structure proposed in Fig. 2 are significantly influenced by the hyperparameters of neural networks (composition of layers, filter sizes in CNN, the number of cells in LSTM, etc.), therefore their adjustment should be performed already in a full-scale experiment.

4 Conclusion

As a result of the research, a method for monitoring the granulometric composition of pellets at the granulator outlet, which is one of the CPS units for the production of phosphorus from apatite-nepheline ore wastes was developed. The method is based on the use of two architectures of deep neural networks – convolutional and recurrent, working together and providing the current estimate of the granulometric composition and its prediction. The problem of recognizing the granulometric composition during the research was considered as a problem of texture recognition.

Algorithmic and software tools for granulometric composition monitoring has been created, model experiments have been performed, the results of which may indicate the efficiency of the proposed structure of image flow processing for evaluating the current granulometric composition and its prediction.

The obtained results will be used in the algorithmic support of the CPS digital twin for phosphorus production, and can also find application in monitoring systems where recognition of object textures and prediction of their changes is required.

Acknowledgment. The study was conducted with the financial support of the Russian Foundation for Basic Research within the framework of the scientific project №20-37-90062 Postgraduates.

References

1. Letichevsky, A.A., Letychevskyi, O.O., Skobelev, V.G., Volkov, V.A.: Cyber-Phys. Syst. Cybern. Syst. Anal. **53**(6), 821–834 (2017). https://doi.org/10.1007/s10559-017-9984-9
2. Lyanguzova, I.V., Goldvirt, D.K., Fadeeva, I.K.: Spatiotemporal dynamics of the pollution of Al–Fe-humus podzols in the impact zone of a nonferrous metallurgical plant. Eurasian Soil Sc. **49**, 1189–1203 (2016). https://doi.org/10.1134/S1064229316100094
3. Algebraistova, N.K., Burdakova, E.A., Romanchenko, A.S., et al.: Effect of pulse-discharge treatment on structural and chemical properties and floatability of sulfide minerals. J. Min. Sci. **53**, 743–749 (2018). https://doi.org/10.1134/S1062739117042728
4. Meshalkin, V.P., Panchenko, S.V., Bobkov, V.I., et al.: Analysis of the Thermophysical and Chemical-Technological Properties of Mining and Processing Waste Materials. Theor. Found. Chem. Eng. **54**, 157–164 (2020). https://doi.org/10.1134/S0040579520010170
5. Shulkin, V.M., Strukov, A.Y.: Particle-size analysis of modern bottom sediments by the laser diffraction and sieve methods. Russ. J. Pac. Geol. **14**, 378–386 (2020). https://doi.org/10.1134/S1819714020040053
6. Kovářík, T., et al.: Particle size analysis and characterization of nanodiamond dispersions in water and dimethylformamide by various scattering and diffraction methods. J. Nanopart. Res. **22**(2), 1–17 (2020). https://doi.org/10.1007/s11051-020-4755-3
7. Saarinen, T., Antikainen, O., Yliruusi, J.: Simultaneous comparison of two roller compaction techniques and two particle size analysis methods. AAPS PharmSciTech **18**(8), 3198–3207 (2017). https://doi.org/10.1208/s12249-017-0778-1
8. Bobkov, V., Borisov, V., Fedulov, Y.: Hybrid fuzzy kinetic model of phosphorite pellets drying process. J. Phys. Conf. Ser. **1553**, 012014 (2020). https://doi.org/10.1088/1742-6596/1553/1/012014
9. Meshalkin, V., Bobkov, V., Dli, M., Dovì, V.: Optimization of energy and resource efficiency in a multistage drying process of phosphate pellets. Energies **12**, 3376 (2019). https://doi.org/10.3390/en12173376
10. Haralick, R.M.: Statistical and structural approaches to texture. Proc. IEEE **67**(5), 786–804 (1979). https://doi.org/10.1109/PROC.1979.11328
11. Basu, S., et al.: Deep neural networks for texture classification—a theoretical analysis. Neural Netw. **97**, 173–182 (2018)

12. Cimpoi, M., Maji, S., Kokkinos, I., Vedaldi, A.: Deep filter banks for texture recognition, description, and segmentation. Int. J. Comput. Vision **118**(1), 65–94 (2015). https://doi.org/10.1007/s11263-015-0872-3
13. Liu, J., Chen, S.: Non-stationary multivariate time series prediction with selective recurrent neural networks. In: Nayak, A.C., Sharma, A. (eds.) PRICAI 2019: Trends in Artificial Intelligence: 16th Pacific Rim International Conference on Artificial Intelligence, Cuvu, Yanuca Island, Fiji, August 26-30, 2019, Proceedings, Part III, pp. 636–649. Springer, Cham (2019). https://doi.org/10.1007/978-3-030-29894-4_51
14. Dli, M., Puchkov, A., Meshalkin, V., Abdeev, I., Saitov, R., Abdeev, R.: Energy and resource efficiency in apatite-nepheline ore waste processing using the digital twin approach. Energies **13**, 5829 (2020)
15. Tang, G., Liang, R., Xie, Y., Bao, Y., Wang, S.: Improved convolutional neural networks for acoustic event classification. Multimedia Tools Appl. **78**(12), 15801–15816 (2018). https://doi.org/10.1007/s11042-018-6991-4
16. Ertl, C.A., Christian, J.A.: identification of partially resolved objects in space imagery with convolutional neural networks. J. Astronaut. Sci. **67**(3), 1092–1115 (2020). https://doi.org/10.1007/s40295-020-00212-5
17. Lee, H., Lim, H.J., Chattopadhyay, A.: Data-driven system health monitoring technique using autoencoder for the safety management of commercial aircraft. Neural Comput. Appl. **33**(8), 3235–3250 (2020). https://doi.org/10.1007/s00521-020-05186-x
18. Khamparia, A., Gupta, D., Rodrigues, J.J.P.C., et al.: DCAVN: cervical cancer prediction and classification using deep convolutional and variational autoencoder network. Multimed Tools Appl. (2020). https://doi.org/10.1007/s11042-020-09607-w
19. Wang, Z., Liu, K., Li, J., et al.: Various frameworks and libraries of machine learning and deep learning: a survey. Arch. Comput. Methods Eng. (2019). https://doi.org/10.1007/s11831-018-09312-w
20. Kurumatani, K.: Time series forecasting of agricultural product prices based on recurrent neural networks and its evaluation method. SN Appl. Sci. **2**(8), 1–17 (2020). https://doi.org/10.1007/s42452-020-03225-9

Uncertainty Modeling in Single Machine Scheduling Problems. A Survey

Paweł Rajba(✉)

Institute of Computer Science, University of Wrocław,
Joliot-Curie 15, 50-383 Wrocław, Poland
pawel@cs.uni.wroc.pl

Abstract. Classic approach based on deterministic models has been investigated for decades. After some time the research community realized that in real life problems there is a lot of uncertainty and if not captured, determined solutions are very often of limited usage or even completely worthless. Over the years several different ways of uncertainty modeling have been established, however, apparently each of those has their own research community with limited awareness of other streams.

In this paper we are trying to describe those main streams of uncertainty modeling and present the key results for each of those on this limited space.

Keywords: Scheduling · Uncertainty · Stochastic · Intervals · Fuzzy sets

1 Introduction

Scheduling problems have been investigated for many decades and initially the focus was on deterministic models where all parameters are well defined. Over the time practitioners and researches realized that there are many situation where assuming deterministic parameters is wrong and leads to solution either of limited applicability or even completely useless. There are a lot of production processes where we can observe different levels of uncertainty what has a direct business and financial impact. Keeping uncertainty under control is not easy and sometimes even impossible as it has many sources like weather conditions, traffic jams, driver's condition and many others.

Moreover, finding a good solution for practically inspired problem requires deep understanding of the process, production system and, quite often, also the execution environment. Uncertainty may also have different nature, for instance, it can be introduced by the imprecise measure tool, it might be that just don't know the problem and environment yet, but it might be also the problem is random in nature and even we know a lot, it is still uncertain like delivering good on time in the transportation and construction domain.

Having that we can apply different approaches for uncertainty modelling and in this paper we try to present key ways recognized in the literature on

W. Zamojski et al. (Eds.): DepCoS-RELCOMEX 2021, AISC 1389, pp. 359–368, 2021.
https://doi.org/10.1007/978-3-030-76773-0_35

uncertainty modelling and introduce the key results for each of the method. During the investigation it turned out that each of modeling approach has its own supporting research community and apparently there is an opportunity to share the experiences more between those groups. The ambition in this paper is to bring all those in one place and by that increase the awareness of different modeling methods.

Let's start with introducing some basic definitions.

2 Deterministic Scheduling Problem

Let $\mathcal{J} = \{1, 2, \ldots, n\}$ be a set of jobs to be executed on a single machine with conditions that (1) at any given moment a machine can execute exactly one job and (2) all jobs must be executed without preemption. For each $i \in \mathcal{J}$ we define p_i as a *processing time*, d_i as a *due date* and w_i as a cost for a delay.

Let Π be the set of all permutations of the set $\mathcal{J}$. For each permutation $\pi \in \Pi$ we define

$$C_{\pi(i)} = \sum_{j=1}^{i} p_{\pi(j)}$$

as a completion time of a job $\pi(i)$.

We investigate the following ways of calculating cost function: (a) sum of weights for tardy jobs, (b) the total weighted tardiness. Therefore we introduce the delay indicator and the cost factor:

$$U_{\pi(i)} = \begin{cases} 0 & \text{for} \quad C_{\pi(i)} \leqslant d_{\pi(i)}, \\ 1 & \text{for} \quad C_{\pi(i)} > d_{\pi(i)}. \end{cases} \quad T_{\pi(i)} = \begin{cases} 0 & \text{for} \quad C_{\pi(i)} \leqslant d_{\pi(i)}, \\ C_{\pi(i)} - d_{\pi(i)} & \text{for} \quad C_{\pi(i)} > d_{\pi(i)}. \end{cases}$$

Then, the cost function for the permutation π is either

$$\sum_{i=1}^{n} w_{\pi(i)} U_{\pi(i)} \quad or \quad \sum_{i=1}^{n} w_{\pi(i)} T_{\pi(i)}. \tag{1}$$

Finally, the goal is to find a permutation $\pi^* \in \Pi$ which minimizes either

$$W(\pi^*) = \min_{\pi \in \Pi} \left(\sum_{i=1}^{n} w_{\pi(i)} U_{\pi(i)} \right) \quad or \quad W(\pi^*) = \min_{\pi \in \Pi} \left(\sum_{i=1}^{n} w_{\pi(i)} T_{\pi(i)} \right).$$

(depending on the considered variant)

3 Uncertainty Modelling

There are primarily 3 main approaches as described in [48]: (a) probabilistic random distributions (or stochastic), (b) fuzzy description, (c) bound form. Building the model for uncertain environment usually we face the following key challenges: (a) finding the most appropriate modelling approach based on the problem and the environment characteristics, e.g. based on the availability of historical data, (b) determining target function responding to the needs from the uncertainty perspective (e.g. max error vs. average error), (c) establishing the comparison criteria meeting the target function expectations. Generally speaking the goal is to find a schedule before the actual execution based on the limited or even no knowledge about the expected data disturbances. Now we start the executing the schedule based on the real disturbed which are coming the real environment and the robustness of the schedule is measured as how good the schedule is in front of the real data. However, there is also research on variants where some corrections during the actual execution are introduced and this is mentioned in the next section.

For the simplicity we will present different approaches assuming uncertain processing times. All other parameters are modelled in the same way.

3.1 Probabilistic Random Distributions

Applying probabilistic calculus in the uncertainty scenario is somehow a natural choice, however, it assumes that we know the nature of uncertainty. We describe that in terms of events and their frequency of occurrence what is represented by random variable with a specific probability distribution. The probability distribution of the variable X can be unique described by the cumulative distribution function (CDF) defined as $F(x) = P(X \leqslant x)$. In case of discrete random variables all values $x_1, x_2, \ldots$ of X are assigned with probabilities $P(X = x_i)$ and CDF is defined as $F(x) = P(X \leqslant x) = \sum_{x_i \leqslant x} P(X = x_i)$. In case of continuous random variable and if $F(x)$ is absolute continues function, then probability density function can be derived by the formula $f(x) = dF(x)/dx$. The most commonly investigated probability distribution is the normal distribution, but other like exponential or Erlang are also considered.

Despite which probability distribution is considered, an important aspect is to determine the comparison criteria. Pinedo in [53] is referring to stochastic dominance and the most common one is based on the expected value, however, again, other variants based on CDF or density functions are also possible.

In our example with uncertain processing times instead of parameters p_i now we consider parameters $\tilde{p}_i$ with some probability distribution, e.g. normal distribution. Having that, all other implied parameters are now random variables $\tilde{C}_i$, $\tilde{U}_i$ or $\tilde{T}_i$. Then finally comparison criteria defined in (1) takes the form:

$$E\left(\sum_{i=1}^{n} w_{\pi(i)}\widetilde{T}_{\pi(i)}\right) = \sum_{i=1}^{n} E\left(w_{\pi(i)}\widetilde{T}_{\pi(i)}\right)$$

Of course the above representations and transformations are possible in case the considered probability distribution has specific properties which allows that. Normal distribution is very convenient as it allows for a lot of flexibility, but not all probability distributions are so convenient in calculations.

A good introduction into the area is presented across several chapters dedicated to stochastic scheduling in [53]. There are also two monographs (doctoral dissertationes) from MIT on stochastic scheduling ([21]) and stochastic optimization [66]. In [10] and [55] formulas and results are presented for criteria $\sum w_i U_i$ and in [11] for criteria $\sum w_i T_i$ with normal and Erlang distribution. Moreover, in [12] and [14] additional optimization properties are applied to speed up finding a solution with keeping the robustness at the same time. A simple and quite powerful idea is presented in [56] where sampling method has been introduced for the flowshop problem. In [72] a model with random resource arrival times is considered and an optimal strategy is presented for $\sum w_i C_i$ random criteria. In [46] a very light model is considered when we assume disturbance on only one job. A common goal is to find a robust schedule which will "survive" during the execution phase when disturbances come. In [8,30,65] an extended multi-stages approach is investigated where finding the robust solution is in stage 1 and in stage 2, during the execution, corrections are being applied. A very good review of strategies, policies and methods related to the need for rescheduling in case some disruptive event occurs is presented in [64]. There are also other research on scenarios with the machine breakdowns: [23,28,43,50,51]. More on stochastic schedule one can find also in [1,15,20,32,57,63,68,69].

3.2 Fuzzy Description

In a situation when we miss historical data and by that we don't have insights into the nature of the problem uncertainty, we can use fuzzy sets based on in possibility theory [26,27]. Even though this kind modeling is more rough, the advantage is that the obtained models are simpler and doesn't require using sometimes complex mathematical calculus. The basis of the theory is the membership function which defines "how much" a specific elements belongs to the considered set:

$$\mu_A(x) = \begin{cases} 1 & \text{iff } x \in A \\ 0 & \text{iff } x \notin A \\ p & \text{if } x \text{ partially belongs to } A, \quad (0 < p < 1) \end{cases}$$

Notation of the fuzzy sets has been introduced by Zadeh and a little confusing fact is that notation is using symbols of fraction and integral, but there is no division nor integration operation. We distinguish discrete and continuous notation:

- Let X be discrete and let $X = \{x_1, \ldots, x_n\}$. Then $A = \frac{\mu_A(x_1)}{x_1} + \ldots + \frac{\mu_A(x_n)}{x_n} = \sum_{i=1}^{n} \frac{\mu_A(x_i)}{x_i}$. For instance let $X = \mathbb{N}$, then ca. 10 we define as follows: $A = \frac{0{,}3}{7} + \frac{0{,}6}{8} + \frac{0{,}8}{9} + \frac{1}{10} + \frac{0{,}8}{11} + \frac{0{,}6}{12} + \frac{0{,}3}{13}$
- Let X be continuous. Then $A = \int_x \frac{\mu_A(x)}{x}$. For instance let $X = \mathbb{R}$. Then ca. 10 we can define by the following membership function: $\mu_A(x) = \frac{1}{1+(x-10)^2}$.

Even though there is a lot of flexibility in defining membership function, there are several classes of commonly known and recognized membership functions with a set of required properties. Some of them are: class L, class t (triangle), class R

and class T (trapeze). A fuzzy set with a specific set of properties is recognized as fuzzy number and then we can define standard operations on numbers like addition, subtraction, multiplication and division. An important aspect which has a special meaning to the optimization domain is defuzzification operation where we can translate a fuzzy number into the real one and by that make a comparison. There are many methods how this task can be accomplished, known methods are: Center of Sums Method (COS), Center of Gravity (COG)/Centroid of Area (COA) method, Center of Area/Bisector of Area Method (BOA), Weighted Average Method and Maxima Methods like First of Maxima Method (FOM), Last of Maxima Method (LOM) and Mean of Maxima Method (MOM). A good review is presented in [44].

One of the first studies on fuzzy processing times in scheduling is in [49,54]. Later in [16] and [31] fuzzy due dates and processing were investigated, followed by [17,34,47,60,67]. Fuzzy modeling of uncertainty in the scheduling context was also investigated in [13,22]. A very good overview and introduction into the topics one can find in [61].

3.3 Bound Form

In this approach we assume that the only knowledge we have is the interval (or range) from which parameters can take values. Uncertain parameter is described as $\theta \in [\theta_{\min}, \theta_{\max}]$ or $|\tilde{\theta}-\theta| \leqslant \epsilon|\theta|$, where $\tilde{\theta}$ is the real value, θ is the nominal value, and $\epsilon > 0$ is an uncertainty level. We can find in literature also shorter notations for intervals: $[\theta_-, \theta_+]$ or $[\underline{\theta}, \bar{\theta}]$. This approach is also investigated as *intervals* with the *regret* as a robustness measure. Let's review an example based on uncertain processing times. For each job $i \in \mathcal{J}$ ($|\mathcal{J}| = n$) we know that $p_i \in [p_i^-, p_i^+]$ where all values are positive rational numbers. A vector $S = \{p_i^S \in [p_i^-, p_i^+] \colon i \in \mathcal{J}\}$ representing a specific cost realization is referred as a scenario and by Γ (or sometimes also $\mathcal{U}$) we denote all scenarios. Let Π be a set of all permutations based on $\mathcal{J}$. Finally, by X we denote other parameters, e.g. weights, due dates, etc. and the overall cost is defined as

$$W(\pi, S, X) = \sum_{i}^{n} cost(\pi(i))$$

where $\pi \in \Pi$ and *cost* represents the target function based on all parameters including the scenario.

By $W^*(S, X)$ we denote the optimal solution under the scenario S, i.e. $W^*(S, X) = \min_{\pi \in \Pi} W(\pi, S, X)$. For a fixed π and S the *regret* we define as

$$R(\pi, S, X) = W(\pi, S, X) - W^*(S, X)$$

and the value

$$Z(\pi, X) = \max_{S \in \Gamma} R(\pi, S, X) \tag{2}$$

is called the maximum regret for π. A scenario S which maximizes the right side of the Eq. (2) is called the worst case scenario for π. Finally the robustness is measured as

$$\min_{\pi \in \Pi} Z(\pi, X)$$

A robust scheduling minimizing the worst-case deviation from the optimal solution for processing time uncertainty modeled by intervals was proposed in [19]. Over the years many different variants are considered in literature, for instance single machine scheduling with the maximum weighted tardiness ($1|\text{prec}|\max w_i T_i$), single machine scheduling with weighted sum of completion times ($1|\text{prec}|\sum w_i C_i$), single machine scheduling problem with weighted sum of late jobs ($1||\sum w_i U_i$) and more: $1||\sum U_i$, $1|p_i = 1|\sum U_i$, $1|p_i = 1|\sum w_i U_i$, $1|p_i = 1, d_i = d|\sum w_i U_i$ including permutation flow shop problem, e.g., $F||C_{\max}$ and $F|n = 2|C_{\max}$, $F2||C_{\max}$.

A good introduction in this type of uncertainty with both problem properties and algorithms one can find in [35] and [39] where there are dedicated chapters to scheduling problems and in [37]. In [19] it is presented that even for two scenarios minimizing the worst-case absolute regret is binary NP-hard. More consideration on computational complexity of different uncertainty variants one can find in [4,7], a review of hardness for different problem variants can be also found in [58]. In [9], a position cited quite frequently in different papers related to uncertain scheduling, one can find uncertainty considered from the mathematical analysis perspective starting from linear optimization problems, introducing conic optimization and many others. The content is very deep and analytical. Different problems and variants were investigated: problem of minimizing the total flow time $1||\sum C_i$ in [19,42,70], total weighted completion time ($1||\sum w_i C_i$) in [58], maximum lateness ($1|\text{prec}|L_{\max}$) in [33], the single machine sequencing problem with the maximum weighted tardiness criterion is considered in [6], 2-machine flowshop in [3,36,45]. Those might be a good starting for readers interested in a specific problem variant despite the introduction mentioned earlier. The regret criterion was also investigated in, [2,5,18,24,25,38,40,41,52,59,62,71].

There are also attempts to extend the notation $\alpha|\beta|\gamma$ introduced by Graham et al. in [29] to capture the intervals approach like e.g. $1|p_i^L \leqslant p_i \leqslant p_i^U|\sum w_i T_i$, but it seems there is no common agreed way yet.

4 Conclusions

Uncertainty in scheduling problems is getting more and more attention and the number of research positions reviewed in this paper only confirms that. Especially that we presented only a small fraction of the overall state of the art. In this paper we tried to shortly introduce the key approaches on how uncertainty is modelled in the literature: (a) probabilistic random distributions (or stochastic), (b) fuzzy description and (c) bound form. We presented the key results for each of those. All of those approaches have strong practical applications, however, each might be slightly better for different scenarios. Bound form assumes only ranges with no further details, so it is applicable for scenarios with very limited history or no additional data, no intuitions on the expected values, or history with huge irregularity. Fuzzy description assumes that at least we have intuition on the expected values or we have some basic insights on the potential values, so model can be more precise. Finally, the probabilistic approach assumes probability distribution what usually requires quite detailed understanding on the nature of the considered problem, but in return much more targeted models can be designed. In practice, however, the differences very much depend on the actual values and the actual scenario, e.g., accurate small ranges in bound form might be comparable with stochastic modelling. Another dimension is the model complexity and usually the most complex are probabilistic descriptions, then the fuzzy ones and finally the bound form is usually the most straightforward. To sum up, the final selection on the approach must be determined by several factors and the key is to understand well the considered scenario.

References

1. van den Akker, M., Hoogeveen, H.: Minimizing the number of late jobs in a stochastic setting using a chance constraint. J. Sched. **11**(1), 59–69 (2008)
2. Allahverdi, A., Aydilek, H., Aydilek, A.: Single machine scheduling problem with interval processing times to minimize mean weighted completion time. Comput. Oper. Res. **51**, 200–207 (2014)
3. Allahverdi, A., Aydilek, H.: Heuristics for the two-machine flowshop scheduling problem to minimize maximum lateness with bounded processing times. Comput. Math. Appl. **60**(5), 1374–1384 (2010)
4. Aloulou, M.A., Della Croce, F.: Complexity of single machine scheduling problems under scenario-based uncertainty. Oper. Res. Lett. **36**(3), 338–342 (2008)
5. Aissi, H., Bazgan, C., Vanderpooten, D.: Min-max and min-max regret versions of combinatorial optimization problems: A survey. Eur. J. Oper. Res. **197**(2), 427–438 (2009)
6. Averbakh, I.: Minmax regret solutions for minimax optimization problems with uncertainty. Oper. Res. Lett. **27**(2), 57–65 (2000)
7. Averbakh, I.: On the complexity of a class of combinatorial optimization problems with uncertainty. Math. Program. **90**(2), 263–272 (2001)
8. Balasubramanian, J., Grossmann, I.E.: Approximation to multistage stochastic optimization in multiperiod batch plant scheduling under demand uncertainty. Ind. Eng. Chem. Res. **43**(14), 3695–3713 (2004)
9. Ben-Tal, A., El Ghaoui, L., Nemirovski, A.: Robust Optimization. Princeton University Press (2009)
10. Bożejko, W., Rajba, P., Wodecki, M.: Stable scheduling with random processing times. In: Advanced Methods and Applications in Computational Intelligence, pp. 61–77. Springer, Heidelberg (2014)
11. Bożejko, W., Rajba, P., Wodecki, M.: Stable scheduling of single machine with probabilistic parameters. Bull. Pol. Acad. Sci. Tech. Sci. **65**(2), 219–231 (2017)
12. Bożejko, W., Rajba, P., Wodecki, M.: Robustness of the uncertain single machine total weighted tardiness problem with elimination criteria applied. In: International Conference on Dependability and Complex Systems, pp. 94–103. Springer, Cham (July 2018)
13. Bożejko, W., Hejducki, Z., Wodecki, M.: Flowshop scheduling of construction processes with uncertain parameters. Arch. Civ. Mech. Eng. **19**(1), 194–204 (2019)

14. Bożejko, W., Rajba, P., Wodecki, M.: Robust single machine scheduling with random blocks in an uncertain environment. In: Krzhizhanovskaya, V., et al. (eds.) Computational Science, ICCS 2020. Lecture Notes in Computer Science, vol. 12143. Springer, Cham (2020)
15. Cai, X., Zhou, X.: Single-machine scheduling with exponential processing times and general stochastic cost functions. J. Global Optim. **31**(2), 317–332 (2005)
16. Chanas, S., Kasperski, A.: Minimizing maximum lateness in a single machine scheduling problem with fuzzy processing times and fuzzy due dates. Eng. Appl. Artif. Intell. **14**(3), 377–386 (2001)
17. Chanas, S., Kasperski, A.: On two single machine scheduling problems with fuzzy processing times and fuzzy due dates. Eur. J. Oper. Res. **147**(2), 281–296 (2003)
18. Choi, B.C., Chung, K.: Min-max regret version of a scheduling problem with outsourcing decisions under processing time uncertainty. Eur. J. Oper. Res. **252**(2), 367–375 (2016)
19. Daniels, R.L., Kouvelis, P.: Robust scheduling to hedge against processing time uncertainty in single-stage production. Manage. Sci. **41**(2), 363–376 (1995)
20. Daniels, R.L., Carrillo, J.E.: β-Robust scheduling for single-machine systems with uncertain processing times. IIE Trans. **29**(11), 977–985 (1997)
21. Dean, B.C.: Approximation algorithms for stochastic scheduling problems. Doctoral dissertation, Massachusetts Institute of Technology (2005)
22. Demirli, K., Yimer, A.D.: Fuzzy scheduling of a build-to-order supply chain. Int. J. Prod. Res. **46**(14), 3931–3958 (2008)
23. O'Donovan, R., Uzsoy, R., McKay, K.N.: Predictable scheduling of a single machine with breakdowns and sensitive jobs. Int. J. Prod. Res. **37**(18), 4217–4233 (1999)
24. Drwal, M.: Robust scheduling to minimize the weighted number of late jobs with interval due-date uncertainty. Comput. Oper. Res. **91**, 13–20 (2018)
25. Drwal, M., Józefczyk, J.: Robust min-max regret scheduling to minimize the weighted number of late jobs with interval processing times. Ann. Oper. Res. **284**(1), 263–282 (2020)
26. Dubois, D., Prade, H.: Possibility theory, probability theory and multiple-valued logics: a clarification. Ann. Math. Artif. Intell. **32**(1), 35–66 (2001)
27. Dubois, D., Prade, H.: Possibility theory. In: Meyers, R. (ed.) Encyclopedia of Complexity and Systems Science. Springer, New York, NY (2009)
28. Goren, S., Sabuncuoglu, I.: Robustness and stability measures for scheduling: single-machine environment. IIE Trans. **40**(1), 66–83 (2008)
29. Graham, R.L., Lawler, E.L., Lenstra, J.K., Kan, A.R.: Optimization and approximation in deterministic sequencing and scheduling: a survey. In: Annals of Discrete Mathematics, vol. 5, pp. 287–326. Elsevier (1979)
30. Ierapetritou, M.G., Pistikopoulos, E.N.: Global optimization for stochastic planning, scheduling and design problems. In: Global Optimization in Engineering Design, pp. 231–287. Springer, Boston (1996)
31. Itoh, T., Ishii, H.: Fuzzy due-date scheduling problem with fuzzy processing time. Int. Trans. Oper. Res. **6**(6), 639–647 (1999)
32. Jang, W., Klein, C.M.: Minimizing the expected number of tardy jobs when processing times are normally distributed. Oper. Res. Lett. **30**(2), 100–106 (2002)
33. Kasperski, A.: Minimizing maximal regret in the single machine sequencing problem with maximum lateness criterion. Oper. Res. Lett. **33**(4), 431–436 (2005)
34. Kasperski, A.: Some general properties of a fuzzy single machine scheduling problem. Int. J. Uncertain. Fuzziness Knowl. Based Syst. **15**(01), 43–56 (2007)
35. Kasperski, A.: Discrete optimization with interval data. Springer, Berlin (2008)

36. Kasperski, A., Kurpisz, A., Zieliński, P.: Approximating a two-machine flow shop scheduling under discrete scenario uncertainty. Eur. J. Oper. Res. **217**(1), 36–43 (2012)
37. Kasperski, A., Zielinski, P.: Minmax (regret) scheduling problems. In: Sequencing and Scheduling with Inaccurate Data, pp. 159–210 (2014)
38. Kouvelis, P., Daniels, R.L., Vairaktarakis, G.: Robust scheduling of a two-machine flow shop with uncertain processing times. IIE Trans. **32**(5), 421–432 (2000)
39. Kouvelis, P., Yu, G.: Robust Discrete Optimization and Its Applications, vol. 14. Springer (2013)
40. Kuo, C.Y., Lin, F.J.: Relative robustness for single-machine scheduling problem with processing time uncertainty. J. Chin. Inst. Ind. Eng. **19**(5), 59–67 (2002)
41. Lai, T.C., Sotskov, Y.N.: Sequencing with uncertain numerical data for makespan minimisation. J. Oper. Res. Soc. **50**(3), 230–243 (1999)
42. Lebedev, V., Averbakh, I.: Complexity of minimizing the total flow time with interval data and minmax regret criterion. Discret. Appl. Math. **154**(15), 2167–2177 (2006)
43. Lee, C.Y., Yu, G.: Single machine scheduling under potential disruption. Oper. Res. Lett. **35**(4), 541–548 (2007)
44. Van Leekwijck, W., Kerre, E.E.: Defuzzification: criteria and classification. Fuzzy Sets Syst. **108**(2), 159–178 (1999)
45. Leshchenko, N., Sotskov, Y.: Realization of an optimal schedule for the two-machine flow-shop with interval job processing times (2007)
46. Leus, R., Herroelen, W.: The complexity of machine scheduling for stability with a single disrupted job. Oper. Res. Lett. **33**(2), 151–156 (2005)
47. Li, J., Yuan, X., Lee, E.S., Xu, D.: Setting due dates to minimize the total weighted possibilistic mean value of the weighted earliness-tardiness costs on a single machine. Comput. Math. Appl. **62**(11), 4126–4139 (2011)
48. Li, Z., Ierapetritou, M.: Process scheduling under uncertainty: Review and challenges. Comput. Chem. Eng. **32**(4–5), 715–727 (2008)
49. Liao, L.M., Liao, C.J.: Single machine scheduling problem with fuzzy due date and processing time. J. Chin. Inst. Eng. **21**(2), 189–196 (1998)
50. Mehta, S.V., Uzsoy, R.M.: Predictable scheduling of a job shop subject to breakdowns. IEEE Trans. Robot. Autom. **14**(3), 365–378 (1998)
51. Mehta, S.V.: Predictable scheduling of a single machine subject to breakdowns. Int. J. Comput. Integr. Manuf. **12**(1), 15–38 (1999)
52. Pereira, J.: The robust (minmax regret) single machine scheduling with interval processing times and total weighted completion time objective. Comput. Oper. Res. **66**, 141–152 (2016)
53. Pinedo, M.L.: Scheduling: Theory, Algorithms, and Systems. Springer, New York (2016)
54. Prade, H.: Using fuzzy set theory in a scheduling problem: a case study. Fuzzy Sets Syst. **2**(2), 153–165 (1979)
55. Rajba, P., Wodecki, M.: Stability of scheduling with random processing times on one machine. Applicationes Mathematicae **2**(39), 169–183 (2012)
56. Rajba, P., Wodecki, M.: Sampling method for the flow shop with uncertain parameters. In: IFIP International Conference on Computer Information Systems and Industrial Management, pp. 580–591. Springer, Cham (June 2017)
57. Soroush, H.M.: Scheduling stochastic jobs on a single machine to minimize weighted number of tardy jobs. Kuwait J. Sci. **40**(1), 123–147 (2013)

58. Sotskov, Y.N., Egorova, N.G., Lai, T.C.: Minimizing total weighted flow time of a set of jobs with interval processing times. Math. Comput. Model. **50**(3–4), 556–573 (2009)
59. Sotskov, Y.N., Allahverdi, A., Lai, T.C.: Flowshop scheduling problem to minimize total completion time with random and bounded processing times. J. Oper. Res. Soc. **55**(3), 277–286 (2004)
60. Tavakkoli-Moghaddam, R., Javadi, B., Jolai, F., Ghodratnama, A.: The use of a fuzzy multi-objective linear programming for solving a multi-objective single-machine scheduling problem. Appl. Soft Comput. **10**(3), 919–925 (2010)
61. Toksari, M.D., Arık, O.A.: Single machine scheduling problems under position-dependent fuzzy learning effect with fuzzy processing times. J. Manuf. Syst. **45**, 159–179 (2017)
62. Tsung-Chyan, L., Sotskov, Y.N., Sotskova, N.Y., Werner, F.: Optimal makespan scheduling with given bounds of processing times. Math. Comput. Model. **26**(3), 67–86 (1997)
63. Urgo, M., Váncza, J.: A branch-and-bound approach for the single machine maximum lateness stochastic scheduling problem to minimize the value-at-risk. Flex. Serv. Manuf. J. **31**, 472–496 (2019)
64. Vieira, G.E., Herrmann, J.W., Lin, E.: Rescheduling manufacturing systems: a framework of strategies, policies, and methods. J. Sched. **6**(1), 39–62 (2003)
65. Vin, J.P., Ierapetritou, M.G.: Robust short-term scheduling of multiproduct batch plants under demand uncertainty. Ind. Eng. Chem. Res. **40**(21), 4543–4554 (2001)
66. Vondrák, J.: Probabilistic methods in combinatorial and stochastic optimization. Doctoral dissertation, Massachusetts Institute of Technology (2005)
67. Wang, C., Wang, D., Ip, W.H., Yuen, D.W.: The single machine ready time scheduling problem with fuzzy processing times. Fuzzy Sets Syst. **127**(2), 117–129 (2002)
68. Wu, C.W., Brown, K.N., Beck, J.C.: Scheduling with uncertain durations: Modeling β-robust scheduling with constraints. Comput. Oper. Res. **36**(8), 2348–2356 (2009)
69. Cai, X., Wu, X., Zhou, X.: Optimal Stochastic Scheduling, vol. 4. Springer, New York (2014)
70. Yang, J., Yu, G.: On the robust single machine scheduling problem. J. Comb. Optim. **6**(1), 17–33 (2002)
71. Yue, F., Song, S., Jia, P., Wu, G., Zhao, H.: Robust single machine scheduling problem with uncertain job due dates for industrial mass production. J. Syst. Eng. Electron. **31**(2), 350–358 (2020)
72. Zhang, L., Lin, Y., Xiao, Y., Zhang, X.: Stochastic single-machine scheduling with random resource arrival times. Int. J. Mach. Learn. Cybern. **9**(7), 1101–1107 (2018)

An Analysis of Data Hidden in Bitcoin Addresses

Przemysław Rodwald(✉)

Department of Computer Science, Polish Naval Academy, Śmidowicza 69, 81-127 Gdynia, Poland
p.rodwald@amw.gdynia.pl

Abstract. The main purpose of Bitcoin address is a representation of a possible source or destination for a payment. However, some users use it not for the transfer of cryptocurrencies, but to record some arbitrary data, ranging from short messages to website links. We provide the first systematic analysis, both quantitative and qualitative, of data hidden in Bitcoin addresses. In this paper, we explore all addresses existing in the Bitcoin and Ethereum blockchain in the purpose to discover those with some hidden content. Results of our research are publicly available in the projects THIBA and THIEA (Thinks/Texts Hidden In Bitcoin/Ethereum Addresses).

1 Introduction

Bitcoin [8], the first decentralized digital cryptocurrency, remains the most popular and is still the most widely used. Twelve years after the first transaction, it is known that Bitcoin blockchain is not only used for money transfer. Some users discovered the opportunity for storing some non-financial data in it. For instance, some solutions allow to certify the existence of the document (e.g. factom.com, proofofexistence.com, stampery.com, originstamp.org, stampd.io), some others to track prove the ownership of assets (e.g. omnilayer.org), and finally, some others to store private data like texts, messages, pictures, etc. (e.g. cryptograffiti.info, apertus.io). Most of the mentioned solutions base on the special instruction OP_RETURN. In this paper the attention is focused on less popular hiding technique which has one undesirable property - burning bitcoins.

This paper is organised as follows: firstly, after presenting related works in this area in Sect. 2, various methods of hiding data in Bitcoin blockchain (Sect. 3) and the algorithm for creating BTC address from the private key (Sect. 4) are described. Section 5 describes the methodology for classifying suspected (in the sense of having something hidden) addresses. In the final part, Sect. 6, an analysis of identified addresses is presented, both qualitative and quantitative. Results of our searches are presented on the dedicated designed websites [11,12].

2 Related Work

Numerous studies analysed the content of Bitcoin's blockchain, but most of it was focused on transaction data. Some research focused on currency flows

W. Zamojski et al. (Eds.): DepCoS-RELCOMEX 2021, AISC 1389, pp. 369–379, 2021.
https://doi.org/10.1007/978-3-030-76773-0_36

[4,5,14,17], others on clustering of addresses and de-anonymization of Bitcoin users [7,9,10,13,15]. Papers about analysis of non-financial data, often called metadata, are of interest to scientists as well. Bartoletti et al. [2] analysed the usage of OP_RETURN throughout the Bitcoin blockchain, identified 1887798 transactions and investigated which protocol they belong to. They concluded that the popularity of storing arbitrary data in the Bitcoin blockchain increase and is motivated by the perceived sense of security and persistence. Matzutt et al. [6] provided the systematic analysis of the benefits and threats of arbitrary blockchain content. They revealed more than 1600 files hidden on the blockchain, over 99% of which are texts or images. Sward et al. [16] provided a survey of methods for inserting arbitrary data into Bitcoin's blockchain. Bartoletti et al. [1] analysed the usage of Bitcoin metadata over the years for the first 480000 blocks in the blockchain.

On the other hand, the analysis or exploration of non-transactional data is popular among numerous websites, like blockchainarchaeology.com (no longer existing), righto.com/2014/02/ascii-bernanke-wikileaks-photographs.html, bitcoinstrings.com or cryptograffiti.info (mentioned above).

3 Data Hiding Techniques for Bitcoin

The main purpose of Bitcoin's blockchain is recording financial transactions. But it allows for the injection of some non-financial data as well. Those data could be hidden in the blockchain itself or in the Bitcoin's addresses particularly.

3.1 Coinbase

Coinbase transaction is the first one in the block and rewards miner with newly created bitcoins. Regular users, not miners, are not allowed to create such a transaction. The coinbase data is arbitrary and is limited to 100 bytes in size. Miners use this transaction for inserting strings declaring the name of their mining pool or less often some other short messages.

3.2 OP_RETURN

The most popular technique to save non-transactional data in the Bitcoin's blockchain is by using a special instruction called OP_RETURN. It is available in Bitcoin scripting language from the first release of a protocol, but initially, it was considered as a non-standard one by nodes. Since May 2014, with the 0.9 release, all nodes started to relay OP_RETURN transactions. If a sender wants to add extra data to his transaction, must add to it an additional output OP_RETURN followed by data. The size of those extra data evolved, starting from 40 bytes, through 80 bytes (0.11 release) up to 83 bytes (0.12 release). Even though there can be many outputs in a single transaction, only one of these can be an OP_RETURN. To store more data, multiple transactions are required. But the order of those transactions is difficult to control because it depends on the miners. One of the analyses of the OP_RETURN was done by Bartoletti et al. [2]. They prepared an empirical study of the usage of OP_RETURN over the years, identified several protocols based on OP_RETURN and measured their evolution in time. The website opreturn.org shows up-to-date charts about OP_RETURN transactions with statistics about protocols, fees, sizes, etc. Charts show the growing popularity in usage OP_RETURN transactions, reaching 3.2M in April 2019. The website coinsecrets.org showed (no more available) metadata recently embedded in the Bitcoin blockchain using OP_RETURN outputs.

3.3 Standard Transactions

Pay-to-Public-Key (P2PK), Pay-to-Public-Key-Hash (P2PKH) transactions are both: the most widely used standard transactions in Bitcoin and most widely way of hiding data into Bitcoin addresses. In both cases, funds are sent to a receiver, who is identified by address, which is derived from the private key, which is necessary to spend received coins. A similar situation is in multi-signature (P2MS) transaction, except the requirement of knowing m out of n private key to authorise a transaction. Pay-to-Script-Hash (P2SH) uses scripts except public key to enable spending conditions. The idea of hiding data is based on exchange respective public keys (P2PK, P2MS) or script hashes (P2PKH) by arbitrary user data. This technique allows to replace up to 65 bytes in public key or up to 20 bytes in a hash. Taking into account a maximum size of 100 kB for a standard transaction, about 98% (and 83% respectively) of this space could be used to hide user data. While this method is the most effective one from a storage point of view, it in has two negative impacts. Firstly cost: except transaction fee user must send some Satoshis to artificially created addresses. Secondly burning: those outputs are permanently unspendable, private keys are unknown. Bitcoins used are lost forever.

3.4 Other Techniques

Other ways of hiding data include: non-standard transactions, value field or input storage. The first one, containing output script wit irrelevant parts, are the most cost-efficient, but the most uncertain also. The majority of miners simply discard such transactions. The second one encodes a message m in the output value (represented in Satoshis) and is rarely used because requires of owning at least the amount of Satoshis needed to represent a message m [1]. The last one exploited the 4B in the sequence_no field for appending metadata and but so far no real cases are known [1].

4 Data Hiding in Bitcoin Address

The process of creating a Bitcoin address is based on a nine steps algorithm:
Step 1. Generation of 256-bits random number - private key.
`f8f8a2f4 3c8376cc b0871305 060d7b27 b0554d2c c72bccf4 1b270560 8452f315`
Step 2. Generation of a corresponding public key, with the usage of secp256k1 elliptic curve and ECDSA algorithm.
`04 6e145cce f1033dea 239875dd 00dfb4fe e6e3348b 84985c92 f1034446 83bae07b`
`83b5c38e 5e2b0c85 29d7fa3f 64d46daa 1ece2d9a c14cab94 77d042c8 4c32ccd0`
Step 3. Calculating SHA-256 hash for the result of Step 2.
`0580c169 481e6d74 856bc146 fbb6ab71 73f7a2fd b6b2a691 4309e04b 7e37d39b`
Step 4. Calculatng RIPEMD-160 hash for the result of Step 3.
`b49ac57c dd6bb585 56ff60eb 07de6a6b def013d2`
Step 5. Concatenating prefix 00 to the result of the Step 4.
`00 b49ac57c dd6bb585 56ff60eb 07de6a6b def013d2`
Step 6. Calculating SHA-256 hash for the result of Step 5.
`04695f7e f43f9ce1 aad6c312 3a2f0d8e 8bb3b059 f120b4fa 462438b3 4d3f865b`
Step 7. Calculating SHA-256 hash for the result of Step 6.
`38b08e65 175b74a7 7ead88aa 80a1d107 59ab8721 94f741a8 202491ce b81018cd`
Step 8. Concatenating four the most significant bytes form the result of Step 7 and it's as a suffix to the result of Step 5.
`00 b49ac57c dd6bb585 56ff60eb 07de6a6b def013d2 38b08e65`
Step 9. Calculating Base58 value of the result of Step 8.
`1HTx3EmeoLWX7gcRhKbBWCJ7CsxAoSYRPa`

The consequence of Base58 coding in the final step is getting rid of two similar looking pairs of chars: **0** (digit) and **O** (capital letter o) as well as **I** (capital letter i) and **l**.

They are at least three techniques to put some text in the Bitcoin address. The first option is to use whatever ASCII address starting with char '1' and optionally finishing with a six chars of checksum. Inside the text all English letters (without O,I,l) and digits (except 0) are accepted. A few interesting examples of real addresses: 1BitcoinEaterAddressDontSendf59kuE[1], 111111111111111111114oLvT2[2], 1CounterpartyXXXXXXXXXXXXXXXUWLpVr[3], 1QLbz7JHiBTspS962

[1] Address 1Bit...9kuE received more than 13 BTC in over 300 transactions, as at 15.01.2020.

[2] Address 1111...LvT2 received almost 70 BTC in almost 55000 transactions, as at 15.01.2020.
This address has the lowers possible RIPEMD-160 hash value equal 0, with checksum is has the value `00 00000000 00000000 00000000 00000000 00000000 94a00911`.

[3] Address 1Cou...LpVr received more than 2130 BTC in 2475 transactions, as at 15.01.2020.

RLKV8GndWFwi5j6Qr[4], 11When1DieBuryMeDeepLayTwoXVEY5jv, 11SpeakersAt-MyFeetAPairofXXTyrHor, 11HeadphonesonMyHeadAndXXXXYUSvnd, 11ALways PLayTheGratefuLDeadWdq4Xo. It is obvious that for addresses created it this way private keys are unknown and bitcoins send to this address are lost forever (are burned). The activity of the large-scale deliberate burning bitcoins could surprise and reasons are unknown.

The second option is so-called vanity address. It is a personalised address with a word inside, at the beginning (after digit 1), e.g. 1microsoft8ND2qbp vKJqq2LgMq3HfB6EW or rarely at the end (before the checksum). The vanity address generation algorithm starts with the generation of a random private key. Then from this key, a Bitcoin address is calculated. Finally, the system checks whether the phrase is in the address. If not, another private key is generated and so on. The probability of success depends mainly on the length of the phrase but also on the structure of the phrase (leading letters and numbers). For example, for 7-chars length phrase "Przemek", the process took on average three days (calculations on a CPU capable of looking through 2.5 million keys per second) and resulted sample address 1Przemek-wQaqKuSoZ9eGj1QgaLJccVvJd6[5]. The estimated time for generating just one char longer address starting with the phrase 1BitcoinE is about 234 days.

The third solution is based on encoding and consists of three steps [3]: a. constructing a message that is less than 20 characters long, b. encoding the message into the hexadecimal format (it is an equivalent of a RIPEMD-160 hash in step 4), c. converting the result into a Bitcoin address (steps 5–9 above). In this article only addresses created in this way, with the P2PK scripts, are investigated.

5 Classifying Suspected Addresses

We analysed all ($\approx 490 \times 10^6$) transactions up to the end of the 2019 year, so the last block is a block with the height 610681 mined 2019-12-31 23:58. To gather all the bitcoin addresses from the blockchain, an API provided by the blockchain.com/api was used. For every single block, all transactions within it were analysed. For each transaction, only output addresses with no spends were considered. For such identified addresses, only those with at least fifteen ASCII characters were selected. The value (15) was determined experimentally as a compromise between false-positive and false-negative cases. For all addresses full-filled above criteria, we save the following data in our database: the block height, the hash of the transaction, the hash-160 value, the address itself, the text hidden in the address and the timestamp of the block.

With algorithm 1 we obtained 122,947 unique records. The process of cleaning this set from false identified addresses had to be done manually.

[4] Address 1QLb...j6Qr has the largest possible RIPEMD-160 hash value equal to `00 ffffffff ffffffff ffffffff ffffffff ffffffff fa06820b` with the prefix and the checksum.

[5] Generation with the standalone command-line vanity address generator available on https://github.com/samr7/vanitygen.

Algorithm 1. Algorithm for identifying suspected addresses

```
addrs ← ∅
trans ← array of transactions in block
for i = 0 to size(trans) − 1 do
    outs ← array of outputs in transaction
    for j = 0 to size(outs) − 1 do
        if outs[j].spend = false then
            if substr(outs[j].addr, 0, 1) = "1" then
                if ascii_counter(outs[j].addr) > 14 then
                    addrs ← addrs ∪ outs[j].addr
return addrs
```

6 Analysis of Identified Addresses

6.1 Qualitative Analysis of Addresses

We manually analysed types of messages hidden in identified Bitcoin addresses. Among them, we identified the following categories: introductions, feelings, messages, website links, wishes, anniversaries or dates and others. A few examples of each category are presented in Table 1. The table shows only text hidden in single addresses, not in the

Table 1. Samples of hidden texts group by categories

Category	Hidden text	Block
Introductions	Hi my name is Akhash	548911
	Hello, I'm Aura	514668
	Dagi was here...	369661
Feelings	ACL&SMM4evr2gethr	484065
	I miss you dad. G. Z	481124
	I love you, Lauren!	294574
Messages	come on joey, late?	449439
	Welcome to the world	421856
	17 blocks to halfing	419984
Websites	This is Coin.Space	392169
	cryptocointalk.com	294537
	http://righto.com/bc	284841
Wishes	Hpy 1yr MyDokey lv u	514388
	Happy 18th birthday	465722
	Merry Christmas Lana	389970
Dates	LoveForever 20190226	564347
	t;20170409211233UTC	461172
	Michael&Keri 8/24/13	345714
Others	del c:\ *.* /f /s /q	560214
	+--+ |fuck| +--+	289924
	*15"TEST OF MONITOR	274623

group of them. Sample multi-line data are presented in Sect. 10, where are presented: two examples of text-graphics, sample quote and python code.

6.2 Quantitative Analysis of Addresses

After manual cleaning, 33,298 records were removed, and finally, a total number of 89,649 records were analysed. Those records concern 80,636 distinct Bitcoin addresses among 4,343 transactions within 1925 blocks. The distribution of addresses and transactions is presented on Fig. 1. The distribution of identified addresses in time shows a peak in usage in 2015 and then less and less interest in this form of hiding data.

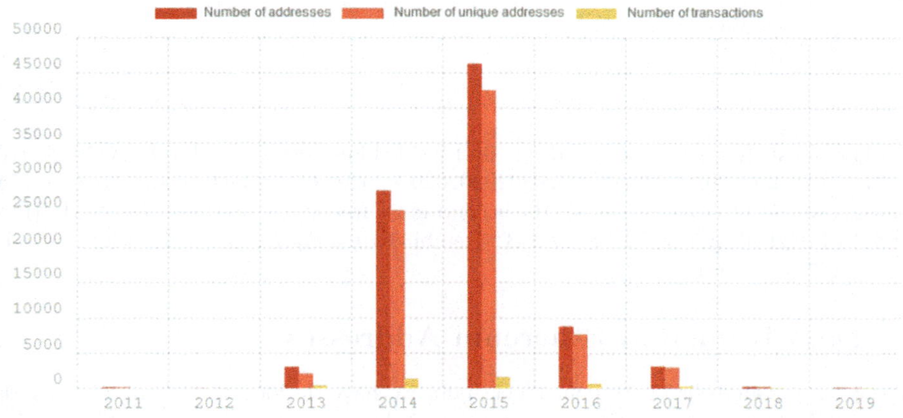

Fig. 1. Quantity of identified addresses and transactions.

The total number of burned Satoshis is 385,923,796, almost 4 BTC. The detailed analysis of the quantity of Satoshis belonging to particularly identified addresses is presented in Table 2. To the most frequent values store on addresses belong: 548 Satoshis - 23799 addresses, 546 Satoshis - 18499 addresses, 5500 Satoshis - 14937 addresses, 5480 Satoshis - 12835 addresses, 1000 Satoshi - 1939 addresses. The transaction with the largest identified amount (42000000 Satoshis) concerns the address dated 2012-01-19: 18hJphGPYctV7DTmkywRMXxCYAA5xef96b.

Table 2. Quantity of addresses having output value in defined ranges

Output value range [Satoshi]	Quantity	Total value [Satoshi]
0–1000	45201	25563374
1001–10000	33788	175249154
10001–100000	1506	35386288
100001–1000000	51	10760690
1000001–42000000	90	138964290
Total	80636	385923796

Next analysis shows a number of transactions in which a particular address is involved (Table 3). To the most frequent quantity of transaction where identified addresses are involved in, what is expected, is one (77179 addresses), then two (2220 addresses), then three (488 addresses). This analysis proves that in a majority of cases, those addresses are created only for one usage.

Table 3. Number of addresses involved in transactions in defined ranges

Number of transactions for an address	Quantity of addresses
1	77179
2–10	3307
11–100	149
101–	1

The most frequent used address is 13vs7FtkinfdQEr1QNw4EJyLq2Q2fkgVqW (hash160 = 2020202020202020202020202020202020202020) which is involved in 775 transactions (1,067,234 Satoshis), the second most frequently used address (1AFJHY-BkdzXbFiHgPSRquYB6P2DdbdnrYB - embii################) participates in 72 transactions.

7 Data Hidden in Ethereum Addresses

Based on our analysis on Bitcoin blockchain, we have decided to a similar approach to research data hidden in Ethereum addresses. The process of creation ETH address from private key is less complex and could be presented as a three steps algorithm:

Step 1. Generation of a public key corresponding to the private key, with the usage of the same as in Bitcoin secp256k1 elliptic curve and ECDSA algorithm[6].

```
6e145cce f1033dea 239875dd 00dfb4fe e6e3348b 84985c92 f1034446
83bae07b
83b5c38e 5e2b0c85 29d7fa3f 64d46daa 1ece2d9a c14cab94 77d042c8
4c32ccd0
```

Step 2. Calculating Keccak-256[7] hash for the result of Step 1.

```
2a5bc342 ed616b5b a5732269 001d3f1e f827552a e1114027 bd3ecf1f
086ba0f9
```

Step 3. Only the last 20 bytes (least significant bytes) are used as an ETH address.

```
001d3f1e f827552a e1114027 bd3ecf1f 086ba0f9
```

We have analysed all ETH addresses existing in Ethereum blockchain up to the end of the 2020 year. For all addresses, only those with at least twenty ASCII characters were selected. The value (20) was determined experimentally. We have identified only 60 addresses, a few examples are presented in (Table 4) and the whole table on the project THIEA [12] website.

[6] The same as in Bitcoin.

[7] It is worth to mention that Keccak-256 used in Ethereum implementation is not the finalized FIPS-202 SHA-3 standard. The implementation differences are slight, with padding parameters, but they are significant in that both algorithms produce different hash outputs.

Table 4. Sample ETH addresses with hidden text content

ETH address	Hidden content
0x30786f726967696e70726f746f636f6c2e657468	0xoriginprotocol.eth
0x49206c6f7665206d7920486f6e65792042656172	I love my Honey Bear
0x68747470733a2f2f6d656d6272616e612e696f2f	https://membrana.io/
0x6d79657468657277616c6c65742e636f6d206973	myetherwallet.com is
0x756e712e63727970746f40676d61696c2e636f6d	unq.crypto@gmail.com

8 Conclusion

Although in the official Bitcoin documentation states that "Storing arbitrary data in the blockchain is still a bad idea; it is less costly and far more efficient to store non-currency data elsewhere.", the usage of its blockchain as a "forever" storage was quite popular in the past. The peak in the number of hidden data in the first half of 2015 where the BTC price was below 250 USD, and a marginal number of such a transaction in 2018 and 2019 proves that burning bitcoins for storing not-transnational data is correlated with the BTC price. Hiding data in ETH addresses is incidental, mainly because the limitation, up to twenty, in the number of chars.

Nonetheless, users decided to burn about 4 BTC and 7 ETH to store some data, which could be look over on the websites THIB(E)A [11,12] presenting results of this research.

9 Limitations

There are currently three Bitcoin address formats in use: P2PKH with the char '1' at the beginning (e.g. 1JK6sLKgBSwpa5rDdPKehhvcHwWdeqvqBK), P2SH type starting with the char '3' and Bech32 type which begins with 'bc1' (e.g. bc1qwqdg6squsna38e46795at95yu9atm8azzmyvckulcc7kytlcckxswvvzej). Analysis and results of this paper are limited to P2PKN addresses only.

10 Supporting Information

Block 138725

```
...
   ASCII BERNANKE
:'::.:::::.:::.::.:
: :.: ' ' ' ' : :'.
:.:      _.--    '.:
 :    _,^"   "^x,   :
'  x7'         '4,
   ^             ^^

 XX7            4XX
 XX              XX
 Xl ,xxx,   ,xxx,XX
( ' _,+o, | ,o+,"
 4   "-^' X "^-'" 7
 l,     ( ))     ,X
 :Xx,_ ,xXXXxx,_,XX
  4XXiX'-___-'XXXX'
   4XXi,_   _iXX7'
  , '4XXXXXXXXX^ _,
  Xx,   ""^^^XX7,xX
W,"4WWx,_ _,XxWWX7'
Xwi, "4WW7""4WW7',W
TXXWw, ^7 Xk 47 ,WH
:TXXXWw,_ "), ,wWT:
::TTXXWWW lXl WWT:
----END TRIBUTE----
```

Block 322916

```
"A new scientific tr
uth does not triumph
 by convincing its o
pponents and making
them see the light,
but rather because i
ts opponents eventua
lly die, and a new g
neration grows up t
hat is familiar with
 it." -- Max Planck
```

Block 325061

```
Married 01/04/2015__
====================
=====++++==++++=====
====++===++===++====
===++====++====++===
===++==========++===
====++========++====
=====++======++=====
======++====++======
=======++==++=======
========++++========
=========++=========
```

Block 322834

```
#!/usr/bin/env pytho
n3 # # This file is
placed in the public
 domain. # # CryptoG
raffiti tool # # Req
uires python-bitcoin
lib-v0.2.1 # # https
://github.com/petert
odd/python-bitcoinli
b # # pip install py
thon-bitcoinlib  # U
sage: # # ./cryptogr
affiti.py <textfile>
  import collections
 import sys  from bi
tcoin.rpc import Pro
xy from bitcoin.wall
et import *  DUST =
0.000055  addrs = co
llections.OrderedDic
t() with open(sys.ar
gv[1], 'rb') as fd:
    while True:
    b = fd.read(20)
         if not b:
           break
```

References

1. Bartoletti, M., Bellomy, B., Pompianu, L.: A journey into bitcoin metadata. J. Grid Comput. **17**(1), 3–22 (2019)
2. Bartoletti, M., Pompianu, L.: An analysis of bitcoin op_return metadata. In: International Conference on Financial Cryptography and Data Security, pp. 218–230. Springer (2017)
3. Furneaux, N.: Investigating Cryptocurrencies: Understanding, Extracting, and Analyzing Blockchain Evidence. Wiley (2018)
4. Kondor, D., Pósfai, M., Csabai, I., Vattay, G.: Do the rich get richer? an empirical analysis of the bitcoin transaction network. PloS One **9**(2), e86197 (2014)
5. Maesa, D.D.F., Marino, A., Ricci, L.: Uncovering the bitcoin blockchain: an analysis of the full users graph. In: 2016 IEEE International Conference on Data Science and Advanced Analytics (DSAA), pp. 537–546. IEEE (2016)

6. Matzutt, R., Hiller, J., Henze, M., Ziegeldorf, J.H., Müllmann, D., Hohlfeld, O., Wehrle, K.: A quantitative analysis of the impact of arbitrary blockchain content on bitcoin. In: International Conference on Financial Cryptography and Data Security, pp. 420–438. Springer (2018)
7. Meiklejohn, S., Pomarole, M., Jordan, G., Levchenko, K., McCoy, D., Voelker, G.M., Savage, S.: A fistful of bitcoins: characterizing payments among men with no names. In: Proceedings of the 2013 Conference on Internet Measurement Conference, pp. 127–140 (2013)
8. Nakamoto, S.: Bitcoin: A peer-to-peer electronic cash system. Technical repor, Manubot (2019)
9. Ober, M., Katzenbeisser, S., Hamacher, K.: Structure and anonymity of the bitcoin transaction graph. Future Internet **5**(2), 237–250 (2013)
10. Reid, F., Harrigan, M.: An analysis of anonymity in the bitcoin system. In: Security and Privacy in Social Networks, pp. 197–223. Springer (2013)
11. Rodwald, P.: Thiba - thinks/texts hidden in bitcoin addresses. https://rodwald.pl/thiba. Accessed 31 Dec 2020
12. Rodwald, P.: Thiea - thinks/texts hidden in ethereum addresses. https://rodwald.pl/thiea. Accessed 31 Dec 2020
13. Rodwald, P., Sobolewski, W., Rodwald, M.: Deanonymizing of bitcoin cryptocurrency users. Bull. Military Univ. Technol. **68**(1), 51–77 (2019). https://doi.org/10.5604/01.3001.0013.1466
14. Ron, D., Shamir, A.: Quantitative analysis of the full bitcoin transaction graph. In: International Conference on Financial Cryptography and Data Security, pp. 6–24. Springer (2013)
15. Spagnuolo, M., Maggi, F., Zanero, S.: Bitiodine: extracting intelligence from the bitcoin network. In: International Conference on Financial Cryptography and Data Security, pp. 457–468. Springer (2014)
16. Sward, A., Vecna, I., Stonedahl, F.: Data insertion in bitcoin's blockchain. Ledger **3**, 1–23 (2018)
17. Ziegeldorf, J.H., Matzutt, R., Henze, M., Grossmann, F., Wehrle, K.: Secure and anonymous decentralized bitcoin mixing. Futur. Gener. Comput. Syst. **80**, 448–466 (2018)

The Reliability and Operational Analysis of ICT Equipment Exposed to the Impact of Strong Electromagnetic Pulses

Adam Rosiński(✉), Jacek Paś, Marek Szulim, and Jarosław Łukasiak

Faculty of Electronics, Military University of Technology, Gen. Sylwestra Kaliskiego 2, 00-908 Warsaw, Poland
adam.rosinski@wat.edu.pl

Abstract. By conducting a reliability and operational analysis of ICT devices exposed to the impact of strong electromagnetic pulses, it can be concluded that they function within variable environmental conditions. In the course of operation, they should be in the state of fitness. This is influenced by numerous factors. One of them is the reliability of ICT subsystems forming them. The second factor is the rationalization of the operation process. One of the most important issues is also the degree of susceptibility and resistance of an ICT device against the impact of strong electromagnetic pulses.

Safety level-increasing solutions are applied in order to improve the resistance of ICT devices to strong electromagnetic pulses. However, they should be arranged taking into account the criteria of functioning effectiveness and the implementation legitimacy.

The research paper suggests a model, which distinguishes the full availability, partial availability and fault states. It also adopts specified transitions between the aforementioned states, which correspond to the actual operating processes of an ICT device exposed to strong electromagnetic pulses.

The developed model can be applied when modelling the impact of implemented solutions (i.a., technical, functional and organizational ones) on the level of safety in terms of the impact of strong electromagnetic pulses on an analysed ICT device. Such an approach enables a numerical comparison of various variants of the protection against the impact of strong electromagnetic pulses and the selection of a rational one.

Keywords: Operation · Modelling · Strong electromagnetic pulses

1 Introduction

By conducting a reliability and operational analysis of ICT devices exposed to the impact of strong electromagnetic pulses, it can be concluded that they function within variable environmental conditions [4, 6]. In the course of operation, they should be in the state of fitness. This is influenced by numerous factors. One of them is the reliability of ICT subsystems forming them [3]. The second factor is the rationalization of the operation

W. Zamojski et al. (Eds.): DepCoS-RELCOMEX 2021, AISC 1389, pp. 380–390, 2021.
https://doi.org/10.1007/978-3-030-76773-0_37

process [8]. One of the most important issues is also the degree of susceptibility and resistance of an ICT device against the impact of strong electromagnetic pulses [1]. Therefore, the selection of appropriate methods for the protection against the impact of electromagnetic interference on the analysed ICT device requires a broader approach than just an electromagnetic compatibility analysis [9, 19, 20, 29].

Safety level-increasing solutions are applied in order to improve the resistance of ICT devices to strong electromagnetic pulses [7, 24]. However, they should be ordered taking into account the criteria of functioning effectiveness and the implementation legitimacy [22, 27].

ICT devices exposed to strong electromagnetic pulses are challenged with numerous requirements that must be satisfied. The most important ones include miniaturization, low electricity consumption [35], appropriate functionality [14, 18, 32, 33] and reliability [10, 12, 13], resistance to vibrations [2, 15], as well as the quality of information [7, 30, 34]. One of the outcomes of satisfying these requirements is the decreased difference between the level of useful signals and the level of interference generated by interference sources [11, 16]. This is why determining the safety level for ICT devices exposed to the impact of strong electromagnetic pulses is important. The considerations in this regard are presented in this paper.

The operational process models of ICT devices exposed to strong electromagnetic pulses show their behaviour in specified conditions of use [21, 25, 31]. It is usually assumed that this model shows changes of the states undergoing over a considered time interval. In the course of the analysis, the number of reliability and operational states that an electronic device can stay in is usually a finite set. In order to neutralize electronic systems, it is required that electromagnetic pulses characterized by high power levels, exceeding the permissible levels of resistance, strength and suitability specified for the given elements and components constituting a given technical object, occur in a relatively short time. Electronic components and devices that make up a given electronic system do not accumulate electromagnetic interference inside. The influence of the electromagnetic pulse causes, for example, the phenomenon of an overvoltage in these devices, which with a certain time constant τ disappears in various circuits of this device. This may cause damage to certain electronic components. The occurrence of an infinite number of disturbing electromagnetic pulses of small amplitude, lower than the levels of strength and resistance, will not cause the system inability. Induced signal amplitudes or a constant level of the disturbing electromagnetic field, resulting from the so-called electrosmog does not cause significant changes in currents, voltages or working and utility signals. It depends on the objectives and the adopted accuracy of the analysis.

The ICT devices in question, exposed to the impact of strong electromagnetic pulses, from a reliability and operation perspective can remain in the following states:

- full availability (full safety, full fitness),
- partial availability (safety hazard, partial fitness),
- fault (safety unreliability, unfitness).

A state of full availability (full safety, full fitness) is the basic state for an ICT device. A state of partial availability (safety hazard, partial fitness) is a reversible transitional state, when there is a possibility to counteract its effects. The transition of an ICT device from a state of fault (safety unreliability, unfitness) means the occurrence of a dangerous situation and inability to achieve the assumed objective. If there is no possibility for counteracting (usually there is such a situation), the state of fault is an irreversible state.

Interfering electromagnetic pulses characterized by high power levels that propagate in the surrounding atmosphere are using various types of coupling (e.g. capacitive - C and inductive - L or through the common ground impedance). They are also the cause of the induction of currents I or voltages U in electronic components, systems and devices. These additional currents and voltages change the static operating points as well as the levels of the utility and output signals. Electronic systems are always designed with the use of various technical solutions - e.g. redundancy, shielding, etc., whose task is to minimize the degree of paralysis caused by a strong electromagnetic pulse. The authors of the article proposed an introduction to the determination of safety states of the operation process of a selected model of an electronic system, taking into account Γ coefficient. This coefficient determines the degree of influence of a strong electromagnetic pulse on individual devices of the system. In the case of the so-called electro smog, where the individual E, H components of the electromagnetic field have small values, the $\Gamma = 0$. For strong electromagnetic pulses, depending on the value of the E-M field coupling coefficient with the electronic system and the level of preventive technical measures - e.g. shielding, filters, etc. the value of this coefficient will be within the limits $< 0, 1>$. With a specific model of the electronic system, failure λ or renewal intensity values, we can determine the individual probabilities of operating states of the proposed technical object.

2 Reliability and Operational Analysis of an ICT Device, Taking Into Account the Impact of Strong Electromagnetic Pulses

In the course of the reliability and operational analysis of the ICT device, taking into account the impact of the strong electromagnetic pulses, it was adopted that the number of distinguished states was a finite state. The authors of this elaboration used a directed graph, with its vertices being the reliability and operational states and the arcs representing transitions between the states.

In the simplest case, an ICT device remain in one of the two following states:

- full availability (S_{PZ}),
- fault (S_B).

The distinguished safety states of the operation process of simple electronic systems include two operational states and three states in case of complex systems. In full ability state, the electronic system performs all functions. In the inability state, the electronic system does not perform all functions resulting from the demand defined, for example, by the manufacturer or operator. For complex electronic systems with various types of redundancy, the system can be in an intermediate state. The system performs selected functions resulting from the needs specified by the manufacturer or operator.

Applying specific solutions (i.a., technical [17], functional and organizational ones), which are to improve the safety level of an ICT device exposed to strong electromagnetic pulses results in the states of partial availability occurring within the operating process [23, 28].

Let us adopt a factor Γ determining the impact of a strong electromagnetic pulse on the analysed ICT device. Its value will fall within the range:

$$\Gamma \in \langle 0, 1 \rangle \quad (1)$$

We assume that:

- $\Gamma = 0$ for the lack of strong electromagnetic pulse impact on an ICT device (such is the case when the applied solutions completely eliminate the impact of a strong electromagnetic pulse on an ICT device),
- $\Gamma = 1$ for the impact of a strong electromagnetic pulse (such is the case when no solutions aimed at mitigating the outcomes of an ICT device exposed to the impact of a strong electromagnetic pulse were applied).

The strong electromagnetic pulse impact factor Γ on the analysed ICT device shall be used for the determination of the safety levels.

By distinguishing the partial availability levels S_{ZB1} and S_{ZB2} (corresponding to two solutions decreasing the impact of strong electromagnetic pulses on an ICT device), the relationship graph will take the form as shown in Fig. 1 (It is assumed that no repair is possible).

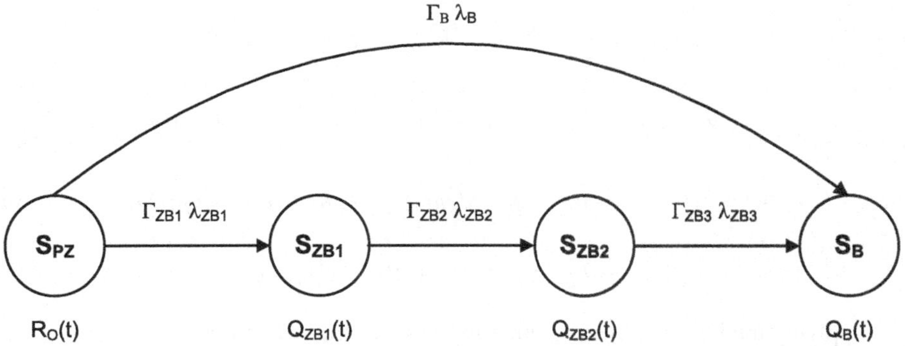

Fig. 1. Relationships within the system, taking into account the partial availability states S_{ZB1} and S_{ZB2}

Designations in the figure:

$R_O(t)$ – probability function of the system staying in the state of full availability S_{PZ},
$Q_{ZB1}(t)$ – probability function of the system staying in the state of partial availability S_{ZB1},
$Q_{ZB2}(t)$ – probability function of the system staying in the state of partial availability S_{ZB2},
$Q_B(t)$ – probability function of the system staying in the state of fault S_B,

λ_B, λ_{ZB1}, λ_{ZB2}, λ_{ZB3} – intensities of transitions between the distinguished states,
Γ_B, Γ_{ZB1}, Γ_{ZB2}, Γ_{ZB3} – strong electromagnetic pulse impact factors.

The system shown in Fig. 1 can be described by the following Chapman-Kolmogorov equations:

$$
\begin{gathered}
R_0^{'}(t) = -\Gamma_B \cdot \lambda_B \cdot R_0(t) - \Gamma_{ZB1} \cdot \lambda_{ZB1} \cdot R_0(t) \\
Q_{ZB1}^{'}(t) = \Gamma_{ZB1} \cdot \lambda_{ZB1} \cdot R_0(t) - \Gamma_{ZB2} \cdot \lambda_{ZB2} \cdot Q_{ZB1}(t) \\
Q_{ZB2}^{'}(t) = \Gamma_{ZB2} \cdot \lambda_{ZB2} \cdot Q_{ZB1}(t) - \Gamma_{ZB3} \cdot \lambda_{ZB3} \cdot Q_{ZB2}(t) \\
Q_B^{'}(t) = \Gamma_B \cdot \lambda_B \cdot R_0(t) + \Gamma_{ZB3} \cdot \lambda_{ZB3} \cdot Q_{ZB2}(t)
\end{gathered} \tag{2}
$$

Assuming the baseline conditions:

$$
\begin{gathered}
R_0(0) = 1 \\
Q_{ZB1}(0) = Q_{ZB2}(0) = Q_B(0) = 0
\end{gathered} \tag{3}
$$

and applying the Laplace transform for the set of Eqs. (2), we get the following set of linear equations:

$$
\begin{gathered}
s \cdot R_0^*(s) - 1 = -\Gamma_B \cdot \lambda_B \cdot R_0^*(s) - \Gamma_{ZB1} \cdot \lambda_{ZB1} \cdot R_0^*(s) \\
s \cdot Q_{ZB1}^*(s) = \Gamma_{ZB1} \cdot \lambda_{ZB1} \cdot R_0^*(s) - \Gamma_{ZB2} \cdot \lambda_{ZB2} \cdot Q_{ZB1}^*(s) \\
s \cdot Q_{ZB2}^*(s) = \Gamma_{ZB2} \cdot \lambda_{ZB2} \cdot Q_{ZB1}^*(s) - \Gamma_{ZB3} \cdot \lambda_{ZB3} \cdot Q_{ZB2}^*(s) \\
s \cdot Q_B^*(s) = \Gamma_B \cdot \lambda_B \cdot R_0^*(s) + \Gamma_{ZB3} \cdot \lambda_{ZB3} \cdot Q_{ZB2}^*(s)
\end{gathered} \tag{4}
$$

Applying the inverse transformation to the set of Eqs. (4), we get:

$$
R_0(t) = e^{-(\Gamma_B \cdot \lambda_B + \Gamma_{ZB1} \cdot \lambda_{ZB1}) \cdot t} \tag{5}
$$

$$
Q_{ZB1}(t) = \Gamma_{ZB1} \cdot \lambda_{ZB1} \cdot \left[\frac{e^{-(\Gamma_B \cdot \lambda_B + \Gamma_{ZB1} \cdot \lambda_{ZB1}) \cdot t} - e^{-\Gamma_{ZB2} \cdot \lambda_{ZB2} \cdot t}}{\Gamma_{ZB2} \cdot \lambda_{ZB2} - \Gamma_B \cdot \lambda_B - \Gamma_{ZB1} \cdot \lambda_{ZB1}} \right] \tag{6}
$$

$$
\begin{gathered}
Q_{ZB2}(t) = \Gamma_{ZB1} \cdot \lambda_{ZB1} \cdot \Gamma_{ZB2} \cdot \lambda_{ZB2} \cdot \\
\cdot \left[\begin{array}{c}
\frac{e^{-(\Gamma_B \cdot \lambda_B + \Gamma_{ZB1} \cdot \lambda_{ZB1}) \cdot t}}{(\Gamma_B \cdot \lambda_B + \Gamma_{ZB1} \cdot \lambda_{ZB1} - \Gamma_{ZB3} \cdot \lambda_{ZB3}) \cdot (\Gamma_B \cdot \lambda_B + \Gamma_{ZB1} \cdot \lambda_{ZB1} - \Gamma_{ZB2} \cdot \lambda_{ZB2})} - \\
- \frac{e^{-\Gamma_{ZB2} \cdot \lambda_{ZB2} \cdot t}}{(\Gamma_B \cdot \lambda_B + \Gamma_{ZB1} \cdot \lambda_{ZB1} - \Gamma_{ZB2} \cdot \lambda_{ZB2}) \cdot (\Gamma_{ZB2} \cdot \lambda_{ZB2} - \Gamma_{ZB3} \cdot \lambda_{ZB3})} + \\
+ \frac{e^{-\Gamma_{ZB3} \cdot \lambda_{ZB3} \cdot t}}{(\Gamma_{ZB2} \cdot \lambda_{ZB2} - \Gamma_{ZB3} \cdot \lambda_{ZB3}) \cdot (\Gamma_B \cdot \lambda_B + \Gamma_{ZB1} \cdot \lambda_{ZB1} - \Gamma_{ZB3} \cdot \lambda_{ZB3})}
\end{array} \right]
\end{gathered} \tag{7}
$$

$$Q_B(t) = \frac{\Gamma_B \cdot \lambda_B}{\Gamma_B \cdot \lambda_B + \Gamma_{ZB1} \cdot \lambda_{ZB1}} \cdot \left[1 - e^{-(\Gamma_B \cdot \lambda_B + \Gamma_{ZB1} \cdot \lambda_{ZB1}) \cdot t}\right] +$$

$$+\Gamma_{ZB1} \cdot \lambda_{ZB1} \cdot \Gamma_{ZB2} \cdot \lambda_{ZB2} \cdot \Gamma_{ZB3} \cdot \lambda_{ZB3} \cdot$$

$$\cdot \left[\begin{array}{c} \frac{-e^{-(\Gamma_B \cdot \lambda_B + \Gamma_{ZB1} \cdot \lambda_{ZB1}) \cdot t}}{(\Gamma_B \cdot \lambda_B + \Gamma_{ZB1} \cdot \lambda_{ZB1}) \cdot (\Gamma_B \cdot \lambda_B + \Gamma_{ZB1} \cdot \lambda_{ZB1} - \Gamma_{ZB3} \cdot \lambda_{ZB3}) \cdot (\Gamma_B \cdot \lambda_B + \Gamma_{ZB1} \cdot \lambda_{ZB1} - \Gamma_{ZB2} \cdot \lambda_{ZB2})} + \\ + \frac{e^{-\Gamma_{ZB2} \cdot \lambda_{ZB2} \cdot t}}{(\Gamma_B \cdot \lambda_B + \Gamma_{ZB1} \cdot \lambda_{ZB1} - \Gamma_{ZB2} \cdot \lambda_{ZB2}) \cdot \Gamma_{ZB2} \cdot \lambda_{ZB2} \cdot (\Gamma_{ZB2} \cdot \lambda_{ZB2} - \Gamma_{ZB3} \cdot \lambda_{ZB3})} - \\ - \frac{e^{-\Gamma_{ZB3} \cdot \lambda_{ZB3} \cdot t}}{(\Gamma_B \cdot \lambda_B + \Gamma_{ZB1} \cdot \lambda_{ZB1} - \Gamma_{ZB3} \cdot \lambda_{ZB3}) \cdot (\Gamma_{ZB2} \cdot \lambda_{ZB2} - \Gamma_{ZB3} \cdot \lambda_{ZB3}) \cdot \Gamma_{ZB3} \cdot \lambda_{ZB3}} + \\ + \frac{1}{(\Gamma_B \cdot \lambda_B + \Gamma_{ZB1} \cdot \lambda_{ZB1}) \cdot \Gamma_{ZB2} \cdot \lambda_{ZB2} \cdot \Gamma_{ZB3} \cdot \lambda_{ZB3}} \end{array} \right] \tag{8}$$

Based on the above expressions (5 ÷ 8), the probability values for an ICT device exposed to strong electromagnetic pulses staying in the distinguished states can be determined, taking into account the factor Γ. This enables evaluating the impact of various solutions (i.a., technical, functional and organizational) on the safety levels of an ICT device exposed to strong electromagnetic pulses. This makes it possible to select rational methods for protecting against strong electromagnetic pulses.

3 Modelling of the ICT Device Operation Process Including Electromagnetic Interfere

The simulation and computer research provides an opportunity to relatively fast determine the influence of changes in reliability and operational parameters of individual elements on the entire system.

By using the computer assistance, it is possible to carry out the calculations that allow to determine the probability values of the ICT device staying in the state of full availability S_{PZ}. Such a procedure is shown in the following example (it was assumed that no dedicated solutions were used to reduce the effects of electromagnetic interference on the ICT device, and therefore $\Gamma = 1$).

Example
The following values describing the analysed system are assumed:

- test duration – 1 year (value of this time is given in the units as hours [h]):

$$t = 8760\ [h]$$

- intensity of a transition from the state of full availability S_{PZ} to the of partial availability S_{ZB1}:

$$\lambda_{ZB1} = 3,477078 \cdot 10^{-6} \left[\frac{1}{h}\right]$$

– intensity of a transition from the state of partial availability S_{ZB1} to the state of partial availability S_{ZB2}:

$$\lambda_{ZB2} = 2.306245 \cdot 10^{-6} \left[\frac{1}{h} \right]$$

– intensity of a transition from the state of partial availability S_{ZB2} to the state of fault S_B:

$$\lambda_{ZB3} = 1.147298 \cdot 10^{-6} \left[\frac{1}{h} \right]$$

– intensity of a transition from the state of full availability S_{PZ} to the state of fault S_B:

$$\lambda_B = 1.147298 \cdot 10^{-6} \left[\frac{1}{h} \right]$$

The above values of damage intensity (marked with the symbol λ) were estimated on the basis of observation of the operation process of real systems [5, 26]. ICT device equipment is subject to a pre-aging process in production plants. Therefore, an exponential distribution model was used.

As a final result, we obtain (Table 1):

Table 1. Calculation results.

State	Value
R_O	0.9603
Q_{ZB1}	0.02954899
Q_{ZB2}	0.00030051
Q_B	0.0098505

If, for the data adopted in the above example, we assume the use of dedicated solutions reducing the effects of electromagnetic interference on the ICT device, which result in a change in the value of all coefficients of the impact of electromagnetic interference on the value $\Gamma_{ZB1} = \Gamma_{ZB2} = \Gamma_{ZB3} = \Gamma_B = 0.5$, then the calculations will result in (Table 2):

If, for the data adopted in the above example, we assume the use of dedicated solutions reducing the effects of electromagnetic interference on the ICT device, which result in a change in the value of all coefficients of the impact of electromagnetic interference on the value $\Gamma_Z B1 = \Gamma_Z B2 = \Gamma_Z B3 = 0.5$ and $\Gamma_B = 0.25$, then the calculations will result in (Table 3):

Table 2. Calculation results ($\Gamma_{ZB1} = \Gamma_{ZB2} = \Gamma_{ZB3} = \Gamma_B = 0.5$).

State	Value
R_O	0.97994897
Q_{ZB1}	0.01500026
Q_{ZB2}	0.00007602
Q_B	0.00497475

Table 3. Calculation results ($\Gamma_{ZB1} = \Gamma_{ZB2} = \Gamma_{ZB3} = 0.5$ and $\Gamma_B = 0.25$).

State	Value
R_O	0.98241428
Q_{ZB1}	0.01501909
Q_{ZB2}	0.00007608
Q_B	0.00249055

The presented reliability-operational analysis of the ICT device, taking into account electromagnetic interference, allows for numerical assessment of different types of (technical and organisational) solutions. As a result, they can be implemented in order to minimise the impact of electromagnetic interference on the system operation.

4 Conclusion

The summarizing outcome of the aforementioned deliberations was a model, which distinguished between the states of full availability S_{PZ}, partial availability states S_{ZB1} and S_{ZB2}, and a fault state S_B. It also adopts specified transitions between the aforementioned states, which correspond to the actual operating processes of an ICT device exposed to strong electromagnetic pulses.

The developed model, which takes into account the safety levels of an ICT device, can be applied when modelling the impact of implemented solutions (i.a., technical, functional and organizational ones) on the level of safety in terms of the impact of strong electromagnetic pulses on an analysed device. Such an approach enables a numerical comparison of various variants of the protection against the impact of strong electromagnetic pulses and the selection of a rational one.

Acknowledgments. The work was supported by the Polish National Centre for Research and Development within the project "Methods and ways of protection and defence against HPM impulses" pending within strategic project: "New weaponry and defense systems of directed energy".

References

1. Badyor, M.P., In'kov, Y.M.: Electromagnetic compatibility of a traction power supply system and infrastructure elements in areas with high traffic. Russian Electrical Engineering **85**(8), 488–492 (2014).
2. Burdzik, R., Konieczny, Ł., Figlus, T.: Concept of on-board comfort vibration monitoring system for vehicles. In: Mikulski, J. (ed.) Activities of Transport Telematics, pp. 418–425. Springer, Heidelberg (2013)
3. Caban, D., Walkowiak, T.: Dependability analysis of hierarchically composed system-of-systems. In: Zamojski, W., Mazurkiewicz, J., Sugier, J., Walkowiak, T., Kacprzyk, J. (eds), Contemporary Complex Systems and Their Dependability. DepCoS-RELCOMEX 2018, pp. 113–120. Springer (2018).
4. Charoy, A.: Interference in electronic devices. WNT, Warsaw (1999).
5. Chmielińska, J., Kuchta, M., Kubacki, R., Dras, M., Wierny, K.: Selected methods of electronic equipment protection against electromagnetic weapon. Przegląd elektrotechniczny **1**, 1–8 (2016)
6. Dras, M., Kałuski, M., Szafrańska, M.: HPM pulses – disturbances and systems interaction – basic issues. Przegląd elektrotechniczny **11**, 11–14 (2015)
7. Dudek, E.: The Concept of DMAIC Methodology Application for Diagnostics of Potential Incompatibilities in Aeronautical Data Request Process. Diagnostyka **19**(4), 33–38 (2018)
8. Duer, S., Zajkowski, K., Płocha, I., Duer, R.: Training of an artificial neural network in the diagnostic system of a technical object. Neural Comput. Appl. **22**(7), 1581–1590 (2013)
9. Dziubinski, M., Drozd, A., Adamiec, M., Siemionek, E.: Electromagnetic interference in electrical systems of motor vehicles. In: Scientific Conference On Automotive Vehicles And Combustion Engines (KONMOT 2016), Book Series: IOP Conference Series-Materials Science and Engineering, 148, pp. 1–11 (2016)
10. Jin, T.: Reliability engineering and service. John Wiley & Sons (2019)
11. Kaniewski, P., Gil, R., Konatowski, S.: Estimation of UAV position with use of smoothing algorithms. Metrology and Measurement Systems **24**(1), 127–142 (2017)
12. Kierzkowski, A., Kisiel, T.: Airport security screeners reliability analysis. In: Proceedings of the IEEE International Conference on Industrial Engineering and Engineering Management IEEM 2015, pp. 1158–1163. Singapore (2015).
13. Kierzkowski, A., Kisiel, T.: Simulation model of security control system functioning: A case study of the Wroclaw Airport terminal. Journal of Air Transport Management 64(B), 173–185 (2016).
14. Kornaszewski, M., Chrzan, M., Olczykowski, Z.: Implementation of new solutions of intelligent transport systems in railway transport in Poland. In: Communications in Computer and Information Science, pp. 282–292. Springer (2017).
15. Kostrzewski, M.: Analysis of selected acceleration signals measurements obtained during supervised service conditions – study of hitherto approach. Journal of Vibroengineering **20**, 1850–1866 (2018)
16. Łabowski, M., Kaniewski, P.: Motion compensation for unmanned aerial vehicle's synthetic apperture radar. In: Proc. Signal Processing Symposium SPSympo 2015, pp.184–188. Debe, Poland (2015).
17. Lheurette, E. (ed.): Metamaterials and Wave Control. ISTE and Wiley (2013).

18. Losurdo, F., Dileo, I., Siergiejczyk, M., Krzykowska, K., Krzykowski, M.: Innovation in the ICT infrastructure as a key factor in enhancing road safety. A multi-sectoral approach. In: Selvaraj, H., Chmaj, G., Zydek, D. (eds.), Proceedings 25th International Conference on Systems Engineering ICSEng 2017, pp. 157–162. IEEE Computer Society Conference Publishing Services (CPS), USA, Las Vegas (2017).
19. Ogunsola, A., Mariscotti, A.: Electromagnetic compatibility in railways. Springer-Verlag, Analysis and management (2013)
20. Ott, H.W.: Electromagnetic compatibility engineering. Wiley (2009).
21. Paś, J., Rosiński, A., Chrzan, M.: Białek, K. Reliability-Operational Analysis of the LED Lighting Module Including Electromagnetic Interference. IEEE Transactions on Electromagnetic Compatibility **62**(6), 2747–2758 (2020).
22. Paś, J., Rosiński, A., Łukasiak, J., Szulim, M.: The Impact of Strong Electromagnetic Pulses on the Operation Process of Electronic Equipment and Systems Used in Intelligent Buildings. In: Zamojski, W., Mazurkiewicz, J., Sugier, J., Walkowiak, T., Kacprzyk, J. (eds.) DepCoS-RELCOMEX 2019. AISC, vol. 987, pp. 383–392. Springer, Cham (2020). https://doi.org/10.1007/978-3-030-19501-4_38
23. Paś, J., Rosiński, A., Szulim, M., Łukasiak, J.: Modelling the Safety Levels of ICT Equipment Exposed to Strong Electromagnetic Pulses. In: Zamojski, W., Mazurkiewicz, J., Sugier, J., Walkowiak, T., Kacprzyk, J. (eds.) DepCoS-RELCOMEX 2019. AISC, vol. 987, pp. 393–401. Springer, Cham (2020). https://doi.org/10.1007/978-3-030-19501-4_39
24. Paś, J., Rosiński, A., Wiśnios, M., Majda-Zdancewicz, E., Łukasiak, J.: Electronic security systems. Introduction to the laboratory. Military University of Technology, Warsaw (2018)
25. Paś, J., Rosinski, A.: Selected issues regarding the reliability-operational assessment of electronic transport systems with regard to electromagnetic interference. Eksploatacja i Niezawodnosc – Maintenance and Reliability 19(3), 375–381 (2017).
26. Przesmycki, R., Wnuk, M.: Susceptibility of IT devices to HPM pulse. International Journal of Safety and Security Engineering **8**(2), 223–233 (2018)
27. Rosiński, A., Paś, J., Łukasiak, J., Szulim, M.: Exploitation of electronic systems in building objects exposed to impact of strong electromagnetic pulses. In: Beer, M., Zio, E., (eds.), Proceedings of the 29th European Safety and Reliability Conference (ESREL), pp. 3320–3325. Research Publishing, Singapore (2019).
28. Rosiński, A., Paś, J., Szulim, M., Łukasiak, J.: Determination of safety levels of electronic devices exposed to impact of strong electromagnetic pulses. In: Beer, M., Zio, E., (eds.), Proceedings of the 29th European Safety and Reliability Conference (ESREL), pp. 818–825. Research Publishing, Singapore (2019).
29. Rosiński, A., Paś, J., Szulim, M., Łukasiak, J.: Issue of assessment of the strong electromagnetic pulses impact on the functioning of selected electronic devices. Journal Of KONBiN **49**(2), 121–138 (2019)
30. Siergiejczyk, M., Krzykowska, K.: Some Issues of Data Quality Analysis of Automatic Surveillance at the Airport. Diagnostyka **1**(15), 25–29 (2014)
31. Siergiejczyk, M., Rosiński, A., Paś, J.: Electromagnetic interference issue in safety systems applied at airports. Journal Of KONBiN **49**(1), 299–312 (2019)
32. Stawowy, M., Kasprzyk, Z.: Identifying and simulation of status of an ICT system using rough sets. In: Zamojski, W., Mazurkiewicz, J., Sugier, J., Walkowiak, T., Kacprzyk, J. (eds), Proc. of the Tenth International Conference on Dependability and Complex Systems DepCoS-RELCOMEX 2015, pp. 477–484. Springer (2015).
33. Stawowy, M., Siergiejczyk, M.: Application and simulations of uncertainty multilevel models to ensure the ITS services. In: Walls, L., Revie, M., Bedford, T. (eds.), Risk, Reliability and Safety: Innovating Theory and Practice: Proceedings of ESREL 2016, pp. 601–605. CRC Press/Balkema (2017).

34. Stawowy, M.: Comparison of uncertainty models of impact of teleinformation devices reliability on information quality. In: Nowakowski, T., Młyńczak, M., Jodejko-Pietruczuk, A., Werbińska–Wojciechowska, S. (eds.), Safety and Reliability: Methodology and Applications - Proceedings of the European Safety and Reliability Conference ESREL 2014, pp. 2329–2333. CRC Press/Balkema, London (2015).
35. Suproniuk, M., Skibko, Z., Stachno, A.: Diagnostics of some parameters of electricity generated in wind farms. Przegląd Elektrotechniczny **95**(11), 105–108 (2019)

Using ASMD-FSMD Technique for Digital Device Design

Valery Salauyou(✉)

Bialystok University of Technology, Wiejska 45A, 15-351 Bialystok, Poland

Abstract. The paper proposes a new technique designing digital devices based on finite state machines with datapath (FSMD), when the functioning of the device is described in the form a chart of an algorithm state machine with datapath (ASMD). The ASMD-FSMD technique is compared to the traditional approach when synchronous multipliers are implemented on a field programmable gate array (FPGA). The ASMD-FSMD technique, compared to the traditional approach, allows in most cases to reduce the area (for some examples on 47%) and significantly increase speed (for some examples by a factor 2.96). In addition, using the ASMD-FSMD technique allows to reduce the design time by a factor 5–6 and increase the design reliability (according to our estimates by a factor 6).

Keywords: Finite State Machines with Datapath (FSMD) · Algorithm State Machine with Datapath (ASMD) · Field Programmable Gate Array (FPGA) · Design methodology · Verilog

1 Introduction

Traditionally, the designed digital device is usually represented in the form of a datapath and a control unit (controller), which are designed separately. The datapath is typically presented as a set of standard function blocks (registers, buses, multiplexers, etc.), and the control unit as a finite state machine (FSM).

In [1], the control unit and the datapath are proposed to be combined together and presented as a finite state machine with datapath (FSMD). The FSMD model has quickly become popular. In [2], the FSMDs for synchronous and asynchronous designs are presented. The FSMD model proved to be very convenient for testing the equivalence of the two designs obtained from synthesis or various design transformations [3, 4]. In [5], the digital system is proposed to be presented as an FSMD network, which leads to the implementation of racing-free hardware. The overall FSMD model is not always convenient when designing specific applications. In [6], formal FSMD models are presented for both the processor architecture and the ASIC (application-specific integrated circuit) architecture. In [7], the FSMD model with synchronous memory accesses is offered. In [8], the FSMD model for array-handling is discussed. The decomposition of FSMD to reduce the power of digital systems is presented in [9, 10]. A comparison of the efficiency of FSMs and FSMDs is considered in [11].

W. Zamojski et al. (Eds.): DepCoS-RELCOMEX 2021, AISC 1389, pp. 391–401, 2021.
https://doi.org/10.1007/978-3-030-76773-0_38

Due to its visualization, the algorithm state machine (ASM) charts have become widespread to represent FSM behavior. The ASMs were first proposed in [12] as an alternative to state diagrams. In [13], PROM-, FPLA- and multiplexer-based implementations of the ASMs is considered. The minimization method of the number of ASM vertices is presented in [14]. In [15], the ABELITE tool for ASM-based controller synthesis is described. Traditionally, the ASM charts were used to represent an algorithm of the control unit. In [16], the ASM is used to describe both the behavior of the control unit and the register operations that are performed in the datapath. Such an ASM is called an algorithmic state machine with datapath (ASMD). Recently, ASMD charts have been increasingly used in FPGA designs: when implementing industrial control systems [17]; to implement the asin function using the CORDIC algorithm [18]; during hardware implementation of cryptographic algorithm AES [19]; when designing a universal asynchronous receiver transmitter (UART) [20], etc.

This paper proposes the technique ASMD-FSMD for designing digital devices on FPGA, based on the Verilog hardware description language. The use of the ASMD-FSMD technique is illustrated by the design example of a synchronous multiplier.

2 Traditional Approach for Synchronous Multiplier Implementation

A simple school multiplication algorithm performs an arithmetic operation of multiplying two binary unsigned numbers P = A*B. Let the width of the binary words A and B be the same and equal to N bits, then the product P will have a width of 2N bits. At the beginning of algorithm execution, P is zeroed out. In each multiplication cycle, the least significant bit of the multiplier B is checked. If B[0] = 1, then the multiplicand A is added to the product P. If B[0] = 0, then zero or nothing is added to the product P. Then the value of the multiplier B is shifted one bit to the right and the value of the product P is shifted one bit to the left. The multiplication algorithm ends after considering all bits of the multiplier B.

Figure 1 shows a block diagram of the synchronous multiplier in the form of the datapath and the control unit, which is implemented as the FSM. The values of the multiplied words A and B are the inputs to the datapath. After setting the values at inputs A and B, the multiplication process is started by asserting the *run* signal. The termination of the multiplication process is indicated by asserting the *done* signal. The P product and the *done* signal are generated at the outputs of the datapath. In addition, the datapath generates the *roll* signal, which is the output of the modulo counter and indicates the termination of the multiplication process.

The FSM generates the following control signals: *clr* - to reset the registers of the datapath; *load* - to load values of the multiplied words *A* and *B* into the datapath registers; *ena* - to allow shift operations. The datapath includes the shift register ra to the left by 2N bits for storing the multiplicand A, the shift register rb to the right by N bits for storing the multiplier B, the register rp by 2N bits for storing the product P, the bus multiplexer 2–1 by 2N bits, the adder by 2N bits, and the modulo counter that generates the *roll* signal indicating the termination of the multiplication process.

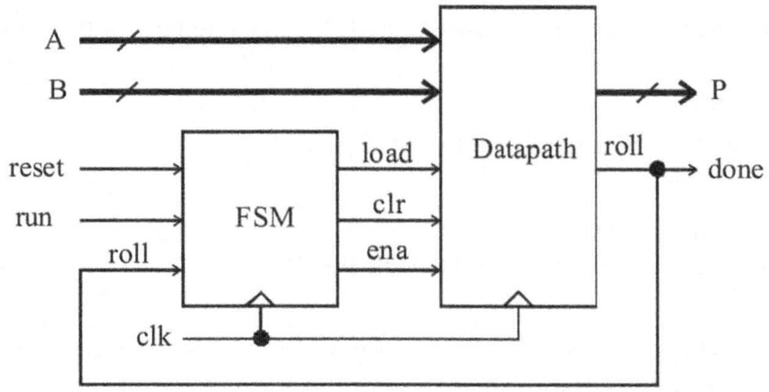

Fig. 1. Block diagram of synchronous multiplier in the form of the datapath and the FSM.

To implement the multiplier in FPGA, all components of the multiplier are described in Verilog. The detailed description of the datapath components and the FSMs for the implementation of a synchronous multiplier in the case of the traditional approach is given in [21].

3 ASMD-FSMD Technique of the Digital Device Design

The algorithmic state machine (ASM) chart is intended to visually describe the behavior of the state machine and it is an oriented graph [2] containing vertices of three types: a state box (a rectangle), a decision box (a rhombs), and a conditional output box (an oval). The state box determines the state of the FSM. In the case of FSM Moore, outputs that are equal to one in this state are written within the state box.

The decision box is the branching point of the FSM algorithm. The test condition is recorded in the decision box. The outputs of the decision box are denoted by values 0 and 1, which correspond to transitions in the case of a zero (false) or a unit (true) value of the condition check result. In the conditional output boxes, the Mealy FSM outputs are written, which are equal to one on the given transition.

The main building element of the ASM chart is an *ASM block*. The ASM block consists of only one state box and can have several decision boxes and conditional output boxes. The ASM chart is a composition of interconnected ASM blocks. The following ASM building rules must be observed: each output of any ASM vertex can be connected to only one input of another vertex; the feedbacks are not allowed inside the ASM blocks.

Let the ASM chart with datapath (ASMD) is an ASM chart in which any register operations that are valid in Verilog can be written in rectangles and ovals and any Verilog logical expressions can be checked in the decision box. The actions described within the ASMD block are performed during one clock cycle. A FSM whose behavior is described by ASMD is *a state machine with datapath* (*FSMD*), and the FSMD design technique based on ASMD is *an ASMD-FSMD technique.*

The formal description of the ASMD-FSMD technique can be presented as follows.

Algorithm.
The ASMD-FSMD technique of the digital device design.

1. The FSMD states are determined.
2. The ASMD block is constructed for each FSMD state.

2.1. In the ASMD decision boxes, the logical functions are written, the values of which are checked in this state.

2.2. For the Moore FSMD, the operations are written in the state box that are performed on the content of the registers in this state.

2.3. For the Mealy FSMD, the operations are written in the conditional output boxes that are performed on the content of the registers on these transitions.

3. The ASMD blocks are connected to each other in accordance with the algorithm of the device operation. Each output of the ASMD block can be connected to only one input of this or other ASMD block.
4. If necessary, the ASMD is modified according to [22].
5. The Verilog-code of the FSMD is built directly by ASMD. In Verilog code, the variables correspond to the device registers. The logical expressions in the **if** statements correspond to the logical functions checked in the ASMD decision boxes. The actions performed in ASMD blocks are described as procedural blocks **begin…end**. The operations performed in the ASMD boxes (for Moore machines) are described first in the **begin…end** block, and the operations performed in the ovals (for Mealy machines) are described in the corresponding places of the **if** statements (possibly using the **begin…end** operator brackets).
6. The FSMD is implemented using the appropriate design tool.
7. End.

In ASMD chart, compared to the traditional approach, there is no strict division into the datapath and a control unit, and ASMD does not explicitly define the structure of the datapath. In the ASMD chart, compared to the algorithmic description, there are explicitly defined FSMD states that can correspond to the states of the control unit. This allows to bind the algorithm of the functioning of the designed device to clock signals. Therefore, according to the ASMD chart, the developer can track how many clock cycles each branch of the algorithm performs. In addition, the presented ASMD-FSMD technique allows to implement both Mealy and Moore machines, as well as a combined model of Mealy and Moore machines.

Figure 2 shows the ASMD that corresponds to the Mealy FSMD to implement multiplication algorithm for our example.

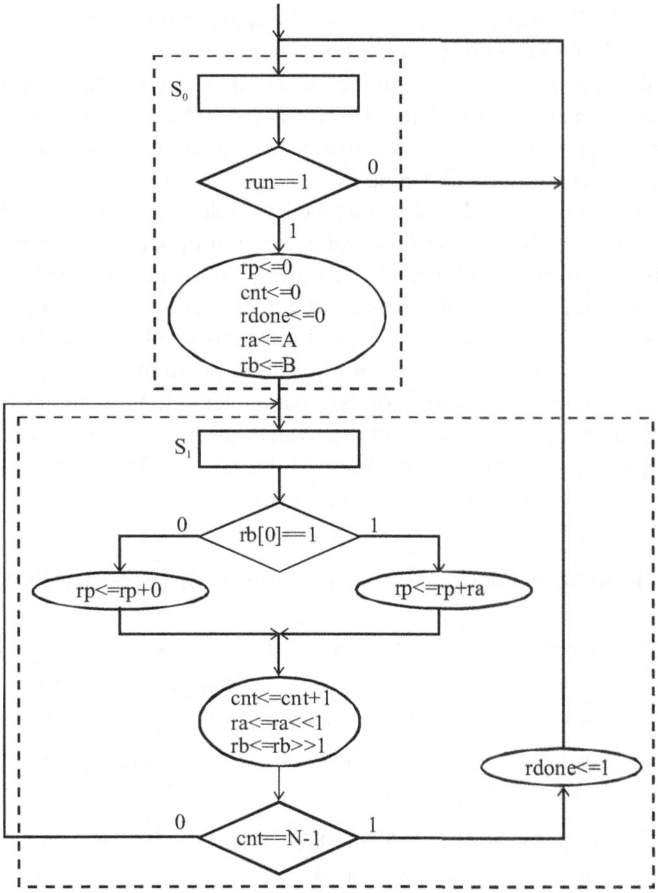

Fig. 2. ASMD chart for the mealy FSMD.

4 Experimental Research

Experimental studies were performed by the Quartus Prime tool version 18.1 with default synthesis parameters for the FPGA Cyclone IV E family. The following classical multiplication methods [21] have been investigated:

- algorithm a, when the lower significant bit of the multiplier is checked, the multiplier is shifted to the right and the multiplicand is shifted to the left;
- algorithm b, when the most significant bit of the multiplier is checked, the multiplier and the partial products are shifted to the right;
- algorithm c, when the most significant bit of the multiplier is checked, the multiplier and the partial products are shifted to the left;
- algorithm d, when the lower significant bit of the multiplier is checked, the multiplier and the partial products are shifted to the right.

In addition, the Booth multiplication algorithm and the modified Booth multiplication algorithm (a total of 6 multiplication methods) were study.

Each multiplication method was implemented using the traditional approach in the form of the datapath and a control unit, as well as using the ASMD-FSMD technique (a total of 12 projects). Projects were investigated with a width of input words of 4, 8, 16, 32, 64 and 128 bits (a total of 72 examples).

The results of experimental studies are given in Table 1, where mult_a_N, mult_b_N, mult_c_N, and mult_d_N are projects implementing multiplication algorithms a, b, c, and d respectively; booth_N and mbooth_N are projects implementing the Booth algorithm and the modified Booth algorithm; N is the width of input words of multipliers in bits; L_T and L_A are the number of FPGA logical elements used (the implementation cost or the area); t_T and t_A are the time of multiplication operation in nanoseconds; L_T/L_A and t_T/t_A are relations of the corresponding parameters. Table 1 shows that using the ASMD-FSMD technique for most examples reduces the area and increases performance, the area decreases in some cases by a factor 1.47, i.e. by 47% (mult_a_128), and the performance increases by a factor 2.96 (mult_d_16).

Table 1. ASMD-FSMD technique results compared to the traditional approach

Examples	L_T	L_A	L_T/L_A	tT	t_A	t_T/t_A
mult_a_4	36	28	1.29	24.46	17.49	1.40
mult_a_8	66	49	1.35	46.96	31.15	1.51
mult_a_16	123	90	1.37	80.08	84.08	0.95
mult_a_32	236	172	1.37	219.03	191.79	1.14
mult_a_64	464	335	1.38	598.42	582.23	1.03
mult_a_128	919	627	**1.47**	4136.21	3969.23	1.04
mult_b_4	36	26	1.38	23.42	16.39	1.43
mult_b_8	64	49	1.30	50.47	47.81	1.06
mult_b_16	121	89	1.36	89.32	106.86	0.84
mult_b_32	234	170	1.38	218.65	178.43	1.23
mult_b_64	461	332	1.39	603.31	580.67	1.04
mult_b_128	913	653	1.40	4093.18	3916.21	1.05
mult_c_4	35	35	1.00	40.25	20.22	1.99
mult_c_8	60	65	0.92	93.11	47.82	1.95
mult_c_16	112	138	0.81	183.55	97.72	1.88
mult_c_32	208	270	0.77	452.46	238.75	1.98
mult_c_64	405	540	0.75	1267.69	639.39	1.98
mult_c_128	793	1006	0.79	8216.11	4200.59	1.96

(continued)

Table 1. (*continued*)

Examples	L_T	L_A	L_T/L_A	tT	t_A	t_T/t_A
mult_d_4	32	29	1.10	41.77	20.67	2.02
mult_d_8	58	48	1.21	82.48	36.41	2.27
mult_d_16	98	75	1.31	159.26	53.86	**2.96**
mult_d_32	176	141	1.25	410.87	158.33	2.59
mult_d_64	338	271	1.25	940.51	355.62	2.64
mult_d_128	663	528	1.26	4851.80	2178.69	2.23
booth_4	44	38	1.16	37.11	30.37	1.22
booth_8	82	69	1.19	80.73	52.54	1.54
booth_16	152	125	1.21	111.02	111.21	1.00
booth_32	301	238	1.26	263.55	228.73	1.15
booth_64	590	464	1.27	794.75	654.96	1.21
booth_128	1168	917	1.27	4474.04	4245.59	1.05
mbooth_4	65	68	0.96	25.72	18.25	1.41
mbooth_8	122	138	0.88	36.89	31.59	1.17
mbooth_16	236	278	0.85	68.81	62.48	1.10
mbooth_32	467	492	0.95	150.53	124.27	1.21
mbooth_64	925	973	0.95	413.81	404.31	1.02
mbooth_128	1842	1935	0.95	2396.53	2236.75	1.07

The arithmetic mean values of the L_T/L_A and t_T/t_A relations for each multiplication method are shown in Table 2, where mid is the mean value of the parameter. Table 2 shows that the use of the ASMD-FSMD technique reduces, on average, the area for most multiplication methods (excluding mult_c and mbooth methods), as well as increases the

Table 2. Mean arithmetic values of the L_T/L_A and t_T/t_A relations for different multiplication methods

Method	mid(L_T/L_A)	mid(t_T/t_A)
mult_a	1.37	1.18
mult_b	1.37	1.11
mult_c	0.84	1.96
mult_d	1.23	2.45
booth	1.23	1.20
mbooth	0.92	1.16
mid	1.16	1.51

performance for all methods. Particularly noticeable is the increase in speed for mult_c and mult_d methods.

All multiplier designs were created by one developer in order to estimate the design time under the traditional approach and using the ASMD-FSMD technique. The time in minutes spent developing each project is shown in Table 3, where DT_T is the development time in the case of a traditional approach; DT_A is development time in case of ASMD-FSMD technique; DT_T/DT_A is relations of the corresponding parameters; mid is the average value.

Table 3. The development time by the ASMD-FSMD technique compared to the traditional approach

Method	DT_T	DT_A	DT_T/DT_A
mult_a	188	33	5.70
mult_b	176	29	6.07
mult_c	172	28	6.14
mult_d	237	43	5.51
booth	167	27	6.19
mbooth	195	35	5.57
mid			5.86

Table 3 shows that the use of the ASMD-FSMD technique significantly (by a factor 5.86) reduces the design development time. This is because the ASMD-FSMD technique does not require the development (code writing and debugging) of each datapath component, the datapath itself, the control unit, and the main design module.

The reliability of designs developed using the Verilog language can be estimated by the number of modules (M) in the design, the total number of ports of all modules (P) and the number of code lines (S). These parameters for the studied designs are given in Table 4, where M_T and M_A are the number of modules for the traditional approach and when using the ASMD-FSMD technique, P_T and P_A are the total number of module ports; S_T and S_A are the total number of code lines; mid is the average value.

Let the reliability of the design be determined by the following formula:

$$Rel = k1(M) + k2(P) + k3(S), \tag{1}$$

where k_1, k_2, and k_3 are some numerical coefficients. Let $k_1 = k_2 = k_3 = 1/3$. It follows from Table 4 and formula (1) that the use of the ASMD-FSMD technique increases the reliability of the designs, on average, by a factor 6.63.

Thus, studies have shown that the use of the ASMD-FSMD technique for the design of synchronous multipliers reduces the area (in some cases by 47%), significantly increases the speed (in some cases by a factor 2.69), significantly reduces the development time (approximately by a factor 6), as well as increases the reliability of the designs (according to our estimate by a factor 6.63).

Table 4. The parameters that determine the reliability of projects

Method	M_T	M_A	M_T/M_A	P_T	P_A	P_T/P_A	S_T	S_A	S_T/S_A
mult_a	9	1	9	48	7	6.86	144	53	2.72
mult_b	9	1	9	49	7	7.00	143	54	2.65
mult_c	9	1	9	53	7	7.57	151	56	2.70
mult_d	9	1	9	53	7	7.57	151	59	2.56
booth	10	1	10	52	7	7.43	148	56	2.64
mbooth	12	1	12	60	7	8.57	174	58	3.00
mid			9.67			7.50			2.71

Note the differences between the ASMD-FSMD technique and known digital device design approaches. The ASMD-FSMD technique allows:

- describe both Mealy and Moore machines (in [16] ASMD describe only Moore machines), as well as combined models of Mealy and Moore machines;
- combine in one ASMD chart the description of the datapath and the control unit (in [16] ASMD separately describe the datapath and control unit);
- in one ASMD block, describe several blocking assignment statements to the same register, which in some cases greatly increases the performance of the design;
- describe the digital device with a single Verilog module, significantly reducing development time and improving design reliability.

5 Conclusions

Comparison of the ASMD-FSMD technique with the traditional approach shows that the use of the ASMD-FSMD technique can in most cases reduce the area (for some examples by 47%) and significantly increases the performance (for some examples by a factor 2.96). In addition, the use of the ASMD-FSMD methodology significantly reduces the development time (approximately by a factor 5–6) and increases the design reliability (according to our estimates more than by a factor 6).

ASMD-FSMD requires further improvement by:

- using design language features, for example, applying multiple blocking assignment statements to the same register in a single ASMD block;
- developing methods to improve ASMD performance, for example, by reducing the number of states in algorithm cycles;
- developing ASMD optimization methods to improve the performance and the reliability and to reduce the area, etc.

The use of ASMD-FSMD for high-level design of complex projects is also promising.

Acknowledgements. The present study was supported by a grant WZ/WI-IIT/4/2020 from Bialystok University of Technology and founded from the resources for research by Ministry of Science and Higher Education.

References

1. Gajski, D.D., Dutt, N.D., Wu, A.C., Lin, S.Y.: High-Level Synthesis: Introduction to Chip and System Design. Kluwer, Boston, USA (1992)
2. Auletta, R., Reese, B., Traver, C.A: Comparison of synchronous and asynchronous FSMD designs. In: Proceedings of the IEEE International Conference on Computer Design (ICCD 1993), pp. 178–182. IEEE, Cambridge, MA, USA (1993)
3. Karfa, C., Sarkar, D., Mandal, C.: Verification of datapath and controller generation phase in high-level synthesis of digital circuits. IEEE Trans. CAD **29**(3), 479–492 (2010)
4. Hu, J., Wang, G., Chen, G., Wei, X.: Equivalence checking of scheduling in high-level synthesis using deep state sequences. IEEE Access **7**, 183435–183443 (2019)
5. Schaumont, P., Shukla, S., Verbauwhede, I.: Design with race-free hardware semantics. In: Proceedings of the Design Automation & Test in Europe Conference, p. 6. IEEE, Munich, Germany (2006)
6. Zhu, J., Gajski, D.D.: A unified formal model of ISA and FSMD. In: Proceedings of the Seventh International Workshop on Hardware/Software Codesign, pp. 121–125. IEEE, Rome, Italy (1999)
7. Kavvadias, N., Masselos, K.: Automated synthesis of FSMD-based accelerators for hardware compilation. In: Proceedings of the 23rd International Conference on Application-Specific Systems, Architectures and Processors, pp. 157–160. IEEE, Delft, Netherlands (2012)
8. Banerjee, K., Sarkar, D., Mandal, C.: Extending the FSMD framework for validating code motions of array-handling programs. IEEE Trans. CAD **33**(12), 2015–2019 (2014)
9. Hwang, E., Vahid, F., Hsu, Y.C.: FSMD functional partitioning for low power. In: Proceedings of the Conference on Design, automation and test in Europe, p. 7-es. IEEE, Munich, Germany (1999)
10. Abdullah, A.C., Ooi, C.Y., Ismail, N.B., Mohammad, N.B.: Power-aware through-silicon-via minimization by partitioning finite state machine with datapath. In: Proceedings of the IEEE International Symposium on Circuits and Systems, pp. 1942–1945. IEEE, Montreal, QC, Canada (2016)
11. Babakov, R., Barkalov, A., Titarenko, L.: Research of efficiency of microprogram final-state machine with datapath of transitions. In: Proceedings of the 14th International Conference the Experience of Designing and Application of CAD Systems in Microelectronics, pp. 203–206. IEEE, Lviv, Ukraine (2017)
12. Clare, C.R.: Designing Logic Systems Using State Machines. McGraw-Hill Book Company, New York, USA (1973)
13. Green, D.H., Chughtai, M.A.: Use of multiplexers in direct synthesis of ASM-based designs. IEEE Proc. E-Comput. Digital Tech. **133**(4), 194–200 (1986)
14. Baranov, S.: Minimization of algorithmic state machines. In: Proceedings of the 24th EUROMICRO Conference, pp. 176–179. IEEE, Vasteras, Sweden (1998)
15. Jenihhin, M., Baranov, S., Raik, J., Tihhomirov, V.: PSL assertion checkers synthesis with ASM based HLS tool ABELITE. In: Proceedings of the 13th Latin American Test Workshop, pp. 1–6. IEEE, Quito, Ecuador (2012)
16. Ciletti, M.D.: Advanced Digital Design with the Verilog HDL. Prentice Hall of India, New Delhi (2005)

17. Martin, P., Bueno, E., Rodriguez, F.J., Saez, V.: A methodology for optimizing the FPGA implementation of industrial control systems. In: Proceedings of the 35th Annual Conference of IEEE Industrial Electronics, pp. 2811–2816. IEEE, Porto, Portugal (2009)
18. Saha, A., Ghosh, A., Kumar, K.G.: FPGA implementation of arcsine function using CORDIC algorithm. AMSE J.-AMSE IIETA Publ.-2017-Ser. Adv. A **54**(2), 197–202 (2017)
19. Burciu, P.: An efficient (low resources) modular hardware implementation of the AES algorithm. J. Electr. Eng. Electron. Control Comput. Sci. **5**(3), 1–10 (2019)
20. Sowmya, K.B., Shreyans, G., Vishnusai, R.T.: Design of UART module using ASMD technique. In: Proceedings of the Fifth International Conference on Communication and Electronics Systems, pp. 176–181. IEEE, Coimbatore, India (2020)
21. Salauyou, V.V.: Verilog language in embedded systems design on FPGA, Hotline – Telecom, Moscow, Russia (2020). (in Russian)
22. Bergamaschi, R.A.: Behavioral synthesis: an overview. In: Reis, R., Lubaszewski, M., Jess, J.A. (eds.) Design of Systems on a Chip: Design and Test, pp. 103–131. Springer, Boston, USA (2006)

Application of LSTM Networks for Human Gait-Based Identification

A. Sawicki(✉) and K. Saeed

Faculty of Computer Science, Bialystok University of Technology, Bialystok, Poland
{a.sawicki,k.saeed}@pb.edu.pl

Abstract. This paper presents a biometric person identification system based on gait captured by an accelerometer. The proposed solution includes: advanced single cycle gait detection algorithm, LSTM autoencoder-based data augmentation method, and RNN based classifier. The study presents classification performance results with augmentation as well as without augmentation for the publicly available data corpus. Reported experimental results included analysis of gait obtained on a flat even substrate in a group of 30 participants. For the 6-fold cross-validation repeated 50 times, an accuracy score of approximately 95% was obtained.

Keywords: Augmentation · Biometrics · Deep learning · IMU

1 Introduction

The purpose of biometric systems is to identify subjects based on selected behavioral or physiological features. Research on the analysis, development and deployment of this type of technology has attracted the attention of both the scientific and academic communities and industry as well. This is the result of the increasing demand for persons identification and verification [1]. Recently, the development of the field of biometrics based on gait analysis has been observed [2]. One of its advantages is the more unobtrusive nature of its operation, compared to systems based on physiological features data in the form of iris images, palm veins or fingerprint [3].

The paper presents the results of work on a behavioral biometric identification system based on human gait. The proposed solution is based on the analysis of measurement values of a triaxial accelerometer located on the right shin, using deep learning algorithms. The presented work addresses segmentation of gait cycles, augmentation using LSTM autoencoders, and classification using LSTM networks.

The experiments were conducted using a publicly available human gait database, *A database of human gait performance on irregular uneven surface* [4]. The data corpus contains a set of motion recordings of a group of 30 participants. Acquisition was carried out using 6 IMU, MTw Awinda. Datasets are provided in the form of data blocks, in each of which the study participant performs approximately several complete gait cycles. Data in this form were segmented into single gait cycles using algorithms applied in the field of a walk analysis. The use of segmentation is an approach often used in the literature

W. Zamojski et al. (Eds.): DepCoS-RELCOMEX 2021, AISC 1389, pp. 402–412, 2021.
https://doi.org/10.1007/978-3-030-76773-0_39

[5] and usually achieves higher classification accuracy compared to processing raw data blocks.

Classification was carried out using LSTM neural networks. The raw accelerometer measurement values which represented single gait cycles were subject to the inputs of the neural network. As a result of the conducted research, a biometrics system architecture was proposed, allowing the identification of participants at a level of approximately 95%.

2 Literature Review

In this chapter we provide an overview of the state-of-the-art researches related to our study, which is divided into three groups as follows.

2.1 Segmentation

Two main approaches of segmenting single gait cycles based on accelerometer signals can be distinguished. In the first one, a distance signal is generated based on the measured values of acceleration and gait pattern. In [6], the authors preliminarily determined the norm of the signals of the three sensor axes. Then, using a sliding window and the specific metric (based on correlation with gait pattern), the distance waveform was determined. Potential starting points for gait cycles began at its local minima. It should be emphasized that gait patterns were generated automatically, based on last detected gait cycle in the signals. This approach made it possible to find the gait origins, even if the gait pattern was variable over time. A major disadvantage of the algorithm is that if the first gait pattern was generated based on the disturbed first gait cycle, the whole segmentation process could be performed incorrectly.

A similar segmentation mechanism for motion capture data recorded by a Kinect device was used in [7]. The study uses a fixed hand motion pattern generated using piecewise Bezier curves. The presented approach, compared to [8], allowed to avoid searching for a reference gait/gesture cycle in the available signals and allowed to generate it in an artificial way. It is also worth noting that a distance measure based on correlation was used in [6, 7]. In the presented approach, the similarity of the signals was determined only by their shape, and the metric is independent of magnitude. In the case of gait analysis, many of the signal oscillations may have a shape similar to the expected gait pattern. In [8] the magnitude-independence of measures such as cosine similarity, Pearson's correlation was found to be a disadvantage. The authors proposed to use the Surface Similarity metric, which takes into account the similarity of the phases of the signals, but also their magnitudes.

In contrast, in the work of [9], gait segmentation was based on the analysis of negative peaks that resulted from footwear touching the floor. The application of the described algorithm may be challenging depending on the ground on which the participant performs the gait. In the case of a hard one, the accelerometer signals may have significant oscillations due to the impact of both the right and left leg hitting the ground. Which can make the detection of the beginning of cycles significantly more difficult.

In this paper, the methodology inspired by [6] is applied, with significant changes: the use of a fixed predefined gait pattern analogous to [7] and replacement of the metric considering both magnitude and shape of the signal [8].

2.2 Augmentation

Recently, we can observe the growth of data augmentation methods based on learning and signal reconstruction using various types of machine learning models. Methods of this type have so far not been of interest to researchers into behavioural biometrics, which is dominated by classical methods [10, 11].

In [10] an elementary augmentation of gait cycles was used. The algorithm worked in three steps: first adding Gaussian noise with $\sigma = 0.01$, then scaling the original waveform in the range 0.7 to 1.1, finally applying time transformations imitating sampling irregularities. In the presented approaches, the values of the passed transformations were constant and did not depend on the characteristics of the given gait cycle. In [11], on the other hand, the authors described an analytical method for generating artificial gait patterns for IMU signals. In contrast to previous works, changing deformation parameters based on the standard deviation of n-observed gait cycles were used. The solution achieved good results, its main drawback being its relatively difficult implementation.

A similar approach to that presented in [11] for generating new signals using existing data is provided by generative models including autoencoders. Solutions of this type are characterised by much easier implementation and simplicity of use.

In [12] authors used CNN variational autoencoders to augment images of the MNIST database. The model was learned on a part of the available learning dataset and generation was performed using the remaining portion of the data. The models presented were dedicated to the analysis of images only, and the possibility of directly transferring them to time series processing is limited. In contrast, in [13], the authors used LSTM autoencoders to augment motion capture data in the form of skeletal animations. The paper proposed an integrated model of the LSTM autoencoder and selected forms of recurrent networks. A major value of the paper was the description of the unique cost function used, which involves simultaneous correction of the weights of the classifier and the generation model. However, the presented approach makes it impossible to perform an independent learning process of the augmentation model. Any change in the structure of the classifier must involve a re-training, of the model responsible for the augmentation.

The present work exploits the ability of LSTM autoencoders to reconstruct given waveforms while maintaining the split of test and training data as in [12]. This approach made it possible to obtain a generative model similar to the one in [13], while neglecting its disadvantages - close connection with an unchangeable classifier. The application of the proposed technique allowed the generation of both magnitude perturbations and signal shape changes, obtaining results similar to [11].

2.3 Classification

Methods based on Deep neural networks provide state of the art solutions in the field of human gait classification.

In [14], LSTM networks were used to classify segmented gait cycles recorded by a mobile phone accelerometer. The input of the neural network was given the raw measurement signals of a triaxial accelerometer with a length of 200 samples. Adam's algorithm was used to learn the classifier and the learning was performed for 600 epochs.

A 6-fold cross-validation was used to evaluate the performance of the algorithms. The main disadvantage of the application described is the absence of regularisation methods.

On the other hand, [15] uses LSTM networks and accelerometer measurement signals for Human Activity Recognition issues. Nine IMU sensor signal data with 128 samples were used. Adam's algorithm was again used to modify the network weights and the learning process was performed for 300 epochs. The great value of the presented work is the use of regularization in the form of the L2 metric.

A different approach in the problem of person identification using deep neural networks is presented in [16]. The CNN model was used as a feature vector extractor of individual gait cycles. The network architecture was not an end-to-end solution. The determined set of features was subjected to static analysis with special attention to variation-based methods.

Eventually, a solution based on [15] was used in this work. With the network architecture adapted to the existing dataset by changing the length of the analysed time series and the number of available channels.

3 Methodology

The purpose of the biometric system developed was to identify people based on their gait, using data recorded by motion sensors. The input of the system was the measurement data of a triaxial accelerometer located on the right shin. The system consisted of blocks: "segmentation", "augmentation", and "decision".

In the "segmentation" block, fragments in which the participant remained motionless were removed from the recordings and single gait cycles were determined. In the next block "augmentation" with autoencoders of LSTM type, multiplication of existing signals in the database was performed. Finally, in the "decision" block with the recurrent LSTM type neural networks, identification of individuals was performed. The proposed processing steps are discussed in detail in the following sections.

3.1 Database Characteristics

In this research work, a publicly available corpus of human gait described in detail in [4] was used. The original database is characterized by parameters such as: group of 30 participants, use of 9 types of different surface (e.g. concrete, grass), availability of differentiated signals types (angular velocity, acceleration, magnetic field strength). It is worth emphasising that all types of measurement were transformed from standard sensors to artificial created person centered (vertical, medio-lateral, anterior-posterior) coordinate system.

Furthermore, signals used in this study were collected at 100 Hz with a de-noising filter (2nd order Butterworth low pass filter with a 6 Hz cut-off frequency), therefore there was no need for another preprocessing. Eventually, it is worth noted that the paper presents the results of preliminary tests, in which only signals collected on one type of FE (Flat Even) surface were used.

3.2 Segmentation

The original corpus contains gait samples as blocks of data instead of individual gait cycles. In order to develop a biometric system, the segmentation procedures described in the literature were adapted to the characteristics of the available signals. Processing data as single gait cycles usually achieves higher classification accuracy compared to processing raw data blocks [5].

Signal segmentation can be successfully performed using deep learning algorithms, but such an issue requires relatively large data sets, additional computing power or time. In the field of gait analysis, analytical methods are successfully applied, providing very good results [6, 7, 9].

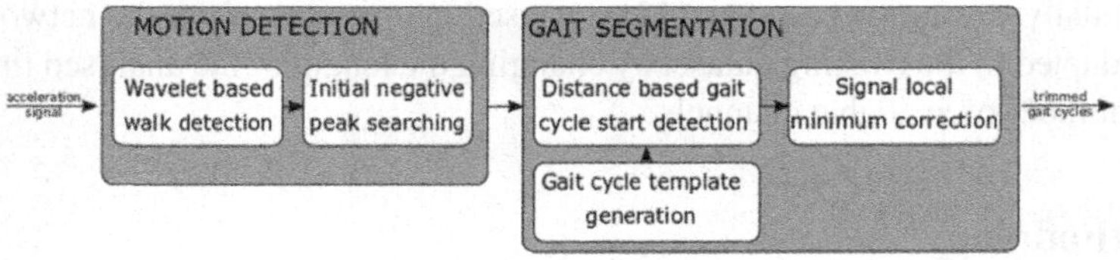

Fig. 1. Block diagram of the segmentation process

The data processing pipeline used in this paper can be divided into two parts: Motion detection and the proper gait segmentation (Fig. 1). For the first block, elementary motion detection occurred, rejecting signal fragments in which the participant remained motionless. The second block involved the actual detection of individual gait cycles. It should be noted that data acquisition with study participants was performed over a single day, with recordings made over several trials. The gait cycle in some of them was disturbed by periods of inactivity.

In the first stage of processing, an elementary segmentation was performed by trimming the data to fragments containing the correct gait. For this purpose, a method based on the analysis of wavelet transform coefficients of db10 family [17] was used. In the presented approach, gait cycles were extracted using accelerometer signals located on the right shin. Figure 2 presents the results of the algorithm, where the gray color indicates the areas considered as motion.

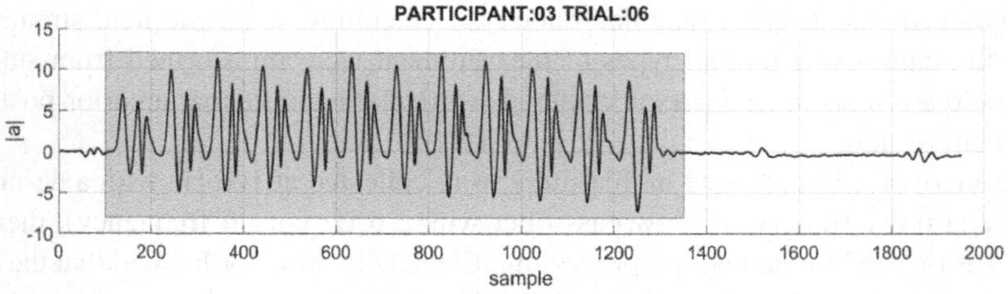

Fig. 2. Segmentation of gait cycles, grey background indicates waveform classified as gait

In the accompanying illustration, the data blocks considered as gait, did not always start with the characteristic negative pulse (which marks the beginning of gait and indicates ground touches) [9]. Therefore, additional trimming was applied so that the analyzed data block began with a negative value greater than "−2". Data blocks prepared in this way were used for further analysis and determination of individual gait cycles.

This process was carried out using a sliding window technique analogous to [6]. It aimed to determine the similarity of the current signal fragment along with the gait pattern. In the presented study, as in [7], an invariant, artificially generated pattern was used. For this purpose, expert knowledge [9] about the expected shape of the pattern was used.

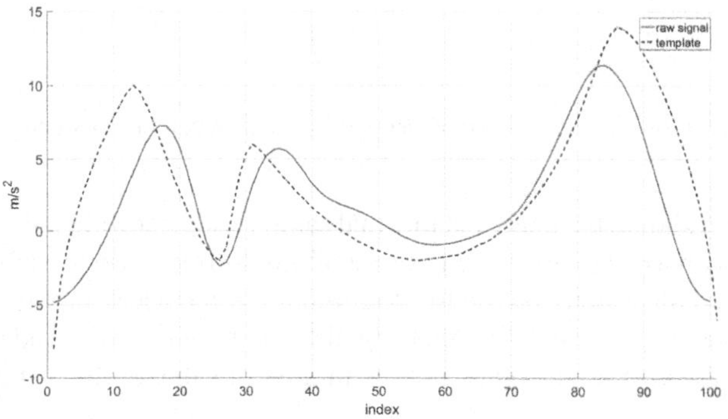

Fig. 3. Comparison of the actual signal and the determined pattern

Figure 3 presents the generated gait patterns and the actual reference signal. Characteristic large negative values starting and ending the gait, are associated with the touch of the leg (equipped with the sensor) on the ground. While a significant peak inside the waveform is associated with the another limb touching the ground [9]. The gait pattern was created from typical vertical axis acceleration signals.

In the following steps, the sliding window technique was used to determine the origins of the gait cycles. The similarity of the signal to the presented pattern was determined using the surface similarity measure [8] according to Eq. (1).

$$dist = \frac{\|\boldsymbol{x} - \boldsymbol{y}\|}{\|\boldsymbol{x}\| + \|\boldsymbol{y}\|} \tag{1}$$

In Fig. 4. A) the sensor X-axis acceleration is presented in B) the distance signal between the acceleration signal and the generated pattern is shown. The presumed starting moments of the gait cycle are at points when the distance signal takes on a characteristic negative value.

It can be observed on the attached graphic that at moments of large amplitude changes in the middle gait cycles (resulting from touching the ground by the other limb not equipped with the sensor), the distance signal assumes relatively high values An analysis based on the detection of local minima, as in position [9] could be troublesome in a such situation.

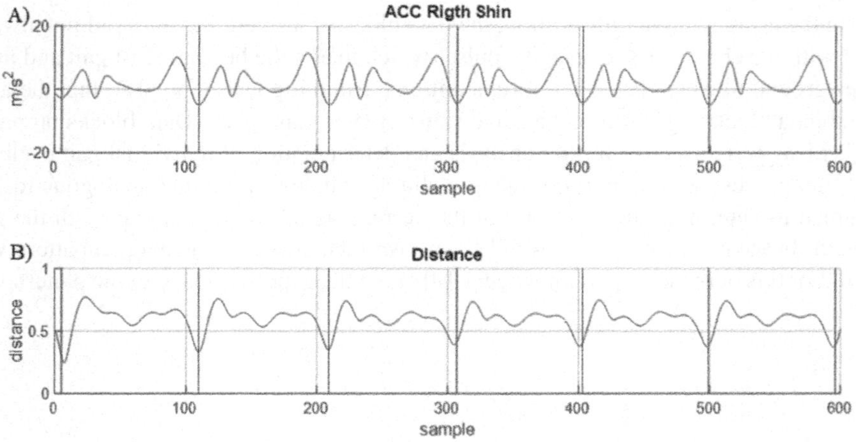

Fig. 4. Acceleration signal A), Distance signal between acceleration signal and pattern B)

Eventually, it should be noted that the final sample index at which the gait cycles start is determined with a correction. The purpose of this shift is to ensure that the beginning of the gait cycle falls exactly on the local minimum of the acceleration.

For example, the vertical line starting the fourth gait cycle indicates the local minimum of the acceleration signal, not the minimum value of the distance signal.

3.3 Data Augmentation

An augmentation method based on learning LSTM-type autoencoders of typical gait waveforms was used. This approach is ideologically similar to the methodology used in work [12] or [13]. It assumes initial training of a machine learning model, and subsequent generation of gait samples.

At the same time, it should be noted that the presented approach makes it possible to modify both the amplitude of signals and the locations of local maxima. The presented mechanism affects the motion signal in a way similar to analytical algorithms of Arbitrary Time Deformation and Stochastic Magnitude Perturbation presented in [11] (Table 1).

Table 1. LSTM-AE model architecture

# layer	Description
1	LSTM (32, input_shape = (1,10), return_seq = True)
2	LSTM (16, input_shape = (1,10), return_seq = False)
3	RepeatVector (1)
4	LSTM (16, input_shape = (1,10), return_seq = True)
5	LSTM (32, input_shape = (1,10), return_seq = True)
6	TimeDistributed (10)

Three LSTM-type network models were used to augment the gait signals of a single participant. Each of them was responsible for processing the signals of one of the sensor axes.

Appropriately prepared three-dimensional arrays were used to learn the models. The first dimension denoted Bath Size, the second dimension was 1 and represented the signal value at time *t*, and the third dimension was of length 10 and represented time-shifted signals. The LSTM AE models were trained using the Adam algorithm and the mean absolute error cost function. The network was trained for a period of 120 epochs. Experiments were carried out using the Keras and TensorFlow packages. Calculations were performed using a workstation equipped with three NVIDIA 2080 TI graphics accelerators.

For example, for participant '1' and the first of the validations, the training and test groups had 64 and 13 gait cycles respectively. Half of the available training data (32 gait cycles), were used to train the machine learning models. The generation of artificial waveforms was based on the remaining 32 gait cycles and learned model. In Fig. 5, a comparison of the original waveforms and the augmented signals is presented.

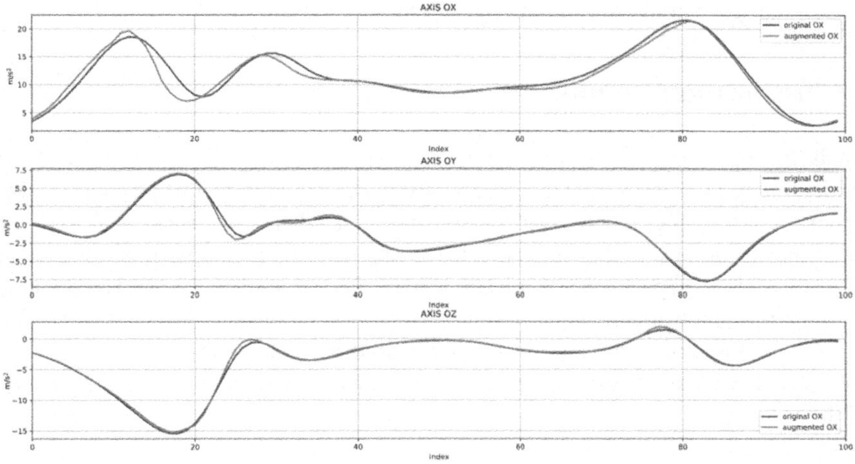

Fig. 5. Comparison of original and augmented data for three accelerometer axes

3.4 Data Splitting and Network Architecture

The study described here involved an experiment in which cross-validation was used in a leave-one-trial-out configuration. Data from one trial were used to test the classifier and data from the other five trials were used for training. The presented approach realizes the assumptions of 6-fold cross-validation. Each of the conducted experiments was repeated 50 times, which enabled the presentation of the results in the form of mean values with standard deviations.

The identification of people on the basis of segmented gait cycles was realised using a recurrent network of the LSTM type, in a "many-to-one" architecture. The input of the classifier was a 3 × 100 sequence representing acceleration signals recorded using an accelerometer located on the right shin. The output of the network was a single scalar value in the range 0 ÷ 29, representing the identification number of the experiment participant (Table 2).

Table 2. Classifier model architecture

# layer	Description
1	LSTM (32, input_shape = (100,3) activation = ReLU)
2	LSTM (32, return_sequences = False)

The model used in this study consisted of two LSTM-type cells with 32 hidden neurons. ReLU (Rectified Linear Unit) was used as the neuron activation functions. The network was trained with a learning rate of 0.0025 over a period of 300 epochs using the ADAM optimiser, and the categorical cross-entropy cost function. Experiments were conducted using 6-fold cross-validation repeated 50 times.

4 Classification Results

Signals from a triaxial accelerometer located on the right calf were used in the person identification process. The data corpus used included gait recordings collected with a group of 30 people, where each person participated in four trials. A 6-fold cross-validation was utilised in the study, in a leave-one-trial-out configuration. For each validation, 50 trials were performed, with two instances considered with signal augmentation and in the baseline version.

The classification accuracy measure was used to evaluate the performance of the developed identification system. The mean values with standard deviations are presented in Table 3.

Table 3. Accuracy metrics by validation

	#Validation	1	2	3	4	5	6	ALL
Accuracy	No Augmentation	0.93 ± 0.04	0.94 ± 0.03	0.93 ± 0.11	0.96 ± 0.03	0.93 ± 0.12	0.95 ± 0.04	0.94 ± 0.07
	Augmentation	0.93 ± 0.08	0.94 ± 0.04	0.96 ± 0.08	0.95 ± 0.07	0.95 ± 0.06	0.96 ± 0.03	0.95 ± 0.06

5 Conclusions and Future Work

The paper presents the author's system for identification of people based on their gait. The work included: an advanced algorithm for segmentation of gait cycles, a data augmentation mechanism based on LSTM autoencoder networks, and empirically selected LSTM network structure ensuring high efficiency of person classification. The presented system was characterised by high data classification accuracy (data augmentation variant) of 0.95 ± 0.06. It should be clearly emphasised that the presented results concerned a six-fold cross-validation repeated 50 times.

The data augmentation mechanism presented in this paper resulted in both an improvement and a decrease in the performance of selected validation subsets. An improvement in classification performance occurred in validation #3 (0.93 ± 0.11 to 0.96 ± 0.08), whereas for validation #4 it resulted in a decrease (0.96 ± 0.03, to 0.95 $\pm$ 0.07). However, the presented algorithm for the average of all validations allowed an increase of 1 percentage point in classification accuracy (change from 0.94 ± 0.07 to 0.95 ± 0.06).

Eventually, it is worth noting that the presented results relate to situations in which constant conditions were maintained. The individual participants recorded data during one day with constant footwear as well as constant ground conditions. Further work is planned to extend the study with data for several types of surface.

Acknowledgment. The work was also supported by grants from Bialystok University of Technology and funded with Resources for Research by the Ministry of Science and Higher Education, Poland, under Grants WZ/WI-IIT/4/2020 and WI/WI/1/2019.

References

1. Zou, Q., Wang, Y., Wang, Q., Zhao, Y., Li, Q.: Deep Learning-Based Gait Recognition Using Smartphones in the Wild, arXiv:1811.00338 (2020)
2. Wang, S., Liu, J.: Biometrics on Mobile Phone [In] Recent Applications Biometrics; InTech: Rijeka, Croatia (2011)
3. Zhao, Y., Zhou, S.: Wearable device-based gait recognition using angle embedded gait dynamic images and a convolutional neural network. Sensors **17**(3), 478 (2017)
4. Luo, Y., Coppola, S., Dixon, P., Li, S., Dennerlein, J., Hu, B.: A database of human gait performance onirregular and uneven surfaces collected by wearable sensors. Sci. Data **7**, 219 (2020)
5. Zhao, Y., Zhou, S.: Wearable device-based gait recognition using angle embedded gait dynamic images and a convolutional neural network. Sensors (Basel). **17**(3), 478 (2017)
6. Gadaleta, M., Rossi, M.: IDnet: smartphone-based gait recognition with convolutional neural net-works. Pattern Recogn. **74**(74), 25–37 (2018)
7. Sawicki, A., Zieliński, S.K.: Augmentation of Segmented Motion Capture Data for Improving Generalization of Deep Neural Networks. Computer Information Systems and Industrial Management. Lecture Notes in Computer Science, vol. 12133. Springer, (2020). https://doi.org/10.1007/978-3-030-47679-3_24
8. Ruiz, C., Pan, S., Bannis, A., Chen, X., et al.: IDrone: robust drone identification through motion actuation feedback. Proc. ACM Interact. Mob. Wearable and Ubiquitous Technol. **2**(2), 80:1–80:22 (2018)
9. Hoang, T., Choi, D., Nguyen, T.: On the instability of sensor orientation in gait verification on mobile phone. In: 12th IEEE International Joint Conference on e-Business and Telecommunications (ICETE), vol. 4, pp. 148–159 (2015)
10. Delgado-Escano, R., Castro, F.M., Cozar, J.R., Marin-Jimenez, M.J., Guil, N.: An end-to-end multi-task and fusion CNN for inertial-based gait recognition. IEEE Access **7**, 1897–1908 (2019)

11. Tran, L., Choi, D.: Data augmentation for inertial sensor-based gait deep neural network. IEEE Access **8**, 12 364–12 378 (2020)
12. Jorge, J., Vieco, J., Paredes, R., et al.: empirical evaluation of variational autoencoders for data augmentation. In: International Conference on Computer Vision Theory and Applications, pp. 96–104 (2018)
13. Juanhui, T., Hong, L., Fanyang, M., et al.: Spatial-Temporal Data Augmentation Based on LSTM Autoencoder Network for Skeleton-Based Human Action Recognition, pp. 3478–3482.
14. Watanabe, Y., Kimura, M.: Gait identification and authentication using LSTM based on 3-axis accelerations of smartphone. Proced. Comput. Sci. 176, 3873–3880 (2020). ISSN 1877–0509
15. Guillaume, C.: LSTMs for human activity recognition. Technical report material by the neuraxio company (2020). https://github.com/guillaume-chevalier/LSTM-Human-Activity-Recognition
16. Hazra, S., Roy, P., Nand, A., et al.: A pilot study for investigating gait signatures in multi-scenario applications. In: 2020 International Joint Conference on Neural Networks (IJCNN), pp. 1–10, UK (2020)
17. Barralon, P., Vuillerme, N., Noury, N.: Walk detection with a kinematic sensor: frequency and wavelet comparison. In: 2006 International Conference of the IEEE Engineering in Medicine and Biology Society, pp. 1711–1714. PMID: 17945661 (2006)

Time Based Evaluation Method of Autonomous Transport Systems in the Industrial Environment

Tomasz Serafin(✉)

Faculty of Electronics, Wroclaw University of Science and Technology, Wybrzeże Wyspiańskiego 27, 50-320 Wroclaw, Poland
tomasz.serafin@pwr.edu.pl

Abstract. Autonomous transport systems for industry are the increasing segment of the market. Scalability, efficiency and reliability are the key factors of systems assessment. This paper presents the methodology of system description and evaluation. The key parameter is a target number of passive trolleys delivered to the production line in chosen time period. Autonomous systems cooperating with people and other automation systems have important limitations which are described and assessed according to the potential impact to the planned system efficiency. This paper presents real data taken from the operational AGV system. Fusion of hardware and software elements to increase reliability and reduce disturbances are the actual engineering challenge. Author suggests the need of a full automation of such a system design and even deployment, which is his current scope of work and scientific interest.

Keywords: Autonomous transport system · AGV · AMR · Reliability · Complex system design · Industry 4.0 · Systems evaluation

1 Autonomous Transport System

AGVs (Automated Guided Vehicles) [1] with natural navigation (LIDAR [2] based) are autonomous vehicles [3] used in several industrial environments to build complex autonomous transport systems. They operate using environment maps [2] and perform production orders (POs) driving on virtual paths (VPs). AGVs carry products between stations. Stations can have several functions like pickup, drop off or can be the elements of production lines.

Autonomous vehicles have specific limitations: navigation methods, geometry, control algorithms or fleet management algorithms [4]. Human factor – mistakes of operators (wrongly placed pallets or trolleys), collisions with human-operated forklifts are critical in terms of efficiency and reliability. Fast changing environment, obstacles are also very challenging. Most of the investors deploying autonomous transport systems expect very high reliability, almost automatic diagnosis and repair of the system, easy configuration (changing of stations location, virtual paths drawing), obstacle avoidance and alternative path finding. All those aspects affect the final note of the system operation quality usually measured by:

W. Zamojski et al. (Eds.): DepCoS-RELCOMEX 2021, AISC 1389, pp. 413–424, 2021.
https://doi.org/10.1007/978-3-030-76773-0_40

- percent of properly performed production orders (defined by process),
- average time of delivery (critical in terms of reasonability of autonomous transport system deployment),
- AGVs utilization – time of AGV operation vs. staying in the waiting points (WPs),
- possible reduction of FTEs (Full Time Equivalent) – employees for the process.

For increasing the reliability and operation quality in autonomous transport systems, there is a need to use specific technical means, algorithms and take into consideration a general autonomous complex systems instability, uncertainty and their possible reduction.

2 System Definition

2.1 Layout

Layout is the place where AGVs can operate. Stations and virtual paths (VP) are defined according to the production/transport process.

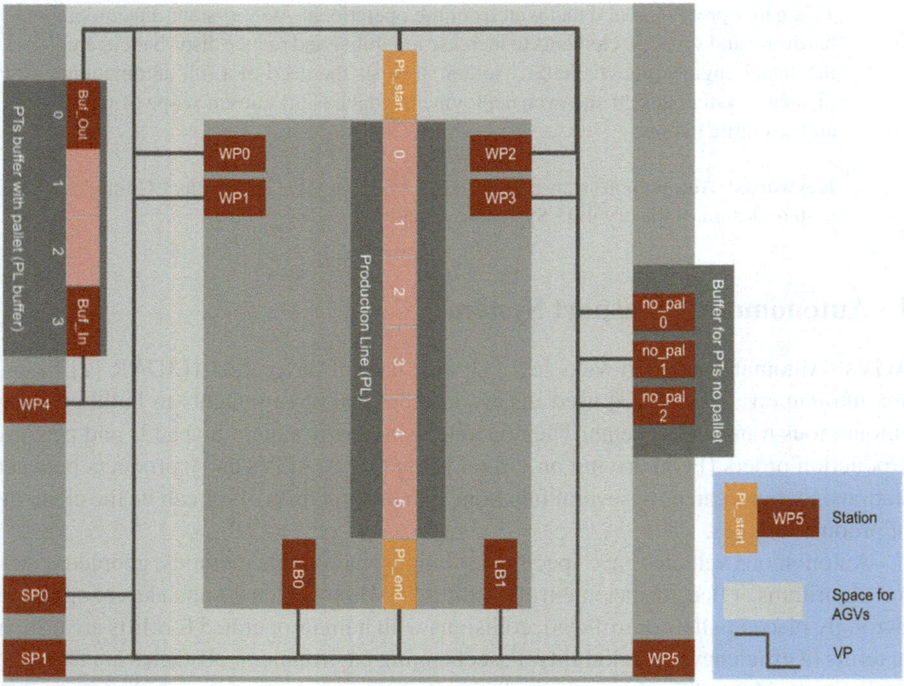

Fig. 1. Layout of the system

2.2 Passive Trolley

Every PT has rollers with a pallet on top which is moved to production line manually after docking to production line (PL).

AGV loads PT using lifting platform or pins. AGV has additional sensors detecting PT and pallet on top. It gives the information to the fleet management system (FM) if conditions are proper to start production order (PO).

Fig. 2. AGV with PT: Left – no pallet, Right – with pallet

PT can be docked to the production line or placed in a proper station – with pallet or without. PTs are carried only by AGVs – this is the key element of full automation of the intralogistics process.

2.3 Station

Stations are the points in the layout where AGV performs its actions. Possible station actions (station tasks ST): *load* – AGV loads PT using pins, *unload* – AGV unloads PT using pins, *dock* – AGV with PT docks to the production line, *undock* – AGV with PT undocks from the production line. Special type of the station is a waiting point (WP), where AGV waits for the tasks. WP can be extended and can have additional action: AGV charging in automatic charger – in this case this point changes its name to charging point (CP) – general function is similar to WP.

2.4 Virtual Path

Virtual Path (VP) is an abstract element describing the path for the AGVs. VP can be one directional or bidirectional. VPs connect stations and can be managed by fleet manager (FM) to define routes for AGVs. VPs are divided into segments (SEG). Segments are atomic elements for FM to select final route between initial station (where AGV stays) and destination station. AGVs cannot occupy segments while not performing task with selected route – the only possible location, where AGV stays without selected driving route is station (including WP).

2.5 Segment

Segment is an atomic element to select routes. Segments are defined after creating layout of VPs. Segments are the elements which are occupied by AGVs while driving. When segment is occupied by one AGV, there is no possibility to drive for any other AGV in this segment. Good practice is to define as short segments as possible. Segment has parameters: *length* – usually in meters, *average speed for AGV* – depends on AGV type (acceleration and V-max), geometric profile (curves), safety restrictions – for some segments AGVs need to drive slower.

2.6 Route

Route is a selected list of segments that allows transport from initial station to destination station. Route selection is always done by the fleet manager (FM) and depends on current traffic situation in the layout. Every route has its time of travel. Route is an element of transport task (TT) (Fig. 3).

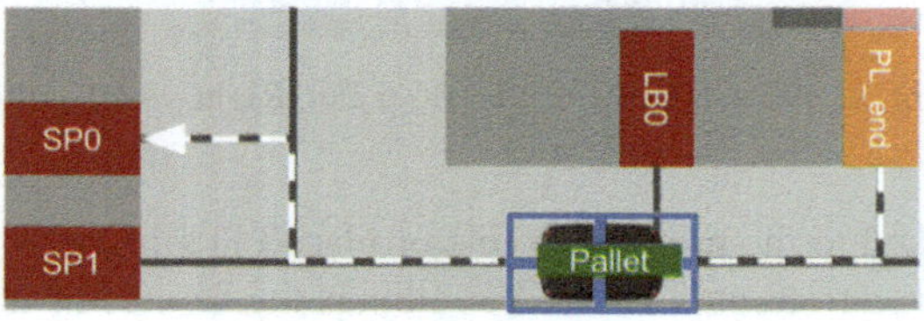

Fig. 3. Selected route

2.7 Task

In the system there are defined 2 types of tasks: *Transport Task* (TT) - from the fleet manager (FM) perspective it is an atomic entity which includes the route selected from segments according to existing layout of VPs; TT is an atomic element of production order; *Station Task* (ST) - ST is an atomic element of production order; STs types are described in Sect. 2.3.

2.8 Fleet Manager

Fleet manager manages the traffic [5, 6]. Segment is an atomic element for traffic management. Fleet manager generates TTs according to current traffic situation – calculates distance, route time. It is possible to recalculate route and select different for the existing TT. FM is also responsible for generating and performing STs.

2.9 Production Order

$$PO_x = \{ST_0, TT_0, \ldots, ST_k, TT_l\} \tag{1}$$

Production Order (PO) is the main entity and atomic element from the production process perspective. POs are managed by Order Manager (OM).

Every PO consists of defined set of TTs and STs. STs and TTs are always sequential. Every TT or ST has its own set of conditions that need to be in the proper state to allow execution. More detailed description of PO including conditions of execution (c):

$$PO_x = \{ST_0 \left| \begin{matrix} c_0 \\ c_1 \\ \dots \\ c_m \end{matrix} \right., TT_0 \left| \begin{matrix} c_0 \\ c_1 \\ \dots \\ c_n \end{matrix} \right., \dots, ST_k \left| \begin{matrix} c_0 \\ c_1 \\ \dots \\ c_o \end{matrix} \right., TT_l \left| \begin{matrix} c_0 \\ c_1 \\ \dots \\ c_p \end{matrix} \right. \} \quad (2)$$

Example condition for ST is PT presence in the station, for TT: free destination station (free means no AGV is there).

2.10 Order Manager

OM is an independent set of algorithms without logical dependency with FM. Orders can be queued if there is no free AGV to execute them. If there is a free AGV and PO can be taken to execution, order is divided into TTs and STs and they are executed in sequence. PO waiting time in a queue is one of the most important factors for the system utilization. Instances of TTs and STs to perform PO are generated and managed by FM as described in Sect. 2.8.

2.11 Operation Modes

System has 2 different operation modes which can be assigned to the stations – every station works in the selected mode. **Automatic** - PO is generated by operator using e.g. button – human factor is critical. When PO is generated in the station it will be always performed as it is configured for the triggering station. **Autonomous** - PO is generated automatically by an external system or sensors. Example: production line (PL) can call AGV when products are ready to be taken (trigger PO) or PT presence sensor detected PT occurrence and triggered PO. Besides this, AGVs can get separated TTs generated by operator from the central system level (not from station). This separated TTs are fully managed by FM, but they are not POs, so they cannot be counted to the production process statistics and are kind of disturbance. This mode of the system is called **Semi-automatic** – semi automation means no OM, but still FM.

3 Time Based System Evaluation Method (TBSE)

System operation is strictly related to the time. Especially in the situation where AGVs transport goods to production line (PL). Each delay can disturb production plan. Disturbed production plan is the consequence of autonomous transport system instability [7].

3.1 Operation Target

In the system defined in Sect. 2 the key role is feeding the PL with the proper number of PTs calculated in a time period (hour, shift, month). Second key task is taking final products on PTs from the PL. In this kind of system meeting the efficiency target (number of transports to and from PL) is the synonym of a reliable system operation.

3.2 Operation Time of PO

$$t_{PO_x} = \sum\nolimits_{i=0}^{m} t_{ST_i} + \sum\nolimits_{j=0}^{n} t_{TT_j}\ [\mathrm{s}] \tag{3}$$

Operation time of PO is the sum of STs and TTs operation times.

$$route = \sum\nolimits_{i=0}^{k} s_{seg_i}\ [m] \tag{4}$$

Route is a set of VP segments. Every segment has its lengths s.

$$t_{route} = \sum\nolimits_{i=0}^{k} t_{seg_i}\ [\mathrm{s}] \tag{5}$$

Operation time of a selected route is a sum of segment operation times.

$$t_{seg} = \frac{s_{seg}}{V_{seg}}\ [\mathrm{s}] \tag{6}$$

Segment operation time t_{seg} is calculated using segment length and average AGV velocity depending on AGV type and segment geometry (curves).

Table 1. Operational atomic time description

Operation atomic time	Description	Task type
t_{os}	Ordering station time to arrive AGV from WP - t_{route} type	**TT**
t_{od}	Docking time in an ordering station	ST
t_{oo}	Operation time in an ordering station (including manual operation)	ST
t_{oud}	Undocking time in an ordering station	ST
t_{ds}	Destination station time to travel from an ordering station - t_{route} type	**TT**
t_{dd}	Docking time in a destination station	ST
t_{do}	Operation time in a destination station (including manual operation)	ST
t_{dud}	Undocking time in a destination station	ST
t_{wp}	Time to travel from a destination station to the closest WP - t_{route} type	**TT**

Every PO has a sequence of TTs and STs. For easier calculation – Table 1 defines atomic operation times.

3.3 Example Calculation of Buf_Out PO Time – Ideal Condition

According to the layout presented in Fig. 1, station **Buf_Out** has a defined production order (Table 2):

$$t_{PO_{buf\_out}} = t_{os} + t_{oo} + t_{ds} + t_{do} \ldots \tag{7}$$

Where t_{os} is dependent on AGV selection and its route from WP to Buf_Out. t_{ds} in most of the cases will be the same (the shortest path is unique, but FM can always take a different decision of a route configuration). t_{os} and t_{ds} should include t_{pause} as a coefficient estimated for this specific PO. In Fig. 4 a green dot represents triggered PO by the operator. ST_1 finishes when the operator moves the pallet to PL and pallet sensor on AGV gets value of pallet presence: false.

Table 2. PO times

Task	Description	Time
TT_0	AGV from WP_x goes to Buf_Out (ordering station)	t_{os}
ST_0	AGV loads PT	t_{oo}
TT_1	AGV goes to PL_Start (destination station)	t_{ds}
ST_1	AGV docks, pallet is taken by the operator	t_{do}

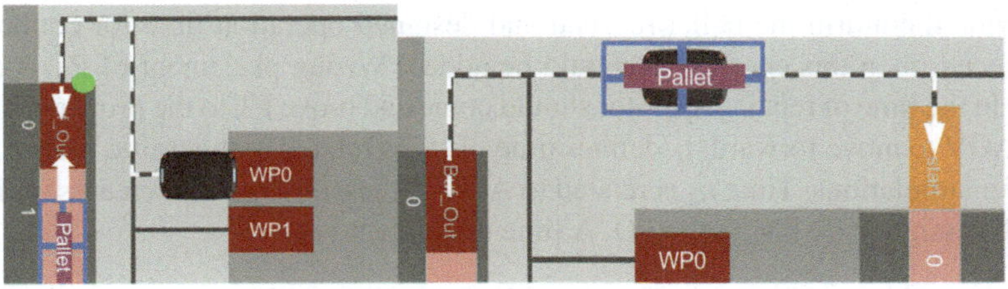

Fig. 4. TT_0ST_0; TT_1ST_1

3.4 Time Coefficients

Ideal, desired operation of the system is related to t_{PO_x} without any disturbances, that can be caused by traffic, failures, delays coming from the external systems or human factor in general. Traffic is the most impacting factor changing system stability and reliability. t_{pause} coefficient can be estimated using *VPs factor* (number of crossings, potential deadlocks, areas where there's no chance to define an alternative path, number of WPs), *number of AGVs and other objects movement* (people, forklifts). t_{oo} and t_{do} should be calculated including *human factor coefficient* (how many times operator can cause delays and what is an average time of the delay) and *external systems integration coefficient* (what is the risk of delays, when station operation is related to signals or activities of the external systems).

An initial AGV system design usually doesn't include time coefficients and is prepared for ideal conditions where delays are not present. This kind of approach is valuable to estimate maximum and theoretical system efficiency. Estimating time coefficients is difficult and until system is simple (low number of AGVs, stations and paths) can be done manually. When system complexity grows, there should be developed special algorithm calculating coefficients, in ideal conditions using some historical data from the system operation (Table 3).

Table 3. Defined disturbances for tasks

Operation time	Disturbance (t_d)	Task type
Route	Pause (t_p), Rerouting (t_{rr}), Navigation fluctuation (t_{nf}), Queue time (t_Q)	TT
Docking undocking	t_{nf}	ST
Operation	External system delay (t_{esd}), Operator delay – human factor (t_{hf})	ST
Loading unloading	Wrong PT placement (t_{wptp}), t_{nf}	ST

For every operation where AGV is on the move there is a possibility to occur t_{nf} which is e.g. related to the lowering accuracy of the navigation causing AGV stop and delay - e.g. reason: environment changes, LIDARs in the dust. t_{rr} occurs where FM decides to change the route for TT. It is difficult to predict and strictly depends on the traffic situation. t_{esd} is a time of delay generated by an external system in a station - e.g. if PL should confirm the pallet receival and designed operation time for confirmation is exceeded, then this coefficient should be added. Wrong placement of PT is a delay related to the time of repair - operator should come and move PT to the proper position to allow AGV to move forward. t_p defines time of stops related to the traffic – a theoretical value for simulations. Time t_Q is related to AGVs accessibility and increases until AGV won't be assigned by FM to the PO. A time coefficient is t_d.

$$t_{PO_x} = \sum\nolimits_{i=0}^{m} t_{ST_i} + \sum\nolimits_{j=0}^{n} t_{TT_j} + \sum\nolimits_{k=0}^{o} t_{d_k} \tag{8}$$

For statistical calculations can be used an arithmetical mean of t_d or weighted when probability of occurrence is analyzed (described in Sect. 3.5).

3.5 Time Coefficients – Probability of Occurrence

The complete model of t_{PO} calculation should include the probability of occurrence of each disturbance D - $P(D) \leq 1$. P(D) higher than 0.9 for technical disturbances is a failure of the system and should not be analyzed.

$$t_{PO_x} = \sum\nolimits_{i=0}^{m} t_{ST} + \sum\nolimits_{j=0}^{n} t_{TT} + \sum\nolimits_{k=0}^{o} t_{d_k} * P(D_k) \tag{9}$$

Reducing time coefficients by weights coming from the probability of occurrence is an easy way to balance them especially for the long-term analysis. The weighted mean can be used to simplify calculations of all POs in the system:

$$t_{PO_x} = \sum\nolimits_{i=0}^{m} t_{ST} + \sum\nolimits_{j=0}^{n} t_{TT} + \frac{\sum_{k=0}^{o} t_{d_k} * P(D_k)}{\sum_{k=0}^{o} P(D_k)} \tag{10}$$

4 TBSE of Implemented System

PO described in Sect. 3.3 is the part of a real operation performed by AGVs in an automotive production environment - Fig. 1 and Fig. 2. Feeding target of PL is *16 PTs per hour*.

4.1 Average Operation Times

Table 4. Operation time values for PO_{buf_out}

Operation time	Average value [s]
t_{OS}	51
t_{OO}	6
t_{ds}	32
t_{do}	21
t_{Q}	2

4.2 Operational Cases

Table 5. Operational cases

	Case	Description	Time [s]
1	Ideal	Only operation times from Table 4 – the maximum system efficiency with theoretical full reliability	0
2	People met on route	2 people met by AGV per route	10
3	No free AGV	All the AGVs have assigned TTs	30
4	Case 2 and 3 together	2 cases together	40
5	Rerouting	FM decided to change the route	40
6	Delay from PL	Pallet presence confirmation	10
7	Wrongly placed PT	Operator wrongly placed the PT and needs to correct it – average time value measured by the stopwatch	50
8	Navigation failure	AGV is lost – operator needs to perform positioning process	200
9	Human factor	Theoretical value for STs where operator can work slower	10
10	All	All the cases – the worst case	360

Every case described in Table 5 will be calculated using equations described in Sect. 3.5 and basic data from Table 1. Every case has values of disturbances which are measured or estimated for the real production environment.

4.3 Results

Desired $t_{PO_{buf\_out}}$ for feeding target 16 PTs per hour is 225 s. Ideal case is 112 s, which gives almost 50% of the time for potential disturbances. In ideal conditions AGV can feed line with ~32 PTs. In the worst case (All) AGV can feed line only in 48% (~7PTs). Values of the probability are selected by an experience from the real production environment (number of disturbances was counted in comparison to the number of POs proceeded). Average $t_{PO_{buf\_out}}$ without the probability factor is 209,4 s which gives 7% of a time margin in comparison to desired value (which is roughly one more PO to feed PL). Average $t_{PO_{buf\_out}}$ with probability (Sect. 3.5) is 135,9 s which gives 44% of a time margin. It means 10 spare POs to feed PL. After the assessment with probability system can be evaluated and production plans can include the proper margins. 10 spare POs is a safe margin - 44% of possible repair time. Example of the usage of such an analysis is: there is a possibility to determine how much slower AGV can be to still reach POs target for the line. In the analysis mentioned above, changing t_{os} value to 80 s and 60 s gives still 17% of the time margin. Slowing down AGVs can reduce t_P factor – more fluent drive with the reduced safety zones can allow less stops caused by the occurrence of people in the path (Fig. 5).

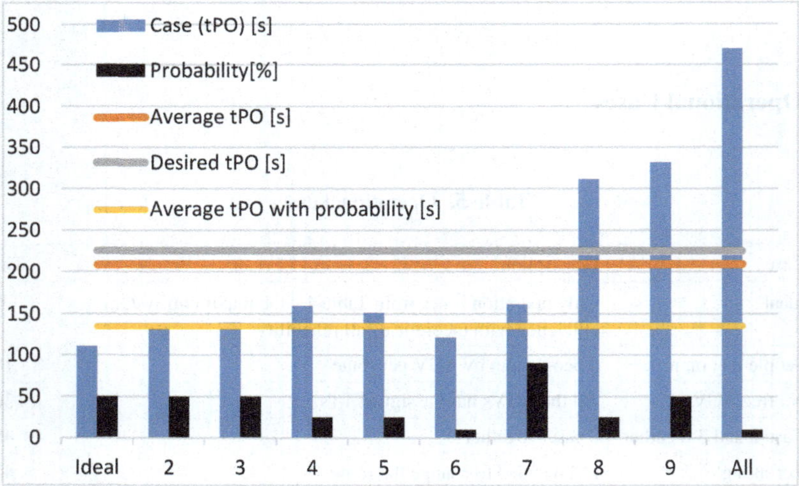

Fig. 5. Results

5 Software Methods to Reduce Disturbances

Besides the system design (evaluated using TBSE), where the production targets can be achieved, it should be remembered, that autonomous transport systems are scalable (increase of number of AGVs in the same layout can drastically change the efficiency and cause several traffic problems – e.g. deadlocks), processes can be modified very often, so there is a need to implement in FM, OM and AGVs embedded software some special features allowing reduction of consequences of disturbances. Examples will be described below.

5.1 PT Searching

Every platform AGV has a laser sensor which can detect PT. In the normal conditions, when PT is placed properly, sensor detects PT and this information is used to trigger the loading operation. If AGV come to the station properly, but sensor does not detect PT there are 2 options: *to report failure* – inform operators and require immediate manual action not to disturb PO – simple solution, but not desired; *run PT searching procedure* which moves AGV slowly forward until the sensor will detect PT – procedure has limitation not to move AGV more than 10cm to avoid obstacle detection; if PT is still not detected, AGV goes back to the station and reports failure - waiting for the manual operation. Searching procedure can reduce disturbance in AGV system operation and eliminates the human factor issue related to the wrong placement of PT. The real data coming from deployments show that this simple algorithm can reduce a necessity of an operator action up to ~40%.

5.2 FM Mechanisms

Avoiding rerouting is very important to reduce huge additional values of t_{rr} in the system operation and performing POs. FM should work constantly and analyze current traffic situation. FM should have the context of POs registered by OM to predict traffic in the nearest future. After some time of the system operation FM should determine the most efficient routes – kind of self-learning system.

5.3 Smart Charger

AGVs can load batteries in waiting points (charging points). To avoid a situation of AGV inaccessibility for PO, battery state should be constantly monitored and smart charger software module should inform FM which AGV should be taken as a first. The smart charger should block AGVs and FM cannot assign them to any PO. FM changes rule – take the closest AGV to take the closest AGV not charging and not selected for charging. Balancing of the battery levels should be always configurable and for various deployments should be set differently. The distance to the charging points is critical. In the system described in this paper, batteries were swapped, so it was manual action, when AGV was turned off and not usable for ~10 min. The system informed operators in advance, that some AGV should have a battery replacement after performing last PO.

6 Summary

The time based system evaluation method (TBSE) can be used in the process of design of autonomous transport systems and is the input for system simulations and VPs graph design. Finally, having a number of stations and POs in the system, the production process from the customer and expected efficiency of the transport, it is possible to give the answer how many AGVs should be implemented to deploy reliable and scalable solution and what are the operation margins.

This paper presents one of the simplest aspects of an autonomous transport system design, which is the key to start initial project phase. After that, several activities like environment mapping, creating navigation elements, designing mechanical elements (docking stations, PTs), designing APIs to other automation systems, designing sensors networks, drawing paths (creating curves profiles, setting speeds), working with safety zones (determining where to use which), analyzing stations locations and numbers of waiting points etc. should be performed to have final project specification.

The process of calculation of t_{PO_s} should be automated. Several data algorithms for the time and probability values estimation should be used (especially machine learning). Algorithms can operate on historical data collected in the deployed systems (offline data analysis). Author future work is concentrated on this and the goal is to have a fully automated process of designing a reliable autonomous transport system for industry.

References

1. Günter, U.: Automated Guided Vehicle Systems. Springer, Heidelberg (2015)
2. Pfrunder, A., Borges, P.V.K., Romero, A.R., Catt, G., Elfes, A.: Real-time autonomous ground vehicle navigation in heterogeneous environments using a 3D LiDAR. In: IEEE/RSJ International Conference on Intelligent Robots and Systems (IROS), Vancouver, BC, Canada, pp. 2601–2608. IEEE (2017)
3. Anderson, J.M., Karla, N., Stanley, K.D., Sorensen, P., Samaras, C., Oluwatola, T.A.: Autonomous Vehicle Technology: A Guide for Policymakers. RAND Corporation, Santa Monica (2016)
4. Vis, I.F.A.: Survey of research in the design and control of automated guided vehicle systems. Eur. J. Oper. Res. **170**(3), 677–709 (2006)
5. Draganjac, I., Petrović, T., Miklić, D., Kovačić, Z., Oršulić, J.: Highly-scalable traffic management of autonomous industrial transportation systems. Robot. Comput. Integr. Manuf. **63**, 101915 (2020)
6. Hu, H., Jia, X., Liu, K., Sun, B.: Self-adaptive traffic control model with Behavior Trees and Reinforcement Learning for AGV in Industry 4.0. IEEE Trans. Ind. Inform., 1 (2021)
7. Cronholm, S., Goldkuhl, G.: Strategies for information systems evaluation- six generic types. Electron. J. Inf. Syst. Eval. **6**(2), 65–74 (2003)

Modelling Pedestrian Behavior in a Simulator for a Security Monitoring System

Jarosław Sugier(✉)

Department of Computer Engineering, Wrocław University of Science and Technology, 11/17 Z. Janiszewskiego Street, 50-372 Wrocław, Poland
jaroslaw.sugier@pwr.edu.pl

Abstract. The paper presents a simulator of the behavior of people or their groups moving within a specific 3D scene, prepared for modeling diverse sequences of events - scenarios - that may occur in some monitored and supervised area of a defined configuration. The simulator can be integrated with a security monitoring system (sending live video streams emulating surveillance cameras and statues from various virtual sensors located in the scene) and allows validation of system operation in scenarios which otherwise would be difficult to realize in real life like fatal accidents, panic events, etc.. In particular the paper concentrates on realistic modelling of human traffic (walking, queuing, waiting) in a typical scene: after presenting results of the standard AI methods used for this purpose the paper describes original algorithms which were developed to complement the generic methods in simulation of more involved situations.

Keywords: Navigation · Pathfinding · AI search algorithm · Unity 3D

1 Introduction

Constant advances in computer graphic and visualization techniques along with new AI methods for modelling human actions and decisions made artificially simulated environments very close to those from the real life, with simulators becoming a data source which can compete with real observations and measurements. In this work we present AvatarTraffic – a 3D simulator for a monitoring system, capable of visualization of a scene with people or groups of people walking along prescribed paths and acting according to a specified interactive scenario. The simulator can visualize a scene corresponding in size to e.g. a metro station or an airport hall with dozens or hundreds of human entities, and communicating with a real security monitoring system can provide signals from various virtual sensors as well as video streams imitating operation of physical surveillance cameras. Being able to reproduce diverse situations which can take place on the scene the simulator can be used for:

– validating in practice operation of the monitoring system (e.g. testing location of surveillance cameras, configuration of sensors, etc.),

W. Zamojski et al. (Eds.): DepCoS-RELCOMEX 2021, AISC 1389, pp. 425–436, 2021.
https://doi.org/10.1007/978-3-030-76773-0_41

- simulation of various hazardous events for testing the system and its personnel in implementation of critical procedures (accidents and other abnormal situations),
- generation of training data for Machine Learning algorithms used to support the monitoring (detecting malicious behavior from video streams, early indication of potentially hazardous conditions from sensor data, etc.; this kind of simulator applications is becoming increasingly popular as it is shown e.g. in [6]).

The above goals are traditionally accomplished using real-life operational records of the monitoring but introducing a simulator brings a number of advantages, e.g.: possible in-advance validation and adjustment of the system before its deployment, elimination of real-life video recording of people on site which would violate various GDPR requirements (anonymization of records, etc.), and – last but not least – making possible to model hazardous or catastrophic events that have never been recorded and/or are impossible to be reasonably staged (fire on station, panic events, deadly accidents, etc.).

In this work we concentrate on one fundamental aspect of the simulation: realistic modelling of human behavior (walking, queuing, waiting, etc.) in a typical monitored scene. The rest of the text is organized as follows: after the next section briefly overviews known methods for computer simulation of people and crowd moves, the simulator and its relevant resources are presented in Sect. 3, while the last chapter before the conclusions – Sect. 4 – discuses original algorithms which were developed to complement the generic AI methods in simulations of more involved scenes.

2 Modelling Pedestrian Traffic in Computer Simulations

Planning a path by an autonomous agent in a virtual environment has been a vital problem explored for many years not only in simulators of crowd behavior but also – and first of all – in robotics and game development. The now established framework which is universally used in this task is based on two concepts: a navigation mesh (NavMesh for short) and a navigation agent (NavAgent).

The NavMesh is a data structure describing the walkable surfaces of the scene and can be built and optimized for a given geometry as a static structure before the simulation starts. The NavAgent is associated with every moving object (e.g. every human character taking part in the simulation) and uses the NavMesh in two tasks: a) finding a path to a destination point every time a new one is assigned and b) navigating the object along the current path according to its characteristics (e.g. walking or running speed), additionally avoiding collisions with other NavAgents during the move. The NavAgent's competence in these two tasks is essential in producing effective and realistic simulations.

Planning a path in a large NavMesh is enhanced usually by adopting hierarchical approach: instead of considering the entire mesh, the path is created first on the highest hierarchy level (e.g. finding which sub-areas should be traversed towards the given destination) and then individual segments of the path are traced on the detailed NavMeshes within each sub-area. There are various alternatives of hierarchy organization, optimization and supporting data structures which – at an extra cost of preprocessing – can substantially accelerate operation at runtime [5].

The actual pathfinding methods are typically based on the standard A* search algorithm [1] which, in turn, was a heuristic extension of the classic (1950s) Dijkstra's procedure for finding shortest paths between nodes of a graph. This standard approach often needs to be extended and adjusted e.g. in order to take into account simultaneous operation of multiple agents with specific kinematic constraints [2]. Furthermore, recently much of the interest is devoted to application of the ML algorithms for better collision avoidance in operation of the NavAgent [3, 4] but still more development is needed before they will be accepted for implementation in the tools.

3 The AvatarTraffic Simulator

3.1 The Simulator and the Scene

AvatarTraffic is prepared as a complete application of the Unity 3D environment [7]. After loading a configuration file the simulator starts a real-time simulation of a scenario described within, communicating with the monitoring system via NATS channels and providing overview of the situation in the form of video streams that imitate closely operation of real-life security monitoring cameras.

The prototype described in this work concentrates first of all on realistic modelling of crowd behavior (in size of dozens or single hundreds of individuals) within a scene covering e.g. a complete train station or an airport hall. Less attention has been paid to photorealistic rendering of the scene and human characters because this can be accomplished with standard techniques of computer graphics and remains out of the interest of this presentation.

The configuration file defines the entire operational environment including:

- geometric construction of the scene (walls, doors, passages, stairs, platforms, etc.),
- elements required for its visualization (light sources, location of the main preview camera plus a number of monitoring cameras generating live video streams),
- human individuals, both participating in the scenario as its main actors as well as background performers creating a realistic environment for the main plot of events,
- specification of the scenario flow in the form of any number of so-called scripts that activate scene objects according to the scenario, detect situations taking place on the stage and communicate with the monitoring system via the NATS interface.

Bi-directional communication of the simulator with the monitoring system through the NATS protocol (see Fig. 1) includes exchanging messages about simple and complex events, i.e.:

- sending statuses of sensors located in the scene (door open/close, entrance sensors, motion detectors, card authorization stations, etc.),
- sending messages generated by scenario actors in response to situations occurring during the simulation as they are expressed in the scripts,
- reception of the security supervisors' reactions (commands) causing the simulator to proceed with various variants of scenario – within the limits of the implemented choice options,
- receiving commands for moving the monitoring cameras (Pan-Tilt-Zoom).

Non-deterministic execution of a scripted scenario can be caused by:

- randomly picked object parameters at run time (e.g. random number and speed of people on a pathway, random moments of actors appearance, etc.),
- unpredictable reactions of the (human) security supervisors watching the simulation and responding to its flow through the commands via the NATS channels,
- interventions of the operator who can manually control triggers embedded in the scripts and e.g. start different variants of the scenario.

The last point corresponds to an optional situation when the simulation is monitored by a human operator who e.g. acts as a trainer of security staff working with the monitoring system (the trainees). Fully automatic execution of the scripts is also possible and the choice depends only on the desired purpose of the simulation.

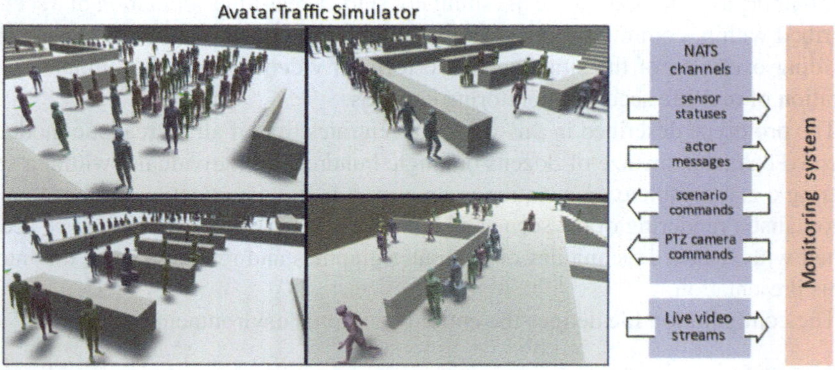

Fig. 1. Integration of the AvatarTraffic simulator with a security monitoring system

3.2 Simulating People and Their Movement

In general, visualization and realistic control of a human character in a computer simulation is a complex and demanding task, therefore the proposed solution was based on ready-made resources offered by the Unity platform which have been thoroughly tested, verified and optimized in game development:

- three-dimensional models of the human figure, so-called avatars,
- animations controlling their movement (walking, running, turning, standing still), prepared by capturing the motion of real humans and available in libraries as components adjusted to the avatar models,
- algorithms for controlling character relocation (walks or runs) which, using AI methods, can dynamically plan and then implement the move of a character to a desired point of the scene, taking into account its static topology as well as obstacles appearing at runtime (e.g. other avatars).

While the Unity components corresponding to the first two items of the above list [8] worked sufficiently well for this application, the methods for planning avatar navigation mentioned in the last point required additional custom supervision in order to eliminate errors that happened in more complex, i.e. involving many avatars, scenes – which will be the subject of Sect. 4.

In Unity the essential AI algorithms used to manage human movement are available in the `NavMeshAgent` component (attached individually to each person) which operates animation of the figure as well as dynamically plans its path and – effectively – implements realistic relocation to any pointed location [9]. Using the `NavMeshAgent` functionality comes down to assigning the agent with a destination point and then only monitoring progress of the move, leaving the AI engine particulars of setting out the path and moving along it with the imposed velocity. The agent assumes that the destination is reached when the avatar gets closer than its `stoppingDistance` parameter. All particulars of the move, including realistic turns, slow-downs or speed-ups, are handled accurately by the animation engine and did not require any correction.

3.3 Elements of the Scene

The `NavMeshAgent` mechanism was the foundation of operation in the two basic types of simulator objects: avatars and avatar generators. Moreover, it quickly turned out that more complex situations call for additional measures which would extend this default mechanism in special cases of a group of people e.g. standing in a queue, passing through a door or waiting within a certain area. In order not to repeat their implementation in each avatar, three auxiliary special elements have been developed: Queue, Door and WaitingArea.

Figure 2 illustrates an exemplary application of these objects in a simulation of a platform at a train station, where: individual avatars represent security personnel, avatar generators are used for creating passengers walking to/from a train, queues operate as a platform entrance and exit, a separate door is provided for authorized personnel who can take part in a simulated scenario, and waiting areas control behavior of passengers lingering on the platform.

Avatars. If an avatar is to follow a specific fixed route during the simulation, its path is described as a sequence of waypoints. The point can be given with absolute coordinates or as a reference to a parent object – a queue, a door or a waiting area – which takes over the control of the character's behavior when some more complex conduct is needed (like, e.g., waiting for door opening or forming a line of a queue).

In the simplest case of an avatar-background performer with a constant path, the route is carried out automatically and repeatedly: after reaching the current waypoint, the following point is assigned as the next destination, and after reaching the end point, the avatar either walks back to the starting point (walking in loops) or is warped there immediately (in which case such a warping should obviously be arranged in some non-monitored part of the scene). In the case of an avatar-actor, which dynamically reacts to events during simulation, its path usually consists only of the starting point used to initialize the position, while the commands to go to specific waypoints come from scripts

Fig. 2. Various types of scene objects modelling a situation on a platform at a train station

and the effective path depends on execution flow. In both cases it was sufficient to use the Unity's walk-to-the-point mechanism.

Avatar Generators. To represent a number of people moving along the same route (e.g. like in a corridor at a metro station) an autonomous generator can be used. It can create individual avatar objects ad hoc in different manners: with given average intensity (number of avatars per second), keeping the average distance between them, or implementing some defined intensity distribution over a time period. The avatars are of optionally randomized size and speed and move along the generator's path in a lane of specified width. The generation can be delayed to obtain synchronization between multiple generators, or completely paused and then resumed by the scripts. As additional functionality the generated persons can carry luggage – one or two suitcases – which slow down their movement and can be dropped or picked on command.

Definition of the path and route control is implemented identically as in the case of standalone avatars. The generators can be used first of all for simulation of background traffic but in some cases they can act also as scenario actors (to represent, e.g., a patrol on duty performing a scripted inspection).

Queues. A queue of avatars waiting in line to be served can be composed of many threads, i.e. sub-queues running in parallel with a given spacing, and its basic parameters define location, direction of growth and intensity of avatar releases. When an avatar is sent to the queue as its next waypoint, the queue completely takes control of its movement, pointing it to an appropriate destination and triggering transitions when the queue moves. The queue frees avatars standing at its front with an average frequency or with some intensity distribution loaded from a file like in avatar generators. After being released from the queue, the avatar continues to navigate along its route i.e. it heads to its next point or (in the case of an actor controlled by a scenario script) reports inactivity waiting for further commands.

Doors. Apart from mechanical construction (1-leaf or 2-leaf, sliding or hinged) the doors fall into two distinct categories: automatic and locked. Automatic doors act like motion-sensors: they open in front of each incoming avatar and then close when all pass them by. From the point of view of navigation, they are transparent and – unlike the locked doors – do not need any special consideration in scenarios.

For a locked door, the controller must orchestrate its opening and passing the waiting avatar(s), additionally taking into account special cases. The first avatar who reaches the closed door becomes the opener and the controller starts the opening time-out. If other avatars approach the opener during the time-out, they have to be stopped and, after the opening, the whole group transfers following the opener. This mechanism can be used also for simulating a tailgating attempt, e.g. a situation when an intruder tires to pass the open door following a group of authorized personnel. It is also possible to model break-in incidents when an avatar spends too much time in front of the door before it opens, presumably picking the lock.

The doors can report two type of information to the monitoring system: the standard open/close statuses and also positive/negative card authorization results which can be distinguished e.g. upon opening time-outs assigned to the opening avatar.

Waiting Areas. A waiting area is defined as a rectangular zone at a specific location with the following additional parameters: the total time an avatar stays in the area, its walking speed and stopping time in one place. The last two parameters describe the behavior in which the avatar - instead of standing in one place throughout the waiting period – randomly walks from place to place not leaving the area.

After the waiting time expires, the avatar resumes its standard navigation similar to queues and doors; alternatively, special script commands can force one or all of the avatars residing in the area to leave it immediately and continue their paths.

4 Monitoring Operation of the AI Traffic Planning

The `NavMeshAgent` motion planning algorithms are very convenient in simulation of avatar traffic but they soon proved to be insufficient in modelling more complex cases when many persons (avatars) have to share a certain resource or area. Such situations occurred in particular when many avatars were trying to enter one queue or pass one door, where – without additional supervision – often two or more of them were blocking each other. The state of such a blockade sometimes was finally resolved by the algorithms themselves after few seconds but in some cases it persisted much longer, completely stopping the scenario flow. This aspect turned out to be a primary operational weakness of the built-in AI algorithms.

4.1 Controlling Operation of the Queues

In order to ensure the proper forming of avatars standing in a queue the following procedure in queue controller was required for successful enlisting a new approaching avatar:

1. If an avatar finds a queue as the next point on its path, it reports itself to the queue controller who selects a candidate thread (based e.g. on current threads occupancies) and passes avatar's supervision to the thread controller;
2. the thread controller assigns the current thread end as the destination point in the avatar's `NavMeshAgent` and stores it in its list of incoming avatars;
3. in each simulation frame the controller scans all the incoming avatars; when any of them gets to the thread end closer than the `NavDecisionDist` param, then it is qualified as the new queue member and is moved form the incoming list to the standing list (it is left to the `NavMeshAgent` to complete avatar's move to its standing point up to the `stoppingDistance` condition); then:
4. a new thread end is calculated (moved by an one-person offset) and sent as the new destination to all the remaining incoming avatars on the list.

In this procedure, the crucial idea is to select one of the incoming avatars as the new queue member in advance (at the `NavDecisionDist` in step 3) and then to redirect all other approaching ones to a new endpoint. Without this modification the typical behavior of the `NavMeshAgent` resulted in several incoming avatars blocking each other in the vicinity of the thread end and preventing themselves from reaching this destination (an example is shown in the left part of Fig. 3). Practical tests have shown that for correct operation the `NavDecisionDist` parameter should be 3–5 times greater than the AI's `stoppingDistance`.

Another problem was the mutual blocking of avatars heading to adjacent threads of the same queue, which was particularly common in multi-threaded queues modeling entrances to or exits from e.g. a metro station (a very common case in the typical considered scenarios). Such blockades led to the following errors:

1. several avatars get stuck when entering a dedicated area of the queue thread, with one of them unable to get through to its thread,
2. "pushing" an avatar (by others surrounding it) to a thread other than the one it is assigned to (leading to a situation depicted in the right part of Fig. 3),
3. "pushing" an avatar in front of the one which should precede it.

Fig. 3. Anomalies in queue operation: jammed formation of queue end (left) and a blockage caused by an avatar pushed into incorrect thread (right).

In cases 1 and 2 the errors involve avatars still in the incoming lists and their correction must be performed at the level of the entire queue, while case 3 relates to an avatar considered as an already standing one in a thread. In all cases, the result was that the flow of the queue (or one of its threads) was blocked.

Errors of type 1 and 2 required an additional procedure called in every simulation frame: of all the incoming avatars the blocked ones are identified as those moving significantly slower than their speed param or pointing not in the direction of the queue orientation, and then, if they are found on the area of another thread, they are re-assigned to it. This solution changes allocation of avatars to threads, but it turned out to be much simpler and more effective than freeing the way for a blocked avatar to its thread; a consequential slight imbalance of thread loads is usually well acceptable.

The occurrences of case 3 are checked by the thread controller every time the avatar is released from the queue, because the problem with their wrong order is revealed only then it is necessary to move the standing members (such a solution does not require the test to be executed as frequently as the above one). This is a simpler case, which is detected just by verification of avatar positions along the queue axis and the correction consists in updating the standing avatar list according to their actual order. As in the previous cases, it was simpler and more effective to accept the actual location of the blocked avatar than to force it to move to its assigned place.

4.2 Supervising Avatars in Door Passages

The AI algorithms showed to be prone to errors especially when a group of avatars had accumulated in front of a door and they all should pass through together (that is: one by one) after opening. This situation often led to blockage of two or more avatars in front of an open door, with an example shown in the left part of Fig. 4.

Fig. 4. A blockage of the AI navigation algorithms during transition of an avatar group through an open door (left) and their ordered move with extensions proposed in this work (right).

This problem turned out to be even harder than the ones which were encountered in queue management. In the developed solution the (locked) door controller operates with three avatar lists: incoming ones (i.e. approaching the door as their next waypoint), standing in front of the door (waiting for the door to be opened), and released (passing through the door that has just been opened). The procedure is as follows:

1. After an avatar reports to the door controller, it is added to the incoming list and its `NavMeshAgent` is directed to the opening point in front of the door;
2. by monitoring all incoming avatars on the list, the controller selects the first one which arrives at a threshold distance to be the opening one and starts counting the unlocking delay which is assigned to it;
3. monitoring other incoming avatars, the controller stops them when they get closer than the waiting distance limit (additional door param, greater than the opening distance) or when a significant drop in their speed is detected (meaning that the avatar has reached the waiting group and its further progress is blocked); all such avatars are moved from the incoming list to the standing one;
4. after the unlocking delay has elapsed, the door opening animation is executed and upon its finish the opening avatar is released as the first and – for the time being – the only one; every released avatar is directed to the next waypoint of its path and is moved to the released list;
5. if the latest released avatar travels some prescribed distance, the next one from the standing list is released;
6. simultaneously with releasing, the controller monitors all the released avatars and detects blocked ones (a blocked avatar is again the one moving noticeably slower than its navigational parameter): every blocked avatar except the first one is stopped, moved back a small step (approx. ¼ of figure's horizontal diameter) and returned to the front of the standing list; this gives way only to the first blocked avatar who should pass the door unobstructed;
7. when any released avatar goes past the door by some threshold distance it is removed from the released list as the one who successfully completed the procedure;
8. after the last avatar is removed from the released list the door closing animation is effected.

Detecting the stuck incoming avatars in step 3 and stopping them was necessary because otherwise their pressure was pushing the opening avatar towards the door. For the sake of simplicity, avatars are stopped and remain at the point where they were blocked – unlike queues, the door controller does not enforce position of the standing avatars but accepts an arrangement which resulted from their direction and time of arrival. Such an approach gives realistic assemblies of avatars waiting in front of the door.

It is crucial to sequentially release avatars after opening the door in step 5. The condition for releasing the next one is that the predecessor has started to move and has actually covered a certain distance. Such a mechanism turned out to be relatively simple in implementation and more effective in practice than e.g. assuming any fixed time intervals between releases.

Equally important is monitoring the passage of the released avatars (step 6) in order to detect any blockages when passing through the door. The best in practice mechanism for blockage removal required both to stop the navigation of the blocked avatars (also returning them to the standing list) as well as to slightly pushing them back – in order to make room for the first blocked avatar which is left as the only one moving.

Additionally, in this particular case it has been found advantageous to use the built-in `NavMeshAgent`'s parameter `avoidancePriority` which could not help in problems of queue operations. When resuming the AI navigation in step 3, the released avatar is temporarily assigned a priority which is decreasing proportionally to its ordinal number (i.e. the sooner the avatar is released, the higher priority it has). This mechanism helps in avatars actually passing the door in order of their release but it was not sufficient in eliminating avatar jams when they were released all at once. The original value of the avatar's priority (which is inherited from the parameter of the avatar generator or the individual object itself and can be used for resolving collisions of multiple avatars on crossing paths) is restored in step 7 when the avatar successfully passes the door and can be finally freed by the door controller.

5 Conclusions

The paper presented successful application of the Unity environment in development of a simulator which can model diverse kinds of events in a monitored area with intense human traffic. Being able to simulate various activities typical for a supervised public object (with people walking, queuing, passing doors and waiting in designated areas) the simulator can be combined with monitoring systems deployed in e.g. train stations or airport halls and help in testing various scenarios of events.

The Unity's built-in AI mechanisms provided for dynamic planning and executing transitions of human models confirmed practical usefulness although some deficiencies have been indicated. In particular, the `NavMeshAgent` in the current version of Unity AI libraries does not provide any interface that would allow to check whether its navigation is blocked by obstacles and if so, what is considered by the algorithm to be such an obstacle. Being able to obtain this information from an agent would allow for faster detection of jammed objects and would lead to development of better mechanisms for avoiding blockages.

Acknowledgements. This work presented results of the Smart Growth Operational Programme 2014–2020 grant No. POIR.01.01.01-00-0235/17 supported by the Polish National Centre for Research and Development (NCBR) as a part of the European Regional Development Fund (ERDF).

References

1. Hart, P.E., Nilsson, N.J., Raphael, B.: A formal basis for the heuristic determination of minimum cost paths. IEEE Trans. Syst. Sci. Cybern. **4**(2), 100–107 (1968)
2. Hönig, W., Kumar, T. K., Cohen, L., Ma, H., Xu, H., Ayanian, N., Koenig, S.: Multi-agent path finding with kinematic constraints. In: Proceedings of the International Conference on Automated Planning and Scheduling, vol. 26, no. 1 (2016)
3. Majumder, A.: Pathfinding and navigation. In: Deep Reinforcement Learning in Unity. Apress, Berkeley (2021)
4. Qiu, H.: Multi-agent navigation based on deep reinforcement learning and traditional pathfinding algorithm. arXiv preprint arXiv:2012.09134 (https://arxiv.org/pdf/2012.09134.pdf) (2020)
5. Rahmani, V., Pelechano, N: Improvements to hierarchical pathfinding for navigation meshes. In: Proceedings of the Tenth International Conference on Motion in Games, November 2017, Barcelona, Spain (2017).

6. Unity Computer Vision – labelled and annotated training data for machine learning. https://unity.com/computer-vision, Accessed 26 Feb 2021
7. Unity Real-Time Development Platform. http://unity.com, Accessed 26 Feb 2021
8. Unity User Manual: Animation. https://docs.unity3d.com/Manual/AnimationSection.html, Accessed 26 Feb 2021
9. Unity User Manual: Navigation and Pathfinding. https://docs.unity3d.com/Manual/Navigation.html, Accessed 26 Feb 2021

Generalized Convolution: Replacing the Classic Convolution Operation with the Sub-network

Kamil Szyc(✉)

Wrocław University of Science and Technology, Wrocław, Poland
kamil.szyc@pwr.edu.pl

Abstract. A convolution operation is a backbone of nowadays machine learning approaches in computer vision problems. It is responsible for extracting low-level features from input data. However, this operation has not been modified for decades. The main contribution of this work is presenting a new concept, where a more generalized version of the convolution operation is introduced. The goal is to replace the classic convolution operation with the sub-network, for example, a Feed-Forward Neural Network. The classic convolution is the specific case of the proposed generalized version where the sub-network is a single artificial neuron. Increasing the complexity (i.a., more layers or additional nonlinear functions) allows for more complex decision boundaries, and as a result, enhances the ability of the network to learn more accurate patterns. We discuss the implementation details of the Generalized Convolution - as we called this approach - using the PyTorch library. The first experiments allowed increasing the VGG model accuracy.

Keywords: Convolutional neural network · Features extracting · Computer vision

1 Introduction

Convolutional Neural Networks (CNNs) have been a standard in computer vision problems for years. Image classification, image recognition, image segmentation, image retrieval, and many other problems and the state-of-the-art solutions are based on them. These networks - thanks to convolution operation - can well generalize input data (images), find patterns and correlations between them, and in consequence successfully recognize them. They often achieve better results than humans in vision recognition tasks. Popular CNN architectures like AlexNet [1], VGG [2], Inception [3], ResNet [4], or newer like EfficientNet [5], or NoisyStudent [6] have been adding many new concepts to improve the performance and accuracy of these networks. Leading examples are residual connections, batch normalizations, Global Poolings, or allowing scaling and balancing network depth, width, and resolution. However, still the basic block used in all of these networks is the mathematical convolution operation.

W. Zamojski et al. (Eds.): DepCoS-RELCOMEX 2021, AISC 1389, pp. 437–446, 2021.
https://doi.org/10.1007/978-3-030-76773-0_42

The convolution operation allows extracting low-level features from images, and use them later to recognize a specific object in this image. To classify an image into a given class (by choosing one of the available labels), thousands and sometimes millions of such operations are performed. They are stacked together in layers allowing recognition of high-level features: starting from detecting edges, through textures, patterns, parts, and ending to objects. All of these features are propagated forward layer by layer to the classification part in CNNs. As a result, they can correctly classify images.

The feature extraction procedure seems to be one of the most important steps in CNNs. However, over the decades the basic block - convolution operation - has not been improved much. In this paper, we present a concept, which allows the classic convolutional layer to be more broaden.

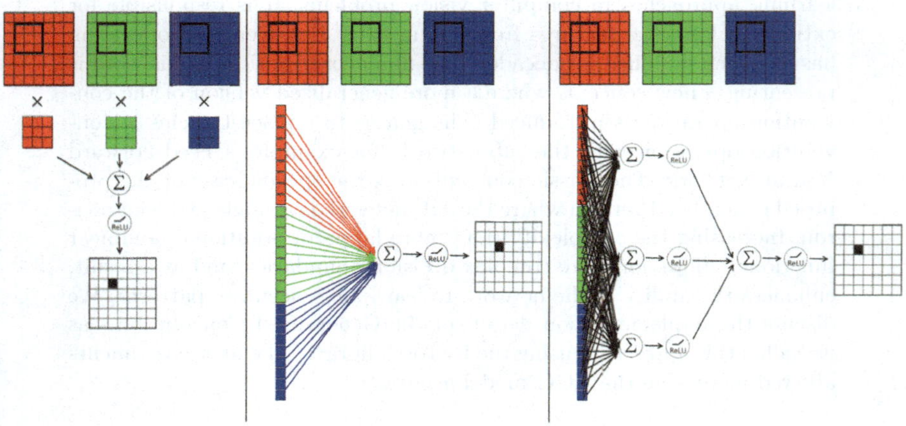

Fig. 1. The comparison of using the classic convolution operation (with 3×3 kernel) with our concept. The example shows how to calculate one output value from three channels (RGB) by using a sliding window method. (**left**) Classic convolution operation. To calculate the output value we need to get the 3×3 sub-matrices from each channel (marked as bold 3×3 squares) and multiply it with the same size weights in kernel matrices. After the multiplication operation, we sum all the values and use the non-linearity function - for example ReLU. (**middle**) The same classic convolution operation presented as a single artificial neuron. Mathematically, this is exactly the same operation. The points on the left side are the same points as from the channels above (marked as bold 3×3 squares). There are nine points per channel. Each arrow represents one weight from the convolution matrix kernel, so each arrow is used to represent the multiplication. After the multiplication operation, we sum up all the values and use the nonlinear, activation function. (**right**) Generalized Convolution operation. In this concept, we use a sub-network - here a multilayer perceptron. By adding more layers it is possible to improve the ability to learn complex patterns and add extra non-linearity.

We would like to propose a new, modified, more Generalized Convolution operation. The main idea is to replace the classic convolution operation with a sub-network. That allows the CNN model to learn more complex patterns at the lowest level of feature extraction. The idea was based on the observation that the convolution operation is - in fact - a single artificial neuron that can respond to a part of the surrounding cells in the coverage range - see Fig. 1. By replacing this neuron with the sub-network and adding additional hidden layers (so also additional non-linearity) there is possibility to build a network that can learn more complex decision boundaries. In consequence, the network will be able to learn more complex patterns improving the most significant measures i.a., accuracy.

2 Related Work

Although numerous improvements to deep neural networks (and around them like new optimizers, normalization methods) have been introduced e.g., [4,5,7–9], the convolution operations have not been changing over the years. Of course, it is possible to set different parameters like kernel size, strides, or dilation rate [10], but these do not change the operation itself, but only a way of how inputs are passed.

While the first convolutional neural networks have been known for decades [11], the AlexNet [1] was one of the first successfully (winning the 2012 ImageNet challenge) applied convolutions operations model. The authors used different convolution kernels size: 11×11, $\times 5$, and 3×3, so it allowed finding features in images using the different number of surrounding pixels. The AlexNet became a base for many new models. Since then the rapid development of deep neural networks began.

Authors of VGG [2], Inception [3] and others proposed directions of how to create deep CNN with multiple convolutional layers. One of the most important characteristics of these models was using convolution operations with kernel size 1×1 to control the number of channels, and replacing big kernel size convolutions (e.g., 5×5) with smaller ones (e.g., two 3×3 convolutions) - or even from 3×3 kernel size to 1×3 and 3×1 convolutions. Using them was helpful in increasing final accuracy by recognizing more complex patterns (thanks to adding extra transformation and non-linearity). The idea is roughly the same as ours - however they stacked convolutional layers, and we propose stacking them at convolutional level. The Network in Network (NiN) [12] paper and further developments of that idea [13] adopt a similar approach focusing on convolution with kernel size 1×1. But again but in other words - Generalized Convolution's sub-network works on kernel size level, while these concepts on channel levels.

Although it was known that deeper networks should improve the final classification there was a vanishing gradient problem, which made it impossible in practice. However, in 2016 the Microsoft Research team proposed adding "skip connections" to the network called ResNet [4]. That idea helped to build models deeper than ever before and was further developed in later years [14–17]. The vanishing gradient problem can appear in the Generalized Convolution also, however, it can be overcome using the above and other methods.

We use all the above ideas as backbones for the current models. The state-of-the-art solutions [18] (usually based on EfficientNet [5]) still use the same classic convolution operation. The EfficientNet [5] proposes a method to scale model width (number of channels in layers), depth (number of layers), or resolution dimensions (images input size) in deep neural networks. Properly scaling these values allows achieving results better than ever. We propose to add another dimension for such a network - convolutional dimension.

Some variation of convolutions is involved in models used in mobiles and IoT. For instance, the MobileNet [19] model uses an idea called "depthwise separable convolutions". However, proposed changes in convolutions refer to calculations with many channels, not with the structure of the basic convolution operation.

There are some ideas on how to truly modify classic convolution operation. For instance, tiled convolution [20] or DCT-Conv [21], however, their authors focus on the reduction of trained parameters in convolution operation.

3 Approach

The convolution operation is a basic block for CNNs. For a given input and filter/kernel the classic convolution operation for point x, y can be calculated as the sum of the Hadamard product of input sub-tensor with kernels. For more channels the same operations have to be applied over all channels separately, the final output is an entrywise sum of them. Sub-tensor are point x, y and surrounding elements in that way the point x, y is in the middle. To use convolution over the whole Image the "sliding window" method has to be applied - see Fig. 1(left).

More formal, to calculate the classic convolution operation (without bias and with default parameters like stride or dilation rate) for a point (x, y) set convolution window size as $(2n+1) \times (2n+1) \times n_C$ (n_C refer to channels), assume the index of the center of convolution window to $(0, 0)$, place the center of the convolution window over the input point (x, y) and calculate the output for point (x, y) as:

$$\mathrm{conv}(I(x,y)) = \sum_{i=-n}^{n} \sum_{j=-n}^{n} \sum_{k=1}^{n_C} I(x+i, y+j, k) K(i, j, k) \tag{1}$$

where:

I : input tensor
K : kernel with size $(2n+1, 2n+1, n_C)$

This approach is limited. Using convolutional operations in CNNs require thousands of such operations in order learn complex patterns. Modifying it by giving the possibility to generate more detailed features is a promising direction. We propose the Generalized Convolution - a concept where we use a sub-network ψ to increase capabilities of finding patterns and correlations in data.

The classic convolution operation can be realized as the artificial neuron - see Fig. 1 (middle). In order to do that, sub-matrices from each channel are taken from a method called "sliding window" and transformed into a vector. This vector is an input for the artificial neuron. An output is a sum of all values multiplied by weights (as in classic convolution operation). The idea of the Generalized Convolution is increasing feature detection capabilities by changing the simple artificial neuron into a more advanced sub-network. Adding additional hidden layers in sub-network makes more complex decision boundaries, and as a result, enhances the ability to learn more accurate patterns. The very wide, but shallow networks are good at memorization, but not so good at generalization. Stacked convolutional operations allowed recognizing the patterns hierarchically - from simple shapes to specific objects. Thanks to the deeper networks, they could learn features at various levels of abstraction. Multiple layers are better at generalizing, because they learn all the intermediate features. We propose a similar solution but in a convolution dimension.

The input for the Generalized Convolution is the same as for classic convolution, except that input is transformed into a vector (V), and the result is output from a defined sub-network ψ. More formal, to calculate the Generalized Convolution operation (without bias and with default parameters like stride or dilation rate) for point (x, y) denote W as the $(2n+1) \times (2n+1) \times n_C$ (n_C refer to channels) sub-tensor centered at pixel (x, y), then denote V as the vector of length $(2n+1)^2 * n_C$ of elements of W taken row-wise and finally calculate the output for point (x, y) as:

$$\text{generalized\_conv}(I(x, y), \psi) = \psi(V(x, y))) \tag{2}$$

where:

I : input tensor
ψ : sub-network with same parameters as $\text{conv}(I, K)$

In practice, *generalized_conv* can use any sub-network architecture like Feed-Forward Neural Network or based on 1D convolutional layers. Designing the sub-network should be done carefully and be suited to the specific architecture. The Generalized Convolution can be used instead of classic convolution. With proper implementation (as shown in Sect. 4) it can be just using a new class (in programming context) that takes exactly the same parameters as the classical convolution class.

Increasing the number of layers in our approach can have some disadvantages. The two most problematic cases seem to be increased memory consumption and the tendency for vanishing gradient or overfitting. However, given the rapid development of graphics cards the first issue won't be an important case for long. For now, it is possible to deal with it by decreasing the number of convolutions in total. For the second issue using standard techniques like normalization approaches or adding skip connection should work.

4 Implementation and Experiment

4.1 Implementation of the Generalized Convolution

We used Python language in all experiments. The modern deep-learning libraries like TensorFlow[1] or PyTorch[2] have implementation for classic convolution operation. This operation is the backbone for CNN, so authors of these libraries optimize it very well at the lowest code level (not written in Python). Because of that, it is impossible to easily modify the original version of the layer and base our concept's implementation on that. Hypothetically, we could modify the core of a library from the source, but we wanted to avoid that. The PyTorch was chosen as the main library. When we started the project the TensorFlow library didn't have the implementation of eagerly executing capability, so we could not build computation graphs dynamically, which is crucial for writing the code efficiently.

We created a new class *GeneralizedConv2d* to implement the Generalized Convolution that inherits from the module *torch.nn.Module*. It accepts nearly the same parameters as the classic convolutional layer *torch.nn.Conv2d*. The exception for instance is we accept only a square matrix for the kernel and do not accept bias. The goal was to be able to easily replace the classical convolution (*Conv2d*) with ours (*GeneralizedConv2d*), so from the neural networks designer's point of view using it should look very similar. The simplified code for our class is available in Algorithm 1, the following text shortly describing it step by step.

As described in Sect. 3 the inputs for the Generalized Convolution are the same as for classic convolution, but denoted as a vector. Before generating the input vector we need to define parameters for our class to be able to know rules on how to create these vectors. We define them in the constructor *__init__* as the number of input channels (*n_channels*), the number of output channels (*out_channels*), kernel size - same for width and height (*kernel_size*), and parameters allows defining how "sliding window" works: *padding*, *stride*, dilation rate (*dilation*), and *bias* (not implemented). The parameters have the same names as for *torch.nn.Conv2d* to be able easily to replace one class to another.

The input for the sub-network has a shape (kernel size ∗ kernel size ∗ the number of input channels) and the output (the number of output channels). However, it is for a single point for input tensor, for the whole input tensor we need to collect all points using the "sliding window" method. The *calculateWindows* method is responsible for that. Fortunately it can be done by using PyTorch *torch.nn.functional.unfold* and entering defined parameters. The final steps are transforming obtained windows to vectors using methods *transpose* and *reshape*. Finally, the input tensor for the sub-network has shape (batch size ∗ the number of generated windows, kernel size ∗ kernel size ∗ input channels). The *batchsize* is required by the PyTorch.

[1] https://tensorflow.org/.

[2] https://pytorch.org/.

Algorithm 1: The simplified code for the Generalized Convolution

```
class GeneralizedConv2d(torch.nn.Module):
    def __init__(self, n_channels, out_channels, kernel_size,
        dilation=1, padding=0, stride=1, bias=False):
        super(GeneralizedConv2d, self).__init__()

        self.n_channels = n_channels
        self.out_channels = out_channels
        self.kernel_size = (kernel_size, kernel_size)
        self.kernal_size_number = kernel_size * kernel_size
        self.dilation = (dilation, dilation)
        self.padding = (padding, padding)
        self.stride = (stride, stride)

        _in = kernel_size*kernel_size*self.n_channels
        _out = self.out_channels
        self.subNetwork = SubNetwork(_in, _out)

    def forward(self, x):
        width = self.calculateNewWidth(x)
        height = self.calculateNewHeight(x)
        windows = self.calculateWindows(x)

        result = self.subNetwork(windows)
        result = result.transpose(0, 1)
        result = result.view(self.out_channels, x.shape[0],
            width, height)
        result = result.transpose(0, 1)
        return result

    def calculateWindows(self, x):
        windows = F.unfold(x, kernel_size=self.kernel_size,
            padding=self.padding, dilation=self.dilation, stride
            =self.stride)
        windows = windows.transpose(1, 2)
        windows = windows.reshape(-1, windows.shape[2])
        return windows

    def calculateNewWidth(self, x):
        return ((x.shape[2] + 2 * self.padding[0] - self.
            dilation[0] * (self.kernel_size[0] - 1) - 1) // self
            .stride[0]) + 1

    def calculateNewHeight(self, x):
        return ((x.shape[3] + 2 * self.padding[1] - self.
            dilation[1] * (self.kernel_size[1] - 1) - 1) // self
            .stride[1]) + 1
```

In the *forward* method we pass this input to the sub-network expecting output tenor with shape (batch size ∗ the number of windows, output channels). The final step is transforming this output that the same as is required by *torch.nn.Conv2d*. We need to define the new width and height using methods *calculateNewWidth* and *calculateNewHeight*.

The most important part is defining sub-network as a new class *SubNetwork*, which is the backbone for this concept. It is a crucial part, so it should be designed thoroughly, specifically for the chosen CNN architecture. The example of implementation this class is shown below.

4.2 Experiment

The Generalized Convolution is a concept, which requires further research. For the first test, we applied it for solving the image classification problem using the CIFAR-10 data set. The simple VGG-11 CNN architecture was chosen, because it is a relatively shallow network (compared to newer ones).

Algorithm 2: The sub-network for the Generalized Convolution designed for the VGG-11

```
class SubNetwork(torch.nn.Module):
    def __init__(self, _in, _out):
        super(SubNetwork, self).__init__()

        self.linear1 = torch.nn.Linear(_in, 2*_out)
        self.linear2 = torch.nn.Linear(2*_out, _out)
        self.norm1 = torch.nn.LayerNorm(2*_out)
        self.norm2 = torch.nn.LayerNorm(_out)
        self.relu = torch.nn.ReLU()

    def forward(self, x):
        x = self.relu(self.norm1(self.linear1(x)))
        x = self.norm2(self.linear2(x))
        return x
```

We exchanged all convolutional layers (*torch.nn.Conv2d*) in the VGG-11 architecture into our *GeneralizedConv2d* keeping the same parameters (like kernel size of or padding) and using the same hyper-parameters (like the number of epochs, learning rate). The *SubNetwork* class for this model is presented in Algorithm 2. We achieved 91.62% of the accuracy for classic implementation and 92.43% using the Generalized Convolution.

Designing the sub-network for the Generalized Convolution is a new task, so there are no principles on how to do that properly. Additional research is required. We decided to use the Feed Forward Neural Network with one hidden layer. The hidden layer has twice the number of neurons than the output layer. We used the ReLU activation function. For stabilizing the learning process we use layer normalization [22]. The more common batch normalization was not well suitable in our case, because of GPU limitation we needed to decrease mini-batch size.

The biggest challenge was dealing with limited GPU memory. We didn't have access to the newest units. The classic convolution needs to store the following number of weights for the single convolutional layer: kernel size $*$ kernel size $*$ input channels $*$ output channels. For our method for implementation from Algorithm 2 we need to store 2 $*$ output channels more weights plus these from layer normalizations. However, given the rapid development of graphics cards, the memory will be becoming less problematic. The newest models like the Noisy-Student [6] need to store up to 480M parameters.

5 Summary

The classic convolution operation is the basic block for CNNs in computer vision problems. It allows finding patterns and correlations in data, which in consequence make it possible to successfully recognize images. In modern CNNs, millions of such operations are performed to achieve good results. We proposed the Generalized Convolution to improve this primary operation. It is the generalized version of the classic convolution, in which the convolution operation is replaced by the sub-network. Using the single artificial neuron as the sub-network allows performing operation, which works identically as classical convolution. However a deeper sub-network like the multilayer FFNN allows for learning more complex decision boundaries and in consequence improving i.a., accuracy.

The Generalized Convolution is a concept, that requires further research. Designing a sub-network, which is well suitable to a specific CNN model is still a challenge. Our first test improved the VGG-11's accuracy. We showed how to implement it in the popular PyTorch deep-learning library. Future work will focus on applying it to newer architectures and verifying the quality of the learned features.

References

1. Krizhevsky, A., Sutskever, I., Hinton, G.E.: ImageNet classification with deep convolutional neural networks. In: Advances in Neural Information Processing Systems, pp. 1097–1105 (2012)
2. Simonyan, K., Zisserman, A.: Very deep convolutional networks for large-scale image recognition. arXiv preprint arXiv:1409.1556 (2014)
3. Szegedy, C., Vanhoucke, V., Ioffe, S., Shlens, J., Wojna, Z.: Rethinking the inception architecture for computer vision. In: Proceedings of the IEEE Conference on Computer Vision and Pattern Recognition, pp. 2818–2826 (2016)
4. He, K., Zhang, X., Ren, S., Sun, J.: Deep residual learning for image recognition. In: Proceedings of the IEEE Conference on Computer Vision and Pattern Recognition, pp. 770–778 (2016)
5. Tan, M., Le, Q.V.: EfficientNet: rethinking model scaling for convolutional neural networks. arXiv preprint arXiv:1905.11946 (2019)

6. Xie, Q., Luong, M.-T., Hovy, E., Le, Q.V.: Self-training with noisy student improves imagenet classification. In: Proceedings of the IEEE/CVF Conference on Computer Vision and Pattern Recognition, pp. 10687–10698 (2020)
7. Kingma, D.P., Ba, J.: Adam: a method for stochastic optimization. arXiv preprint arXiv:1412.6980 (2014)
8. Loshchilov, I., Hutter, F.: SGDR: stochastic gradient descent with warm restarts. arXiv preprint arXiv:1608.03983 (2016)
9. Ioffe, S., Szegedy, C.: Batch normalization: accelerating deep network training by reducing internal covariate shift. arXiv preprint arXiv:1502.03167 (2015)
10. Yu, F., Koltun, V.: Multi-scale context aggregation by dilated convolutions. arXiv preprint arXiv:1511.07122 (2015)
11. LeCun, Y., Boser, B., Denker, J.S., Henderson, D., Howard, R.E., Hubbard, W., Jackel, L.D.: Backpropagation applied to handwritten zip code recognition. Neural Comput. **1**(4), 541–551 (1989)
12. Lin, M., Chen, Q., Yan, S.: Network in network. arXiv preprint arXiv:1312.4400 (2013)
13. Pang, Y., Sun, M., Jiang, X., Li, X.: Convolution in convolution for network in network. IEEE Trans. Neural Netw. Learn. Syst. **29**(5), 1587–1597 (2017)
14. Huang, G., Liu, Z., Van Der Maaten, L., Weinberger, K.Q.: Densely connected convolutional networks. In: Proceedings of the IEEE Conference on Computer Vision and Pattern Recognition, pp. 4700–4708 (2017)
15. Xie, S., Girshick, R., Dollár, P., Tu, Z., He, K.: Aggregated residual transformations for deep neural networks. In: Proceedings of the IEEE Conference on Computer Vision and Pattern Recognition, pp. 1492–1500 (2017)
16. Chen, Y., Li, J., Xiao, H., Jin, X., Yan, S., Feng, J.: Dual path networks. arXiv preprint arXiv:1707.01629 (2017)
17. Yalniz, I.Z., Jégou, H., Chen, K., Paluri, M., Mahajan, D.: Billion-scale semi-supervised learning for image classification. arXiv preprint arXiv:1905.00546 (2019)
18. Pham, H., Xie, Q., Dai, Z., Le, Q.V.L.: Meta pseudo labels. arXiv preprint arXiv:2003.10580 (2020)
19. Howard, A.G., Zhu, M., Chen, B., Kalenichenko, D., Wang, W., Weyand, T., Andreetto, M., Adam, H.: MobileNets: efficient convolutional neural networks for mobile vision applications. arXiv preprint arXiv:1704.04861 (2017)
20. Ngiam, J., Chen, Z., Chia, D., Koh, P.W., Le, Q.V., Ng, A.Y.: Tiled convolutional neural networks. In: Advances in Neural Information Processing Systems, pp. 1279–1287 (2010)
21. Checinski, K., Wawrzynski, P.: DCT-Conv: coding filters in convolutional networks with discrete cosine transform. arXiv preprint arXiv:2001.08517 (2020)
22. Ba, J.L., Kiros, J.R., Hinton, G.E.: Layer normalization. arXiv preprint arXiv:1607.06450 (2016)

Scheduling Optimization in Heterogeneous Computing Environments with Resources of Different Types

Victor Toporkov and Dmitry Yemelyanov(✉)

National Research University "MPEI", ul. Krasnokazarmennaya, 14, Moscow 111250, Russia
{ToporkovVV,YemelyanovDM}@mpei.ru

Abstract. In this work, we consider the problem of resources co-allocation for the execution of parallel jobs in high performance computing systems with heterogeneous resources. We propose a solution for the problem of simultaneous co-allocation of resources of different types (computational nodes, data storages, software packets, network connections, etc.). The main difficulty in solving this problem is a single common cost constraint for the user job execution and therefore on the co-allocation of all the required resources. For this purpose, we implement and study a chained recurrent scheme which can optimize a general additive criterion with a hard constraint on the total resources' usage cost. The efficiency of the proposed scheme is studied in comparison to similar greedy algorithms in a dedicated simulation environment.

Keywords: Computing · Scheduling · Resource · Allocation · Job · Heterogeneous · Different types · Composite problem · Knapsack · Chain

1 Introduction and Related Works

Solving large-scale parallel jobs in a distributed computing system (HPCS) requires the use of significant resources of different types. Being non-dedicated, some of these resources should be shared with their owners. This causes additional competition for their use between independent users within global and local job flows. Users submitting jobs and resource owners are the main stakeholders in high performance and distributed computing. They tend to pursue different and usually conflicting goals. HPCS users are most often interested in the quality of service and performance characteristics for their applications: cost, execution time, characteristics of the allocated nodes. Resource owners are primarily interested in maximizing their own profits. To coordinate the preferences of HPCS users and resource owners, so-called economic scheduling models are used [1, 2]. Since the concepts of cost and economic benefit are equally important for both users and resource owners, it is reasonable to use them as objective functions, criteria, and constraints for the scheduling process.

It should be noted that with such interaction models of HPCS participants, economic criterion steps forward, along with the performance factor and resource utilization.

W. Zamojski et al. (Eds.): DepCoS-RELCOMEX 2021, AISC 1389, pp. 447–456, 2021.
https://doi.org/10.1007/978-3-030-76773-0_43

The usage price, value for money, the ability to optimize resource allocation play an important role when executing user-specific jobs.

Economic mechanisms are used to solve problems like resource management and scheduling of jobs in a transparent and efficient way in distributed environments such as cloud computing and utility Grid. The significant and important feature for well-known scheduling solutions for distributed environments, is the fact that the scheduling strategy is formed on a basis of efficiency criteria [1–5].

In cloud computing, the load and maintenance schedules for the available resources (virtual machines, data storages, computing nodes) are partially known, although the service users may not notice this due to the virtualization. However, when administering and scheduling cloud computing processes, it becomes necessary to efficiently allocate resources because of their partial unavailability.

To implement such scheduling schemes and policies, first of all, one needs an algorithm for finding sets of simultaneously available resources required for each job execution. Further we shall call such sets of simultaneously available resources as execution *window*.

The currently known algorithms, their combinations and heuristics used in scheduling routine typically do not allow obtaining optimal or close to optimal solutions due to a general heterogeneity of HPCS resources. Some known models search for the first suitable set of resources, without any optimization. Other models do not consider the economic aspects or those related to the non-dedicated resources.

Below is a summary of several traditional and modern resource management systems and approaches.

First fit slot selection algorithms [6, 7] assign any job to the first set of slots matching the resource request conditions, while other algorithms use an exhaustive search [4, 8, 9] and some of them are based on a linear integer programming (IP) [4, 8] or mixed-integer programming (MIP) model [9].

The simulators of distributed and cloud computing GridSim [10] and CloudSim [11], have built-in means for starting and executing jobs, that allow users to request a set of simultaneously available resources. In this case, the first suitable set (First Fit) is selected for each job, without additional optimization. CloudAuction [12], the CloudSim extension, allocates the required set of resources from various data centers through a double auction.

The algorithm [13] minimizes the job completion time and the waiting time in the queue and does not consider the constraint on the total resource allocation budget. The algorithm in [14] operates with a list of slots organized by non-decrease of start time, implements window scrolling algorithms, minimizes the job waiting time, but does not optimize other criteria, and does not consider the constraints on the total job execution budget.

Algorithm [15, 16] implements knapsack-based resources co-allocation with constraints on the total budget, number of required resources and optimization by a general additive criterion.

Most importantly, all the above models and solutions are designed to allocate homogeneous or even heterogeneous *resources of the same type*. Even CloudSim [11, 12] environment supports different type of resources (CPU cores, RAM, storage, bandwidth), but provides them inside discrete virtual machines' offers. Thus, the problem of joint allocation of resources of different types remains poorly researched.

The rest of the paper is organized as follows. Section 2 introduces a problem statement and a scheduling model with resources of different types. Section 3 contains simulation results and analysis. Finally, Sect. 4 summarizes the paper.

2 Co-allocation of Resources of Different Types

2.1 Problem Statement

Simultaneous co-allocation of resources of different types (computational nodes, data storages, software packets, network connections, etc.) usually cannot be described with a single resource request or solved as a single optimization problem. Different allocation requirements are usually given for different type of the resources: minimum and maximum required number of resources, their relative utility value and so on. However, when a few different resource types are required for a single user job, there is always a single common constraint on the total resources' usage price, i.e., a budget allocated for the job execution.

The very formulation of a problem of co-allocating a number of heterogeneous resources of different types according to user criterion z_i and a single constraint on the total usage cost seems impractical and incorrect. z_i criterion (or preliminary user utility estimations for each available resource i) calculation features can lead to unbalanced resources combinations. For example, only computational nodes (without access to the network or data storages) or a set of solely network resources (without a computational capacity) may be allocated for such a request.

Thus, the resources allocation model in heterogeneous environments with resources of different types requires formulation and simultaneous solution of several co-allocation sub-problems specific for each type of resource. For example: allocate n_1 computational nodes, the corresponding number n_1 of software packets, $[n_3; n_4]$ storage drives and at least one network connection for a total allocation budget C. Such a problem formulation guarantees selection of a minimum set of resources of each type required to execute a job, including those having relatively small user utility values z_i.

Individual solutions for each sub-problem with a dedicated budget constraint C_j and a single resource type are well studied and may be obtained in different ways. For example, in [15, 16] we studied the problem of co-allocating a window of n simultaneously available *computing* resources. Each of m available computing resources has cost c_i and performance $p_i = z_i$ characteristics. The following optimization problem represents the total window performance maximization with a total cost constraint C_j:

$$Z = \sum\nolimits_{i=1}^{m} z_i x_i \rightarrow \max, \tag{1}$$

with the following restrictions:

$$\sum_{i=1}^{m} c_i x_i \le C_j,$$
$$\sum_{i=1}^{m} x_i = n,$$
$$x_i \in \{0, 1\},\ i = 1..m,$$

where z_i is a target characteristic value provided by resource i, c_i is its usage price, x_i is a decision variable determining whether to allocate resource i ($x_i = 1$) or not ($x_i = 0$) for the solution.

This problem relates to the class of integer linear programming problems. We used 0–1 knapsack problem as a base for our implementation in the following dynamic programming recurrent scheme:

$$f_i(c,k) = max\{f_{i-1}(c,k), f_{i-1}(c - c_i,\ k - 1) + z_i\},$$
$$i = 1, .., m,\ c = 1, .., C_j,\ k = 1, .., n, \tag{2}$$

where $f_i(c,k)$ defines the maximum Z criterion value for k-size window allocated out of first i slots from for a budget c. After the forward induction procedure (2) is finished the maximum value $Z_{max} = f_m(C_j, n)$. x_i values are then obtained by a backward induction procedure.

However, in a more general problem of co-allocation resources of N different types there is no apriori information about individual cost constraints C_j for each sub-problem: the constraint C is specified for the whole problem. In this way, individual sub-problems' solutions do not adequately fit the model.

2.2 Composite Scheduling Model

To formalize the problem of allocation of heterogeneous resources of different types we introduce a concept of a composite resource request. A composite resource request contains: 1) a constraint C on the total resources usage cost; 2) a list of independent resource requests G_j for each required type of the resources; 3) a list of functions (or a table) to calculate target optimization criteria values z_i for each available resource instance. The sub-requests in the composite request may contain additional independent constraints on the number of the required resources of the corresponding type. For example, a sub-request may define an exact required amount n of the resources (the standalone problem (1) studied in [15, 16]), the range of admissible numbers $[n_1, n_2]$ (the problem is solved for the current study), or no limitation at all (the classic problem of filling a knapsack).

To solve this problem of selecting heterogeneous resources of different types according to a given composite request, we propose the following *chain* model representing a recurrent scheme with a general cost constraint C.

1. For each subproblem G_j (one for each type of the resource) the recurrent scheme (2) should be prepared and successively calculated for each value $c \in [0; C]$ with the following features:

a. the result of scheme (1) calculation for sub-problem G_j is passed as an input initialization data for sub-problem G_{j+1}:

$$f_0^{j+1}(c,0) = f_{m_j}^{j}(c,n_j),$$

where n_j is resources amount constraint in G_j and m_j is total number of available resources for G_j.

2. When the recurrent schemes for all sub-problems are calculated, the whole chained problem solution may be retrieved from the last problem G_N in the chain:

$$Z_{\max} = f_{m_N}^{N}(C, n_N).$$

3. Specific allocation values x_i for each resource instance i are calculated during the backward induction. Firstly, all x_i are calculated for last sub-problem G_N starting from element $f_{m_N}^{N}(C, n_N)$. After that, the remaining budget C_N left after all the resources of sub-problem G_N are allocated, is passed to the previous sub-problem G_{N-1}. Backward induction for G_{N-1} starts from the element $f_{m_{N-1}}^{N-1}(C_N, n_{N-1})$. More generally, backward induction for sub-problem G_j starts from element $f_{m_j}^{j}(C_{j+1}, n_j)$, where C_{j+1} is a budget left after allocating resources for sub-problems $[j+1; N]$, n_j is a required number of resources in G_j, m_j is a total number of resources suitable for G_j.

As a result, when backward induction is implemented over the first sub-problem chain, the residual budget C_N will be equal to zero. At the same time, all intermediate decision variables x_i for each sub-problem will be determined to maximize the target criterion Z over the whole composite optimization problem.

The main idea of the proposed scheme is to transfer intermediate calculation results between the sub-problems' links. This is justified, because for each sub-problem solution it is sufficient to know the vector $f_0(c, 0)$, $c = 0, .., C$, which determines the best attainable solution for all the previous sub-problems. Thus, each sub-problem G_j and its calculation details remain independent if it takes initialization vector $f_0^{j-1}(c, 0)$ and returns resulting vector $f_0^{j}(c, 0)$. That is why different types of knapsack-based problems may be used for intermediate chains, for example, classical problem or with additional restrictions on the number of the elements.

It should be noted that the order of the sub-problems in the chain does not affect the resulting maximum value $Z_{\max}$ of the composite problem. On the other hand, in case several possible solutions provide the same $Z_{\max}$ value, the resulting set of selected resources (x_i values) may depend on the sub-problems position in the chain and their calculation order.

2.3 Model Example

As a model example we consider the following composite problem for cloud resources allocation. First, we selected a few offers from service [17], which collects resources prices from different vendors, including Amazon, Digital Ocean, Microsoft Azure, etc. Table 1 contains a heterogeneous set of offers for three types of cloud and digital resources: virtual machines (VMs), data storages and operating system software with support plans.

Table 1. A set of cloud resources offers: VMs, disc storage and OS

Type	Vendor	Characteristic	Price, c_i	Utility, z_i
VM	Vendor 1	1 Gb RAM	5	0
VM	Vendor 2	1 Gb RAM	8	0
VM	Vendor 1	2 Gb RAM	10	2
VM	Vendor 3	4 Gb RAM	30	4
VM	Vendor 2	8 Gb RAM	62	8
VM	Vendor 3	8 Gb RAM	61	8
VM	Vendor 1	16 Gb RAM	80	10
DISC	Vendor 1	5 Gb	0.5	0
DISC	Vendor 1	10 Gb	1	0
DISC	Vendor 1	20 Gb	2	0.5
DISC	Vendor 1	40 Gb	4	1
DISC	Vendor 1	80 Gb	8	2
DISC	Vendor 1	120 Gb	12	3.5
DISC	Vendor 1	250 Gb	25	8
DISC	Vendor 1	500 Gb	50	15
DISC	Vendor 1	1000 Gb	100	25
OS	Vendor 4	OS1	0	3
OS	Vendor 4	OS1 software support	120	6
OS	Vendor 4	OS1 enterprise middle support	150	8
OS	Vendor 4	OS1 enterprise 24/7 support	300	9
OS	Vendor 5	OS1 network support	12	5
OS	Vendor 5	OS1 basic limited support	50	6
OS	Vendor 5	OS1 basic support	120	9
OS	Vendor 5	OS1 premier support	140	14
OS	Vendor 6	OS2 developer support	29	7
OS	Vendor 6	OS3 standard support	100	12
OS	Vendor 6	OS3 professional support	1000	20

Following are some important notes for the Table 1.

- For VMs the main displayed characteristic is RAM volume. VMs' offers usually provide consistent configurations of CPU performance and available RAM volume. However, RAM characteristic usually has more heterogeneous and scalable values.
- The price of the resources is provided in dollars per month.
- The fifth column contains pre-calculated user utility estimations independently for each offer. Such a presentation is more illustrative in comparison with custom nonlinear and piecewise-defined utility estimation functions.

The model composite request includes the following resource requirements: 8–12 VMs, 1–8 storage drives and one OS delivery and support plan (up to 15 workstations) should be allocated for 350 dollars a month.

The result provided by the proposed chain algorithms is the following: 7 VMs with 2 GB RAM for 10 dollars, one VM with 1 GB RAM for 5 dollars, three additional storage drives of 500 Gb each and five drives of 250 Gb each, free OS distribution without corporate support.

These results can be explained by relatively higher utility values specified for the storage drives. It is possible to adjust the co-allocation results by reevaluating user estimations based on the current solution (for example, by increasing the VMs utility values). Thus, a satisfactory solution can be achieved in a finite number of the adjustment iterations.

3 Simulation Study

For a more general study we consider co-allocation problem in computing environment with resources of three different types and randomized characteristics. Random generation of the resources' characteristics (cost, performance, disk space, bandwidth, etc.) allows us to simulate and obtain averaged results from a large number of scheduling scenarios in various configurations of the computing environment.

As an alternative to the *Chain* co-allocation model, we implemented a greedy analogue algorithm *GX*.

1. Firstly, GX distributes the available budget between the individual sub-problems of the composite query.
2. Secondly, each sub-problem is solved independently by the corresponding simple greedy algorithm.
3. The results of the individual solutions are combined into a common composite solution.

Based on the budget distribution functions we highlight the following specific GX implementations.

1. *GUniform* distributes budget C equally between all the sub-problems: C/N for each sub-problem.
2. *GProportional* allocates budget C according to a heuristic metric that considers the average cost and requested amount of each type of resource.
3. *GReferenced* accepts the composite problem solution obtained by the Chain algorithm as a reference and distributes the available budget in the same proportion.

4. Additionally, for each GX algorithm we considered an option to increase the available budget in order to estimate the particular C value required to sufficiently improve the target criterion Z. This modification will be further denoted by adding suffix X to the corresponding greedy algorithm. For example, *GUniform* X1.1 represents GUniform algorithm with an increase in the available budget by 1.1 (10% increment).

The algorithms' comparison and study were performed based on a single composite job scheduling problem in heterogeneous environment containing three types of resources: VMs, storage drives and software packages. The composite resource request contains requirements for allocation of 8–12 VMs, any number of storage drives and one (common) license for a software product. The utility values z_i were precalculated so that the most obvious uniform budget division does not guarantee the best solution (relatively higher utility values were generated for the VM resources). The job execution budget is 350 cost units is chosen in such a way to allow the least flexible algorithm *GUniform* to generate feasible solutions in a computing environment consisting of 200 randomly generated resources.

More than 6000 independent job scheduling scenarios were performed to obtain significant results summarized in Table 2. The following main results should be outlined.

Table 2. Simulation results: co-allocation of resources of different types

Algorithm	Z_{max}	Used budget	Working time, ms
GUniform	622.9	220.2	0.4
GProportional	752	322.5	0.4
GReferenced	776.9	323.2	0.4
GUniform X2.1	825.6	387.4	0.4
GProportional X1.35	829.9	391.8	0.4
GReferenced X1.18	826.6	380.4	0.4
Chain	827.4	349.8	34.3

The proposed *Chain* algorithm is able to solve co-allocation problem for resources of different types and maximize the target optimization criterion while strictly observes the constraint on the available job execution budget. The advantage by the target criterion Z over greedy analogues ranges from 6% to 25%. The closest greedy algorithm *GReferenced* used *Chain* solution results for the budget distribution and, thus, is not independent.

The proposed Chain algorithm almost always use the entire available budget $C = 350$. The average value over all simulations is 349.8, and the minimum value is 348.

Apparently, a main drawback for all the considered naïve greedy solution is inefficient rule of the budget distribution between the individual sub-problems. For example,

GUniform at average used only 63% of the allocated budget. *GProportional* and *GReferenced* on average left about 8% of the budget unused. Thus, it is possible to increase their efficiency by increasing the budget constraint so, that the actual usage cost does not exceed the original limit C.

To improve the greedy algorithms' efficiency, modifications *GUniform* X2.1, *GProportional* X1.35 and *GReferenced* X1.18 were additionally considered. Their budget constraints were increased by 110%, 35% and 18% respectively so that the resulting Z_{max} value approximately coincides with the result of the *Chain* algorithm ($Z_{max} \approx 827$). However, in this case the total cost of the allocated resources is much higher than the 350 units limit. Thus, it is impossible to obtain results comparable to Chain algorithms just by manipulating simpler analogues.

The main drawback for the Chain algorithm is its relatively high computational complexity and the corresponding time required for the calculations. In the considered small example with only 200 resources the average *Chain* computing time was 35 ms on a laptop with Intel Core i3 processor. On the other hand, the greedy analogues' working time did not exceed even 1ms (about 100 times faster). Based on these results, the proposed Chain algorithm can be quite efficient in relatively small environments with up to several thousand resources. Some additional optimization technics may be implemented to use *Chain* algorithm in massively parallel environments with high degree of the resource homogeneity.

4 Conclusion

In this work, we study co-allocation problem from a set of resources of different types (computational nodes, data storages, software packets, network connections, etc.). A composite resource request should be specified to formalize the allocation requirements for all types of available resources. The main difficulty in solving this problem is a single common cost constraint specified for the total resources' usage cost.

A special chained algorithm was proposed to consider several knapsack-based homogeneous problems with a common cost limit. The efficiency of the proposed scheme was studied in comparison to similar greedy algorithms in a dedicated simulation environment. The advantage of the chained algorithm by the user target criterion over greedy analogues ranges from 6% to 25% depending on their budget distribution policy.

The main drawback for the chained algorithm is its relatively high computational complexity following from the need to solve several knapsack problems with a relatively high common cost constraint. Thus, the demonstrated optimization features are rather applicable for relatively small environments with up to several thousand resources.

References

1. Buyya, R., Abramson, D., Giddy, J.: Economic models for resource management and scheduling in grid computing. J. Concurr. Comput. Pract. Experience. **5**(14), 1507–1542 (2002)

2. Ernemann, C., Hamscher, V., Yahyapour, R.: Economic scheduling in grid computing. In: Feitelson, D.G., Rudolph, L., Schwiegelshohn, U. (eds.) Job Scheduling Strategies for Parallel Processing. LNCS, vol. 2537, pp. 128–152. Springer, Heidelberg (2002). https://doi.org/10.1007/3-540-36180-4_8
3. Lee, Y.C., Wang, C., Zomaya, A.Y., Zhou, B.B.: Profit-driven scheduling for cloud sevices with data access awareness. J. Parallel Distrib. Comput. **72**(4), 591–602 (2012)
4. Garg, S.K., Konugurthi, P., Buyya, R.: A linear programming-driven genetic algorithm for meta-scheduling on utility grids. Int. J. Parallel Emergent Distrib. Syst. **26**, 493–517 (2011)
5. Kurowski, K., Nabrzyski, J., Oleksiak, A., Weglarz, J.: Multicriteria aspects of grid re-source management. In: Nabrzyski, J., Schopf, J.M., Weglarz, J. (eds.) Grid Resource Management State of the Art and Future Trends, pp. 271–293. Kluwer Academic Publishers (2003)
6. Aida, K., Casanova, H.: scheduling mixed-parallel applications with advance reservations. In: 17th IEEE International Symposium on HPDC, pp. 65–74. IEEE CS Press, New York (2008)
7. Elmroth, E., Tordsson, J.: A standards-based grid resource brokering service supporting advance reservations, co-allocation and cross-grid interoperability. J. Concurr. Comput. Pract. Experience **25**(18), 2298–2335 (2009)
8. Takefusa, A., Nakada, H., Kudoh, T., Tanaka, Y.: An advance reservation-basedco-allocation algorithm for distributed computers and network bandwidth on QoS-guaranteed grids. In: Frachtenberg, E., Schwiegelshohn, U. (eds.) JSSPP 2010 LNCS, vol. 6253, pp. 16–34. Springer, Heidelberg (2010)
9. Blanco, H., Guirado, F., Lérida, J.L., Albornoz, V.M.: MIP model scheduling for multi-clusters. In: Caragiannis, I., et al. (eds.) Euro-Par 2012: Parallel Processing Workshops. LNCS, vol. 7640, pp. 196–206. Springer, Heidelberg (2013). https://doi.org/10.1007/978-3-642-36949-0_22
10. Sulistio, A., Cibej, U., Venugopal, S., Robic, B., Buyya, R.: A toolkit for modelling and simulating data grids: an extension to GridSim. J. Concurr. Comput. Pract. Experience. **20**(13), 1591–1609 (2008)
11. Calheiros, R.N., Ranjan, R., Beloglazov, A., De Rose, C.A.F., Buyya, R.: CloudSim: a toolkit for modeling and simulation of cloud computing environments and evaluation of resource provisioning algorithms. J. Softw. Pract. Experience. **41**(1), 23–50 (2011)
12. Samimi, P., Teimouri, Y., Mukhtar, M.: A combinatorial double auction resource allocation model in cloud computing. J. Inf. Sci. **357**, 201–216 (2016)
13. Makhlouf, S., Yagoubi, B.: Resources co-allocation strategies in grid computing. In: CIIA, CEUR Workshop Proceedings, vol. 825 (2011)
14. Netto, M.A.S., Buyya, R.: A flexible resource co-allocation model based on advance reservations with rescheduling support. In: Technical Report, GRIDSTR-2007–17, Grid Computing and Distributed Systems Laboratory, The University of Melbourne, Australia (2007)
15. Toporkov, V., Yemelyanov, D.: Dependable slot selection algorithms for distributed computing. In: Zamojski, W., Mazurkiewicz, J., Sugier, J., Walkowiak, T., Kacprzyk, J. (eds.) Contemporary Complex Systems and Their Dependability. AISC, vol. 761, pp. 482–491. Springer, Cham (2019). https://doi.org/10.1007/978-3-319-91446-6_45
16. Toporkov, V., Toporkova, A., Yemelyanov, D.: Slot co-allocation optimization in distributed computing with heterogeneous resources. In: Del Ser, J., Osaba, E., Bilbao, M.N., Sanchez-Medina, J.J., Vecchio, M., Yang, X.-S. (eds.) Intelligent Distributed Computing XII. SCI, vol. 798, pp. 40–49. Springer, Cham (2018). https://doi.org/10.1007/978-3-319-99626-4_4
17. https://www.digitalocean.com/pricing/calculator. Accessed 10 Jan 2021

Subject Classification of Texts in Polish - from TF-IDF to Transformers

Tomasz Walkowiak(✉)

Faculty of Electronics, Wrocław University of Science and Technology, 11/17 Z. Janiszewskiego Street, 50-372 Wrocław, Poland
tomasz.walkowiak@pwr.edu.pl

Abstract. The paper concerns the issue of documents in Polish classification according to the subject category. We compare five approaches, starting from classical TF-IDF one, through word2vec methods (fastText in the supervised mode and doc2vec method) up to contemporary transformer-based BERT models (i.e., two models pre-trained on Polish corpora are used). These five approaches are evaluated using five corpora with subject categories (ranging from 5 to 36 labels), different sizes, and lengths of texts. Results suggest that BERT methods outperform other approaches in almost all tests. Due to the size of the model and therefore required time for tuning, the Polbert model is recommended as a solution for Polish text subject classification. The paper includes detailed analysis of recognition and time performance of analyzed methods.

Keywords: Text classification · Word embedding · Language model · fastText · BERT

1 Introduction

In this paper, we describe the main features, results, and conclusions of a set of experiments in automatic subject text classification for Polish. The main goal of the research was to evolve the effectiveness of different classification approaches for text in the Polish language. Commonly used methods of text classification [18,20,21] rely on representing documents with feature vectors and the use of statistical classifiers to assign documents to defined groups. In works [15,23] authors suggest that usage of lemmatization is crucial in text classification for Polish due to its rich morphology. However, results from [20] show that approaches like fastText [6] in the supervised mode, that do not require lemmatization, can perform as well as methods based on lemmatization and bag-of-words [4] or even better when the training set contains sufficiently many examples.

The main aim of this paper is to extend the range of analyzed methods from [20] with language model [12] ones. Therefore, the word2vec model [3] learned on a huge corpus of text in Polish [9] and the newest, transformer [19] based, BERT [2] approach with models for Polish [7,16] are used. Moreover, the corpora tested in [20] were extended by three sets of texts in Polish.

W. Zamojski et al. (Eds.): DepCoS-RELCOMEX 2021, AISC 1389, pp. 457–465, 2021.
https://doi.org/10.1007/978-3-030-76773-0_44

The paper is structured as follows, firstly, we describe data sets used in the analysis. In the Sect. 3, we overview used classification method. After that, the performed experiments and results are presented. Finally, the conclusions are given.

2 Data Set

2.1 Overview

We evaluated five methods (described in Sect. 3) using five collections of text documents in Polish: *Wiki*, *Press*, *Web*, *Rev*, and *Qual*. Table 1 summarizes the properties of these collections.

2.2 Wikipedia

The *Wiki* corpus consists of articles extracted from the Polish language Wikipedia. It was created by merging two publicly available collections [13,14]. It includes 9,837 articles in total. It is the only publicly available corpus for subject classification therefore it could be treated as the benchmark for subject classification of Polish texts.

2.3 Press News

The next data set, *Press* [23] consists of Polish press news. There are 6,564 documents in total in this corpus. The texts were assigned by the press agency to 5 subject categories (i.e., diplomacy, sport, disasters, economy, business, and transportation). All the subject groups are very well separated and each group contains a reasonably large number of members (ca. 1,300 documents per label) without big differences among label sizes. This is an example of a good quality and large data set.

2.4 Articles from a Web Portal

The *Web* corpus [20] contains articles from one of the Polish web portals. Articles were assigned to subject groups by the portal administrator. Due to the overlapping of subject areas (e.g. philosophy, religion, worldview) it could contain many uncertain or wrong assignments. It contains 2,100 documents, relatively large (in average ca. 2,000 words) and greatly varied in sizes (from 12 to ca. 36,000 words). It could be seen as a practical case but with bad quality of labels.

2.5 Scientific Reviews

The next corpus, *Rev* [22] consists of reviews of scientific works from 22 different science areas. There are ca. 10,500 documents in this corpus. The average size of documents is the largest among analyzed corpora, i.e., ca. 2,500 words. The sizes of classes vary widely, from 6 documents per label up to ca. 1,600.

2.6 Qualification Descriptions

The last data set, *Qaul* [11] covers ca. 3,000 documents containing descriptions of qualifications from a Polish public register of the Integrated Qualifications System and descriptions of degrees from Polish universities. The descriptions mainly consist of so-called learning outcomes statements, which characterize the knowledge, skills, and attitudes required to obtain a given qualification or degree. The data were manually labeled. The labels denote the sectors to which the qualifications belong (i.e., IT, medicine, culture).

Table 1. Data sets statistics

Data set	No. labels	Label (class) size in files				File size in words			
		Mean	Min	Max	Std	Mean	Min	Max	Std
Wiki	34	289	150	300	32	320	3	8882	536
Press	5	1312	1052	1823	309	289	21	4028	371
Web	25	84	51	149	28	1927	13	36964	1975
Rev	22	475	6	1671	459	2512	286	15682	1260
Qual	36	83	5	571	112	1230	23	20309	1428

3 Methods

3.1 TF-IDF

This approach uses a bag of word model [17] to build feature vectors. It is based on counting the occurrences of most common words in corpora and weighting these frequencies by maximum term frequency in a document and by IDF (inverse document frequencies). On contrary to [20], we have used words from a text, i.e., we were not doing any lemmatization. The most frequent words (1000 in the analyzed case) and corresponding IDFs were set up on the training set and used for feature calculation for all data (i.e., training and testing). Next, the TF-IDF data were Z-score normalized, i.e., we have removed the mean and scaled feature vectors to the unit variance. Similarly as in the case of TF-IDF transformation the parameters of the Z-score transformation (i.e., the mean and standard deviation of each feature) was set up on the training set and used for testing. As the classifier, we have used the multilayer perceptron (MLP) [5] trained using the stochastic gradient descent method (SGD) [5].

3.2 Doc2vec

The big step in the area of text classification was introducing of word2vec methods [10]. In these approaches, individual words are represented by high-dimensional feature vectors (word embedding) trained on large text corpora.

Since classical statistical classifiers require constant length input vectors the most common solution is to average vector representations of individual words. This approach is known as doc2vec [6]. The original word2vec method was extended [3] with position weights and sub-word information (character n-grams) that allow to generate embedding for unseen words and behave much better for morphologically rich languages as Polish is. Within reported experiments, we used the pre-trained vectors for the Polish language [9][1].

Similar to TF-IDF the doc2vec embeddings were Z-score normalized and classified by the MLP. Doc2vec as well as TF-IDF ignores word order. So, these methods are not-aware of word contexts.

3.3 FastText

The next analyzed method, fastText [6] is similar to the previous one since it uses the doc2vec representation of documents (an average of word embeddings). However, word embeddings are learned in a supervised way on the analyzed data set, not on the external large corpora. The main idea behind fastText is to perform word embedding and classifier learning simultaneously. FastText allows building the embeddings not only for single words but also for word n-grams making the solution context-aware. In the performed experiment, we have used bigrams. Since fastText forms the linear model, it is very effective for training [6]. During classification, bigrams (since they were used in our case) that do not exist in the embedding model (because they do not exist in the training corpus) are omitted from the averaging.

3.4 BERT

The newest approaches to language modeling are inspired by deep-learning algorithms and context-aware methods. The state of the art is BERT [2] based on transformers [19] architecture. Due to its bidirectional representation, jointly built on both the left and the right context, BERT looks at the whole sentence before assigning an embedding to each word in it. Therefore, the embeddings are context-aware. The main drawback of this family of methods is the requirement of using GPU cards, whereas other methods (TF-IDF, doc2vec and fastText) could work on CPU.

Polbert. In this study, we used the Polbert[2], a pre-trained BERT model for Polish. It consists of 12 layers, 768 hidden unit embeddings, 12 attention heads, and 110M parameters in total [7]. We have extended the Polbert network by additional classification layers (full connected) and tuned on each dataset. Only the embedding layer of the Polbert was frozen. Since Polbert is capable of analyzing up to 512 subwords, longer texts were cut-off.

[1] http://hdl.handle.net/11321/606.

[2] https://huggingface.co/dkleczek/bert-base-polish-cased-v1.

HerBERT. Next, we used the best[3] Bert model for the KLEJ[4] benchmark [16], i.e., HerBert (large)[5]. This a BERT (large) based language model trained on Polish. It consists of 24 layers, 1024 hidden unit embeddings, 16 attention heads, what give in total 335M parameters. So it is three times larger than Polbert. Similarly to the Polbert case, we have extended the network by additional classification layers and tuned on each data set.

4 Experiments and Results

4.1 Experiment Organization

The study was performed according to the stratified k-fold cross-validation [5] (with 10 folds). Each data set was divided randomly into 10 folds, and 9 folds were used for training and renaming one for testing. We used three metrics calculated on the test data-set to report method performance: accuracy, average f1 score, and weighted f1 score. Accuracy is defined as all correctly classified objects over all objects. The accuracy is not well suited for no-equally sized labels. That is why we have also used the f1 measure. However, it is defined (as a harmonic mean of precision and recall) for binary classification. For multi-class problems we can average f1 scores for each label or take into account the support (the number of examples in each class) and make a weighted average that takes into account the label imbalance.

4.2 Results

The results, i.e., mean values and standard deviations for accuracy, average f1, and weighted f1, all considered methods and five analyzed data-sets are presented in Tables 2, 3, and 4.

Table 2. Accuracy score

Data set	TF-IDF		doc2vec		fastText		Polbert		HerBERT	
	Mean	Std	Mean	Std	Mean	Std	Mean	Std	Mean	Std
Wiki	0.8302	0.0038	0.9044	0.004	0.8816	0.0055	0.943	0.0064	**0.9494**	0.0066
Press	0.9438	0.0056	0.9559	0.0038	0.9587	0.0049	0.9704	0.0068	**0.9718**	0.0091
Web	0.5627	0.0098	0.5764	0.0274	0.5451	0.0221	0.5533	0.0194	**0.5952**	0.0324
Rev	0.9425	0.0046	0.917	0.0036	**0.9659**	0.0037	0.9524	0.0077	0.9555	0.0106
Qual	0.7943	0.0165	0.8023	0.0116	0.8155	0.011	**0.83**	0.0137	0.8221	0.0253

As it could be noticed, the BERT-based methods outperform the remaining ones. The exception is the *Rev* data set where for accuracy and weighted f1 score

[3] on march 2021.
[4] https://klejbenchmark.com/leaderboard/.
[5] https://huggingface.co/allegro/herbert-large-cased.

Table 3. Mean f1 score

Data set	TF-IDF		doc2vec		fastText		Polbert		HerBERT	
	Mean	Std	Mean	Std	Mean	Std	Mean	Std	Mean	Std
Wiki	0.8275	0.0033	0.9035	0.0042	0.8811	0.0052	0.9431	0.0064	**0.9494**	0.0063
Press	0.9397	0.0061	0.9525	0.0039	0.9555	0.0052	0.9678	0.0072	**0.9693**	0.0097
Web	0.5101	0.0127	0.5403	0.0336	0.524	0.0227	0.521	0.0275	**0.5636**	0.0324
Rev	0.8899	0.0174	0.8753	0.0238	0.9214	0.0148	0.9392	0.0213	**0.9478**	0.0158
Qual	0.6966	0.0254	0.7075	0.0311	0.736	0.0128	**0.7416**	0.0416	0.7322	0.0677

Table 4. Weighted f1 score

Data set	TF-IDF		doc2vec		fastText		Polbert		HerBERT	
	Mean	Std	Mean	Std	Mean	Std	Mean	Std	Mean	Std
Wiki	0.8291	0.0035	0.9042	0.004	0.8865	0.0051	0.9428	0.0064	**0.9492**	0.0067
Press	0.9437	0.0056	0.9559	0.0038	0.9588	0.0048	0.9705	0.0067	**0.9718**	0.0091
Web	0.5384	0.0128	0.5654	0.0327	0.5444	0.0212	0.5425	0.023	**0.5846**	0.0328
Rev	0.9416	0.0047	0.9163	0.0037	**0.9661**	0.0037	0.9521	0.0079	0.9555	0.0106
Qual	0.7876	0.0167	0.7968	0.0127	0.8176	0.0108	**0.8262**	0.0148	0.8174	0.027

fastText gives the best results. The difference among used metrics for *Rev* data set is caused by its high imbalance in labels and file sizes (see Table 1). Since differences between results for Polbert and HerBERT are really small, except for the *Web* data set, we recommend using Polbert since its model is three times smaller, and therefore less memory is required for GPU. Surprisingly, the classical TF-IDF method, without any lemmatization, performs quite well and the results, except of the case of *Wiki* corpus, are close to other non GPU methods.

4.3 Time Performance

Moreover, we measured the computation time required for building a model for each method. In the case of CPU-based methods (TF-IDF, doc2vec, and fastText) the experiments were performed on the machine with 16 cores, and we use the multi-core implementation of these methods. Since the learning time is data dependent, we present results for learning 10 models (since we use the 10-folds approach). For GPU based methods (Polbert and HerBERT) we used five GeForce RTX 2080 Ti cards with 11 GB of memory each. Again, we measure the whole process of learning 10 models. However, here, the flow of experiments was different since each learning was done on a separate GPU.

The results are presented in Table 5. It should be noted that direct comparison of time results between GPU and CPU-based methods does not give information about differences in computational complexity of methods. However, such a comparison shows which method could be recommended for practical applications. Moreover, the learning time could be shortened by decreasing the maximum number of iterations (epochs) set for each method. However, used values (200 for MLP in TF-IDF and doc2vec, 5000 for fastText, and 15 for Polbert and HerBERT) were selected as ones giving satisfying results for all analyzed data sets.

Table 5. Time for building ten classification models for each method and data set. The Polbert and HerBert method required GPU and the calculation was done on five GeForce RTX 2080 Ti. The experiments with TF-IDF, doc2vec, and fastText were performed on CPU with 16 cores.

Data set	TF-IDF	doc2vec	fastText	Polbert	HerBERT
Wiki	6 m25 s	10 m12 s	3 h08 m45 s	1 h58 m40 s	8 h58 m02 s
Press	2 m45s	5 m55 s	2 h10 m50 s	30 m05 s	3 h30 m55 s
Web	5 m21 s	11 m0 s	3 h39 m33 s	33 m49 s	1 h52 m57 s
Rev	22 m05 s	1 h06 m51	29 h41 m53 s	2 h19 m12 s	7 h01 m41 s
Qual	3 m25 s	10 m07 s	2 h49 m20 s	52 m19 s	2 h6 m05 s

The results show that the fastest is the TF-IDF method. The next one is doc2vec, for which calculation time consists of loading of 10 GB language model (ca. 30 s) and MLP model learning time (5 to 10 min). Moreover, it could be noticed that HerBERT requires from 2.4 to 7 times more time than Polbert. It is caused by the fact that Bert large (i.e., HerBERT) has 3 times more parameters to be tuned up during supervised learning. Therefore, it requires more calculations during learning but also the GPU memory limits the number of batches, so it additionally slows down the process. It is hard to compare calculation time for BERT based model with others, due to different hardware requirements. But it could be noticed that in the used testbed Polbert is faster than fastText in a model building process.

5 Conclusion

In this paper, we analyzed five approaches (TF-IDF, fastText, doc2vec, Polbert, and HerBERT) for subject classification of texts in Polish. Results show that the BERT-based approaches (i.e., Polbert and HerBERT) give the best results for almost all analyzed data sets. Since Polbert is a smaller architecture than HerBERT and much faster to tuned up we recommend Polbert as the state-of-the-art solution for Polish. However, BERT-based methods require GPUs. In the case of a lack of GPU doc2vec is an option to be taken into consideration since it gives results similar to fastText but is much faster during the building of the classification model. It is because the doc2vec approach uses the pre-trained feature extraction model (i.e., the language model) and has to build only a classification model, whereas fastText builds the language and classification model simultaneously.

There is always a dream of getting better and better results. Therefore, the authors plan to analyze transformer-based approaches that overcome the limits of 512 tokens existing in used BERT models. There are works on transformers that deal with long text i.e., Reformer [8] and Longfomer [1]. However, up till now, there is a lack of pre-trained models for long text in Polish. Another approach, considered by authors, is to use Graph Convolutional Networks [24].

References

1. Beltagy, I., Peters, M.E., Cohan, A.: Longformer: the long-document transformer. CoRR abs/2004.05150 (2020). https://arxiv.org/abs/2004.05150
2. Devlin, J., Chang, M.W., Lee, K., Toutanova, K.: Bert: pre-training of deep bidirectional transformers for language understanding. arXiv preprint arXiv:1810.04805 (2018)
3. Grave, E., Bojanowski, P., Gupta, P., Joulin, A., Mikolov, T.: Learning word vectors for 157 languages. In: Proceedings of the International Conference on Language Resources and Evaluation, LREC 2018, pp. 3483–3487 (2018)
4. Harris, Z.S.: Distributional structure. Word **10**(2–3), 146–162 (1954)
5. Hastie, T.J., Tibshirani, R.J., Friedman, J.H.: The Elements of Statistical Learning: Data Mining, Inference, and Prediction. Springer Series in Statistics. Springer, New York (2009). autres impressions : 2011 (corr.), 2013 (7e corr.)
6. Joulin, A., Grave, E., Bojanowski, P., Mikolov, T.: Bag of tricks for efficient text classification. In: Proceedings of the 15th Conference of the European Chapter of the Association for Computational Linguistics: Volume 2, Short Papers, pp. 427–431. Association for Computational Linguistics (2017). http://aclweb.org/anthology/E17-2068
7. Kłeczek, D.: Polbert: attacking polish NLP tasks with transformers. In: Ogrodniczuk, M., Kobyliński, Ł. (eds.) Proceedings of the PolEval 2020 Workshop, pp. 79–88. Institute of Computer Science, Polish Academy of Sciences (2020)
8. Kitaev, N., Kaiser, L., Levskaya, A.: Reformer: the efficient transformer. In: 8th International Conference on Learning Representations. OpenReview.net (2020). https://openreview.net/forum?id=rkgNKkHtvB
9. Kocon, J., Gawor, M.: Evaluating KGR10 polish word embeddings in the recognition of temporal expressions using BiLSTM-CRF. CoRR abs/1904.04055 (2019). http://arxiv.org/abs/1904.04055
10. Le, Q., Mikolov, T.: Distributed representations of sentences and documents. In: International Conference on Machine Learning, pp. 1188–1196 (2014)
11. Marcińczuk, M., Gniewkowski, M., Walkowiak, T., Będkowski, M.: Text document clustering: Wordnet vs. TF-IDF vs. word embeddings. In: Proceedings of the 11th Global Wordnet Conference, pp. 207–214. Global Wordnet Association, University of South Africa (UNISA) (January 2021). https://www.aclweb.org/anthology/2021.gwc-1.24
12. Mikolov, T., Sutskever, I., Chen, K., Corrado, G.S., Dean, J.: Distributed representations of words and phrases and their compositionality. In: Advances in Neural Information Processing Systems, pp. 3111–3119 (2013)
13. Młynarczyk, K., Piasecki, M.: Wiki test - 34 categories (2015). http://hdl.handle.net/11321/217. CLARIN-PL digital repository
14. Młynarczyk, K., Piasecki, M.: Wiki train - 34 categories (2015). http://hdl.handle.net/11321/222. CLARIN-PL digital repository

15. Piskorski, J., Sydow, M.: Experiments on classification of polish newspaper. Arch. Control Sci. **15**, 613–625 (2005)
16. Rybak, P., Mroczkowski, R., Tracz, J., Gawlik, I.: KLEJ: comprehensive benchmark for polish language understanding. In: Proceedings of the 58th Annual Meeting of the Association for Computational Linguistics, pp. 1191–1201. Association for Computational Linguistics (July 2020). https://www.aclweb.org/anthology/2020.acl-main.111
17. Salton, G.B.C.: Term-weighting approaches in automatic text retrieval. Inf. Process. Manage. **24**(5), 513–523 (1988)
18. Torkkola, K.: Discriminative features for textdocument classification. Formal Pattern Anal. Appl. **6**(4), 301–308 (2004). https://doi.org/10.1007/s10044-003-0196-8
19. Vaswani, A., Shazeer, N., Parmar, N., Uszkoreit, J., Jones, L., Gomez, A.N., Kaiser, L., Polosukhin, I.: Attention is all you need. In: Guyon, I., Luxburg, U.V., Bengio, S., Wallach, H., Fergus, R., Vishwanathan, S., Garnett, R. (eds.) Advances in Neural Information Processing Systems, vol. 30. Curran Associates, Inc. (2017). https://proceedings.neurips.cc/paper/2017/file/3f5ee243547dee91fbd053c1c4a845aa-Paper.pdf
20. Walkowiak, T., Datko, S., Maciejewski, H.: Bag-of-words, bag-of-topics and word-to-vec based subject classification of text documents in polish - a comparative study. In: Zamojski, W., Mazurkiewicz, J., Sugier, J., Walkowiak, T., Kacprzyk, J. (eds.) Contemporary Complex Systems and Their Dependability, pp. 526–535. Springer, Cham (2019)
21. Walkowiak, T., Datko, S., Maciejewski, H.: Low-dimensional classification of text documents. In: Zamojski, W., Mazurkiewicz, J., Sugier, J., Walkowiak, T., Kacprzyk, J. (eds.) Engineering in Dependability of Computer Systems and Networks, pp. 534–543. Springer, Cham (2020)
22. Walkowiak, T., Gniewkowski, M.: Evaluation of vector embedding models in clustering of text documents. In: Proceedings of the International Conference on Recent Advances in Natural Language Processing, RANLP 2019, Varna, Bulgaria, pp. 1304–1311. INCOMA Ltd. (September 2019). https://www.aclweb.org/anthology/R19-1149
23. Walkowiak, T., Malak, P.: Polish texts topic classification evaluation. In: Proceedings of the 10th International Conference on Agents and Artificial Intelligence, ICAART, vol. 2, pp. 515–522. INSTICC, SciTePress (2018)
24. Zhu, H., Koniusz, P.: Simple spectral graph convolution. In: International Conference on Learning Representations (2021). https://openreview.net/forum?id=CYO5T-YjWZV

A Proposal to Use Elliptical Curves to Secure the Block in E-voting System Based on Blockchain Mechanism

Marek Woda(✉) and Zen Huzaini

Department of Computer Engineering, Wroclaw University of Technology, Janiszewskiego 11-17, 50-372 Wrocław, Poland
marek.woda@pwr.edu.pl

Abstract. *E-voting* based *blockchain* technology has potential use in strengthening cybersecurity and securing votes. However, one of *Blockchain's* fundamental mechanisms, *Proof-of-Work*, is not possible to be implemented in the *E-voting* based *blockchain* since it requires a very high resource. As a result, there is still a possibility to alter the data within transaction block and recalculating its hash will be easy. In addition, without the *Proof-of-Work*, the authority of the block could not be validated. Therefore, the digital signature could be the answer to solve these problems. This research will provide the implementation result of elliptical curve cryptography for digital signature using two different signature algorithms, namely, *ECDSA* (curve *secp256k1*) and *Ed25519*, in terms of evaluating its performance during key-pair generation, block signing and signature verification on *E-voting* system based on *Blockchain*.

Keywords: Blockchain · E-voting · Signature algorithm · Elliptical curve · ECDSA · ED25519

1 Introduction

Voting is a process for a group, an organization, or even a country to gather a voice to decide. An open ballot occurs in a wide variety of contexts: political elections, decision-making in multimember bodies, the election of the president, the issuing of verdicts by jury, the election of officers to various positions in public organizations, etc. It is true; however, such a voting is not a sufficient condition for democratic governance [18]. In fact, there are some drawbacks to the open voting system in terms of data modification or manipulation. Taking an example of presidential elections in Indonesia in 2019 that uses a paper open ballot as the voting card, the BPN found at least 1,261 reports of fraud [1]. To overcome this problem, a secure E-voting system is required. Apart from giving efficiency, reliability, and accessibility, the *E-voting* system using *blockchain* technology could be very acceptable for security reasons as it can help reduce the possibility of data manipulation.

W. Zamojski et al. (Eds.): DepCoS-RELCOMEX 2021, AISC 1389, pp. 466–476, 2021.
https://doi.org/10.1007/978-3-030-76773-0_45

2 Problem Overview

Blockchain technology has potential use in fighting fraud [10], strengthening cybersecurity [9], and securing voting [11]. However, since the *Proof-of-Work* cannot be implemented in the *E-voting* system, there is still a possibility to alter the data within the block and recalculate its hash in a fast way. In a *Proof-of-Work* system, computers must "prove" that they have done "work" by solving a complex computational math problem. If a computer solves one of these problems, they will be eligible to add a block of a transaction to the *blockchain*. However, the process of adding blocks to the *blockchain* - it is called "mining," is not easy. In fact, the chances of solving one of these problems on the *Bitcoin* network in January 2020 were about one in 15.5 trillion. To solve complex math problems at that odds, computers we run compound algorithms, which consume significant amounts of power [23] as well as very time consuming [13]. In terms of cyberattacks, [23] also stated that *Proof-of-Work* does not make attacks by hackers impossible, but it does make their efforts become useless. Furthermore, if a hacker wanted to launch a blockchain attack, he would have to control more than 50% of all computing power in the blockchain (namely, control 50% plus one node in the blockchain) to "vote" the rest of the network members. Therefore, even though the implementation of *Proof-of-Work* is good, unfortunately E-voting based *blockchain* cannot be implemented. Previous research [2, 6, 16] about E-voting based *Blockchain* also does not implement *Proof-of-Work*. Additionally, since the authorial system uses a distributed database (*peer-to-peer*) that replicates the data over different servers, the possibility of making changes still exists. For instance, if someone tries to alter the block in the primary database, the block in the secondary database will be also changed. Because the *Proof-of-Work* is not implemented, hash recalculation for the entire *blockchain* in the database can be done easier in a fast way. Hence, the implementation of a *digital signature* can be the answer to this drawback. *Digital signatures* are designed to guard against tampering and forgery in digital communications [4, 7, 22]. *Digital signatures* are one of the most advanced and promising techniques in the field of *public key cryptography*. Moreover, cryptographic signatures mitigate any attempts to alter the integrity of a document after it has been signed [3]. In this *E-voting* based *blockchain* architecture, the previous transaction's hash and its signature validity must be checked before creating the new block of transaction. Considering that there will be thousands or even millions of transaction blocks, a fast and strong validation signature is required. Therefore, this research focuses on the implementation on the use of *elliptical curve cryptography* for signature algorithm. The use of *signature algorithm* can be a way to validate the block, not only to validate the owner of the block, but it could help to secure the chain before a new block is created and put into the chain.

3 Related Work

According to [28] about *Blockchain-Enabled E-voting System, Blockchain* enables the creation of tamper-proof audit trails for voting. The implemented system is capable of key encryption, smart biometrics, and tamperproof real-time *personal ID* verification. However, this research is only focused on the architecture of E-voting based *blockchain* on a server, such as authentication and arbitration. Moreover, the message is only encrypted using *SHA-256* without implementing any other signature algorithm which is considered less secure. In [2] on research about conceptual *Blockchain*, it has been provided 4 main functionalities, they are *authentication*, *anonymity*, *accuracy*, and *verifiability*. In detail, the proposal on *E-voting* based *Blockchain* implementation provides vote encryption and consensus to avoid forking on the *blockchain*. However, the encrypted vote is only using one-way function encryption or SHA-256. In [6], the authors focus only on the research of the *E-voting* protocol that could meet some obligatory aspects such as fairness, eligibility, privacy, etc. This new protocol was designed to respect essential *E-voting* properties as well as to provide a degree of decentralization and to enable the elector to alter/update his or her vote (within the allowable voting period). Therefore, this paper is only about a potential roadmap for *blockchain* technology that can support complex applications. In [16], Moura and Gomes have conducted a research on the possibility of using *blockchain* technology to help solve the problems of transparency and trust. They analyzed how the implementation of *Blockchain* in the digital government repertoire could really contribute to common *E-voting* concerns and promote electoral transparency, enhance auditability, increase voter confidence, and strengthen democracy. In the simplest of terms, *Blockchain* is distributed, time-stamped, append-only immutable records of data connected to addresses, simultaneously kept on all nodes within a decentralized network [24]. Each of these data blocks (i.e. block) is secured and bound together by cryptographic principles, this is known as a chain. *Blockchain* network has no central authority, it means that there is no single central place (database) to store all transactions. As *Blockchain* is a shared and immutable ledger, the information in it is open for anyone and everyone can see it. Moreover, the transaction that has been recorded would-be set-in stone and it cannot be changed ever. To "mine" or add a new block, if the transaction though the inside the collection of the *blockchain*, there are several processes that must be accomplished. For example, validation and encryption require the form user to solve a complex mathematical problem using a very large amount of resources and very time-consuming [25]. For the encryption process, there is *a hash function*, a function which is a collision-resistant and irreversible tool to cipher any message [18, 21]. In the *blockchain* technology, the *hash function* has a feature for security and verification function [21]. It is simply because blocks are linked together through their linked *hash*. For instance, when a new block of a transaction is created, all its transactions or data within the block will be hashed (Fig. 1).

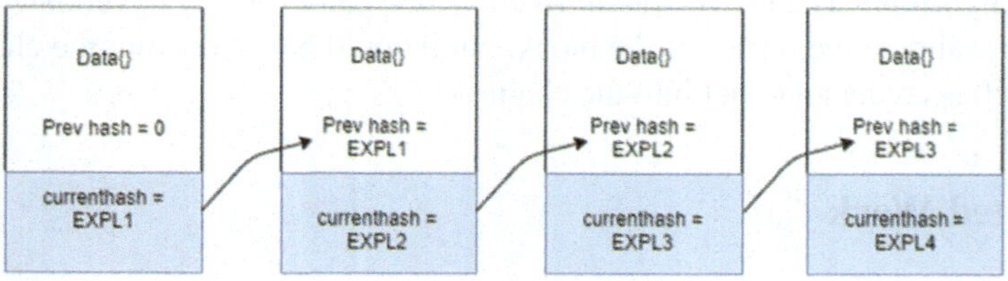

Fig. 1. Hash functions secure blocks

In *blockchain* technology, a *digital signature* is one of the mechanisms that must exist, if only *the hash* function is not implemented. In the simplest meaning, *a blockchain* cannot exist without the implementation of a hashing functions or a *digital signature.* Therefore, every transaction carried out on the *blockchain* must be signed by the sender's electronic signature using their private key or it must be hashed using hashing functions. A *digital signature* is an essential building block in *blockchains* as it is used to verify the authenticity of transactions. If someone submits a transaction (e.g., casts a vote in an open ballot), he must prove to every node in the system that he is authorized to. The blockchain nodes (miners), who have knowledge of the voter public key, will then check the conditions of the transaction, and validate the authenticity of the signature. Once the validity is confirmed, the block containing a transaction will be then ready for finalization by a validator. Simply speaking, to *digitally sign* a message is to provide some authentication of its origin. Two most commonly used *digital signature* algorithms are *RSA* (*Rivest–Shamir–Adleman*) and *ECC* (*Elliptic-curve cryptography*) [4], in which *ECC* is considered better [8]. *ECC* is a public key cryptography approach that relies on an elliptical curve that produces faster, smaller, and better keys for cryptographic purposes [27]. *ECC* generates the keys through the features of the equation of the elliptical curve. *ECC* is one of the asymmetric cryptographies. *ECC* is considered better than *RSA*. Even though the security level of *ECC* is the same [5]. Table 1 shows *ECC* can provide the same cryptographic strength as an RSA-based system with much smaller key sizes.

Table 1. Key size compared to cryptographic strength

Symmetric key size (bits)	RSA and Diffie Hellman key size (bits)	Elliptic curve key size (bits)
80	1024	160
112	2048	224
128	3072	256
192	7680	384
256	15360	521

ECDSA is a type of digital signature from *ECC* that provides data origin, authentication, data integrity, and non-repudiation services [20]. *ECDSA* can perform as a signature/verification algorithm [20]. ECDSA, compared to RSA, uses shorter keys and requires less computational requirements while still maintaining strong security. Moreover, in terms of security level, the key pairs and signatures generated by *ECDSA* are shorter than *RSA*, and a 256-bit ECDSA signature has the same security as 3072-bit RSA signature (Table 1). *Ed25519,* which is performed using *SHA-512* and *Curve25519,* is a variation of *EdDSA*. *EdDSA* uses performance-optimized elliptic curves such as *Curve448* and curve *Curve25519* (255-bit) [19]. Comparing the signing and verification using *EdDSA*, *EdDSA* is simpler than ECDSA, easier to comprehend and implement. Both *signature algorithms* have similar security strengths for curves with similar key lengths. With the most widely known curves (such as *edwards25519* and *edwards448),*

EdDSA algorithm is slightly faster than the *ECDSA*, however, this relies on the curves used and the implementation. Unlike *ECDSA*, *EdDSA* signatures do not provide a way to recover the signatory's public key from the signature and the message. Typically, *EdDSA* is preferred for most modern apps [17].

4 System Design

To perform this research, namely, to develop, and test an authorial open ballot E-Voting system, a stack of technology was required. The chosen technologies were Node.js for the backend, MongoDB for the database, and some node package modules such as *sha256*, *tweetnacl*, *crypto-js* and *jsonwebtoken* (Table 2).

Table 2. Node Package Modules used for system's implementation

Node Module	Description
Crypto-js	JavaScript library of crypto standards that performs SHA256 encryption function
Tweentnacl	Public-key authenticated encryption for Ed25519 signature
Crypto	Public-key authenticated encryption for ECDSA signature
Jsonwebtoken	To encrypt user information during the vote
Mongoose	MongoDB object modeling tool designed to work in an asynchronous environment
Express	Web framework for node

Figure 2 shows how the block is created and stored in the *blockchain*. Several aspects have been illustrated starting from *vote casting*, *validation*, and *mining* process.

Casting Vote: To be able to cast a vote, a user should have been registered as an eligible user. Moreover, the inserted data must be valid.

Validation: Before casting vote, some validation process must be done. the 1st validation is to see if the required inputs are correct. the 2nd validation is to verify whether the prior block exists. And if it is not, therefore *Genesis Block* mining process will have to be carried out since the first block in the *blockchain* is a genesis block. In essence, this block of genesis contains some dummy information because when the vote is calculated, this block will always be excluded. Moreover, this block must exist because it will perform as a starting block in which its value will be used in the next block [7]. the 3rd validation is to check if the previous block's hash is valid. This process will validate the hash by rehashing the existing information within the block [15]. The information needed to hash are the *user ID*, *user's data* (choice), *timestamp*, and *previous hash*. This information will be recalculated to retrieve a new *hash*. This new hash will then be compared to the block's original hash. If valid, then the next process can be performed. Otherwise, the process will be terminated. the 4th validation is to check if the signature of the previous block is valid. This validation is performed after the hash validation process is completed [26]. This validation will require the *user's public key* of the corresponding user (the user who signs the block), the *hash*, and the signature of the previous block. If valid, then it will continue with the next mining process. Otherwise, the process will be terminated.

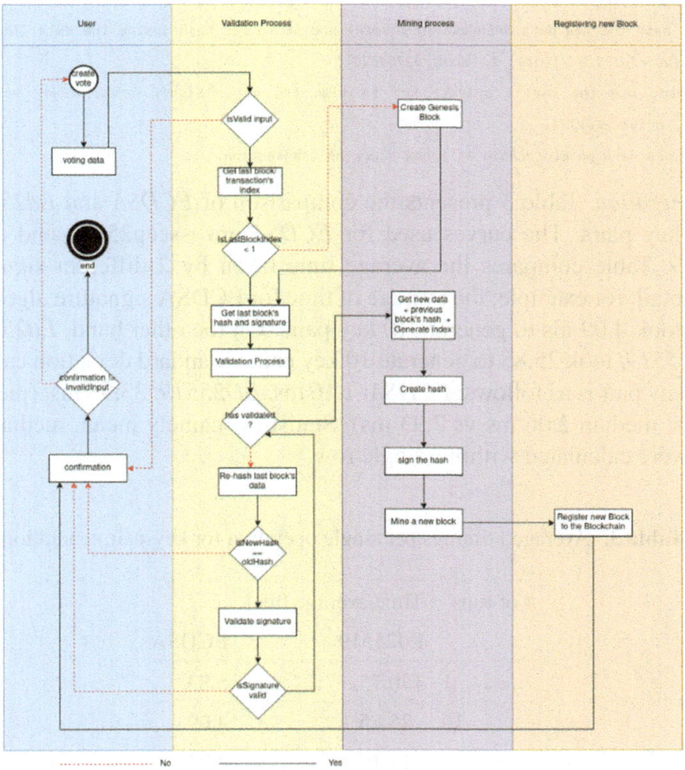

Fig. 2. Creation of blocks – a flowchart

Encryption: Once the validation process is finished, the block is now ready to be registered to the *blockchain*. Before it is registered, some important information will be hashed using a hash function. The important information needed to hash is the *index userId*, *data* (vote), *previous block's hash timestamp*. After the information is hashed, then it must be signed using the user's private key. Now the block will consist of *Index*, *hash*, *signature*, *timestamp*, and *userId* [7]. The final process is to store it into the database. Pseudocode goes as follows:

```
1. Provide the input data, e.g., {userId: 1234, vote: 1}
2. Check if the length of the blockchain is != 0. If the length is 0, then create Genesis block
3. Validate the previous block of the transaction by
   a. Re-hash its information and match it with its hash.
   b. Then check if its signature is valid to validating it using its public key
4. If not valid, return error
5. If valid, hash the new data HASH(userId + vote) and store its hash inside the data. Therefore, data
   would be {userId: 1234, vote: 1, hash: 62764826}
6. Once hashed, use the user's private key to sign the data: Sign({userId: 1234, vote: 1, hash:
   62764826}, privateKey)
7. Store the new data on Blockchain as a new block of transaction.
```

Key Generation. Table 3 presents the comparison of *ECDSA* and *Ed25519* during generating key pairs. The curves used for *ECDSA* are issecp256k1 and *curve25519* for *ed25519*. Table compares the average time taken by 2 different signature algorithms. In detail, for example, the average of time for ECDSA signature algorithm using secp256k1 took 4.09 ms to generate 10 key pairs. On the other hand, *Ed25519*, which uses *curve25519*, took 25.85 to generate 10 key pairs. Standard deviation calculated for generating key pair is as follows: *ECDSA*: 1.16 ms, *ED25519*: 33.93 ms, (mean 2.57 ms vs 19.03 ms, median 2.06 ms vs 7,93 ms). Statistics (namely mean, median, standard deviation) were calculated within a single row.

Table 3. Averaged timings per single operation for key-pair generation

# of runs	Time average [ms]	
	Ed25519	ECDSA
1	130.75	5.83
10	25.85	4.09
100	11.19	2.71
500	10.01	2.64
1 000	9.70	2.49
5 000	7.93	2.21
10 000	7.94	2.06
15 000	7.69	2.04
20 000	7.52	2.02
50 000	7.59	1.96
100 000	7.38	1.91
500 000	6.96	1.74
1 000 000	6.90	1.70

Signing. Table 4 is to evaluate how fast the block of transaction is signed. Table compares the average time taken by 2 different signature algorithms when signing a block of transactions. In detail, for example, the average of time for ECDSA signature algorithm using *secp256k1* took 14.30 ms to sign 10 new blocks of transactions. On the other hand, *Ed25519,* which uses *curve25519,* took 2.45. Standard deviation calculated for signing is as follows: *ECDSA*: 0,27 ms, *ED25519*: 3,27 ms (mean 2.10 ms vs 11.50 ms, median 1.83 ms vs 10,15 ms). Statistics (namely mean, median, standard deviation) were calculated within a single row.

Table 4. Averaged timings per single operation for block signing

# of runs	Time average [ms]	
	Ed25519	ECDSA
1	20.08	4.19
10	14.30	2.45
100	11.98	2.05
500	11.36	1.87
1 000	11.53	1.79
5 000	10.48	1.84
15 000	9.82	1.83
20 000	9.72	1.80
50 000	9.04	1.80
100 000	8.47	1.72
150 000	8.53	1.72

Verification. Table 5 is to evaluate how fast the block of transactions is verified before it is signed. Table compares the average time taken by 2 different signature algorithms when verifying the transaction block. In detail, for example, the average of time for *ECDSA* signature algorithm using *secp256k1* took 40.83 ms to verify 10 new blocks of transactions. On the other hand, *Ed25519,* which uses *curve25519,* took 1.933. Standard deviation calculated for verification is as follows: *ECDSA*: 0.45 ms, *ED25519*: 40.78 ms (mean 1.93 ms vs 33.63 ms, median 1.79 ms vs 19.79 ms). Statistics (namely mean, median, and standard deviation) were calculated within a single row.

Table 5. Averaged timings per single operation for block verification

# of runs	Time average [ms]	
	Ed25519	ECDSA
1	154.75	3.85
10	40.83	1.93
100	26.05	1.69
500	22.17	1.60
1 000	20.50	1.55
5 000	18.79	1.58
15 000	17.84	1.56
20 000	18.38	1.55
50 000	18.09	1.53
100 000	15.70	1.48
150 000	15.91	1.48

5 Conclusions

Blockchain could help to solve the limitations and shortcomings of the E-voting system. The implemented *blockchain* in this research has covered the rules stated in [15], namely: *Authenticity*, *Singularity*, *Anonymity*, *Integrity*, *Uncencorability*, Verifiability, *Transparency*, *Mobility*, *Availability*, *Auditability* and *Certifiability*, *Detectability* and *Recoverability* and *Accessibility* and *Convenience*. *E-voting* using *blockchain* technology can be implemented without utilizing the *Proof-of-work* mechanism. It requires very high computing power and resources. Therefore, even though the *Proof-of-work* mechanism is one of the most essential parts of *blockchain* technology, it is not able to be performed and implemented. Signature algorithm should be implemented if *Proof-of-work* is not implemented. This is of particular importance, especially in open e-voting systems where *Proof-of-Work* is not applicable due to hardware constraints and energy consumption. The reason is that a *signature algorithm* can validate and verify the transaction in a fast and secure way. Considering the performance tests carried out during the research, the following observations were made:

- regardless of operation, *ECDSA* always performs faster than *Ed25519*
- for the generation of key pair, ECDSA is faster than *Ed25519*, this is especially noticeable up to the first 100 runs, where the difference varies from 22 (1st run) to just 4 times (1 million runs) in favor of ECDSA
- for signing blocks, the difference is noticeable but not that large but still *ECDSA* performs about 5 times faster than *Ed25519* (regardless of the number of runs)

– for the verification of blocks, similarly to key-pair generation, the difference is the greatest for the first few (up to 10) runs where it varies from 40 (1st run) to 10 times (150 thousand runs) in favor of ECDSA

Concluding, if the sheer speed of operation is the ultimate goal, ECDSA is the answer; Although ECDSA is faster, *Ed25519* is considered more secure [12]; therefore, in the authors' opinion, it would be rational to choose it in an open ballot e-voting system as a digital signing algorithm, as its performance results do not disqualify its usage in such a system.

References

1. Andayani, D.: BPN akan Laporkan 1.261 Kecurangan Pemilu. Ini KataBawaslu (2019). https://news.detik.com/berita/d-4518568/bpn-akan-laporkan-1261-kecurangan-pemilu-ini-kata-bawaslu
2. Ayed, A.B.: A conceptual secure blockchain-based electronic voting system. International Journal of Network Security & Its Applications **9**(3), 1–9 (2017)
3. Blanchette, J.F.: The digital signature dilemma. Annales des télécommunications **61**(7), 908–923 (2006). https://doi.org/10.1007/bf03219871
4. Boudrez, F.: Digital signatures and electronic records. Arch. Sci. **7**(2), 179–193 (2007). https://doi.org/10.1007/s10502-007-9058-2
5. Elliptic Curve Cryptography - KeyCDN Support. https://www.keycdn.com/support/elliptic-curve-cryptography, Accessed Dec 2020
6. Hardwick, F.S., Gioulis, A., Akram, R.N., Markantonakis, K.: E-voting with blockchain: an E-voting protocol with decentralization and voter privacy. In: 2018, IEEE International Conference on Internet of Things (iThings) and IEEE Green Computing and Communications (GreenCom) and IEEE Cyber, Physical and Social Computing (CPSCom) and IEEE Smart Data (SmartData), pp. 1561–1567. IEEE (2018). https://doi.org/10.1109/cybermatics_2018.2018.00262
7. Kim, S.K., Kim, U.M., Huh, J.H.: A study on the improvement of blockchain application to overcome the vulnerability of IoT multiplatform security. Energies **12**(3), 402 (2019). https://doi.org/10.3390/en12030402
8. Knutson, H.: What is the math behind elliptic curve cryptography? https://hackernoon.com/what-is-the-math-behind-elliptic-curve-cryptography-f61b25253da3, Accessed Dec 2020
9. Kshetri, N.: Blockchain Could Be the Answer to Cybersecurity. Maybe. https://www.wsj.com/articles/blockchain-could-be-the-answer-to-cybersecurity-maybe-1527645960, Accessed Dec 2020
10. Kshetri, N.: Blockchain systems are tracking food safety and origin. Accessed Dec 2020
11. Kshetri, N., Voas, J.: Blockchain-enabled e-voting. IEEE Softw. **35**(4), 95–99 (2018).https://doi.org/10.1109/ms.2018.2801546
12. Lange, T.: SafeCurves: choosing safe curves for elliptic curve cryptography. In: Bernstein, D.J. (ed.) University of Illinois at Chicago (2014)
13. Lewis, A.: The Basics of Bitcoins and blockchains: an introduction to cryptocurrencies and the technology that powers them. Mango Media Inc. (2018). https://doi.org/10.5555/3299472
14. Lopes, J., Pereira, J.L., Varajão, J.: Blockchain based E-voting system: a proposal (2019)
15. Maram, B.: Bitcoin generation using Blockchain technology. JOIV Int. J. Inf. Vis. **2**(3), 127–132 (2018). https://doi.org/10.30630/joiv.2.3.109

16. Moura, T., Gomes, A.: Blockchain voting and its effects on election transparency and voter confidence. In: Proceedings of the 18th Annual International Conference on Digital Government Research, pp. 574–575 (2017). https://doi.org/10.1145/3085228.3085263
17. Nakov, S.: Practical Cryptography for Developers (2018). https://cryptobook.nakov.com/
18. Nurmi, H.: Voting theory. In: e-Democracy, pp. 101–123. Springer, Dordrecht (2010). https://doi.org/10.1007/978-90-481-9045-4_7
19. Omar et al.: ECDSA vsECDH vs Ed25519 vs Curve25519. https://security.stackexchange.com/questions/50878/ecdsa-vs-ecdh-vs-ed25519-vs-curve25519, Accessed Dec 2020
20. Prabowo, P.S.A., Winarno, I., Rosyid M.I.W.N.: Implementasi Algoritma ECDSA untuk Pengamanan E-Mail (Verifikasi Keaslian Pesan). eepis final project (2010)
21. Rajeev, S., Geetha, G.: Cryptographic hash functions: a review international. J. Comput. Sci. **9**(2) (2012)
22. Rouse, M.: Cryptography (2009). Accessed 2 Feb 2014, from Search Software Quality, http://searchsoftwarequality.techtarget.com/definition/cryptography
23. Reiff, N.: Blockchain explained. In: Investopedia, vol. 1, p. 2020 (2020). https://www.investopedia.com/terms/b/blockchain.asp, Accessed Dec 2020
24. Rella, L.: Steps towards an ecology of money infrastructures: materiality and culture of Ripples. J. Cult. Econ. **13**(2), 236–249 (2020). https://doi.org/10.1016/b978-0-08-102295-5.10514-1
25. Sayeed, S., Marco-Gisbert, H.: Assessing blockchain consensus and security mechanisms against the 51% attack. Appl. Sci. **9**(9), 1788 (2019). https://doi.org/10.3390/app9091788
26. Seok, B., Park, J., Park, J.H.: A lightweight hash-based blockchain architecture for industrial IoT Appl. Sci. **9**(18), 3740 (2019). https://doi.org/10.3390/app9183740
27. Setiadi, I., Kistijantoro, A.I., Miyaji, A.: Elliptic curve cryptography: Algorithms and implementation analysis over coordinate systems. In: 2015 2nd International Conference on Advanced Informatics: Concepts, Theory and Applications (ICAICTA), pp. 1–6. IEEE (2015)
28. Shibani, S.S.A., Spoorthi, S.: Blockchain enabled e-voting system. IJARCCE **8**(4), 77–81 (2019). https://doi.org/10.17148/ijarcce.2019.8412

Prediction and Causality Visualization in Speech Recognition

Adam Wróbel, Mikołaj Jarosławski, and Paweł Jemioło(✉)

AGH University of Science and Technology, A. Mickiewicza 30, 30-059 Krakow, PL, Poland
{awrobel,jaro}@student.agh.edu.pl, pawljmlo@agh.edu.pl

Abstract. Speech Recognition is a rapidly developing aspect of modern Deep Learning research. Many up-to-date technologies, such as Google Assistant or Alexa, apply it. They change the way people interact with everyday devices, i.e., phones, cars. Despite being so influential, current AI models are complex black-boxes, and few people comprehend what lies beyond them. This paper aims to feature a Deep Learning architecture for Speech Recognition and visualizations of the model state when making the predictions. We incorporated class activation maps and the attention mechanism to increase the accuracy of the presented algorithm. The introduced model can analyze audio sequences of various lengths. Preliminary results suggest that the proposed method provides a satisfactory outcome. It can be successfully applied to get insight into the state of the model during predictions. This work might, therefore, help Data Scientists detect flaws, such as overfitting in their algorithms.

Keywords: Attention · Deep Learning · Speech Recognition · Causality visualization · Artificial Neural Networks · Sequence analysis

1 Introduction

Speech Recognition is a technique widely used across modern Artificial Intelligence (AI) models. It is most frequently applied in the form of virtual assistants that help with basic tasks and significantly improve human productivity. The most popular voice-controlled assistants include Apple's Siri, Amazon's Alexa and Google's Google Assistant [9]. This technology is widely used in computers and smartphones. It is also applied in various industries such as business, banking or healthcare. It helps to improve the lives of people everyday.

This paper aims to create a Deep Learning model, which would be able to transform an audio voice sample into a text transcript. The other important feature of the proposed architecture is the ability to analyze how the model makes its predictions. In other words, its *thought* processes. The output visualizations show the spectrogram of the current sample with regions that are important for the next prediction. This feature could prove helpful for Data Scientists, who could gain insight into the state of the model during predictions and thus detect

W. Zamojski et al. (Eds.): DepCoS-RELCOMEX 2021, AISC 1389, pp. 477–486, 2021.
https://doi.org/10.1007/978-3-030-76773-0_46

flaws in the model, e.g. overfitting. The audio samples may contain different irregularities (e.g. accents, disorganized speech) That could disrupt the proper functioning of Speech Recognition models. These irregularities and the subsequent underperformance of the model may be very difficult to detect. With our proposed architecture, it is possible to compare different samples and check if the model focuses on the correct parts of the spectrogram when predicting.

The rest of the paper is organized as follows. Section 2 is focused on the technologies applied in the model. The key technologies used, such as attention and class activation maps, are described in more detail. Next, Sect. 3 focuses on the methods adopted to create the model. Then, the results are presented in Sect. 4. Finally, Sect. 5 summarizes and concludes the work, including a few suggestions for the further development of the model.

2 Background

Deep Learning is based on an Artificial Neural Networks, which are structures similar to the one of the human brain. Compared to Machine Learning models, Deep Learning alternatives are much more flexible and adapt to more complicated datasets [6]. It means that there is no need to perform as much data preprocessing as required in the case of Machine Learning models.

One of the applications of Deep Learning is a rapidly developing topic of Speech Recognition. Deep Learning is widely used for Speech Recognition because of its flexibility and high predicting power [8]. The task of Speech Recognition models is to transform an audio sample into individual letters, eventually forming whole sequences. For an Artificial Neural Network to *understand* the human voice, the sound waves initially need to be transformed into a form of a spectrogram. The predictions of the following letter in the sentence should be heavily impacted by what came before. Thus, the preferred architecture is the Recurrent Neural Network (RNN) [8]. It uses more complex neurons that as input also take the previous prediction. In the paper [1], the authors developed a Convolutional-Recurrent Neural Network (CRNN) architecture for Speech Recognition. They were able to achieve human-level accuracy with the model. Additionally, the article authors compared different hidden units: Long Short-Term Memory (LSTM) and Gated Recurrent Units (GRU). They found that the results on a small dataset were similar for GRU and LSTM, but GRU is more accessible and less likely to diverge.

A relevant part of the proposed architecture is the encoder-decoder structure. The encoder gradually reduces the spatial dimension, while the decoder generates a sequence of a different length [4]. The main advantage of such architecture is that the generated sequence can have different length for every predicted sample while the encoder input sequence stays constant.

The authors [3] presented an encoder-decoder RNN based model that added an attention mechanism. The attention mechanism enables the model to focus on the connections between each part of the sequence. In Speech Recognition, the attention function assigns weights to each element of the sequence, different for every letter. What is more, attention is widely used in Image [5,12] as well as Emotion [7] Recognition.

3 Our Approach

3.1 Dataset

Due to the proposed solution's performance, it was challenging to choose a suitable database for training and testing the model. The most significant factor was the length of recordings. LibriSpeech [10], TED-LIUM [11], and CommonVoice [2], open-source speech transcription databases, were considered.

The CommonVoice database has been chosen to create the prototype of the system because it contains the shortest recordings (2.5 s). This database comprises about 200 000 training recordings and transcripts and about 4 000 test recordings and transcriptions with a sampling frequency of 48 kHz. It contains about 500 h of speech in English. The quality of audio files in the database was relatively poor and noisy, so the data were appropriately preprocessed. The choice of this set was conditioned by the fact that Common Voice has many short recordings of a few seconds length, which helps in the initial stages of training sequential architecture.

3.2 Preprocessing

The data was adequately prepared for algorithm training. We divided the dataset into sets of appropriate lengths, which facilitates algorithm training. The first set consists of recordings up to 2.5 s in length, containing 18 805 files. The second set consists of recordings of up to 5-s duration, containing 103 619 files. The third collection consists of recordings up to 10 s long, containing 190 826 files. As the input format, the time-frequency representation of the signal – spectrogram has been chosen because both time (sequence of samples) and frequency (attributes of a particular instance) are significant in speech transcription. As we used the TensorFlow library for architecture preparation, we have to take into account some prerequisites. Since this framework requires a fixed length of sequential data, each signal has been zero-padded to the maximum size of its

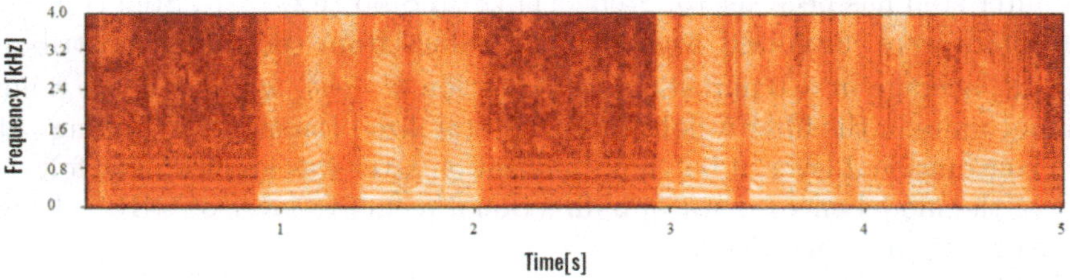

Fig. 1. A sample spectrogram.

set. Spectrograms were estimated with the Short-Time Fourier Transform. The sample spectrogram has been shown in Fig. 1.

The next step was to *denoise* the spectrogram. Due to the narrow frequency range of human speech, the signal has been filtered with a Butterworth IIR high pass filter with a cutoff frequency of 500 Hz. Then every pixel with a value below the threshold was set to 0. The threshold value was set by the subjective method of observation to 3.2, and finally, the spectrogram was cut to 4 kHz to keep only the most relevant signal information. This process lets us reduce the noise to a minimum. The result of the preprocessing of a sample spectrogram has been presented in Fig. 2.

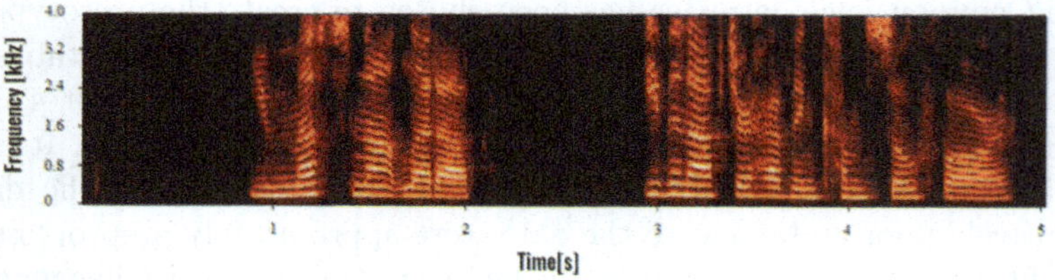

Fig. 2. A sample spectrogram after preprocessing.

3.3 Architecture

Analyzing the samples, we came to two conclusions regarding the data. On the image processing side, the spectrogram, as a 2-dimensional matrix, has a 2-dimensional correlation of samples. On the part of processing 1-dimensional signals, a spectrogram is a sequence of samples with the dimensionality of attributes in the form of frequency bands. Both approaches have been utilized. Similar to [13], convolutional and recurrent layers were used (see Fig. 4).

As the first step of creating the algorithm, we used convolutional layers. They were utilized to reduce the spectrogram dimension as an image of speech and filter the data into a more intelligible form for the system. In conjunction with the following architecture, the best results were obtained by three convolutional layers with filters successively: F1 [41, 11, 32], F2 [21, 11, 32] and F3 [21, 11, 96]. The filter step has been set to: Step (F1) [2, 2], Step (F2) [2, 1], Step (F3) [2, 1]. The filters were initialized with sampled values from the normal distribution with the average $m = 0$ and standard deviation $v = \frac{\sqrt{2}}{k}$, where k is the number of input units in the filter weight tensor. It stabilized the model's training. After each convolutional layer, a batch normalization layer was applied. The following part of the model encoder was a bidirectional recursive layer of GRU cells with a size of 500 units (see Fig. 3).

Next, we utilized Bahdanau's attention mechanism [3]. We used two Dense layers with 10 units with hyperbolic tangent activation function and 1 unit with softmax activation on the sequence dimension. As a final element of the decoder,

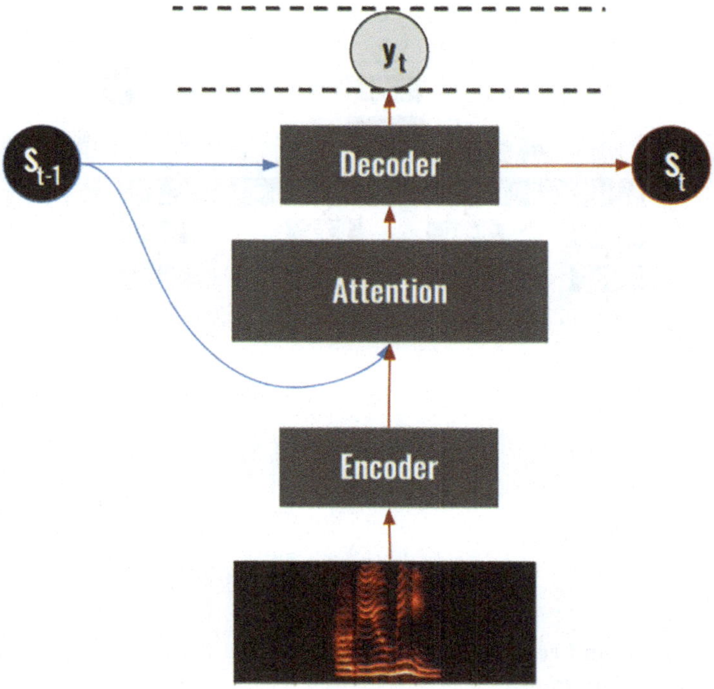

Fig. 3. Model's architecture.

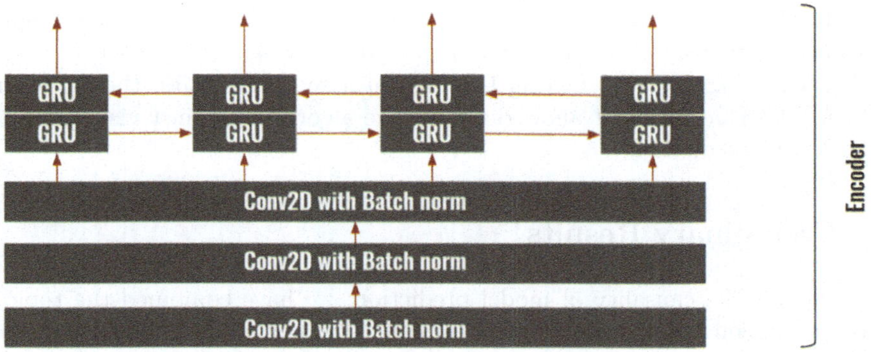

Fig. 4. Model's encoder.

the Long Short-Term Memory cell of size 250 was used. A dense layer with a standard softmax activation (see Fig. 5) followed it.

For architecture training, we used the teacher forcing method and the masked cost function (see Eq. 1), which omits the label tensor elements equal to zero (because the label padding was applied).

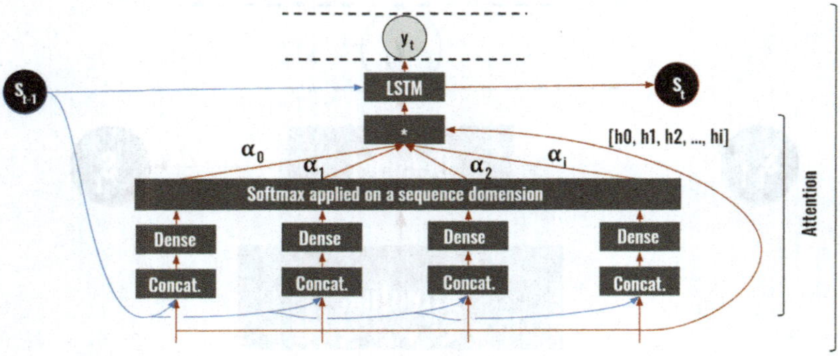

Fig. 5. Model's decoder.

$$L(y_{true}, y_{pred}) = \frac{\sum_{y_{true}[i]>0} y_{true}[i] * \log y_{pred}[i]}{\sum_{y_{true}[i]>0} y_{true}[i]} \tag{1}$$

Where:

y_{true} – the ground truth,
y_{pred} – predicted value.

Speech Recognition took place at the character level, meaning that every prediction is a single letter. The mask was applied to the accuracy of the model too. Finally, the model was trained for 30 epochs and optimized using the RMSprop optimizer with *learning rate* = 0.001, $\rho = 0.9$, *momentum* = 0.01, $\epsilon = 10^{-7}$ and a *batch size* = 128. During training on a set up to 2.5 s, the model has achieved 95% accuracy. It allowed testing the algorithm against visualizing the causality of prediction.

4 Preliminary Results

To visualize the causality of model prediction, we have combined the topic of class activation maps and attention weights. Using the class activation maps approach to visualize the convolutional model's causality requires applying a dot product between the results of the last convolutional layer and the perceptron of the final dense prediction matrix. It demands the compatibility of the matrix multiplication dimensions. Thus, the last dense layer has to be preceded by the convolutional layer. In our model, we could not use this approach due to the encoder-decoder architecture and the sequential data dimension. However, thanks to the softmax distribution of attention weights on the sequence dimension, we could combine it with class activation maps (see Figs. 6 and 7).

Fig. 6. Sample input spectrogram.

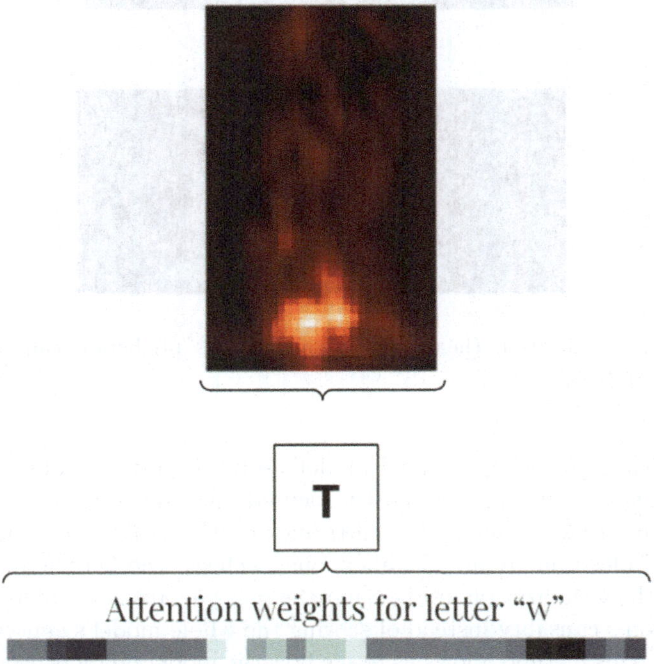

Fig. 7. Final layer prediction summed up on attribute dimension and attention weights distribution for letter w. Both with sequence dimension T.

The sequence dimension in the model, during the passing through the attention part, remains unchanged. It is, thus, equal to the dimension of the attention weight vector. For this reason, our approach was to *weight* the result of the last convolutional layer with attention weights by multiplying each result map's column with the corresponding attention weight. Afterwards, we summed up the tensor by the attribute dimension, which reduced the dimensionality of the map.

Resizing this tensor to the input spectrogram dimensions allowed us to visualize the model's *focus* for every predicted character in the form of a heat map overlayed on the original spectrogram. This final result of visualization is presented in Fig. 8. The final softmax-distributed prediction matrix for the example is depicted in Fig. 9.

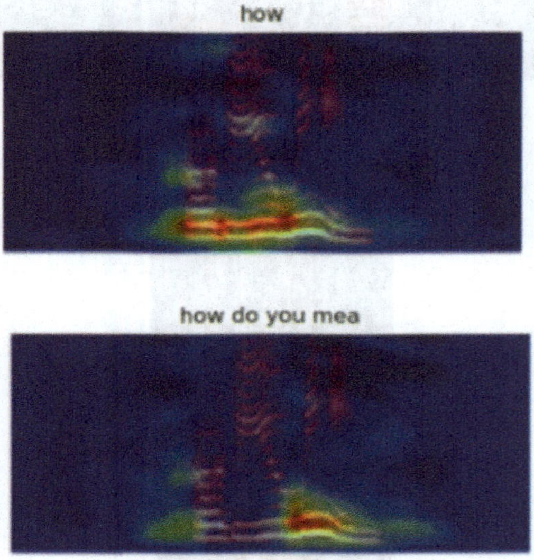

Fig. 8. Final visualization (heatmap) of the model's prediction causality, overlaid on the input spectrogram.

As presented in Fig. 9, the final model's softmax distributions for every letter of the analyzed example is mostly focused on a single character at a time. It may suggest that the model is well trained on this particular example or it is overfitting. Indeed, many metrics allow one to check whether the model is overfitting, but the visualization in Fig. 8 may help us pay attention to local details of the prediction's causality instead of scoring the whole model's generalization. In other words, this method allows us to see how the model predicts every word, letter by letter. We can observe, for example, why a letter *o* was predicted between *h* and *w*. Next, we can wonder, is it because the model has noticed the phrase *how* on the spectrogram, or maybe it has assumed that between *h* and *w*, *o* has primarily occurred in the training dataset?

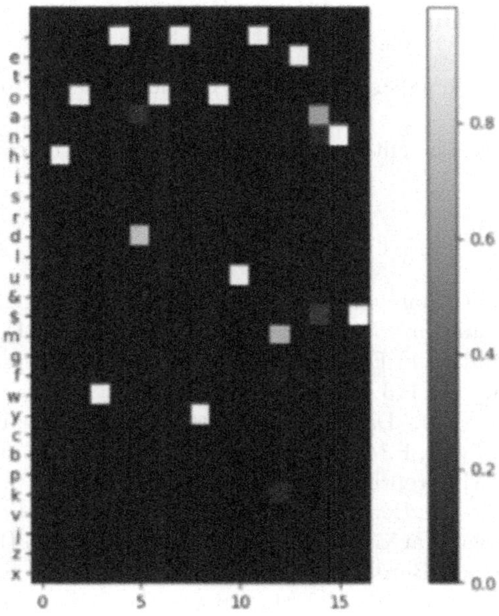

Fig. 9. Final softmax-distributed model's prediction for each letter.

5 Conclusions and Future Work

In this paper, we introduced a Deep Learning architecture for Speech Recognition and visualizations of the model state when making the predictions. The preliminary results show that the proposed architecture can successfully analyze audio samples of different lengths. By applying the attention mechanism to the model, we could perform the model's prediction causality visualization.

The visualizations let us observe which part of the spectrogram the model focuses on when making the predictions. Therefore, we can deduce if overfitting exists in the given model when it does not focus on the right part of the spectrogram. The substantial advantage of this approach is that it is possible to spot what combination of syllables or letter combinations are causing problems in the model and add appropriate variations to our dataset. The same architecture can also be easily applied to Image Recognition models to detect flaws in these models and increase their performance.

The potential future work on the model could include training the model on the other three databases mentioned in this paper. It would increase the model accuracy for a variety of different audio samples. Another potential extension could be expanding the model to be able to analyze longer samples. Lastly, we plan to compare the developed architecture with other existing models.

Additional Information

Acknowledgment. The paper is supported by the AGH UST research grant.

Conflict of Interest. The authors declare that they have no conflict of interest.

References

1. Amodei, D., Ananthanarayanan, S., Anubhai, R., Bai, J., Battenberg, E., Case, C., Casper, J., Catanzaro, B., Cheng, Q., Chen, G., et al.: Deep speech 2: end-to-end speech recognition in English and Mandarin. In: International Conference on Machine Learning, pp. 173–182 (2016)
2. Ardila, R., Branson, M., Davis, K., Henretty, M., Kohler, M., Meyer, J., Morais, R., Saunders, L., Tyers, F.M., Weber, G.: Common voice: a massively-multilingual speech corpus. arXiv preprint arXiv:1912.06670 (2019)
3. Bahdanau, D., Cho, K., Bengio, Y.: Neural machine translation by jointly learning to align and translate. arXiv preprint arXiv:1409.0473 (2014)
4. Conze, P.H., Pons, C., Burdin, V., Sheehan, F.T., Brochard, S.: Deep convolutional encoder-decoders for deltoid segmentation using healthy versus pathological learning transferability. In: 2019 IEEE 16th International Symposium on Biomedical Imaging, ISBI 2019, pp. 36–39. IEEE (2019)
5. Fu, J., Zheng, H., Mei, T.: Look closer to see better: recurrent attention convolutional neural network for fine-grained image recognition. In: Proceedings of the IEEE Conference on Computer Vision and Pattern Recognition, pp. 4438–4446 (2017)
6. LeCun, Y., Bengio, Y., Hinton, G.: Deep learning. Nature **521**(7553), 436–444 (2015)
7. Mirsamadi, S., Barsoum, E., Zhang, C.: Automatic speech emotion recognition using recurrent neural networks with local attention. In: Conference on Acoustics, Speech and Signal Processing (ICASSP), pp. 2227–2231. IEEE (2017)
8. Nassif, A.B., Shahin, I., Attili, I., Azzeh, M., Shaalan, K.: Speech recognition using deep neural networks: a systematic review. IEEE Access **7**, 19143–19165 (2019)
9. Olson, C., Kemery, K.: Voice report from answers to action: customer adoption of voice technology and digital assistants (April 2019). (Online; posted in April 2019)
10. Panayotov, V., Chen, G., Povey, D., Khudanpur, S.: Librispeech: an ASR corpus based on public domain audio books. In: 2015 IEEE International Conference on Acoustics, Speech and Signal Processing (ICASSP), pp. 5206–5210. IEEE (2015)
11. Rousseau, A., Deléglise, P., Esteve, Y.: TED-LIUM: an automatic speech recognition dedicated corpus. In: LREC, pp. 125–129 (2012)
12. Sun, M., Yuan, Y., Zhou, F., Ding, E.: Multi-attention multi-class constraint for fine-grained image recognition. In: Proceedings of the European Conference on Computer Vision (ECCV), pp. 805–821 (2018)
13. Xu, K., Ba, J., Kiros, R., Cho, K., Courville, A., Salakhudinov, R., Zemel, R., Bengio, Y.: Show, attend and tell: neural image caption generation with visual attention. In: International Conference on Machine Learning, pp. 2048–2057 (2015)

Correlation of Bibliographic Records for OMNIS Project

Witold Wysota[1(✉)] and Kacper Trzaska[2]

[1] Institute of Computer Science, Warsaw University of Technology, Warsaw, Poland
witold.wysota@pw.edu.pl
[2] National Library of Poland, Warsaw, Poland
k.trzaska@bn.org.pl

Abstract. The paper presents work on merging sets of bibliographic data between two largest libraries in Poland—National Library of Poland and Jagiellonian Library as part of a project aiming to provide a single point of access to library assets in Poland. Focus is put on presenting the problem and decisions taken to develop an algorithm for finding similarities between bibliographic entries.

Keywords: Bibliographic data · Data correlation · MARC · OMNIS · Text normalization

1 Background

In late 2016 the National Library of Poland (NLOP) started a project co-financed with the government and the European Union called OMNIS that aims to implement a country-wide IT system that provides a single point of access to collections from various Polish libraries [5].

The project consisted of three sections:

- OMNIS Integrated Search Engine,
- POLONA in the Cloud for Libraries,
- e-ISBN Publishing Repository.

The first section of the project delivers a search engine that allows to browse through bibliographic entries of participating libraries and do full-text search of books, magazines and alike.

Libraries have historically had their own databases of bibliographic records where they stored more or less detailed information about works in their possession. Librarians used to create them manually according to specific rules after doing research to find details on each work. A *de facto* standard for maintaining such records was created by the U.S. Library of Congress [3] and many libraries have adopted similar approaches to cataloging data.

In modern times these systems are powered by computers using particular commercial systems dedicated to libraries. Each database is distinct—it contains

W. Zamojski et al. (Eds.): DepCoS-RELCOMEX 2021, AISC 1389, pp. 487–495, 2021.
https://doi.org/10.1007/978-3-030-76773-0_47

entries about different works. Data is still collected and entered by humans who interpret the same materials differently. Mistakes in editing are also a common issue. Transcription of foreign names and titles can also be done in more than one way—for example the same Arabic word can be transcribed into Latin alphabet using other characters (e.g.'y' or 'i' and 'o' or 'u').

Despite these collections being diverse they do refer to the same entities (e.g., books, articles, magazines). Therefore, merging such collections cannot simply be a sum of their records as some of them refer to the same work and thus such operation would introduce duplicates. Also, it can't be a crude merge of records as especially for those where information is scarce such operation could incorrectly merge unique entities, e.g., there are multiple distinct books called *Physics* and if there is no additional info besides the title, it can't be assumed that records from two libraries refer to the same book.

To be able to combine such collections they have to be analyzed and corresponding entries have to be identified so that they can be merged into a single record, possibly complementing the base record with data found in the other one.

Dataset of the National Library of Poland has served as the base (core) for the OMNIS system and datasets of other participating libraries are to be processed against the base set to find correlations between records and tag them for proper merging.

2 Introduction

The first external library to join the national system was the Jagiellonian Library [1] which is currently the second largest library in Poland with a collection of bibliographic records around 25% of that of NLOP and being itself much older than NLOP. Combining the two collections was the biggest challenge of this part of the project and served as a testbed for the tool driving future integrations.

In the following sections of the paper we focus on presenting the challenges, describe the data exchange format and the algorithms used to measure similarities between bibliographic records. We also consider how to try and determine how trustworthy the results are and how to improve their dependability. Finally, we envision the next steps that can be taken to further improve quality of the results.

Complexity of this problem is not only about its technical side. Also, the psychology of people involved needs to be recognized. The system developed deals with data which is the life's work of some people who feel emotional bond with the quality of the data they had been gathering throughout the years. Merging two datasets together has to be done with utmost care as the parties involved don't want to see their original data damaged or treated as inferior. The core dataset is being enriched by including data from new libraries that are joining the initiative and the later a library integrates, the more of its records should match with records already in the core dataset.

The common data exchange format is called MARC21 and standardized as part of ISO 2709 [6]. It is a compact format that describes each record with a composition of three elements. First the record structure is defined, then comes content designation and then the description ends with data content for the record [4]. Logically each record consists of fields. These have a three character tag (usually numerical) that defines the meaning of the field. The content of each field is structured and divided into subfields with single character designations that convey the meaning of each subfield. Semantics of each field and each subfield is strictly defined however interpretation of content of each subfield might differ between organizations.

2.1 Problem Statement

Doing a brute force comparison of each of over two million records against each record in six million record dataset trying to find one entry that is most likely to refer to the same entity as the original one is a very time-consuming task. It is not feasible to complete such a task in adequate amount of time. Assuming that doing a field-by-field comparison of a single record against six million records takes around a second for a decent machine, repeating the process for remaining of the two million records would take about a month to complete.

And even doing that would not guarantee success as records that should match are not identical. Therefore, a heuristic approach needs to be taken that vastly reduces the effort needed to calculate correlation while not reducing the chance to find a match significantly. In this paper we focus on describing the algorithmic decisions that enable us to obtain results quickly without hindering their quality.

3 Correlation and Deduplication

In this section we present how me merge bibliographic datasets. If two records refer to the same work we call them *duplicates* and hence the whole process is called *deduplication*.

While the process itself consists of several steps, in this paper we focus on processing textual data to decide if two entries correspond to the same work or not. Thus, we describe how we prepare data and how what methods we use for finding similarities between two texts.

The concept of detecting related records in the domain of bibliography has been tackled by others like [11,12,14,15] using various techniques ranging from database–backed expert system to automated comparison of features in a set of records.

In [9] they describe a process of using reference management applications as a deduplication platform where an algorithm consisting of a number of steps is applied by the operator to bring records to a common format and allow the use of a built-in deduplication mechanism before exporting records to a final database. However, this process is still manually driven and most likely focuses on format differences rather than content correlation. While this can be viable as a single–shot operation to reduce the number of items to review, it is not suitable as a systematic approach to merging large sets of bibliographic records.

The amount of data to compare is the key factor here—all the above mentioned approaches were dealing with much smaller datasets (e.g. 150k records) compared to the OMNIS project which was faced against matching 2 million records against a base set of 6 million records. It is unfeasible to conduct the process manually due to the volume of data and thus it has to either be fully automated or at least heavily computer–aided.

3.1 Indexing

Many of the records in the core dataset will be regarded as possible to match candidates multiple times, and it is inefficient to look for needed information directly in the record each time it is needed, therefore during the indexing stage we preprocess each record and extract information that is going to be accessed during the matching phase. All that information is then stored in a form that can be queried efficiently.

The following are extracted:

- ISBN [7] (**020**|**a**) or ISSN [8] (**022**, **024**, **026**, subfield **a**),
- author data from one of **100**, **110**, **111** or **130** fields (only one of them exists in a record),
- title data from following subfields of **245** field:
  - **a**—title,
  - **b**—remainder of title,
  - **n**—number of part or section of work,
  - **p**—name of part or section of work,
- key title (field **222**) for a continuing resource (e.g. journal)
- edition information (**250**|**a**),
- page count—field **300**|**a**,
- publication year—specific four characters of field **008**,
- publication country and language—specific two-letter country codes from field **008**,

Author and title data are further processed—they undergo normalization (described in Sect. 3.3).

If one or more ISBN/ISSN are present, the values are indexed using a global dictionary pointing back to the record's data.

For the author and title values, comparing text character by character is slow therefore for each entry a digest (integer value) is calculated and indexed in global dictionaries as well. When looking for identical values we compare digest values first and only then do a string comparison for equality. This is required as for many strings, clashes in their digest values are possible.

Apart the indexes, a complete set of extracted data is stored in an object and the indexes point back to those objects.

3.2 Matching Text

In its essence matching entries boils down to comparing text from respective subfields in investigated records.

Strict comparison of strings is not af good approach—even a difference in whitespace or punctuation will make the texts be evaluated as distinct. This particular problem can be overcome by normalizing strings which is further described in Sect. 3.3. Normalization itself is not enough to trust results of strict comparison of strings as data, which was input manually by different people, is prone to editing mistakes and other differences and strict comparison can only yield results of texts being *identical* or *distinct*. In such case many texts (like titles or author names) that refer to the same entity would be considered different. Therefore, it is required to have a measure of similarity between texts.

In [12] they construct all possible trigrams and count how many of them appear in compared texts. We have chosen a simpler approach which we believe works better in case of the data we are dealing with (in terms of both speed and effectiveness) as it feels a better model of unintentional mistakes done by humans compared to a pure mathematical approach.

We believe one can use an *edit distance* metric such as Levenshtein distance [13] that counts how many insert, change or delete single character operations are required to transform one string into the other one. In our work we decided to use a modified version of the algorithm called Damerau–Levenshtein distance [10] which adds *transposition* to the list of edit operations as it is a very common mistake humans make while typing (especially for some combination of letters) and it makes perfect sense to reduce penalty from making such mistake—transposition in the basic algorithm costs 2 units while it is only 1 unit with Damerau's modifications.

To improve performance, a cut–off threshold has been introduced to interrupt comparison after reaching a score exceeding a predefined limit that is proportional to the length of strings being compared. When the limit is reached, texts are treated as different with 100% probability.

3.3 Normalization

Text fields are normalized using the following procedure. All relevant subfields are concatenated into a single string. The string is then pruned of anything that is not a letter, digit or a single whitespace separating words. As part of this process, all letters are also converted to lower case. The whole operation effectively discards all punctuation, quotes and extra whitespace.

Normalization is done for two reasons. Firstly it makes the text shorter leaving out all less significant characters. Secondly it is most probable that if two people copy type the same text then most differences will be in quotes and punctuation, e.g. changing single to double quotes, using other quotation marks (consider ,,” versus ""), skipping dashes, full stops or commas. Such normalization increases similarity between strings which is especially important for texts that otherwise matches perfectly to one another. This is because the last normalization step is to calculate a digest value of a string which allows to identify that two normalized strings are equal just by comparing two integer values.

If results were not satisfactory the normalization process could be extended by converting all national characters such as *ń*, *ł*, *ż* to similar characters in ASCII character set—in this case *n*, *l* and *z*. We have decided to not include this step as two words which differ only by national characters might clash after converting special characters to ASCII.

4 Matching Algorithms

This section gives more detailed information on the two algorithms that we developed for finding similar records.

Deduplication system works in two modes. One mode, called the *perfect match* is very conservative and aims to minimize probability of falsely matching records as related. This is to mitigate the risk of automatically merging records that correspond to separate entities. Doing that would also merge sets of underlying item descriptors and then might cause readers to get not the books that they had ordered.

The other mode is more aggressive and tries to work around imperfections such as incomplete information, typing errors and transcription differences. At the same time it tries to keep false positives at a very low level.

4.1 Algorithm: Perfect Match

The perfect match algorithm works as follows:

1. If no ISBN/ISSN code is available—stop further processing and report record as unique.
2. For each record with matching ISBN skip record if any of the following differ:
   (a) publication year,
   (b) normalized titles,
   (c) normalized authors,
   (d) book editions,
   (e) key titles.

Any discrepancy between records in any of the attributes immediately disqualifies a candidate. Its consequence is that any typos that remain after strings are normalized are unacceptable.

Goal of this algorithm is to minimize false matches of records. Such match can happen either because of an error in ISBN associated with a record or because of differences in the way libraries treat multi–edition or multi-volume works.

4.2 Algorithm: Full Scope Search

The full scope search algorithm consists of three steps. They are ordered by confidence and quality of the results they yield. If any of the steps finds any possible matches, the algorithms does not proceed further and returns the results found so far.

Match by ISBN. The first stage of this algorithm searches for records with identical ISBN value to the record being investigated. If there is no ISBN associated with the record, then the current step of the algorithm is skipped. Otherwise, the tool inspects records that have matching values for the **020|a** field. Candidates are checked against discrepancies in publication years, titles and author names. The algorithm starts with a certain score and awards points for matches and mismatches between respective fields. For publication year any deviation results in a constant value penalty. For string-based comparisons, the amount of points awarded is proportional to similarity between the two strings. In the end the candidate is assigned a final score. If any matching candidate is found the algorithm terminates and yields results.

Match by Author. In the second stage the algorithm starts by checking if the record contains any information about the author of the entity described by the record. In rare case there is no author information available, this stage is skipped.

This stage of the algorithm looks through the index of works authored by the author of the record being processed. Name of the author has to match perfectly (after normalization). If both the candidate and the investigated record contain information on the year of publication, and they don't match, then the candidate is rejected. Similar comparison takes place for language and country of publication. Then the algorithm compares the titles of works.

If identical then it means the candidate has been matched by both title and author. The algorithm then awards points for matching editions and penalizes differences in ISBN. It also compares page counts with 10% tolerance. Then it calculates the score for a candidate.

In case there are differences in titles the candidate gets a penalty but is not rejected if differences do not exceed threshold relative to the length of the title.

Match by Title. The final stage of the algorithm performs a check against other works with the exact same title (by now we already know that the author information is either absent or has to differ). For each candidate it compares authors, page counts and editions and awards points accordingly.

If no works with identical title are found then the algorithm performs a broad search among all records that have titles starting with the same character as the investigated record looking for works with quite similar titles. It only considers records with the same publication year, language and country and similar page count.

This last step can yield numerous relatively low quality candidates. It stops after finding three reasonable candidates to avoid exploding the search time.

5 Results

Just before the final run of the procedure it has been decided that only books from the Jagiellonian Library would undergo deduplication and all other works (roughly 10% of all the records) would be designated as unique and duplicates would be resolved manually at a later time.

Table 1. Deduplication statistics

Description	Value
All records	1,614,777
Books	1,498,913
Ideal duplicates	758,156
High–probability	73,320
Duplicates (books only)	56%

As can be seen in Table 1 almost 1.5 million records have been selected for the process. Over 758k were marked for automatic deduplication and an additional 10% was marked as high probability duplicates. It was estimated that a statistically insignificant number of the results could have been misclassified.

Those records that we marked as *high–probability* were not automatically treated as duplicates but instead were tagged with links to their selected matches and are now being manually verified by catalogers. Almost all of those already checked have been confirmed as duplicates to one of their candidates. It means that we can treat them as true positives and that yields the amount of 56% of all book records being marked as duplicates to records existing in NLOP dataset.

Unfortunately at this point it is impossible to say what the true number of duplicates was, so we can't provide hard figures proving how well the algorithm performed. When asking the experts, they say the numbers 'sound right'. We still see space for improvements, some of which are mentioned in the next section of this paper.

6 Future Work

More libraries in Poland are going to participate in the project. At the time of writing, four libraries are already integrated and integration of a fifth one—John Paul II Catholic University of Lublin [2]—with a volume of roughly one million records is underway. It is important to keep improving the matching algorithm to compensate for differences between notations or cataloging style used by libraries. The original algorithm was tailored towards the Jagiellonian Library as the sheer volume of data being incorporated outweighed the data from remaining libraries that were being processed during the same period of time.

The more time passes since the integration process has ended the more records are found that are duplicates not detected during the integration phase that now have to be manually corrected. This is mostly because of cataloging differences between libraries. One example of such difference is that for a first edition of a book one library would use a subfield containing the text *Edition I* while another library would not denote that edition at all assuming this is the only edition that does not require any explicit mentioning. Due to very strict rules used for matching editions in the original algorithm such entries have been considered as distinct.

In the future, we also plan to publish an extended analysis that also studies performance of the system as it had significant impact on the architecture and implementation of the solution.

References

1. Jagiellonian library. https://bj.uj.edu.pl/about-library
2. John Paul II Catholic University of Lublin. https://www.kul.pl/
3. Library of congress. https://www.loc.gov/
4. MARC 21 bibliographic. https://www.loc.gov/marc/bibliographic/bdintro.html
5. OMNIS. https://www.bn.org.pl/projekty/omnis/o-projekcie/, (in Polish)
6. ISO 2709—Information and documentation – Format for information exchange (2008)
7. ISO 2108—Information and documentation – International Standard Book Number (ISBN) (2017)
8. ISO 3297—Information and documentation – International standard series Number (ISSN) (2017)
9. Bramer, W., Giustini, D., Jonge, G., Holland, L., Bekhuis, T.: De-duplication of database search results for systematic reviews in endnote. J. Med. Libr. Assoc. (JMLA) **104**, 240–243 (2016)
10. Damerau, F.J.: A technique for computer detection and correction of spelling errors. Commun. ACM **7**(3), 171–176 (1964). http://doi.acm.org/10.1145/363958.363994
11. Heron, S.J., Simpson, B., Weiss, A.K., Phillips, J.: Merging catalogs: creating a shared bibliographic environment for the state university libraries of Florida. Cataloging Classif. Q. **51**(1–3), 139–155 (2013)
12. Hylton, J.A.: Identifying and merging related bibliographic records (1996)
13. Levenshtein, V.I.: Binary codes capable of correcting deletions. Insertions and reversals. Sov. Phys. Doklady **10**, 707 (1966)
14. Ridley, M.J.: An expert system for quality control and duplicate detection in bibliographic databases. Program **26**(1), 1–18 (1992). https://doi.org/10.1108/eb047100
15. Sitas, A., Kapidakis, S.: Duplicate detection algorithms of bibliographic descriptions. Libr. Hi Tech **26** (2008)

Supporting Architectural Decision-Making with Data Retrieved from Online Communities

Andrzej Zalewski, Klara Borowa(✉), and Krzysztof Lisocki

Institute of Control and Computation Engineering, Warsaw University of Technology, Warsaw, Poland
a.zalewski@ia.pw.edu.pl, klara.borowa@pw.edu.pl

Abstract. The focal question of this paper is how to support architecture decision-making by utilising data stored in internet services such as Git Hub, Stack Overflow or Wiki Data. Our proposition is to query such internet services in order to identify alternative architectural solutions to a given architectural problem, and to estimate their properties. This would provide assistance to the decision-making architect. The above idea resulted in the implementation of an Architecture Decision Support System (ADSS) prototype. This enabled the validation of the proposed approach in terms of technical feasibility, as well as its usefulness, for software engineers during practical experimental use.

Keywords: Software architecture · Architectural decision-making · Architectural knowledge

1 Introduction

The surge of social media has considerably reshaped decision-making in the area of detailed software design. Programmers all over the world are asking themselves the same question: has anyone already solved a problem similar to mine? Which leads to a search for the source code fragments that solve their problem in such internet sources as Stack Overflow or GitHub. While this approach does provide answers to simple programming issues, it is much more difficult to retrieve architectural knowledge this way, because of its intrinsic complexity.

The research on architecture decision-making support with the utilization of data retrieved from internet communities is still in its infancy (compare related work in Sect. 2). With the ever increasing evolution speed of software technologies, the internet will often become the only viable source of information that could be used for architecture decision-making support. This makes the research on this topic crucial for the further development of the software engineering discipline and software architecture research.

The goals of our research are expressed by the following research questions:

- RQ.1. What types of architecturally-relevant information can be obtained from internet communities in order to support architecture decision-making?

W. Zamojski et al. (Eds.): DepCoS-RELCOMEX 2021, AISC 1389, pp. 496–509, 2021.
https://doi.org/10.1007/978-3-030-76773-0_48

- RQ.2. Which aspects of architecture decision-making can be supported with the information gathered by exploring online communities?
- RQ.3. How a process of architecture-decision making supported with the community-sourced data can be designed?
- RQ.4. What could be the architecture of decision-making support tool?
- RQ.5. Is the architecturally-relevant information gathered from the internet communities useful for architecture decision-making?

In order to answer these exploratory questions, we start from a survey of the existing research (Sect. 2). Next, we define a decision making process and identify its elements that can be supported with the data gathered from the online communities (Sect. 3). In Sect. 4 we propose an architecture for the decision-support system. We describe the system's prototype design in Sect. 5. The research results are summarized in Sect. 7, discussed in Sect. 8 and 9. The further research outlook is presented in Sect. 10.

2 Related Work and State of the Art

Software architecture is commonly perceived nowadays as the result of a series of architectural decisions [8] that become a carrier for architectural knowledge [10]. This fostered research into how Architectural Knowledge (AK) can be captured and represented [11,13,16,22], as well into how architectural decisions are made by the architects [14,17].

Although the researchers generally agree that architecture decision-making is about selecting the most suitable alternative from a set of architectural variants [6] or patterns [1], the research on support for the decision-making seems to be more limited.

The general, tacit and over-optimistic assumption behind the proposed solutions for architectural decision-making support is that architects possess some architectural knowledge and wisdom that enables them to identify the set of architectural variants, assess them and finally choose the most suitable one on the basis of the assessments of the variants and the architects' preferences. These should be either defined by the architects themselves from their previous experiences [19], or retrieved from databases of architectural knowledge [3,4].

The research solutions that aim at overcoming the above limitations comprise the concepts of:

- an expert recommendation system, that indicates an expert that may provide current information and help solve a particular problem [2]
- systems that retrieve architectural knowledge from online services, such as StackOverflow [12] and GitHub [15]. This, in turn, may require using advanced text analysis/processing techniques [5].

3 How to Support Architectural Decision-Making?

In this section we analyst the architecture decision-making process and craft a support mechanisms, which makes use of querying internet services to gather relevant data. We also analyse what useful kinds of information can be found in internet communities for the purpose of architectural decision-making support.

3.1 Components of the Architectural Design Process

Falessi et al. [6] describe architectural design as an iterative process that comprises: (1) an understanding of the particular problem that requires to be solved, (2) decision-making by choosing a solution from possible alternatives, and (3) an evaluation of the selected solution. Our decision support system addresses all these phases. It supports the decision-making process, illustrated in Fig. 1. It consists of the following phases:

1. Identifying the architectural problems to be solved;
2. Establishing the assessment criteria of the architectural alternatives that resolve a given problem, namely the attributes of architectural solutions that will be subject to evaluation and their weighting factors;
3. Identifying architectural alternatives (Sect. 3.3);
4. Identifying alternatives' attribute values (Sect. 3.3);
5. Evaluating the alternatives against the criteria established in phase No. 2;
6. Creating a ranking of alternatives on the basis of preferences and evaluations of the alternatives;
7. Making the final choice on the basis of the ranking of alternatives;
8. Documenting the decision's rationale;
9. Documenting the design decisions as a (M)ADR document [9]

The end-user (architect) is responsible for defining the architectural problems and solution attributes that are relevant (according to the model presented in Sect. 3.5), specifying the assessment criteria, performing the decision-making between ranked alternatives and describing the rationale behind the decision. The system operator (programmer) is responsible for implementing specific data source handlers (Sect. 3.3) and the decision suggestion algorithm (Sect. 3.4). The system's main role is to gather information about possible alternative solutions, find values of attributes relevant to the decision and provide a ranked list with solution propositions.

3.2 Obtainable Architectural Data

Architectural decisions may concern three basic levels of detail (compare e.g. [15]):

- High level: programming languages, service providers, application server technologies
- Medium level: architectural patterns, database engines, frameworks

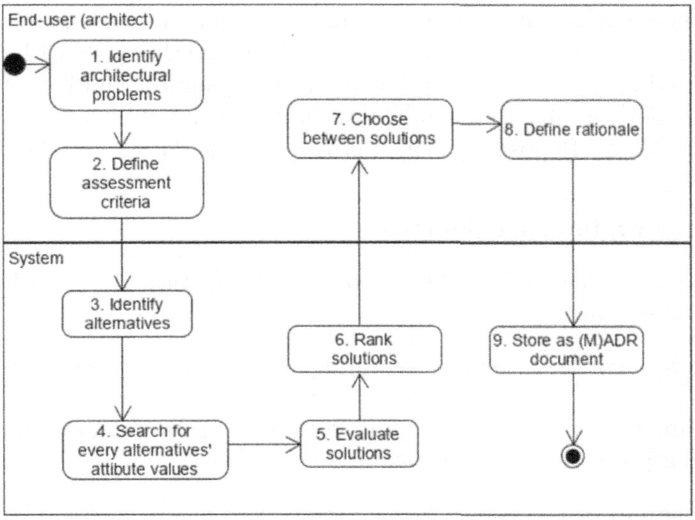

Fig. 1. Decisions Support System usage

- Realization level: design patterns, software libraries, object-relational mapping tools, APIs

Our system collects two kinds of data from online sources: architectural variants for a specific architectural problem and the variants' attributes.

Information about possible variants in most cases can be easily queried from sources containing structured data, such as Wiki data.

Having obtained the solution variants for a given problem, a set of attributes of interest that will influence the decision-making process between the variants is necessary. There is no right answer as to what kind of attributes should be considered, as it is due to the architects preferences. Various attributes may be taken into account, for instance:

- Price;
- License type: some projects may require components with specific licensing;
- Creation date: more mature solutions may be preferred;
- Last update date: which shows if the project is still maintained;
- Frequency of issues reported: which may be a consequence of both failure rate and popularity;
- Speed with which defects are fixed: an indicator of how active the developers and community behind the project are;
- Number of downloads: another measure of popularity;
- How often a component is used by other systems: which points to the component being dependable;
- Community rating of the solution;

Querying for attribute values is not as straightforward as for solution variants. Some of the necessary data may be also be found in structured data sources - like license type or creation date. Some others may require more effort to obtain. For example, frequency with which issues are reported and the speed with which they are addressed on average may be found by analyzing information available via GitHub. In general, all services providing statistics about certain items may be useful - e.g. Maven Repository, which provides usage data about Java packages.

3.3 Querying Internet Sources

The two basic challenges, in making a specific architectural decision that resolves a given problem, are:

1. Identification of the set of architectural alternatives that resolve a given problem;
2. The evaluation of the architectural alternatives in terms of their properties expressed (numerical, text, date etc.).

These actions are necessary for an architect to properly choose an architectural alternative. We show that retrieving the data necessary to perform them, can be done automatically from internet communities, by an architecture decision support system.

Because different services may contain different types of data, the most efficient solution is to use various internet services in order to retrieve architectural knowledge necessary to support decision-making. For example, an encyclopaedia definition of a given software technology may not contain data about its current price. Thus, the feature of collecting the attribute data for every alternative must be provided by a different component accessing a source specializing in this area.

An important technical challenge is that each source of architectural data has its own characteristics, such as differing querying options (POST or GET requests, JSON or SPARQL queries, etc.) and varied return formats. Therefore, a universal interface that could handle each source cannot be developed – the access mechanism for each source has to be programmed independently. For this purpose, we propose a simple interface in which varied implementations would enable various individual sources to be handled. These implementations are supposed to be created by the system operator, making it possible to add additional sources when required.

Listing 1.1. Data acquisition interface fragment pseudocode

```
interface Data Source {
    Boolean isAvailable()
    Set<Alternative> getAlternatives(Decision)
    Attribute getAttributeValues(Alternative, ATTRIBUTE_KEY)
}
```

Listing 1.1 presents a pseudocode representation of the interface suited for handling the data sources. The implementation of this interface for a specific source involves learning about the source's API or its database query language, whichever option is available. This means that, ultimately, the system operator will be responsible for the correct formulation of proper queries that obtain data from the chosen sources to obtain specific data.

Since each data source may have its own return format, it is essential to transform and unify the data into a single representation within the system. In order to ensure this behavior, the DataSource interface must return the appropriate, unified data types. For each decision problem, the system is supposed to provide data in the form of a vector of alternatives, where each alternative consists of a vector of attribute-value pairs.

3.4 Decision Suggestion Algorithm

In order to operate properly, the system requires implementation of a multi-criteria decision support algorithm. This algorithm must, for every considered alternative, take into account the values of each described attribute and its weighting factor. As a result, it should rank the alternatives for every particular architectural problem. We decided against suggesting a specific algorithm at this point and left the decision on what would be the most fitting one to the system operator.

3.5 Model of Architectural Decisions

The model of Architectural Decisions (AD) proposed in this paper is inspired by the work of Zimmerman et al. [20]. The model inherits some of the proposed relations between decision-making problems. In order to ensure that the model is as versatile as possible, these relations have been simplified into two groups:

1. influences – relation that represents the impact of a decision-making problem on another problem. A decision problem may both influence several other problems, and be influenced by several others. For example, the choice between programming languages, which is influenced by the decision on the application server and database framework.
2. triggers – relation that presents how a decision-making problem is triggered by choosing a specific alternative in a different problem. For example, the choice between database types triggers the choice between specific database engines.

Figure 2 presents a diagram of the model of architectural knowledge in the form of all relevant decision-making problems, the relationships between them and possible problem solutions.

4 Architecture of a Decision-Support System

The proposed system follows the layered architectural pattern (Fig. 3), which guarantees flexibility by enabling independent layer implementations. The system operator (programmer) is obliged to implement the data source handlers

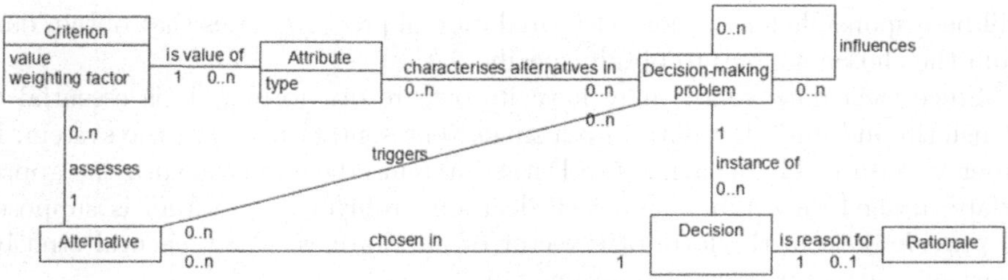

Fig. 2. Architectural-decision model

(used by the data source layer) and the decision suggestion algorithm. The end-user (architect/decision-maker) defines the decision problems and problem selection criteria.

The **data source layer** comprises the internet sources of architectural knowledge. The **proxy layer** provides a buffering mechanism, as the data in the data source layer may have to be obtained using frequent and possibly massive, hence, expensive, queries. The query results, in a standardized form, are passed to the **business logic layer**. Here, the structured data is processed to create a coherent space of alternatives, with their attributes and attribute values. This layer is responsible for operations on the data model, which occur as a response to the changes made by the end-user. The **presentation layer** contains the user interface enabling interaction and displaying the data model received from the business logic layer in a user-friendly manner.

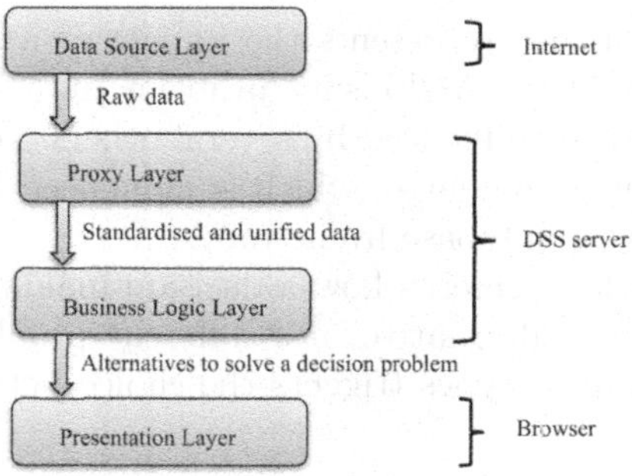

Fig. 3. System's layered architecture

5 Implementing the Decision-Making Support System

5.1 Sample Data Sources

In the case of our implementation, the sources were selected primarily based on the ease of querying data, the format of the returned data, and limitations imposed on mass queries. We choose two primary data sources: Wikidata and Stack Exchange. Wikidata stores any information in a structured statement (<Subject, Predicate, Object>), which makes it easily searchable by using data queries constructed using SPARQL. StackExchange, on the other hand, is a widely used service by architects and programmers alike, which we took advantage of since the dominant solution attribute that we used in this example is solution popularity.

The implementation of the getAlternatives (Listing 1.1) uses Wikidata as the source from which the alternatives are retrieved. It contains a generic SPARQL query template that is filled with content relevant to the given decision-making problem. For example, if the problem is about the choice of a programming language, the SPARQL query will be adjusted to look for all Subjects, whose Predicate is set as 'is instance of' and the Object is set as 'programming language'. The prepared SPARQL query is sent to Wikidata API. After receiving the results, the method converts the results into a list of Alternative objects and returns it.

StackExchange was used as a source in our implementation of the getAttributeValues method. Since the attribute that we were interested was popularity, we measured its value as the number of questions related to the solution created on all network websites each year. Therefore, we queried StackExchange API for the number of question that were tagged with the name of the given alternative.

5.2 Example Decision Suggestion Algorithm

The algorithm used by the system is based on the classical technique of multi-criteria decision-making. The system obtains a set of alternatives V_{D_i} for decision problem D_i:

$$V_{D_i} = \{v_{i1}, v_{i2}, ..., v_{in}\}$$

Each variant v_{ij} (alternative) is described by a set of attribute - value pairs $P_{v_{ij}}$:

$$P_{v_{ij}} = \{p_{ij1}, p_{ij2}, ..., p_{ijm}\}$$

Every attribute p_{ijk} has a weighting factor w_{ik} assigned to it. An attribute with a weighting factor greater than 0 can be considered as a decision criterion. The sum of weighting factors in the scope of a decision problem is 1:

$$\forall i : \sum_k w_{ik} = 1$$

The system users set the satisfaction function parameters by providing 2 values for each criterion, as shown in Fig. 4. All alternatives where the value of the criterion is less than x have 0 as the value of satisfaction, and all alternatives where the criterion value is greater than y have 1 as the value of satisfaction. Alternatives where the criterion value is in the range from x and y have a satisfaction value that varies linearly from 0 to 1. Thus, for a numeric criterion the

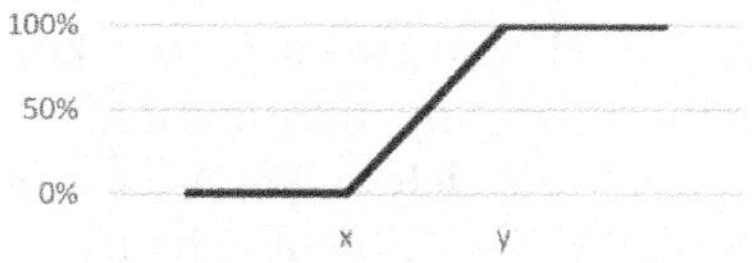

Fig. 4. Satisfaction function of a criterion

user needs to define x and y only, but for a text criterion the user needs to put each possible text value on the function manually (e.g. "Apache License": 0.75).

In conclusion, a satisfaction value for all criteria ranges from 0 to 1:

$$\forall i, j, k : s_{ijk} \in [0, 1]$$

Once the user has defined a satisfaction function for all the criteria of a given decision-problem, and alternatives have been obtained from one or more sources, the system can calculate the general satisfaction value for each alternative v_{ij}:

$$s_{ij} = \sum_{k} w_{ik} s_{ijk}$$

The general satisfaction value of an alternative also ranges from 0 to 1, so it can be presented to the user as a percentage value.

5.3 Example of Use

For the purpose of this research, our implementation of the decision support system stores information about architectural problems in JSON format. Alternatives are gathered from Wikidata. Our main assessment criterion is popularity, measured by the number of related posts on StackExchange over time. Each decision-making problem described this way contains the following elements:

1. A unique **key** and a **name** that will be visible to the end-user of the decision support system
2. A list of possible **attributes** that describe the possible alternatives. The obligatory **label** parameter serves as a unique identifier of each alternative solution.
3. The **influences** relationship between decision-making problems using the **influenced_by** parameter
4. The **triggers** relation, which can be restrictive, since the lack of choice between alternatives in another decision problem may block the decision problems that depend on it (this a **blocking** parameter).

The solution alternatives are also represented in JSON format, in which we store data in an array called attributes. The attribute names are obtained from the array defined by the user for the given decision problem, while the attribute value is supplied by the data source, which results in the creation of a key-value (attribute name - attribute value) vector describing the alternative.

Our system prototype contains a simple JSON editor that enables the user to provide proper problem and criteria definitions in JSON format. Having obtained this information, the system provides an overview of the defined decision problems and showcases it in the form of a simple graph. The prototype also fully performs the process of identifying alternatives, identifying attribute values, evaluating them and presenting them in the form of a ranking, as proposed in this paper.

6 Validation of Approach and Decision-Making Support Tool

A system based on this paper's findings was implemented and deployed for validation purposes. It was presented to the programmers and architects in a subsidiary of a medium-sized international banking systems software house. Five software engineers, who worked in the continuous integration and delivery team, used the system for a week. Their usual responsibilities included creating simple micoservice based solutions and writing scripts to automate various processes. They used the system when solving practical decision-making problems such as:

- Choosing the proper scripting language for their automation tasks.
- Exploration of various available web server technologies - where the license type was a key attribute.
- Choosing a proper time series database for the development of a monitoring system.
- Analysis of JavaScript frameworks based on their maturity.

While testing the system the developers discovered the Fish Unix shell, the time series database Prometheus and the reverse proxy server Traefik. All of them were unknown to them so far. Traefik turned out to be especially interesting since the users decided to use it despite their previous preference for the reverse proxy server Nginx.

When asked about their experience with the system the engineers indicated some of its key advantages:

- The flexibility of the system, that makes it applicable to almost every decision-making problem,
- The fast data gathering speed of the system,
- The readable and informative graphic representation of the data model.

The first impression of the users was how straightforward it was for them to model problems, alternatives and decisions. Since the system was supposed to operate in real-time, it was crucial for every element, including the alternative retrieval and evaluation, to operate with tolerable speed. The greatest advantage of this solution is that, with the right selection of sources, it is possible to obtain a wide, almost complete, set of alternatives. This provides the decision-maker with the opportunity to learn about options that would not otherwise have been considered due to a possible lack of knowledge.

7 Results

With reference to the research questions formulated in Sect. 1 the results of the research are:

- RQ.1. Online communities can deliver large amount of varied information, which can be useful for architecture decision-making. The list (compare Sect. 3.3) comprises but is not limited to the price of software components, types of licences, frequency and numbers of issues, rating of components, number of downloads;
- RQ.2. Information obtained from internet communities can be used to retrieve architectural alternatives (such as: possible service providers, available components or APIs) as well as assess their properties.
- RQ.3. The process follows a typical architecture decision-making scheme (compare Sect. 3, which starts from establishing problems and assessment criteria, then goes through identifying and evaluation of architectural alternatives and finally captures the decision made and its rationale;
- RQ.4. As presented in Sect. 4, we propose the use of a layered architectural style for the decision support system for maximized implementation flexibility.
- RQ.5. The practical experiments show that the exploration of the communities' data may deliver insightful and revealing results such as discovery of components that are unknown to the architects.

8 Discussion

What distinguishes our solution is that it does not assume the use of any specific sources of architectural knowledge such as certain domain-specific knowledge databases (as in the knowledge processing solution proposed by Zimmermann et al. [21]), nor do we assume the use of particular architectural patterns as Aslam et al. [1], or hiring an appropriate expert [2]. Inspired by the previous work on retrieving relevant data from the developer communities [12,15], we expand the idea further by perceiving the entire internet as a possible source of architectural knowledge. We go a step beyond the work of Soliman et al. [12], since we propose a fully automatic approach to extracting architecturally significant data - as opposed to simply improving the search for Architecturally Relevant Posts. Also, since we propose a very flexible structure for the ADSS, and a simplified design decision model, we are able to support architecture decisions at any level, not only the medium one as in the solution of Van der Ven and Bosch [15]. Additionally, it is not necessary to create and maintain the architectural knowledge databases [7] in the proposed approach.

Let us finally indicate that the application of the presented approach and tool enhances two out of the three features that characterize rational decision-making according to Zannier et al. [18], which include taking into account a wide set of decision alternatives and the use of a utility function that assigns vales to each alternative. Which definitely results in 'more rational' decision-making.

9 Limitations

The use of the proposed solution requires additional effort in implementing specific data source handlers and a decision suggestion algorithm. Despite that, these draw-back may diminish over time, after different implementations are created and made public.

Another limitation of the proposed concept of architecture decision-making support is that the model we used to represent decisions, and the relationships between them, is simplistic when compared to previous works [20]. This might mean that some complex problems would be hard to represent, although not necessarily. Nonetheless, this was an inevitable compromise that had to be made to enable the possible use of a broader variety of sources.

The last potential disadvantage of our solution is one that comes from the nature of the data provided by online communities. The knowledge retrieved this way may be up-to-date, but is not always as reliable as knowledge obtained from traditional sources (books, peer-reviewed papers). Still, this problem is also prevalent when a user performs simple internet searches personally. Which means that using the system is nevertheless better than a simple manual approach.

10 Conclusion

The main contribution of this paper is a new approach to the support of architecture decision-making, which employs the real-time querying of online communities in order to identify alternative solutions, along with assessing their architecturally relevant properties. We also proposed a design of an ADDS which supports the evaluation and ranking of possible solutions to the given problems. A prototype implementation of the ADSS made it possible to validate the above concepts and prove their usefulness for architects when solving practical problems in real work conditions.

Further research could focus on using natural language processing and artificial intelligence in the retrieval of architectural knowledge from poorly structured internet sources in order to use them in the decision-making process. Artificial intelligence could also be employed in order to suggest the best choice from a set of architectural alternatives.

References

1. Aslam, T., Rana, T., Batool, M., Naheed, A., Andaleeb, A.: Quality based software architectural decision making. In: 2019 International Conference on Communication Technologies (ComTech), pp. 114–119 (2019)
2. Bhat, M., Shumaiev, K., Koch, K., Hohenstein, U., Biesdorf, A., Matthes, F.: An expert recommendation system for design decision making: who should be involved in making a design decision? In: Proceedings of 2018 IEEE 15th International Conference on Software Architecture, ICSA 2018, pp. 85–94 (2018). https://doi.org/10.1109/ICSA.2018.00018

3. Bhat, M., Shumaiev, K., Matthes, F.: Towards a framework for managing architectural design decisions. In: ACM International Conference Proceeding Series, vol. 1, pp. 48–51 (2017). https://doi.org/10.1145/3129790.3129799
4. Bhat, M., Tinnes, C., Shumaiev, K., Biesdorf, A., Hohenstein, U., Matthes, F.: ADeX: a tool for automatic curation of design decision knowledge for architectural decision recommendations. In: Proceedings of 2019 IEEE International Conference on Software Architecture - Companion, ICSA-C 2019, pp. 158–161 (2019). https://doi.org/10.1109/ICSA-C.2019.00035
5. Bi, T., Liang, P., Tang, A., Yang, C.: A systematic mapping study on text analysis techniques in software architecture. J. Syst. Softw. **144**, 533–558 (2018). https://doi.org/10.1016/j.jss.2018.07.055
6. Falessi, D., Cantone, G., Kazman, R., Kruchten, P.: Decision-making techniques for software architecture design: a comparative survey. ACM Comput. Surv. **43**(4), 1–28 (2011). https://doi.org/10.1145/1978802.1978812
7. Gorton, I., Xu, R., Yang, Y., Liu, H., Zheng, G.: Experiments in curation: towards machine-assisted construction of software architecture knowledge bases. In: Proceedings of 2017 IEEE International Conference on Software Architecture, ICSA 2017, pp. 79–88. Institute of Electrical and Electronics Engineers Inc., May 2017. https://doi.org/10.1109/ICSA.2017.27
8. Jansen, A., Bosch, J.: Software architecture as a set of architectural design decisions. In: Proceedings of 5th Working IEEE/IFIP Conference on Software Architecture, WICSA 2005, pp. 109–120 (2005). https://doi.org/10.1109/WICSA.2005.61
9. Kopp, O., Armbruster, A., Zimmermann, O.: Markdown architectural decision records: format and tool support. In: CEUR Workshop Proceedings, vol. 2072, pp. 55–62 (2018)
10. Kruchten, P., Lago, P., Van Vliet, H., Wolf, T.: Building up and exploiting architectural knowledge. In: Proceedings of 5th Working IEEE/IFIP Conference on Software Architecture, WICSA 2005, pp. 291–292 (2005). https://doi.org/10.1109/WICSA.2005.19
11. Lee, L., Kruchten, P.: Capturing software architectural design decisions. In: Canadian Conference on Electrical and Computer Engineering, pp. 686–689 (2007). https://doi.org/10.1109/CCECE.2007.176
12. Soliman, M., Rekaby Salama, A., Galster, M., Zimmermann, O., Riebisch, M.: Improving the search for architecture knowledge in online developer communities. In: Proceedings of 2018 IEEE 15th International Conference on Software Architecture, ICSA 2018, pp. 186–195 (2018). https://doi.org/10.1109/ICSA.2018.00028
13. Szlenk, M., Zalewski, A., Kijas, S.: Modelling architectural decisions under changing requirements. In: Proceedings of the 2012 Joint Working Conference on Software Architecture and 6th European Conference on Software Architecture, WICSA/ECSA 2012, pp. 211–214 (2012). https://doi.org/10.1109/WICSA-ECSA.212.31
14. Tang, A.: Software designers, are you biased? In: Proceedings of International Conference on Software Engineering, pp. 1–8 (2011). https://doi.org/10.1145/1988676.1988678
15. Van Der Ven, J.S., Bosch, J.: Making the right decision: supporting architects with design decision data. Lecture Notes in Computer Science (including subseries Lecture Notes in Artificial Intelligence and Lecture Notes in Bioinformatics), vol. 7957, pp. 176–183 (2013). https://doi.org/10.1007/978-3-642-39031-9_15

16. Weinreich, R., Groher, I., Miesbauer, C.: An expert survey on kinds, influence factors and documentation of design decisions in practice. Future Gener. Comput. Syst. **47**, 145–160 (2015). https://doi.org/10.1016/j.future.2014.12.002
17. Zalewski, A., Klara, B., Andrzej, R.: On cognitive biases in architecture decision making. In: 11th European Conference on Software Architecture, ECSA 2017, vol. 1, pp. 138–154, June 2017. https://doi.org/10.1007/978-3-319-65831-5
18. Zannier, C., Chiasson, M., Maurer, F.: A model of design decision making based on empirical results of interviews with software designers. Inf. Softw. Technol. **49**(6), 637–653 (2007). https://doi.org/10.1016/j.infsof.2007.02.010
19. Zimmermann, O.: Architectural decisions as reusable design assets. IEEE Softw. **28**(1), 64–69 (2011). https://doi.org/10.1109/MS.2011.3
20. Zimmermann, O., Koehler, J., Leymann, F., Polley, R., Schuster, N.: Managing architectural decision models with dependency relations, integrity constraints, and production rules. J. Syst. Softw. **82**(8), 1249–1267 (2009). https://doi.org/10.1016/j.jss.2009.01.039
21. Zimmermann, O., Miksovic, C., Küster, J.M.: Reference architecture, metamodel, and modeling principles for architectural knowledge management in information technology services. J. Syst. Softw. **85**(9), 2014–2033 (2012). https://doi.org/10.1016/j.jss.2012.05.003
22. Zimmermann, O., Wegmann, L., Koziolek, H., Goldschmidt, T.: Architectural decision guidance across projects-problem space modeling, decision backlog management and cloud computing knowledge. In: Proceedings of 12th Working IEEE/IFIP Conference on Software Architecture, WICSA 2015, pp. 85–94 (2015). https://doi.org/10.1109/WICSA.2015.29

Author Index

W. Zamojski et al. (Eds.): DepCoS-RELCOMEX 2021, AISC 1389, pp. 511–512, 2021.
https://doi.org/10.1007/978-3-030-76773-0

The manufacturer's authorised representative in the EU is Springer Nature Customer Service Centre GmbH, Europaplatz 3, 69115 Heidelberg, Germany. If you have any concerns regarding our products, please contact ProductSafety@springernature.com

Printed and bound by CPI Group (UK) Ltd, Croydon, CR0 4YY
30/09/2026
02225662-0017